T0292758

# Short-range Wireless Communication

# Short-range Wireless Communication

**Alan Bensky**

Newnes is an imprint of Elsevier
The Boulevard, Langford Lane, Kidlington, Oxford OX5 1GB, United Kingdom
50 Hampshire Street, 5th Floor, Cambridge, MA 02139, United States

© 2019 Elsevier Inc. All rights reserved.

No part of this publication may be reproduced or transmitted in any form or by any means, electronic or mechanical, including photocopying, recording, or any information storage and retrieval system, without permission in writing from the publisher. Details on how to seek permission, further information about the Publisher's permissions policies and our arrangements with organizations such as the Copyright Clearance Center and the Copyright Licensing Agency, can be found at our website: www.elsevier.com/permissions.

This book and the individual contributions contained in it are protected under copyright by the Publisher (other than as may be noted herein).

**Notices**
Knowledge and best practice in this field are constantly changing. As new research and experience broaden our understanding, changes in research methods, professional practices, or medical treatment may become necessary.

Practitioners and researchers must always rely on their own experience and knowledge in evaluating and using any information, methods, compounds, or experiments described herein. In using such information or methods they should be mindful of their own safety and the safety of others, including parties for whom they have a professional responsibility.

To the fullest extent of the law, neither the Publisher nor the authors, contributors, or editors, assume any liability for any injury and/or damage to persons or property as a matter of products liability, negligence or otherwise, or from any use or operation of any methods, products, instructions, or ideas contained in the material herein.

**Library of Congress Cataloging-in-Publication Data**
A catalog record for this book is available from the Library of Congress

**British Library Cataloguing-in-Publication Data**
A catalogue record for this book is available from the British Library

**ISBN: 978-0-12-815405-2**

For information on all Newnes publications
visit our website at https://www.elsevier.com/books-and-journals

*Publisher:* Jonathan Simpson Acquisition
*Acquisition Editor:* Tim Pitts Editorial
*Editorial Project Manager:* Ana Claudia A. Garcia
*Production Project Manager:* Kamesh Ramajogi
*Cover Designer:* Mark Rogers

Typeset by SPi Global, India

*To my wife Nurit and daughters Chani, Racheli and Ortal.*

# Contents

# Preface

*Short-range Wireless Communication—Fundamentals of RF System Design and Application* was first published in the year 2000. The following excerpt from its Preface is relevant to the third edition as well.

*Developers, manufacturers and marketers of products incorporating short-range radio systems are experts in their fields—security, telemetry, medical care, to name a few. Often they add a wireless interface just to eliminate wires on an existing wired product. They may adapt a wireless subsystem, which is easy to integrate electrically into their system, only to find that the range is far short of what they expected, there are frequent false alarms, or it doesn't work at all. It is for these adapters of wireless subsystems that this book is primarily intended.*

*Other potential readers are curious persons with varied technical backgrounds who see the growing applications for wireless communication and want to know how radio works, without delving deeply into a particular system or device. This book covers practically all aspects of radio communication including wave propagation, antennas, transmitters, receivers, design principles, telecommunication regulations and information theory. Armed with knowledge of the material in this book, the reader can more easily learn the details of specialized radio communication topics, such as cellular radio, personal communication systems (PCS), and wireless local area networks (WLAN).*

*The technical level of this book is suitable for readers with an engineering education or a scientific background, working as designers, engineering managers, or technical marketing people. They should be familiar with electrical circuits and engineering mathematics. Elementary probability theory is needed in some of the early chapters. Readers without an appropriate background or who need to brush up on probability are advised to jump ahead to Chapter 10* (Chapter 9 in the third edition).

In 2004 the second edition was published. In the four years between the two publications, Wi-Fi became the ubiquitous Standard choice for Wireless LAN while significantly multiplying its spectrum efficiency, Bluetooth rapidly gained acceptance for short-range wireless accessories while other IEEE 802.15.4 personal area network technologies, namely Zigbee and ultra wideband (802.15.4a) began to make their mark on the short-range wireless scene. Fourteen years later, the pace of technological innovation has not let up, and new concepts and technologies along with their acronyms have come to the fore—MIMO, metamaterials, WBAN, WSN, IoT to mention a few. This new, third edition, covers these and significantly expands on RFID, while also introducing location awareness, millimeter and optical communication and energy harvesting. While short-range wireless has taken many new directions, the basics of radio communication have not changed, and the new book devotes most of its pages to descriptions of radio propagation, antennas, modulation, transmitter and receiver design and implementation, as well as an introduction to information theory as a tool for dealing with wireless communication reliability. The chapter on regulations and standards has been updated and expanded to present relevant rules for operating short-range wireless devices in the United States and the European Community. While all chapters that were included in the previous editions were updated and corrected where necessary, this third edition has four new chapters: Wireless Local Area Networks for a comprehensive description of IEEE 802.11 evolution, Wireless Personal Area Networks covering Bluetooth and Bluetooth LE as well as IEEE 802.15.4 and other sub-1000 MHz standards, Radio Frequency Identification, and a new Applications and Technologies chapter containing subjects that were not topical or not addressed in the earlier publications.

In common with the previous editions, Mathcad worksheets are available to aid the understanding and use of the more technical engineering material. The book also has extensive source references at the end of each chapter.

I've tried to give a broad but thorough treatment of short-range wireless communication, connecting theory with implementation. I hope the third edition will serve as a key to understanding advances and new technologies that will inevitably continue to appear, some of them due to the work of readers of this book.

*Alan Bensky*, December 2018

# Engineering Worksheets

A number of radio engineering worksheets accompany this book. Their aim is to help you understand the technical explanations in the book and to serve as a tool for your own work and study. In order to use the worksheets, you need to install a version of PTC Mathcad Prime, which supports *.mcdx files. All of our files run on the free (as this book is written) reduced functionality version, PTC Mathcad Express (https:/ www.ptc.com/en/products/mathcad-express-free-download). It is recommended that you become familiar with Mathcad Prime through tutorials and "Help" in order to benefit from the worksheets. Here are general comments on the style of the worksheets:

- Names of variables on the worksheets are not necessarily the same as in the book text.
- Assignments to variables may not be connected to the text. Change the assignments at will in the worksheets to solve the examples in the text and to suit your own interest and requirements.
- Note that you can easily change the units of measurement of results to those you are most familiar with, for example, from meters to feet or centimeters to inches.
- In many worksheets, yellow marking of assignment terms is user-entered given values, and results are marked in blue.

Following is a synopsis of the individual worksheets. Companion site for Engineering Worksheet is available online in https://www.elsevier.com/books-and-journals/book-companion/9780128154052.

## *Complex_numbers.mcdx*: Complex numbers

This worksheet finds real and imaginary parts and argument of a complex number, for use with Mathcad Express. It shows how to create a constellation diagram on an *x*, *y* plot. This worksheet is not needed for fully functional Mathcad, which has built-in complex number functions and polar plotting.

## *Conversion.mcdx*: Impedance transformations

This worksheet lets you convert capacitances and inductances between their values and their reactances. You can also find resonant frequencies and convert between serial and parallel impedances.

## *diffraction.mcdx*: Diffraction

This worksheet plots the formula for diffraction over a knife-edge barrier and converts from the normalized distance of the barrier tip from the line of sight to the absolute value called height. The first Fresnel zone limit is also calculated.

## *diversity.mcdx*: Transmit diversity

This worksheet demonstrates the use of the Alamouti algorithm for spatial diversity with two transmitter antennas and one receiving antenna. The known channel parameters are chosen with random amplitude and random phase. 8PSK modulation is assumed, with eight possible equal-amplitude symbols. You can choose any two source symbols for the demonstration. Gaussian noise, with user-determined standard deviation, is added to the received signal at each of the two consecutive reception times of the algorithm. The similarity of the resulting estimated transmitted symbols to the true values is easily recognized. You can see how the maximum likelihood decision rule works by comparing the detected symbols to each of the possible transmitted symbols. Finally a constellation diagram shows the estimated symbols in relation to the noiseless constellation.

## *Helical.mcdx:* Helical antenna

This worksheet is used to design a helical antenna over a ground plane. You can specify turns per unit length to calculate the height or the height to get number of turns.

## *Intermodulation.mcdx*: Intermodulation distortion

This worksheet facilitates finding third-order spurious signal frequency and power given interfering signal frequencies and power and third-order intercept. It also finds two-tone dynamic range and third-order intercept of cascaded stages.

## *Loop.mcdx*: Loop antenna

This worksheet is used to design a printed circuit board or air small loop antenna and calculates radiation and loss resistances, inductance, and resonating capacitance. The loss resistance is apt to be greater than specified since board material and surroundings are not taken into account.

## *Loop_match.mcdx*: Coupling loop match

This worksheet facilitates designing coupling loop matching for a small resonant loop antenna with a given area. The worksheet shows design for square loops. For round loops, use the circular perimeter for a given loop area to find inductance instead of the square perimeter as shown. The worksheet calculates the equivalent round coupling loop radius, given the required mutual inductance for the match and distance between loops and skew of loop centers. The fourth step in the worksheet shows adjustment of the radiating loop resonating capacitor to obtain a good match. Residual reactance in the match can be canceled out with an inductor or capacitor in series with the coupling loop.

## *Matching.mcdx*: Impedance matching

This worksheet calculates impedance matching circuit components, with different topographies. Terminations are resistive, but complex impedances can be dealt with by absorbing reactances in adjacent circuit components. Also designs an RF balun.

## *Microstrip.mcdx*: Microstrip transmission lines

This worksheet finds printed microstrip width or impedance according to frequency and board thickness and relative permittivity and also calculates wavelength and effective relative permittivity in board.

## *MIMO.mcdx*: MIMO spatial multiplex

This worksheet demonstrates how, using MIMO antennas, separate transmit data streams can be received without cointerference over a common channel resource. The condition is that the number of data streams is equal to or greater than the least number of antennas between the two terminals. A constellation diagram shows the estimated symbols in relation to the noiseless constellation.

## *Patch.mcdx:* Microstrip patch antenna

This worksheet is used to design a square patch antenna. From frequency and board substrate thickness and dielectric constant, the worksheet finds the length of the sides and feed impedance at the center of the edge. By specifying the desired input impedance, often 50 Ω, the worksheet finds the point on the centerline of the patch for connecting a feed probe, the center wire of coax cable, for example.

## *PLL.mcdx*: Charge pump PLL

This worksheet calculates charge pump phase-lock loop component values and displays loop and filter gain versus frequency plots.

## *Probability.mcdx*: Probability

This worksheet shows how to find the probability of errors in a sequence of bits, dependent on the bit error rate. It also defines average, variance, and standard deviation for a discrete function of a random variable and for a continuous Gaussian random variable.

## *Radiate.mcdx*: Radio wave propagation

This worksheet is used for finding the relationships between transmitting and receiving signal strengths, field strength, antenna gains, and transmission distance in free space.

## *range.mcdx*: Open field path gain

This worksheet plots path gain over an open field (wave reflection only from ground) for given frequency, antenna heights, and polarization. A free space plot is also shown. From the plot, you can find required receiver sensitivity for a given transmitter radiated power and range or estimate range when transmitted and received powers are known.

## *RFID_FF.mcdx*: RFID far field

This worksheet finds the backscattered power to a far-field RFID reader, given the switched source impedances from the tag antenna during tag response modulation. It also calculates modulation and power efficiency factors that indicate link performance.

## *RFID_NF.mcdx*: RFID near field

This worksheet finds the magnetic flux density created by a near-field RFID reader and uses it to calculate the induced voltage in a passive tag. Then, it estimates the reflected voltage in the reader due to tag modulation.

## *Sensitivity.mcdx*: Noise figure and sensitivity

This worksheet calculates the cascaded noise figure, which it uses to find receiver sensitivity, and also calculates sensitivity when the antenna is not at standard room temperature due to environmental noise.

## *Translines.mcdx*: Transmission lines

This worksheet calculates various transmission line parameters at a given frequency. It defines VSWR, reflection coefficient, and return loss and converts between them. It finds input impedance when load impedance is known and load impedance for a given input impedance, taking into account the transmission line loss.

# Abbreviations

| | |
|---|---|
| **AAD** | Additional authentication data |
| **ACK** | Acknowledge |
| **ACL** | Asynchronous connectionless link |
| **ADC** | Analog to digital converter |
| **AES** | Advanced encryption standard |
| **AFA** | Adaptive frequency agility |
| **AGC** | Automatic gain control |
| **AID** | Association identifier |
| **AMP** | Alternate MAC/PHY |
| **AODV** | Ad hoc on-demand distance vector |
| **AP** | Access point |
| **APC** | Adaptive power control |
| **ARQ** | Automatic repeat request |
| **ASK** | Amplitude shift keying |
| **AWGN** | Added white Gaussian noise |
| **BAN** | Body area network |
| **BER** | Bit error rate |
| **BLE** | Bluetooth low energy |
| **BPM** | Burst position modulation |
| **BPSK** | Binary phase shift keying |
| **BR/EDR** | Basic rate/enhanced data rate |
| **BSS** | Basic service set |
| **BT** | Bandwidth time period product |
| **CCA** | Clear channel access; clear channel assessment |
| **CCK** | Complementary code keying |
| **CCM** | Counter mode with CBC-MAC (of AES standard) |
| **CCMP** | Counter mode cipher block chaining message authentication code protocol |
| **CDMA** | Code division multiple access |
| **CE** | European conformity (marking) |
| **CEPT** | Conference of Postal and Telecommunications Administrations |
| **CRC** | Cyclic redundancy check |
| **CSI** | Channel state information |
| **CSMA** | Carrier sense multiple access |
| **CSMA/CA** | Carrier sense multiple access with collision avoidance |
| **CTS** | Clear to send |
| **DAA** | Detect and avoid |
| **DAC** | Digital to analog converter |
| **DCF** | Distributed coordination function |
| **DDC** | Digital down converter |

| | |
|---|---|
| **DDS** | Direct digital synthesizer |
| **DFS** | Dynamic frequency selection |
| **DIAC** | Dedicated inquiry access code |
| **DIFS** | Distributed coordination function interframe space |
| **DL** | Downlink |
| **DMG** | Directional multigigibit |
| **DNG** | Double negative metamaterial |
| **DPSK** | Differential phase shift keying |
| **DSP** | Digital signal processor |
| **DSR** | Dynamic source routing |
| **DSSS** | Direct sequence spread spectrum |
| **DUC** | Digital up converter |
| **EAPoL** | Extensible authentication protocol over LAN |
| **EFC** | Electric field communication |
| **EIRP** | Equivalent (or effective) isotropic radiated power |
| **EMC** | Electromagnetic compatibility |
| **EMI** | Electromagnetic interference |
| **EOF** | End-of-frame |
| **ERC** | European Radiocommunications Committee |
| **ERP** | Effective radiated power |
| **ERP** | Extended rate physical layer |
| **ESS** | Extended service set |
| **ETSI** | European Telecommunications Standards Institute |
| **EU** | European Union |
| **EUT** | Equipment under test |
| **FBE** | Frame-based equipment |
| **FCC** | Federal Communications Commission |
| **FCS** | Frame check sequence |
| **FDM** | Frequency division multiplex |
| **FEC** | Forward error correction |
| **FFD** | Full function device |
| **FFT** | Fast Fourier transform |
| **FHS** | Frequency hop synchronization |
| **FHSS** | Frequency hopping spread spectrum |
| **FRS** | Family radio service |
| **FSK** | Frequency shift keying |
| **FTM** | Fine timing measurement |
| **GFSK** | Gaussian frequency shift keying |
| **GIAC** | General inquiry access code |
| **GMSK** | Gaussian minimum shift keying |
| **HBC** | Human body communication |
| **HCCA** | HCF controlled channel access |
| **HCF** | Hybrid coordination function |

| | |
|---|---|
| **HE** | High efficiency |
| **HRP UWB** | High rate pulse repetition frequency UWB |
| **HT** | High throughput |
| **ICV** | Integrity check value |
| **IEEE** | Institute of Electrical and Electronic Engineers |
| **IETF** | Internet Engineering Task Force |
| **IFFT** | Inverse fast Fourier transform |
| **IMD** | Intermodulation distortion |
| **IP** | Internet protocol |
| **IR** | Impulse radio |
| **IR** | Infrared |
| **IrDA** | Infrared Data Association |
| **ISM** | Industrial scientific medical |
| **ITU** | International Telecommunication Union |
| **IV** | Initialization vector |
| **L2CAP** | Logical Link Control and Adaptation Protocol |
| **LAN** | Local area network |
| **LBE** | Load-based equipment |
| **LBT** | Listen before talk; listen before transmit |
| **LE** | (Bluetooth) low energy |
| **LEACH** | Low-energy adaptive clustering hierarchy |
| **LED** | Light-emitting diode |
| **LLC** | Logical link control |
| **LNA** | Low-noise amplifier |
| **LPF** | Low-pass filter |
| **LPRS** | Low-power radio service |
| **LRP UWB** | Low rate pulse repetition frequency UWB |
| **LTE** | Long-term evolution |
| **M2M** | Machine to machine |
| **MAC** | Medium access control |
| **MANET** | Mobile ad hoc network |
| **MBANS** | Medical body area network systems |
| **MB-OFDM** | Multiband OFDM |
| **MCF** | Mesh coordination function |
| **MDS** | Minimum discernable signal |
| **MIC** | Message integrity check |
| **MICS** | Medical implant communications service |
| **MIMO** | Multiinput multioutput |
| **MPDU** | MAC protocol data unit |
| **MU** | Medium utilization (factor) |
| **MU-MIMO** | Multiuser MIMO |
| **NAV** | Network allocation vector |
| **NCO** | Numerically controlled oscillator |

| | |
|---|---|
| **NRZ** | Non return to zero |
| **NSD** | Noise spectral density |
| **OFDM** | Orthogonal frequency division multiplex |
| **OFDMA** | Orthogonal frequency division multiple access |
| **OLED** | Organic LED |
| **OMI** | Operation mode indication |
| **OOK** | On-off keying |
| **O-QPSK** | Offset quadrature phase shift keying |
| **OWC** | Optical wireless communication |
| **PA** | Power amplifier |
| **PAL** | Protocol adaption layer |
| **PAM** | Pulse amplitude modulation |
| **PBCC** | Packet binary convolutional coding |
| **PCF** | Point coordination function |
| **PHY** | Physical layer |
| **PKC** | Public key cryptography |
| **PKI** | Public key infrastructure |
| **PLL** | Phase locked loop |
| **PMK** | Pairwise master key |
| **PPDU** | PHY protocol data unit |
| **PPM** | Pulse position modulation |
| **PRF** | Pulse repetition frequency |
| **PSK** | Phase shift keying |
| **PTK** | Pairwise transit keys |
| **QAM** | Quadrature amplitude modulation |
| **QoS** | Quality of service |
| **QPSK** | Quadrature phase shift keying |
| **QTP** | Quiet time period |
| **RAW** | Restricted access window |
| **RCRS** | Radio control radio service |
| **RERR** | Route error |
| **RFD** | Reduced function device |
| **RFID** | Radio frequency identification |
| **RLAN** | Radio local area network |
| **RREP** | Route reply |
| **RREQ** | Route request |
| **RSA** | Rivest-Shamir-Adleman algorithm |
| **RSNA** | Robust security network association |
| **RTS** | Request to send |
| **RU** | Resource unit |
| **S2S** | Station to station |
| **SAW** | Surface acoustic wave |
| **SCO** | Synchronous connection oriented (link) |

| | |
|---|---|
| **SDM** | Spatial division multiplex |
| **SFD** | Start frame deliminator |
| **SIFS** | Short interframe space |
| **SIG** | (Bluetooth) special interest group |
| **SKC** | Symmetric key cryptography |
| **SNR** | Signal-to-noise ratio |
| **SOF** | Start-of-frame |
| **SRD** | Short-range devices |
| **STA** | Station |
| **SU-MIMO** | Single-user MIMO |
| **TDMA** | Time division multiple access |
| **TDOA** | Time difference of arrival |
| **TEDS** | Transducer electronic data sheet |
| **TEG** | Thermoelectric energy generator |
| **TKIP** | Temporal key integrity protocol |
| **TOF** | Time of flight |
| **TPC** | Transmit power control |
| **TTDR** | Two-tone dynamic range |
| **TWT** | Target wakeup time |
| **UDP** | User datagram protocol |
| **UL** | Uplink |
| **U-NII** | Unlicensed National Information Infrastructure |
| **UVC** | Ultraviolet communication |
| **UWB** | Ultra-wideband |
| **VFO** | Variable frequency amplifier |
| **VHT** | Very high throughput |
| **VLC** | Visual light communications |
| **WBAN** | Wireless body area network |
| **WBFM** | Wide band frequency modulation |
| **WEP** | Wireless equivalent privacy |
| **WLAN** | Wireless local area network |
| **WMN** | Wireless mesh network |
| **WMTS** | Wireless medical telemetry service |
| **WPA** | Wi-Fi protected access |
| **WSN** | Wireless sensor network |

CHAPTER

# Introduction

1

## 1.1 Historical perspective

A limited number of short-range radio applications were in use in the 1970s. The garage door opener was one of them. An L-C tuned circuit oscillator transmitter and superregenerative receiver made up the system. It suffered from frequency drift and susceptibility to interference, which caused the door to open apparently at random, leaving the premises unprotected. There still may be similar systems in use today, although radio technology has advanced tremendously. Even with greatly improved circuits and techniques, wireless replacements for wired applications—in security systems for example—still suffer from the belief that wireless is less reliable than wired and that cost differentials are too great to bring about the revolution that cellular radio has brought to telephone communication.

Few people will dispute the assertion that cellular radio is in a class with a small number of other technological advancements—including the proliferation of electric power in the late 19th century, mass production of the automobile, and the invention of the transistor—that have profoundly affected human lifestyle in the last century. Another development in electronic communication within the last 30 or so years has also impacted our society—satellite communication—and its impact has come even closer to home with the spread of direct broadcast satellite television transmissions.

That wireless techniques have such an overwhelming reception is not at all surprising. After all, the wires really have no intrinsic use. They only tie us down and we would gladly do without them if we could still get reliable operation at an acceptable price. Cellular radio has been of lower quality, lower reliability, and higher price generally than wired telephone. Even though each generation brings it closer to parity on all counts, its acceptance by the public is nothing less than phenomenal. Imagine the consequences to lifestyle when electric power is able to be distributed without wires!

Considering the ever-increasing influence of wireless systems in society, this book was written to give a basic but comprehensive understanding of radio communication to a wide base of technically oriented people who either have a curiosity to know how wireless works, or who will contribute to expanding its uses. While most chapters of the book will be a gateway, or even a prerequisite, to understanding the basics of all forms of radio communication, including satellite and cellular systems, the emphasis and implementations are aimed at what are generally defined as



**Short-range Wireless Communication.** https://doi.org/10.1016/B978-0-12-815405-2.00001-4
© 2019 Elsevier Inc. All rights reserved.

short-range or low-power wireless applications. These applications are undergoing a fast rate of expansion, in large part due to the technological fall-out of the cellular radio revolution.

## 1.2 Reasons for the spread of wireless applications

One might think that there would be a limit to the spread of wireless applications and the increase in their use, since the radio spectrum is a fixed entity and it tends to be depleted as more and more use is made of it. In addition, price and size limitations should restrict proliferation of wire replacement devices. However, technological developments defy these axioms.

- We now can employ higher and higher frequencies in the spectrum whose use was previously impossible or very expensive. In particular, solid-state devices have been developed to amplify at millimeter wavelengths, or tens of gigahertz. Efficient, compact antennas are also available, such as planar antennas, which are often used in short-range devices. The development of surface acoustic wave (SAW) frequency-determining components allow generation of UHF frequencies with very simple circuits.
- Digital modulation techniques have largely replaced the analog methods of previous years, permitting a multiplication of the number of communication channels that can occupy a given bandwidth.
- We have seen much progress in circuit miniaturization. Hybrid integrated circuits, combining analog and digital functions on one chip, and radio-frequency integrated circuits are to a large part responsible for the amazingly compact size of cellular telephone handsets. This miniaturization is not only a question of convenience, but also a necessity for efficient design of very short-wavelength circuits.

## 1.3 Characteristics of short-range radio

"Short-range" and "low-power" are both relative terms, and their scope must be asserted in order to see the focus of this book. Hardly any of the applications that we discuss will have all of these characteristics, but all of them will have some of the following features:

- RF power output of several microwatts up to 100 milliwatts
- Communication range of centimeters up to several hundred meters
- Principally indoor operation
- Omnidirectional, built-in antennas
- Handheld, mobile terminals
- Simple construction and relatively low price in the range of consumer appliances
- Unlicensed operation

- Noncritical bandwidth specifications
- UHF operation
- Battery-operated transmitter or receiver

Our focus on implementation excludes cellular radios and wireless telephones, although an understanding of the material in this book will give the reader greater comprehension of the principles of operation of those ubiquitous devices.

## 1.4 Short-range radio applications

Table 1.1 lists some short-range radio applications and characteristics that show the focus of this text.

A new direction in short-range applications has appeared in the form of high-rate data communication devices for distances of several meters. This is being developed

**Table 1.1** Short-range radio characteristics

| Application | Frequencies (MHz) | Characteristics |
|---|---|---|
| Security Systems | 300-500, 800, 900 | Simplicity, easy installation |
| Emergency Medical Alarms | 300-500, 800 | Convenient carrying, long battery life, reliable |
| Computer Accessories—mouse, keyboard | UHF | High data rates, very short range, low cost |
| RFID | 100 kHz—2.4 GHz | Very short range, active or passive transponder |
| WLAN | 2.4, 5-6 GHz | High continuous data rates, spread spectrum and OFDM modulation |
| WPAN | 2.4 GHz | Medium data rates, low cost |
| Wireless Microphones; Wireless Headphones | VHF, UHF | Analog high fidelity voice modulation, moderate price |
| Keyless Entry—Gate, car door openers | UHF | Miniature transmitter, special coding to prevent duplication |
| Wireless bar code readers | 900 MHz, 2.4 GHz | Industrial use, spread spectrum, expensive |
| Wireless power meters | 2.4 GHz | Realtime measurement of electricity consumption and instantaneous demand |
| Internet of Things (IOT) | UHF, 2.4 GHz | Wireless monitoring and control of appliances and industrial apparatus with minimum human intervention |

by the Bluetooth consortium of telecommunication and PC technology leaders for eliminating wiring between computers and peripherals, as well as wireless internet access through cellular phones. Several other standards have been developed based on Zigbee physical layer specifications. Mass production brings sophisticated communication technology to a price consumers can afford, and fallout from this development is evident in the applications in the table above, improving their reliability and increasing their acceptance for replacing wiring.

## 1.5 Elements of wireless communication systems

Fig. 1.1 is a block diagram of a complete wireless system. Essentially all elements of this system will be described in detail in the later chapters of the book. A brief description of them is given below with special reference to short-range applications.

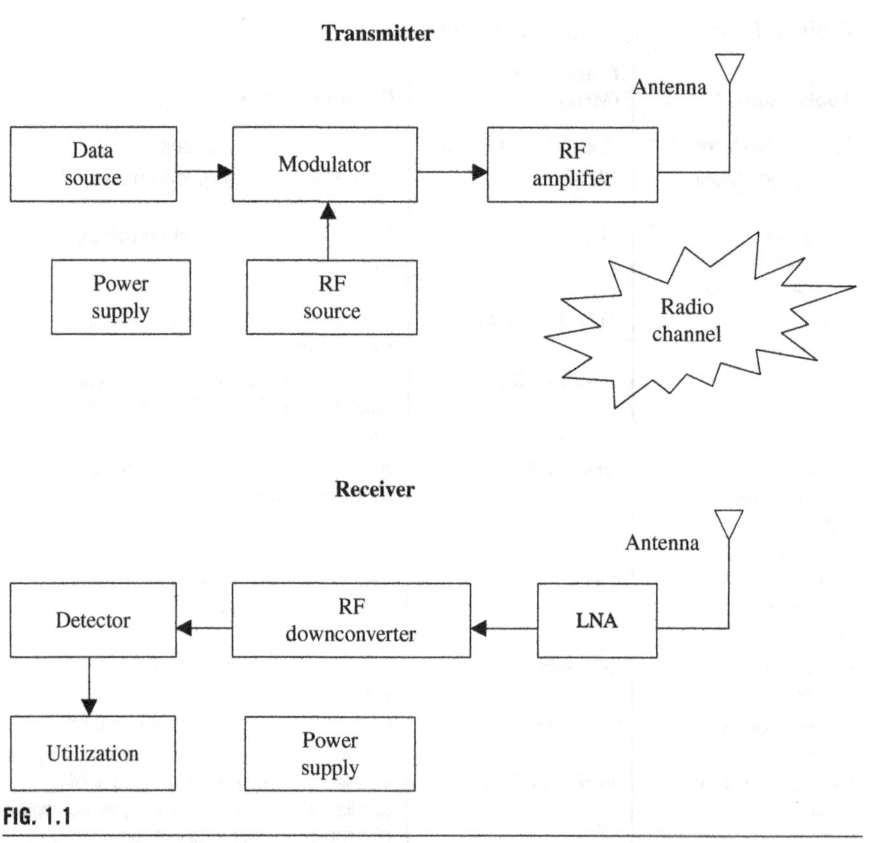


**FIG. 1.1**

The wireless system.

### 1.5.1 Data source

This is the information to be conveyed from one side to the other. Each of the devices listed in Table 1.1 has its own characteristic data source, which may be analog or digital. In many of the cases the data may be simple on/off information, as in a security intrusion detector, panic button, or manually operated remote control unit. In this case, a change of state of the data will cause a message frame to be modulated on an RF carrier wave. In its simplest form the message frame may look like Fig. 1.2. An address field identifies the unit that is transmitting and the data field conveys the specific information in on/off form. A parity bit or bits may be appended for checking message integrity.

Other digital devices have more complex messages. Computer accessories and WLANs send continuous digital data over the short-range link. These data is organized according to protocols that include sophisticated error detection and correction techniques (see Chapter 9).

Audio devices such as wireless microphones and headsets send analog data to the modulator. However, this data must be specially processed for best performance over a wireless channel. For FM transmission, which is widely used for these devices, a pre-emphasis filter increases the high frequencies before transmission so that, in the receiver, deemphasizing these frequencies will also reduce high-frequency noise. Similarly, dynamic range is increased by the use of a compandor. In the transmitter weak sounds are amplified more and strong signals are amplified less. The opposite procedure in the receiver reduces background noise while returning the weak sounds to their proper relative level, thus improving the dynamic range.

A quite different aspect of the data source is the case for RFIDs. Here, the data is not available in the transmitter but is added to the RF signal in an intermediate receptor tag, also called a transducer. See Fig. 1.3. This transducer may be passive or active, but in any case the originally transmitted radio frequency is modified by the transducer and detected by a receiver that deciphers the data added and passes it to a host computer.

### 1.5.2 Radio frequency generating section

This part of the transmitter consists of an RF source (oscillator or synthesizer), a modulator, and an amplifier. In the simplest short-range devices, all three functions may be included in a circuit of only one transistor. Chapter 5 details some of the common configurations. Again RFIDs are different from the other applications in that the modulation is carried out remotely from the RF source.

| Address bits | Data | Parity |
|---|---|---|

**FIG. 1.2**

Message frame.

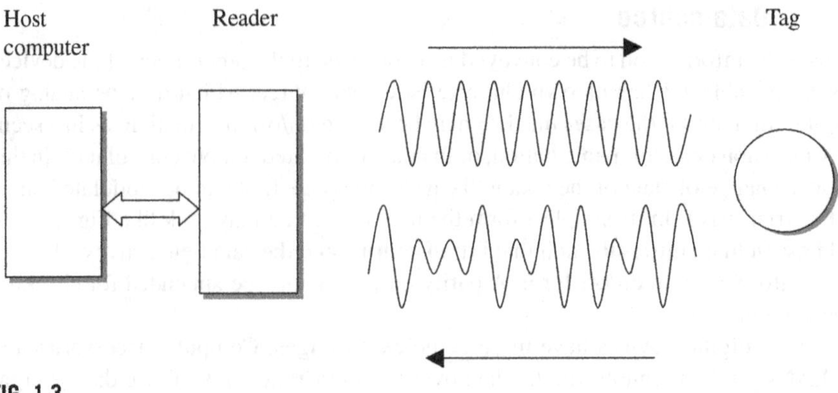


**FIG. 1.3**

RFID.

### 1.5.3 RF conduction and radiation

Practically all short-range devices have built-in antennas, so their transmission lines are relatively short and simple. However, particularly on the higher frequencies, their lengths are a high enough percentage of wavelength to affect the transmission efficiency of the transmitter. Chapter 3 explains antennas and the transmission lines encountered in short-range systems and the importance and techniques of proper matching. The antennas of short-range devices also distinguish them from other radio applications. They must be small—often a fraction of a wavelength—and omnidirectional for most uses.

### 1.5.4 Radio channel

By definition, the radio channel for short-range applications is short, and for a large part the equipment is used indoors. The allowed radio frequency power is relatively low and regulated by the telecommunication authorities. Also, the devices are often operated while close to or attached to a human (or animal) body, a fact which affects the communication performance. Reliable operating range is difficult to predict for these systems, and lack of knowledge of the special propagation characteristics of short-range radio by manufacturers, sellers, and users alike is a dominating reason for its reputation as being unreliable. Short-range devices are often used to replace hard wiring, so when similar performance is expected, the limitations of radio propagation compared to wires must be accounted for in each application. Chapter 2 brings this problem into perspective.

### 1.5.5 Receivers

Receivers have many similar blocks to transmitters, but their operation is reversed. They have an antenna and transmission line, RF amplifiers, and use oscillators in their operation. Weak signals intercepted by the antenna are amplified above the circuit noise by a low noise amplifier (LNA). The desired signal is separated from all the others and is shifted lower in frequency in a down converter, where it may be more effectively amplified to the level required for demodulation, or detection. The detector fulfills the ultimate purpose of the receiver; conversion of the data source which was implanted on the RF wave in the transmitter back to its original form.

While the transmitted power is limited by the authorities, receiver sensitivity is not, so the most obvious way to improve system performance is by improving the sensitivity and the selectivity to reduce interference from unwanted sources. This must be done under constraints of physics, cost, size, and often power consumption. Chapter 7 deals with these matters.

An important factor in low-power system design, and sometimes a controversial one, is the type of modulation to use. In the case of the simpler systems—security and medical alarms, for example—the choice is between amplitude shift keying (ASK), parallel to amplitude modulation in analog systems, and frequency shift keying (FSK), analogous to frequency modulation (FM). In Chapter 4 we look at the pros and cons of the two systems.

### 1.5.6 Power supplies

In most short-range devices, at least one side of the wireless link must be completely untethered—that's what wireless is for! When size is limited, as it is in hand-operated remote control transmitters and security detectors, battery size and therefore energy is limited. The need to change batteries often is not only highly inconvenient but also expensive, and this is an impediment to more widespread use of radio in place of wires. Thus, low current consumption is an important design aim for wireless devices. This is usually harder to achieve for receivers than for transmitters. Many short-range applications call for intermittent transmitter operation, in security systems, for example. Transmitters can be kept in a very low current standby status until data needs to be sent. The receiver, on the other hand, usually doesn't know when data will be sent so it must be alert all the time. Even so, there are techniques to reduce the receiver duty cycle so that it doesn't draw full current all the time. Another way to reduce receiver power consumption is to operate it in a reduced power standby mode, wherein operation goes to normal when the beginning of a signal is detected. This method often entails reduced sensitivity, however.

## 1.6 Overview of chapters

The book is organized as two sections. Chapters 2 through 10 contain material relating to short-range wireless devices in general. Chapters 11, 12, 13, and 14 describe networks and applications.

Chapter 2 discusses radio propagation and factors that affect communication range and reliability. Among the topics covered are link budget, diversity techniques and reliability improvement through multiple antennas (MIMO).

Antennas and transmission lines are the subjects of Chapter 3. Included is a discussion of metamaterials and how their use allows overriding some limitations of size and bandwidth in conventional designs.

Chapter 4 covers the various forms of signals used for information transmission and modulation, and overall wireless system properties.

Transmitter and receiver architectures are presented in Chapters 5 and 6 respectively. Chapter 5 explains signal generation and control including fractional-N synthesizers and direct digital synthesis. Among the types of receivers discussed in Chapter 6 are software defined and cognitive radios.

Radio system design is the subject of Chapter 7. It explains how to find sensitivity and from it, range, and the importance of quantifying noise. Information about power sources and battery types is also included in this chapter.

Examples of components, from chips to modules and complete boards are presented in Chapter 8, some of them designed to specific wireless communication standards.

Chapter 9 is an introduction to information theory. While not specific to wireless communication, the basics of probability, communication capacity and error detection and correction coding is essential for reliable wireless system design.

Regulations and Standards imposed by country and regional regulatory bodies are reviewed in Chapter 10, with emphasis on equipment requirements for non-licensed bands and applications.

The IEEE 802.11, the Wi-Fi wireless interface specification with its various implementations, is described in Chapter 11 whose subject is wireless local area networks. Also explained is Wi-Fi security. Chapter 12 covers wireless personal area networks including the Bluetooth, Zigbee and other specifications, both for the 2.4 GHz band and frequencies below 1000 MHz. Included in this chapter is a description of the ultra-wideband technology.

A comprehensive examination of radio frequency identification is the aim of Chapter 13. Various aspects of RFID are covered including active and passive technologies, protocols, security and NFC.

The final chapter of the book, Chapter 14, is about networks, applications and ancillary issues that aim to give a more complete coverage of short-range wireless communication to complement the theory and implementations discussed in detail in preceding chapters. Among its topics are the Internet of Things and the ad hoc networks and derivatives that support it, optical and millimeter wavelength communication, location awareness, wireless body area networks, and energy harvesting.

## 1.7 Summary

Short-range radio is an expanding and distinct adjunct to wireless communication in general. While its basic operating characteristics are the same as all radio systems, there are many features and specific problems that justify dealing with it as a separate field. Among them are low power, low cost, small size, battery operation, uncertainty of indoor propagation, and unlicensed operation on crowded bands. The rest of this book delves into the operational and design specialties of short-range radio communication from the electromagnetic propagation environment through antennas, receivers and transmitters, regulations and standards, and a bit of relevant information theory. The last four chapters describe in detail current developments that are bringing wireless to the home, office and factory at an unprecedented extent. Electronic worksheets are available and referred to throughout the book that can be used to work out examples given in the text, and to help the reader solve his own specific design problems.

CHAPTER

# Radio propagation

# 2

It is fitting to begin a book about wireless communication with a look at the phenomenon that lets us transfer information from one point to another without any physical medium—the propagation of radio waves. If you want to design an efficient radio communication system, even for operation over relatively short distances, you should understand the behavior of the wireless channel in the various surroundings where this communication is to take place. While the use of "brute force"—increasing transmission power—could overcome inordinate path losses, limitations imposed on design by required battery life, or by regulatory authorities, make it imperative to develop and deploy short-range radio systems using solutions that a knowledge of radio propagation can give.

The overall behavior of radio waves is described by Maxwell's equations. In 1873, the British physicist James Clerk Maxwell published his *Treatise on Electricity and Magnetism* in which he presented a set of equations that describe the nature of electromagnetic fields in terms of space and time. The Appendix gives a brief description of those equations. Heinrich Rudolf Hertz performed experiments to confirm Maxwell's theory, which led to the development of wireless telegraph and radio. Maxwell's equations form the basis for describing the propagation of radio waves in space, as well as the nature of varying electric and magnetic fields in conducting and insulating materials, and the flow of waves in waveguides. From them, you can derive the skin effect equation and the electric and magnetic field relationships very close to antennas of all kinds. A number of computer programs on the market, based on the solution of Maxwell's equations, help in the design of antennas, anticipate electromagnetic radiation problems from circuit board layouts, calculate the effectiveness of shielding, and perform accurate simulation of ultra high frequency and microwave circuits. While you don't have to be an expert in Maxwell's equations to use these programs (you do in order to write them!), having some familiarity with the equations may take the mystery out of the operation of the software and give an appreciation for its range of application and limitations.

## 2.1 Mechanisms of radio wave propagation

Radio waves can propagate from transmitter to receiver in four ways: through ground waves, ionosphere waves, troposphere waves and space waves [1, p. 11.4].



Short-range Wireless Communication. https://doi.org/10.1016/B978-0-12-815405-2.00002-6
© 2019 Elsevier Inc. All rights reserved.

Ground waves exist only for vertical polarization, produced by vertical antennas, when the transmitting and receiving antennas are close to the surface of the earth (see **Polarization** under Section 3.1.4 in Chapter 3). The transmitted radiation induces currents in the earth, and the waves travel over the earth's surface, being attenuated according to the energy absorbed by the conducting earth. The reason that horizontal antennas are not effective for ground wave propagation is that the horizontal electric field that they create is short circuited by the earth. Ground wave propagation is dominant only at relatively low frequencies, up to a few MHz, and generally follows the curvature of the earth well beyond line-of-sight, so it needn't concern us here.

Ionosphere wave propagation, also called sky waves, is dependent on reflection from the ionosphere, a region of rarified air high above the earth's surface that is ionized by sunlight (primarily ultraviolet radiation). The ionosphere is responsible for long-distance communication in the high-frequency bands between 3 and 30 MHz and somewhat beyond. It is very dependent on time of day, season, longitude on the earth, and the multiyear cyclic production of sunspots on the sun. It makes possible long-range communication using very low power transmitters. Most short-range communication applications that we deal with in this book use VHF, UHF, and microwave bands, generally above 40 MHz. There are times when ionosphere reflection occurs at the low end of this range, and then sky wave propagation can be responsible for interference from signals originating hundreds of kilometers away. However, in general, sky wave propagation does not affect the short-range radio applications that we are interested in.

The troposphere is the lower portion of the atmosphere where air temperature increases with height and pressure decreases in height, resulting in changes in the refractive index of the air and consequently bending of electromagnetic waves at VHF and UHF beyond the line-of-sight that otherwise would be limited by the curvature of the earth. This propagation mode is also not relevant in our discussion of short-range low power communication.

Virtually all propagation of short-range signals is via space waves, where radiation travels between transmitter and receiver on multiple direct paths at the speed of light. In almost all cases, the exceptions being millimeter wave and optical short-range communication over a line-of-sight narrow beam path, the received signal is a vector sum of a direct line-of-sight signal and signals from the same source that are reflected, scattered or diffracted off the earth and objects in the vicinity of the transmitter.

Free space signal strength, when there are no reflections from earthly objects or the ionosphere, is a function of the dispersion of the waves from the transmitter antenna. Power density $S$ from an isotropic radiator at a distance $d$ equals the transmitter power $P_t$ divided by the surface area of a sphere with radius $d$:

$$S = \frac{P_t}{4\pi d^2} \tag{2.1a}$$

An isotropic radiator radiates equally in all directions. Thus, in free space, the power density is inversely proportional to the square of the distance. When the radiator has gain $G_t$ and power into the antenna is $P_t$, the power density in the direction of maximum radiation is

$$S = \frac{P_t G_t}{4\pi d^2} \tag{2.1b}$$

In this free-space case the signal field strength $E$ decreases in inverse proportion to the distance away from the transmitter antenna. When the radiated power $P_t G_t$ is known, the field strength is:

$$E = \sqrt{S \cdot \Omega_s} = \frac{\sqrt{30 P_t G_t}}{d} \tag{2.2}$$

where $\Omega_s = 120\pi$ is the impedance of free space. When $P_t$ is in watts and $d$ is in meters, $E$ is in volts/meter.

Power into the receiver $P_r$ given the radiated power $P_t G_t$ and signal wavelength $\lambda$ is

$$P_r = S \cdot A_e = \frac{P_t G_t G_r \lambda^2}{(4\pi d)^2} \tag{2.3}$$

$G_r$ is receiver antenna gain, and $\lambda$ is the wavelength. $A_e$ is the effective antenna area:

$$A_e = \frac{\lambda^2 G_r}{4\pi} \tag{2.4}$$

The effective antenna area is related physically to the area of an aperture antenna, such as a horn antenna or parabolic antenna. It equals the actual aperture area times an aperture efficiency constant that is less than unity. However, Eq. (2.4) is just as relevant for a wire antenna.

Eq. (2.3) is called the Friis free space equation. It is particularly accurate at high UHF and microwave frequencies when high-gain antennas are used, located many wavelengths above the ground. Signal strength between the earth and a satellite, and between satellites, also follows the inverse distance law, but these cases aren't in the category of short-range communication! At microwave frequencies, signal strength is also reduced by atmospheric absorption caused by water vapor and other gases that constitute the air.

Eqs. (2.1), (2.2), and (2.3) are valid in the far field, where the free space impedance is a constant ratio of the electric field strength and the magnetic field strength. The far field distance $d$ satisfies the following conditions:

$$\begin{aligned} d &> \frac{2D^2}{\lambda} \\ d &\gg D \\ d &\gg \lambda \end{aligned} \tag{2.5}$$

where $D$ is the largest linear dimension of the antenna [2, p. 108]. As a rule of thumb, for $D \geq \lambda$ the first condition in Eq. (2.5) holds, whereas for $D < \lambda$ consider the far field as starting from a distance of twice the wavelength.

Mathcad worksheet **"Radiate.mcdx"** is useful for calculations involving free space radio wave propagation and antenna parameters.

## 2.2 Open field propagation

Although the formulas in the previous section are useful in some circumstances, the actual range of a VHF or UHF signal is affected by reflections from the ground and surrounding objects. The path lengths of the reflected signals differ from that of the line-of-sight signal, so the receiver sees a vectorially combined signal with components having different amplitudes and phases. The reflection causes a phase reversal. A reflected signal having a path length exceeding the line-of-sight distance by exactly the signal wavelength or a multiple of it will almost cancel completely the desired signal ("almost" because its amplitude will be slightly less than the direct signal amplitude due to longer path length and absorption by the reflector). On the other hand, if the path length of the reflected signal differs exactly by an odd multiple of half the wavelength, the total signal will be strengthened by "almost" two times the free space direct signal.

In an open field with flat terrain there will be no reflections except the unavoidable one from the ground. It is instructive and useful to examine in depth the way received power varies with distance in this case.

In Fig. 2.1 we see transmitter and receiver antennas separated by distance $d$ and situated at heights $h_1$ and $h_2$. Communication is symmetrical so it doesn't matter which side is the transmitter. Using trigonometry, we can find the line of sight and reflected signal path lengths $d_1$ and $d_2$. Just as in optics, the angle of incidence equals the angle of reflection $\theta$. We get the relative strength of the direct signal and reflected signal using the inverse path length relationship. If the ground were a perfect mirror, the reflected signal strength would be proportional to the inverse of $d_2$. In this case, the reflected signal phase would shift 180 degrees at the point of reflection. However, the ground is not a perfect reflector. Its characteristics as a reflector depend on its conductivity, permittivity, the polarization of the signal and its angle of incidence. In the Mathcad worksheet **"range.mcdx"** we have accounted for these factors to find the reflection coefficient, which approaches $-1$ as the distance from the

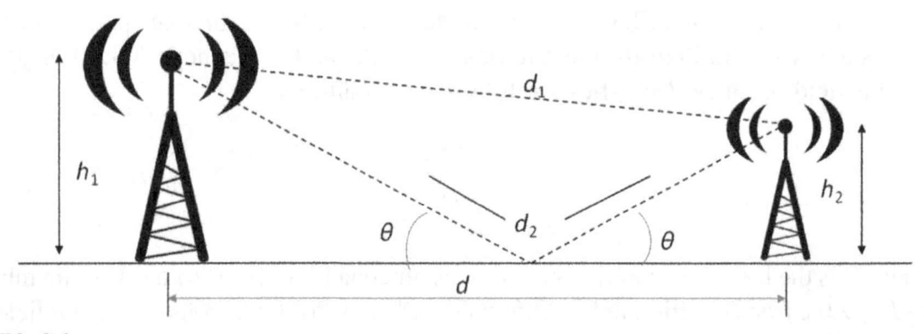


**FIG. 2.1**

Open field signal paths.

transmitter increases and $\theta$ approaches zero. The signal strengths of the direct and reflected waves reaching the receiver are represented as complex numbers since they have both phase and amplitude. The two signals are added vectorially and the resultant is the total open space signal strength at the given distance.

Fig. 2.2 is a plot of the total open field path gain, shown as a solid line curve, versus distance using the following parameters:

Polarity—horizontal
Frequency—300 MHz
Antenna heights—both 3 m
Relative ground permittivity—15

The path gain is the ratio of the power at the input to the receiver antenna over the radiated power from the transmitter. More on this in Section 2.5. The mathematical details are given in the Mathcad worksheet **"range.mcdx."** The received power, not including receiver antenna gain, at a given distance is the path gain in dB plus the radiated power in dBm or dBW.

Also shown is a plot of free space path gain versus distance (dotted line). In the open field plot, signal strength is referenced to the free space field strength at a range of 1 m.

Notice in Fig. 2.2 that, up to a range of around 30 m, there are several sharp depressions of path gain, but the signal strength is mostly higher than it would be

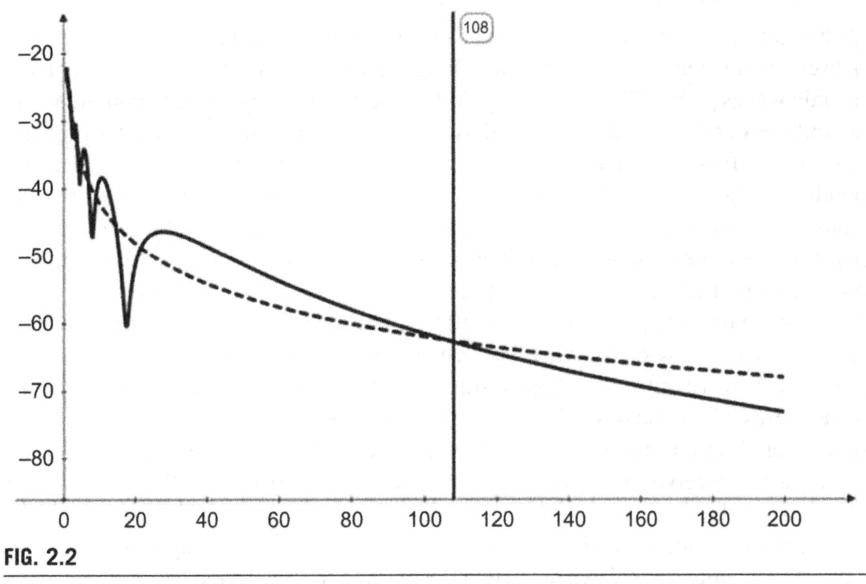


**FIG. 2.2**

Path gain vs. range at 300 MHz. Solid line is open field gain, dashed line is free space gain.

in free space up to around 100 m. Beyond this distance, signal strength decreases more rapidly than for the free space model. Whereas there is an inverse square law for the path gain vs. distance in free space, in the open field beyond around 100 m (for these parameters) the path gain follows an inverse power of 4 law. Increasing the antenna heights extends the distance at which the inverse power of 4 law starts to take effect. This distance $d_m$, at which open space path gain and free space path gain are equal, is shown by a vertical line in the figure. It can be approximated by

$$d_m = 12h_1h_2/\lambda \quad (2.6)$$

where $h_1$ and $h_2$ are the transmitting and receiving antenna heights above the ground and $\lambda$ is the wavelength, all in the same units as the distance $d_m$.

In plotting Fig. 2.2, we assumed horizontal polarization. Both antenna heights, $h_1$ and $h_2$, are 3 m. When vertical polarization is used, the extreme local variations of signal strengths up to around 50 m are reduced, because the ground reflection coefficient is less at larger reflection angles. However, for both polarizations, the inverse power of 4 law comes into effect at approximately the same distance. This distance in Fig. 2.2 where $\lambda$ is 1 m is, from Eq. (2.6): $d_m = 108$ m. In Fig. 2.2 we see that this is the distance where the open-field path gain falls continuously below the free-space path gain.

## 2.3 Diffraction

Diffraction is a propagation mechanism that permits wireless communication between points where there is no line-of-sight path due to obstacles that are opaque to radio waves [2, p. 129]. For example, diffraction makes it possible to communicate around the Earth's curvature, as well as beyond hills, buildings, and other obstructions. It also fills in the spaces around obstacles when short-range radio is used inside buildings. Fig. 2.3 is an illustration of diffraction geometries, showing an obstacle whose extremity has the shape of a knife edge. The obstacle should be seen as a half plane whose dimension is infinite into and out of the paper. The signal strength at a receiving point relative to the free-space signal strength without the obstacle is the diffraction gain. The phenomenon of diffraction is due to Huygen's principle which states that each point on the wave front emanating from the transmitter is a source of a secondary wave emission whose resultant is in the direction of propagation. Thus, at the knife edge of the obstacle, as shown in Fig. 2.3A, there is radiation in all directions, including into the shadow. In Fig. 2.3B, there is a line-of-sight path, but the signal at the receiver also gets vectorial contributions from the diffraction at the obstruction edge.

Diffraction gain $G_r$ can be expressed in a rather complicated way as a function of a parameter $\nu$ that normalizes the height $h$ with the transmitter and receiver distances

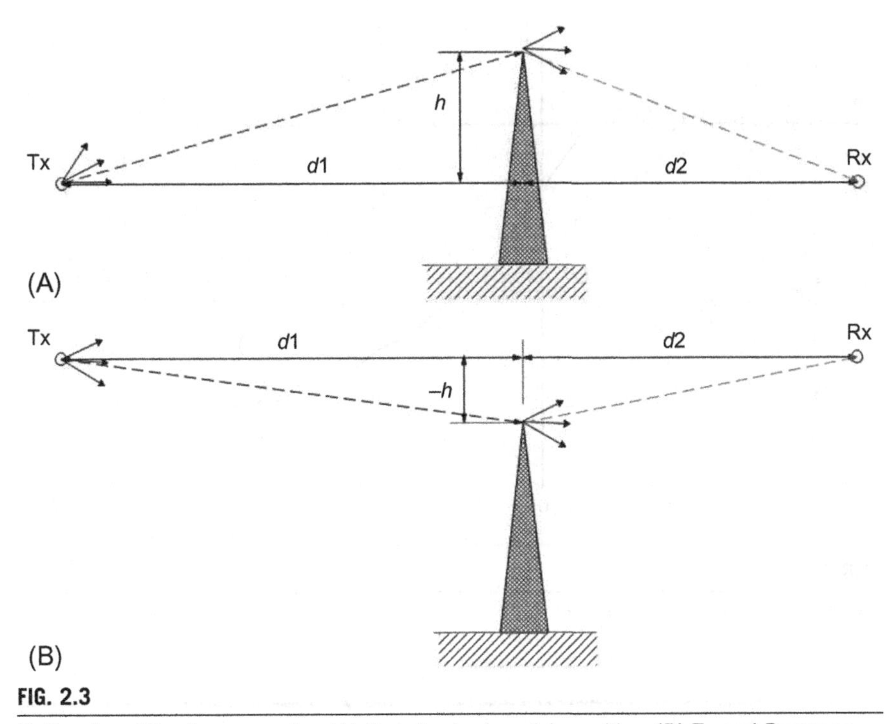


**FIG. 2.3**

Knife-edge diffraction geometry. (A) Rx is in shadow, $h$ is positive. (B) Tx and Rx are on line-of-sight, $h$ is negative.

from the obstacle, $d_1$ and $d_2$, and the wavelength $\lambda$. $E_d$ is the field strength of the diffracted signal at the receiver and $E_0$ is the line-of-sight field strength.

$$G_r = 10\log(|F(\nu)| \tag{2.7}$$

where

$$F(v) = \frac{E_d}{E_0} = \frac{1+j}{2} \int_{v}^{\infty} \exp\left(-j\pi t^2/2\right) dt \tag{2.8}$$

and

$$\nu = h\sqrt{\frac{2(d_1 + d_2)}{\lambda d_1 d_2}} \tag{2.9}$$

Fig. 2.4 shows a plot of the diffraction gain as a function of normalized height $\nu$. It is used by calculating $\nu$ from Eq. (2.9) and extending a vertical line from that point on the abscissa to the curve. Example 2.1 demonstrates this.

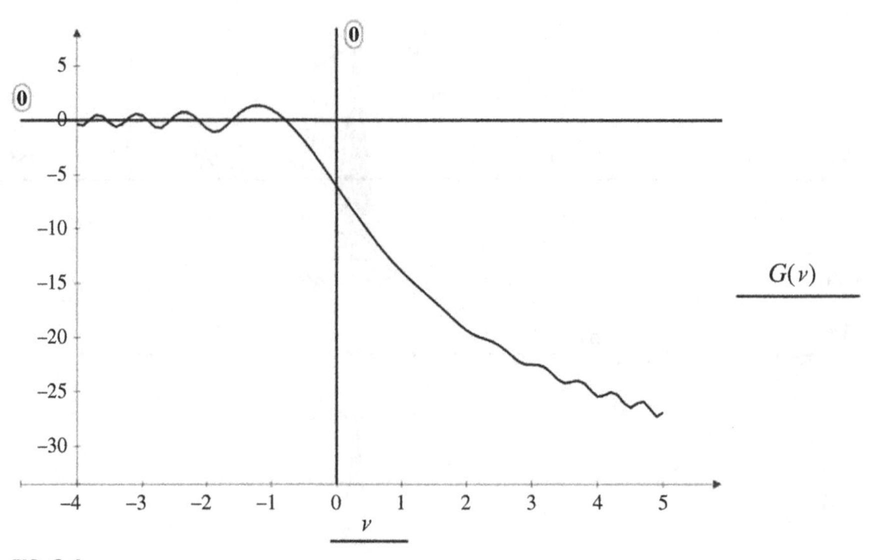


**FIG. 2.4**

Normalized plot of diffraction gain.

## Example 2.1

A long concrete wall 6 m high separates two radio terminals, each 20 m from it. Operating frequency is 434 MHz. The terminal antennas are $x$ meters above the ground. The wall is opaque to radio waves. Find the diffraction gain for the following cases:

**(1)** $x = 3$
**(2)** $x = 6$
**(3)** $x = 8$

**Solution**

Determine $h$ for the three cases and then calculate $\nu$ using Eq. (2.9). Wavelength $\lambda$ is $c/434$ MHz $= 0.691$ m. The diffraction gain is found from Fig. 2.4.

**(1)** $h = 3$, $\nu = 1.6$, $G = -17$ dB
**(2)** $h = 0$, $\nu = 0$, $G = -6$ dB
**(3)** $h = -2$, $\nu = -1.076$, $G = 1.2$ dB

Mathcad worksheet **"diffraction.mcdx"** can be used to conveniently solve this and similar problems.

Note that the barrier affects the received signal strength even when there is a clear line of sight between the transmitter and receiver ($h$ is negative as shown in Fig. 2.3B). When the barrier edge is on the line of sight, diffraction gain is approximately −6 dB, and as the line-of-sight path gets farther from the barrier (to the left of the $\nu = 0$ axis in Fig. 2.4), the signal strength varies in a cyclic manner around 0 dB gain. When the path from transmitter to receiver is blocked by the barrier and gets farther from the barrier edge into the shadow, the signal attenuation increases steeply.

Admittedly, the situation depicted in Fig. 2.3 is idealistic, since it deals with only one barrier of very large extent. Normally there may be several partitions and other obstacles near or between the line of sight path and a calculation of the diffraction gain would be very complicated, if not impossible. However, a knowledge of the idealistic behavior of the diffraction gain and its dependence on distance and frequency can give qualitative insight. The Mathcad worksheet **"diffraction.mcdx"** lets you see how the various parameters affect the diffraction gain.

Virtual propagation paths between and around the line-of-sight between a transmitter and receiver are considered as bounded by specially defined ellipsoid surfaces called Fresnel zones. They form concentric circles as they are cut by a plane surface normal to the line-of-sight. Path lengths touching each Fresnel zone at all points on the ellipsoid are longer than the line-of-sight path by a multiple $n$ of one half wavelength. Referring to the dimensions in Fig. 2.3, the radius of the first Fresnel zone is

$$r = \sqrt{\frac{\lambda \cdot d1 \cdot d2}{d1 + d2}} \tag{2.10}$$

An obstruction that is outside of the first Fresnel zone doesn't cause significant diffraction loss. This rule is relevant even for the case (most cases, in fact) where the obstruction doesn't have a knife edge as assumed here. A rule-of-thumb for microwave links is that a radius of 55% of $r$ should be clear of obstructions to avoid significant diffraction loss [2]. That is, $h < -0.55r$.

## 2.4 Scattering

A third mechanism affecting path loss, after reflection and diffraction, is scattering. Rough surfaces in the vicinity of the transmitter do not reflect the signal cleanly in the direction determined by the incident angle, but diffuse it, or scatter it in all directions. As a result, the receiver has access to additional radiation and path loss may be less than it would be from considering reflection and diffraction alone. The degree of roughness of a surface and the amount of scattering it produces depends on the height of the protuberances on the surface compared to a function of the wavelength and the angle of incidence. The critical surface height $h_c$ is given by

$$h_c = \frac{\lambda}{8\cos\theta_i} \tag{2.11}$$

where $\lambda$ is the wavelength and $\theta_i$ is the angle of incidence. It is the dividing line between smooth surfaces. Protuberances greater than $h_c$ are considered rough and the reflection coefficient has to be modified by a scattering loss factor [3, p. 360].

## 2.5 Path loss

Equivalent isotropic radiated power, EIRP, is the total radiated power from a transmitter antenna times the numerical directivity of the antenna in the direction of the receiver, or the power delivered to the antenna times the antenna numerical gain. The numerical path loss is the ratio of EIRP to the power available at the receiver, which is the output of an isotropic antenna substituted for the receiver antenna. An isotropic radiator is an ideal antenna that radiates equally in all directions and therefore has a gain of 0 dB. Sometimes, for clarity, the ratio is called the *isotropic* path loss. In free space, the isotropic path loss $PL$ is derived from Eq. (2.3), resulting in

$$PL(d) = \frac{\text{EIRP}}{(P_r/G_r)} = \frac{P_t G_t}{(P_r/G_r)} = \left(\frac{4\pi d}{\lambda}\right)^2 \quad (2.12)$$

We have just examined several factors that affect the nonfree space path loss of VHF-UHF signals—ground reflection, diffraction, and scattering. For a given site, it would be very difficult to calculate the path loss between transmitters and receivers, but empirical observations have allowed some general conclusions to be drawn for different physical environments. These conclusions involve expressing the rate of signal attenuation in terms of an exponent and a short reference distance $d_0$, up to which free space propagation holds. We then can write the path loss for $d > d_0$ as dependent on the exponent $n$:

$$PL = PL(d_0)\left(\frac{d}{d_0}\right)^n = \left(\frac{4\pi d_0}{\lambda}\right)^2 \left(\frac{d}{d_0}\right)^n \quad (2.13)$$

Table 2.1 shows the exponent $n$ for different environments.

The path loss in Eq. (2.13) may be expressed in dB, while adding a term to express statistical variation:

$$PL_{\text{dB}} = PL(d_0) + 10n \cdot \log\left(\frac{d}{d_0}\right) + X_\sigma|_{\text{dB}} \quad (2.14)$$

where $X_\sigma$, in dB, is a Gaussian random variable with standard deviation $\sigma$ that expresses the path loss variations in different positions or environments with the

**Table 2.1** Path loss exponents for different environments [3, p. 362]

| Environment | Path gain exponent *n* |
|---|---|
| Free space | 2 |
| Open field (long distance) | 4 |
| Cellular radio—urban area | 2.7-4 |
| Shadowed urban cellular radio | 5-6 |
| In building line-of-sight | 1.6-1.8 |
| In building—obstructed | 4-6 |

same exponent $n$. $n$ is determined from a linear regression of large number of measurements in a given environment.

The inverse path loss ratio is sometimes more convenient to use. This is the path gain and when expressed in decibels is the negative of Eq. (2.14).

## 2.6 Link budget

A link budget is a tabulation of factors that affect signal strength along with transmission power and receiver sensitivity, with the purpose of determining the trade offs needed to maintain a wireless communication link. The following equation is perhaps the simplest expression of a link budget.

$$\mathrm{EIRP}_{\mathrm{dB}} = PL_{\mathrm{dB}} - G_{r\_\mathrm{dB}} + P_{r\_\mathrm{min}} + M \tag{2.15}$$

where $PL_{\mathrm{dB}}$ is path loss in dB, $G_{r\_\mathrm{dB}}$ is receiver antenna gain in dB, $P_{r\_\mathrm{min}}$ is receiver sensitivity and $M$ is the link margin.

In this case, when path loss over a desired communications range and receiver antenna gain and sensitivity for a given link performance is known, then the minimum required transmitter radiated power is estimated. A positive link margin is added to assure a required up-time percentage, or link reliability, when the other parameters, particularly path loss, are not known to the desired precision. Any unknown parameter in the expression can be found, for example receiver sensitivity when transmitted power is known, by rearranging the terms in the equation. All quantities in Eq. (2.15) are in decibels. Note that transmitter radiated power is the equivalent isotropic radiated power EIRP, which is the power delivered to the antenna in dBm or dBW plus the antenna gain related to an isotropic antenna in dBi.

Fig. 2.5 shows example plots of path gain versus distance for free-space propagation, open field propagation, and path gain with exponent 6, a worst case taken from Table 2.1 for "In building—obstructed." Transmitter and receiver heights are 2 m, polarization is vertical, and the frequency is 915 MHz. The reference distance $d_0$ is 5 m, used for the open field curve and the $n = 6$ curve. When there is a known dependence of path loss (or path gain) on distance, then distance can be found from the path loss.

### Example 2.2

Assume that the open field range of a security system transmitter and receiver is 300 m. Transmitter and receiver antennas are 2 m high, both with 0 dB gain. The operating frequency is 915 MHz, transmitter power is 0.5 mW and link margin is 0. Reference distance is 5 m. What range can we expect for their installation in a building with many obstructions?

Solution

Step 1: Find receiver sensitivity. From the open field curve in Fig. 2.5, path gain at 300 m is −87.4 dB. Path loss is the negative of this value. Convert transmitter radiated power to dBm:

Continued

**Example 2.2—Cont'd**

$EIRP_{dBm} = 10\log(EIRP/1\,mW) = -3\,dBm$. Using the link budget expression (2.15), receiver sensitivity equals −90.4 dBm.

Step 2: Using the maximum value of $n$ from Table 2.1, "in building—obstructed," $n = 6$. From Eq. (2.15) path loss $PL_{dB} = -3\ dBm + 90.4\ dBm - 0\ dB = 87.4\ dB$. Using the $n = 6$ curve in Fig. 2.5, the ordinate of −87.4 dB path gain corresponds to a distance of 24.2 m. Thus, a wireless system that has an outdoor range of 300 m may be effective only over a range of around 24 m, on the average, in an indoor installation that has many obstructions.

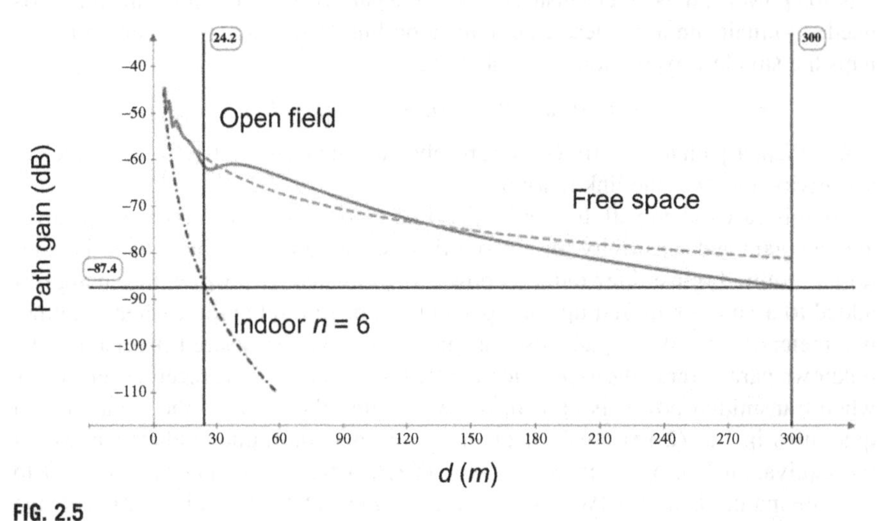


**FIG. 2.5**

Path gain for different propagation environments.

## 2.7 Multipath

We have seen that reflection of a signal from the ground has a significant effect on the strength of the received signal. The nature of short-range radio links, which are very often installed indoors and use omnidirectional antennas, makes them accessible to a multitude of reflected rays. Indoors, for example, signals are reflected from floors, ceilings, walls, and the various furnishings and people that are invariably present near the transmitter and receiver while outdoors reflections are from buildings and vehicles among other common objects. Thus, the total signal strength at the receiver is the vector sum of not just two signals, as we studied in Section 2.2, but of many signals traveling over multiple paths. In many cases indoors, there is no direct line-of-sight path, and all signals are the result of reflection, diffraction and scattering. This is shown in Fig. 2.6.

The communication channel can be characterized observing the channel impulse response. If a short pulse is transmitted on the channel carrier frequency, the output

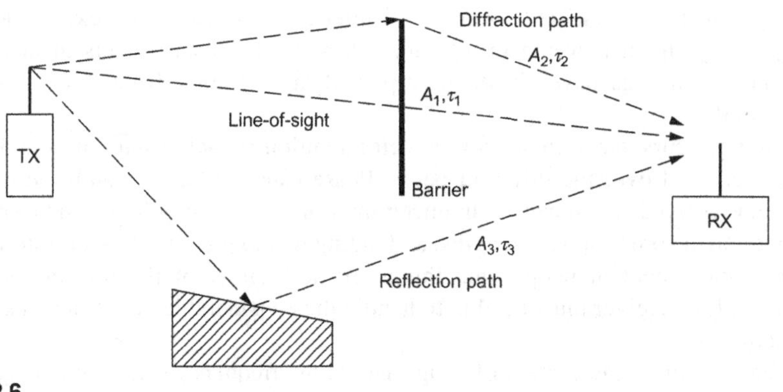


**FIG. 2.6**

Multipath.

of the receiver may look like Fig. 2.7. The transmitted pulse width should be considerably shorter than the differences in reflection times from objects in the vicinity of the link.

Fig. 2.7, a *power delay profile*, shows the effects of the multiple paths of different distances that a single transmitted short pulse takes on its way to the receiver. $\tau_k$ is the propagation time over one of the paths. This is the *excess delay*, measured from the time of arrival of the earliest signal at the receiver. *Maximum excess delay*, $\tau_m$, is the time of the longest path whose signal is not lower than a given amount (say, 10 dB) below the maximum, which isn't necessarily the first, direct path return. Other excess delay parameters are *mean excess delay*, which is the first moment of the power delay profile and the standard deviation from the mean, the *RMS delay*

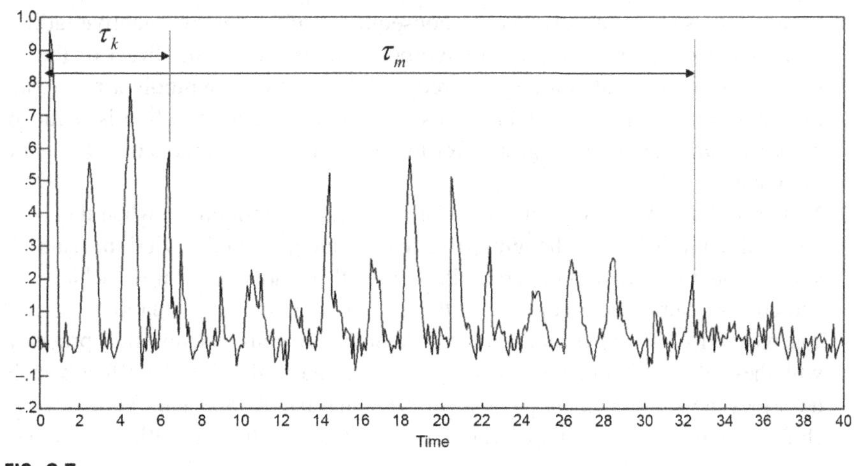


**FIG. 2.7**

Power delay profile.

*spread*, which is calculated from the differences squared of the excess delays weighted by signal power over all paths [2, p. 199]. Measurements in factories and other buildings have shown multipath delays ranging from 40 to 800 ns [3, p. 366].

Fading occurs due to the result of vector addition at each instant of time of the waves received over the different paths. Phase cancellation and addition of the resultant received signal causes an uncertainty in signal strength over small range changes on the order of wavelengths, and fading will express itself over time when there is movement of people and things in the vicinity of the transmitter and receiver. The receiver must be able to handle the considerable variations in signal strength.

There are three categories of fading: flat fading, frequency selective fading, and motion or time fading.

**(a)** Flat fading occurs when symbol length is significantly longer than the maximum excess delay. In this case, signal amplitude remains essentially constant for the duration of each symbol but the value of the amplitude and symbol phase changes over short distances. Flat fading is also described in the frequency domain as being the case where signal bandwidth is smaller than the channel coherence bandwidth. The coherence bandwidth is proportional to the inverse of the RMS delay spread, or roughly to the reciprocal of the maximum excess delay [4, p. 960]. Narrow band signals, then, are susceptible to flat fading. Flat fading is countered by error correction coding and diversity reception.

**(b)** Frequency selective fading occurs when the symbol length is shorter than the delay spread, or equivalently when signal bandwidth is larger than the channel bandwidth. It is therefore the type of multipath interference encountered by high bandwidth, high data rate signals in outdoor areas where excess delays are relatively long. A consequence of frequency selective fading is intersymbol interference where symbols received over the direct or the shortest reflecting paths are interfered with by previous symbols arriving at the same time over longer delay paths. Frequency selective fading is reduced by *equalization*, where a digital filter in the receiver counteracts the effects of the channel.

**(c)** Motion fading. When the transmitter or receiver is in motion, or when the physical environment is changing (tree leaves fluttering in the wind, moving vehicles, and people moving around), there will be fading over time, which can contain amplitude and phase distortion, and time delay fluctuations. The receiver AGC, synchronization and demodulation circuits must deal properly with these effects. Path length change over time causes the Doppler effect, which describes the shifting of frequencies in the signal bandwidth. The frequency shift due to the velocity component of motion along the direct or reflecting path is

$$f_d = \frac{V}{c} f_c \tag{2.16}$$

where $f_d$ is the Doppler shift, $V$ the velocity, $c$ the speed of light, and $f_c$ is the carrier frequency. $f_d$ is positive when the path length is shortened with time and negative when it is lengthened. It is maximum over the direct path or reflecting paths on the extended line-of-sight, for example when transmitter motion relative to receiver is along the line of sight. Doppler shift for movement normal to line-of-sight is zero. The different Doppler shifts along all of the scattering paths create a Doppler spectrum and fading because of the changing of the vectorial combinations of the reflected signals. Motion fading is called slow fading when signal amplitude and phase do not change appreciably during a symbol period and fast fading otherwise.

## 2.8 Rayleigh fading

In describing the variation of the resultant signal amplitude and phase in a multipath environment, we distinguish two cases: (1) There is no line-of-sight path and the signal is the resultant of a large number of randomly distributed reflections. This is Rayleigh fading. (2) The random reflections are superimposed on a signal over a dominant constant path, usually the line of sight. This is Rician fading.

Short-range radio systems that are installed indoors or outdoors in built-up areas are subject to multipath fading essentially of the first case. Our aim in this section is to determine the signal strength margin that is needed to ensure that reliable communication can take place at a given probability. While in many situations there will be a dominant signal path in addition to the multipath fading, restricting ourselves to an analysis of the case where all paths are the result of random reflections gives us an upper bound on the required margin. See Ref. [2, p. 212] for details of Rician fading.

The Rayleigh fading can be described by a received signal $R(t)$, expressed as

$$R(t) = r \cdot \cos(2\pi \cdot f_c \cdot t + \theta) \tag{2.17}$$

where $r$ and $\theta$ are random variables for the peak signal, or envelope, and phase. Their values vary with time, when various reflecting objects are moving (people in a room, for example), or with changes in position of the transmitter or receiver which are small in respect to the distance between them. We are not dealing here with the large-scale path gain that is expressed in Eqs. (2.12), (2.13). For simplicity, Eq. (2.17) shows a CW (continuous wave) signal as the modulation terms are not needed to describe the fading statistics. Received signal power vs. time is shown in Fig. 2.8 under the condition of Rayleigh fading.

The envelope of the received signal, $r$, can be statistically described by the Rayleigh distribution whose probability density function is

$$p(r) = \frac{r}{\sigma^2} e^{\frac{-r^2}{2\sigma^2}} \tag{2.18}$$

where $\sigma^2$ is the average received signal power. This function is plotted in Fig. 2.9. The curve is normalized with $\sigma$ equal to 1. In this plot, the average, or mean, value of

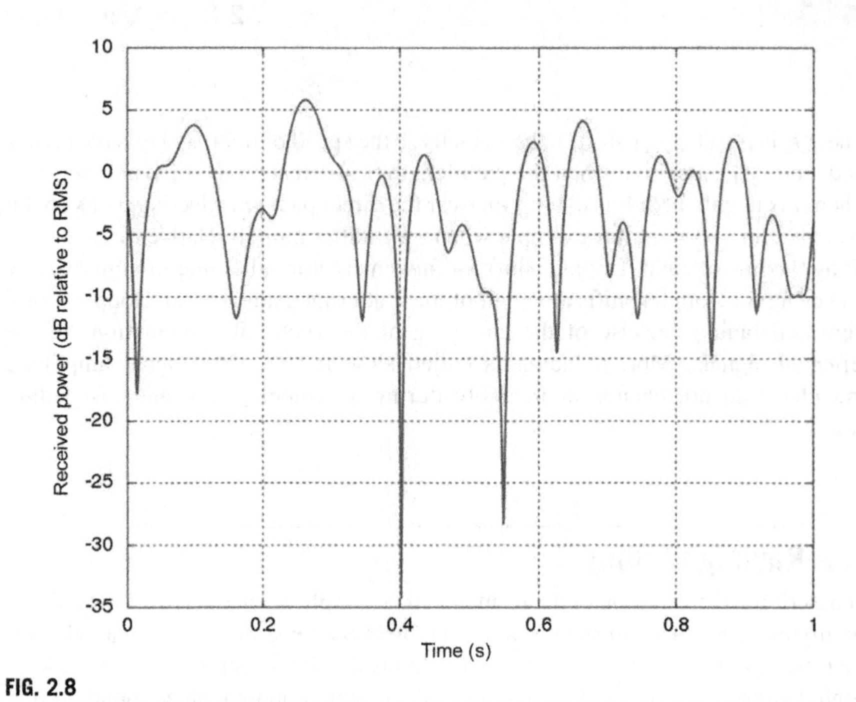


**FIG. 2.8**

Rayleigh fading signal of 10 Hz Doppler.

*Copied from Wikipedia, "Rayleigh fading". Attribution: Splash at the English language Wikipedia.*

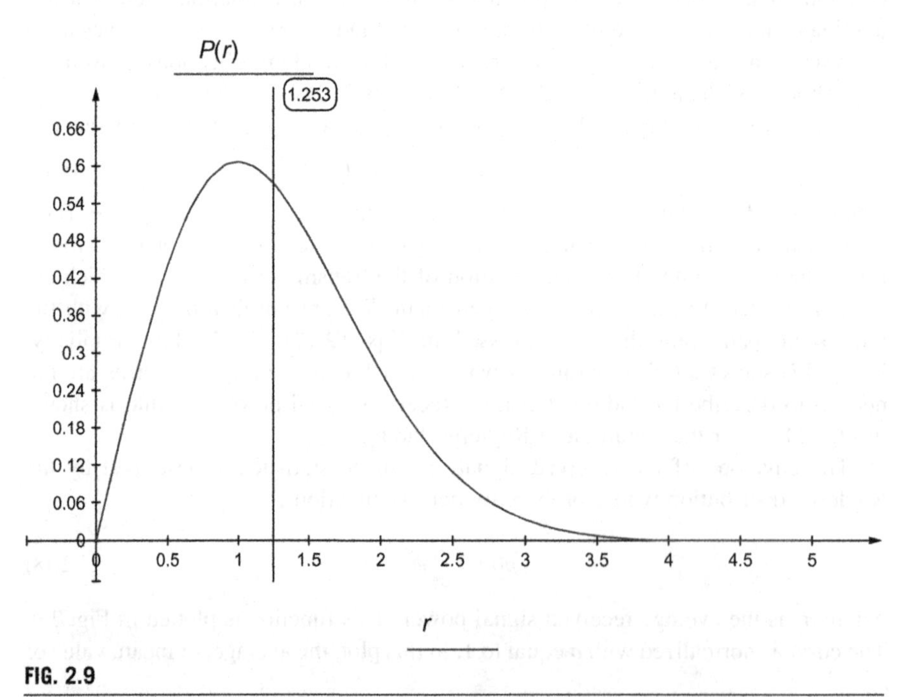


**FIG. 2.9**

Rayleigh probability density function.

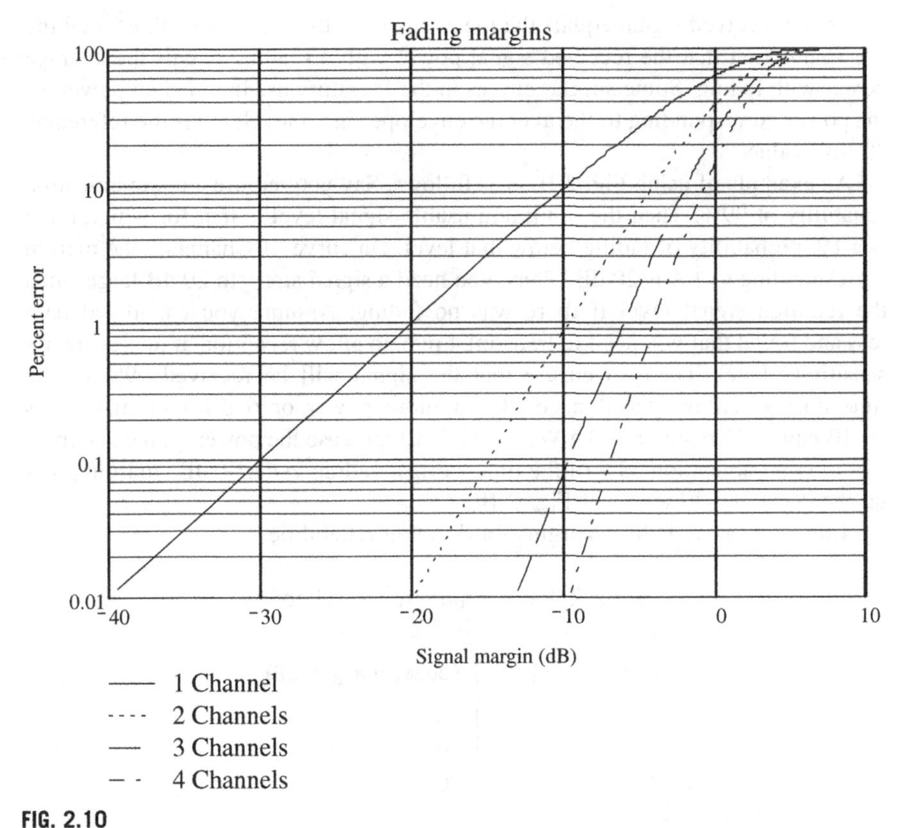


**FIG. 2.10**

Fading margins.

the signal envelope, shown by a vertical line, is 1.253. Note that it is not the most probable value, which is $\sigma = 1$. The area of the curve between any two values of signal envelope strength $r$ represents the probability that the signal strength will be in that range. The average for the Rayleigh distribution, which is not symmetric, does not divide the curve area in half. The parameter that does this is the *median*, which in this case equals 1.1774. There is a 50% probability that a signal will be below the median and 50% that it will be above.

As stated above, the Rayleigh distribution is used to determine the signal margin required to give a desired communication reliability over a fading channel with no line of sight. The curve labeled "1 Channel" in Fig. 2.10 is a cumulative distribution function with logarithmic axes. For any point on the curve, the probability of fading below the margin indicated on the abscissa is given as the ordinate. The curve is scaled such that "0 dB" signal margin represents the point where the received signal equals the mean power of the fading signal, $\sigma^2$, making the assumption that the received signal power with no fading equals the average power with fading. Some similar curves in the literature use the median power, or the power corresponding to the average envelope signal level, $r_a$, as the reference, "0 dB" value.

An example of using Fig. 2.10 is as follows. Say you require a communication reliability of 99%. Then the minimum usable signal level is that for which there is a 1% probability of fading below that level. On curve "1 Channel," the margin corresponding to 1% is 20 dB. Thus, you need a signal strength 20 dB larger than the required signal level if there was no fading. Assume you calculated path loss and found that you need to transmit 4 mW to allow reception at the receiver's sensitivity level. Then, to ensure that the signal will be received 99% of the time during fading, you'll need 20 dB more power or 6 dBm (4 mW) plus 20 dB equals 26 dBm or 400 mW. If you don't increase the power, you can expect loss of communication 63% of the time, corresponding to the "0 dB" margin point on the "Channel 1" curve of Fig. 2.10.

Table 2.2 shows fading margins for different reliabilities.

**Table 2.2** Fading margins vs. reliability from "1 Channel" in Fig. 2.10

| Reliability (%) | Fading margin (dB) |
|---|---|
| 90 | 10 |
| 99 | 20 |
| 99.9 | 30 |
| 99.99 | 40 |

## 2.9 Diversity techniques

Communication reliability for a given signal power can be increased substantially in a multipath environment through diversity reception. If signals are received over multiple, independent spatially separated channels, the largest signal can be selected for subsequent processing and use. The key to this solution is the independence of the channels. The multipath effect of nulling and of strengthening a signal through vectorial combination is dependent on transmitter and receiver spatial positions, on wavelength (or frequency) and on polarity. Let's see how we can use these parameters to create independent diverse channels.

### 2.9.1 Space diversity

A signal that is transmitted over slightly different distances to a receiver may be received at very different signal strengths. For example, in Fig. 2.2 the signal at 17 m is at a null and at 11 m at a peak. If we had two receivers, each located at

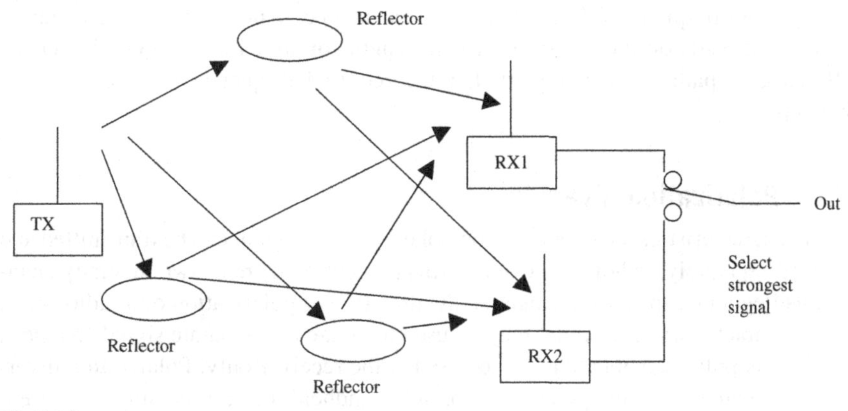


FIG. 2.11
Space diversity.

one of those distances, we could choose the strongest signal and use it. In a true multipath environment, the source, receiver, or the reflectors may be constantly in motion, so the nulls and the peaks would occur at different times on each channel. Sometimes Receiver 1 has the strongest signal, at other times Receiver 2. Fig. 2.11 illustrates the paths to two receivers from several reflectors. Although there may be circumstances where the signals at both receiver locations are at around the same level, when it doesn't matter which receiver output is chosen, most of the time one signal will be stronger than the other. By selecting the strongest output, the average output after selection will be greater than the average output of one channel alone. To increase even more the probability of getting a higher average output, we could use three or more receivers. From Fig. 2.10 you can find the required fading margin using diversity reception having 2, 3, or 4 channels. Note that the plots in Fig. 2.10 are based on completely independent channels. When the channels are not completely independent, the results will not be as good as indicated by the plots.

It isn't necessary to use complete receivers at each location, but separate antennas and front ends must be used, at least up to the point where the signal level can be discerned and used to decide on the switch position.

### 2.9.2 Frequency diversity

You can get a similar differential in signal strength over two or more signal channels by transmitting on separate frequencies. For the same location of transmitting and receiving antennas, the occurrences of peaks and nulls will differ on the different frequency channels. As in the case of space diversity, choosing the strongest channel will give a higher average signal-to-noise ratio than on either one of the channels.

The required frequency difference to get near independent fading on the different channels depends on the diversity of path lengths or signal delays. The larger the difference in path lengths, the smaller the required frequency difference of the channels.

### 2.9.3 Polarization diversity

Fading characteristics are dependent on polarization. A signal can be transmitted and received separately on horizontal and vertical antennas to create two diversity channels. Reflections can cause changes in the direction of polarization of a radio wave, so this characteristic of a signal may be used to create two separate signal channels. Thus, cross-polarized antennas can be used at the receiver only. Polarization diversity is particularly advantageous in a portable handheld transmitter, since the orientation of its antenna will not be rigidly defined. Polarization diversity doesn't allow the use of more than two channels, and the degree of independence of each channel will usually be less than in the frequency and spatial diversity cases. However, it may be simpler and less expensive to implement and may give enough improvement to justify its use, although performance will be less than can be achieved with space or frequency diversity.

### 2.9.4 Diversity implementation

In the descriptions above, we talked about selecting or switching to the channel having the highest signal level. A more effective method of using diversity is called "maximum ratio combining." In this technique, the outputs of each independent channel are added together after the channel phases are made equal and channel gains are adjusted for equal signal levels. Maximum ratio combining is known to be optimum as it gives the best statistical reduction of fading of any linear diversity combiner. In applications where accurate amplitude estimation is difficult, the channel phases only may be equalized and the outputs added without weighting the gains. Performance in this case is almost as good as in maximum ratio combining [3, p. 170, 171].

Space diversity has the disadvantage of requiring significant antenna separation, at least in the VHF and lower UHF bands. In the case where multipath signals arrive from all directions, antenna spacing on the order of $0.5\lambda$ to $0.8\lambda$ is adequate in order to have reasonably independent, or decorrelated, channels. This is at least one-half meter at 300 MHz. When the multipath angle spread is small—for example, when directional antennas are used—much larger separations are required.

Frequency diversity eliminates the need for separate antennas, but the simultaneous use of multiple frequency channels entails increased total power and spectrum utilization. Sometimes data are repeated on different frequencies so that simultaneous transmission doesn't have to be used. Frequency separation must be adequate to create decorrelated channels. The bandwidths allocated for unlicensed short-range use are rarely adequate, particularly in the VHF and UHF ranges (transmitting simultaneously on two separate bands can and has been done). Frequency diversity to reduce the effects of time delay spread is achieved with frequency hopping or direct sequence spread spectrum modulation, but for the spreads encountered in indoor applications, the pulse rate must be relatively high—of the order of several megabits per second—in order to be effective. For long pulse widths, the delay spread will not be a problem anyway, but multipath fading will still occur and the amount of frequency spread normally used in these cases is not likely to solve it.

When polarity diversity is used, the orthogonally oriented antennas can be close together, giving an advantage over space diversity when housing dimensions relative to wavelength are small. Performance may not be quite as good, but may very well be adequate, particularly when used in a system having portable hand-held transmitters, which have essentially random polarization.

Although we have stressed that at least two independent (decorrelated) channels are needed for diversity reception, sometimes shortcuts are taken. In some low-cost security systems, for example, two receiver antennas—space diverse or polarization diverse—are commutated directly, usually by diode switches, before the front end or mixer circuits. Thus, a minimum of circuit duplication is required. In such applications the message frame is repeated many times, so if there happens to be a multipath null when the front end is switched to one antenna and the message frames are lost, at least one or more complete frames will be correctly received when the switch is on the other antenna, which is less affected by the null. This technique works for slow fading, where the fade doesn't change much over the duration of a transmission of message frames. It doesn't appear to give any advantage during fast fading, when used with moving hand-held transmitters, for example. In that case, a receiver with one antenna will have a better chance of decoding at least one of many frames than when switched antennas are used and only half the total number of frame repetitions is available for each. In a worst-case situation with fast fading, each antenna in turn could experience a signal null.

### 2.9.5 Statistical performance measure

We can estimate the performance advantage due to diversity reception with the help of Fig. 2.10. Curves labeled "2 Channels" through "4 Channels" are based on the selection combining technique.

Let's assume, as before, that we require communication reliability of 99%, or an error rate of 1%. From probability theory (see Chapter 9) the probability that two independent channels would both have communication errors is the product of the error probabilities of each channel. Thus, if each of two channels has an error probability of 10%, the probability that both channels will have signals below the sensitivity threshold level when selection is made is 0.1 times 0.1, which equals 0.01, or 1%. This result is reflected in the curve "2 Channels". We see that the signal margin needed for 99% reliability (1% error) is 10 dB. Using diversity reception with selection from two channels allows a reliability margin of only 10 dB instead of 20 dB, which is required if there is no diversity. Continuing the previous example, we need to transmit only 40 mW for 99% reliability instead of 400 mW. Required margins by selection among three channels and four channels are even less — 6 dB and 4 dB, respectively. Remember that the reliability margins using selection combining diversity as shown in Fig. 2.7 are ideal cases, based on the Rayleigh fading probability distribution and independently fading channels. However, even if these premises are not realized in practice, the curves still give us approximations of the improvement that diversity reception can bring.

## 2.10 MIMO

In the previous sections we saw the multipath phenomenon as an obstacle to radio communication and discussed ways to combat it. MIMO—multi-input multi-output—uses multipath to improve communication parameters. In one respect, we can understand this by noting that if signals on multiple communication paths are combined coherently then more power would be available at the receiver relative to the noise power. MIMO takes advantage of multipath to increase S/N and consequently link capacity, thus transforming a usually degrading factor into a benefit. There are other aspects as well to using multipath as an asset rather than an obstruction. MIMO improves range, data rate, reliability, and facilitates multiple communication channels over common resources of frequency and time.

MIMO means that both the transmitter and the receiver have more than one antenna: multi-input (to the channel) multi-output (from the channel). We are mostly familiar with SISO—single input single output. Other multi-antenna configurations are MISO—multi-input single output—and SIMO—single input multiple output, which we discussed above in connection with diversity. There are three ways to take advantage of multiple antennas on one or both sides of the communication link: beam forming, spatial diversity, and spatial multiplex, which are discussed below.

### 2.10.1 Beam forming

The beam forming principle is shown in Fig. 2.12 [5, Chapter 7]. The array elements, three of them in this example labeled A, B, and C, are omnidirectional antennas—a vertical dipole or monopole over a horizontal ground plane. For reception from a target at an angle $\theta$ from bore sight, the phase at an element terminal relative to the phase at A is

$$\beta_i = -2\pi \cdot i \cdot \frac{d}{\lambda} \sin(\theta), \quad i = 0, 1, \ldots, (N-1) \tag{2.19}$$

where $N$ is the number of elements ($N = 3$ here), $d$ is the distance between elements, and $\lambda$ is signal wavelength. Each element output is connected to a phase shift

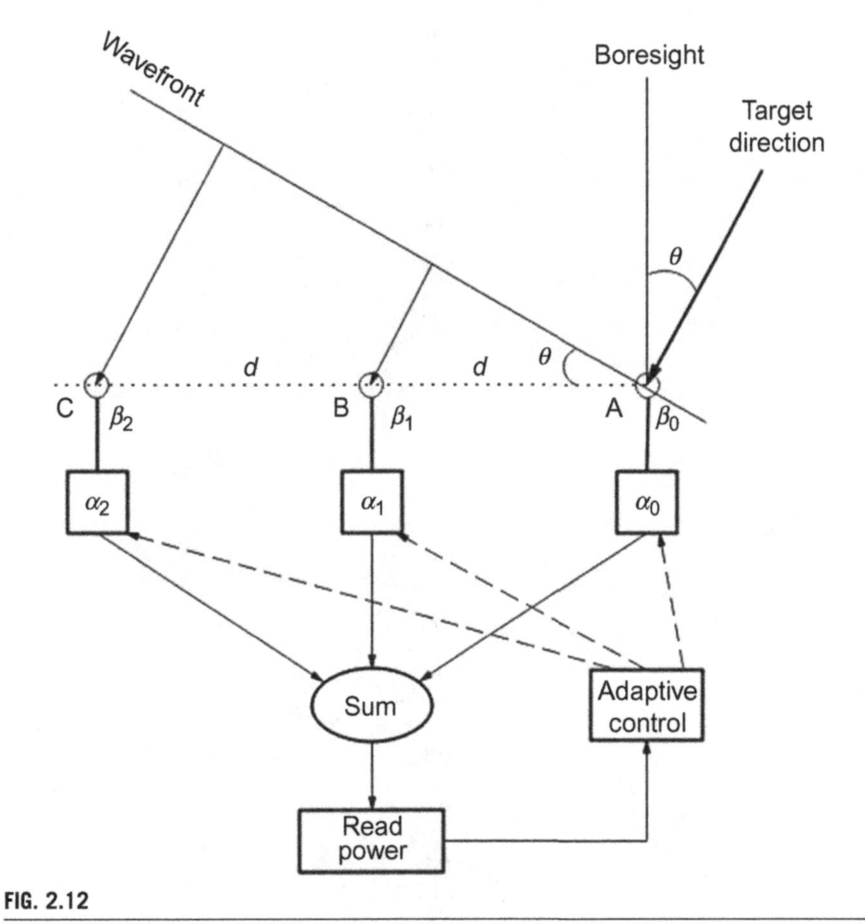


**FIG. 2.12**

Three-element adaptive antenna array.

network which is adjusted by an adaptive control block to determine the antenna pattern. For maximum gain in the direction of $\theta$, phases $\alpha_1$ and $\alpha_2$ are set to cancel out $\beta_1$ and $\beta_2$ while the reference phase shift $\alpha_0=\beta_0=0$. The phase shift network can also be adjusted to null out reception of an interfering signal, for example, from direction $\theta$. The same principle applies for transmission, where the phase shift network is controlled to beam a signal in the desired direction. Antenna patterns created in this way are shown in Fig. 2.13. The range of adjustment is $-90° > \theta < 90°$ but significant side lobes are created at the larger deviations from bore sight. The pattern is the same at the back of the array since the elements are omnidirectional. Narrower beamwidths and reduced side lobes are achieved with a larger number of elements.

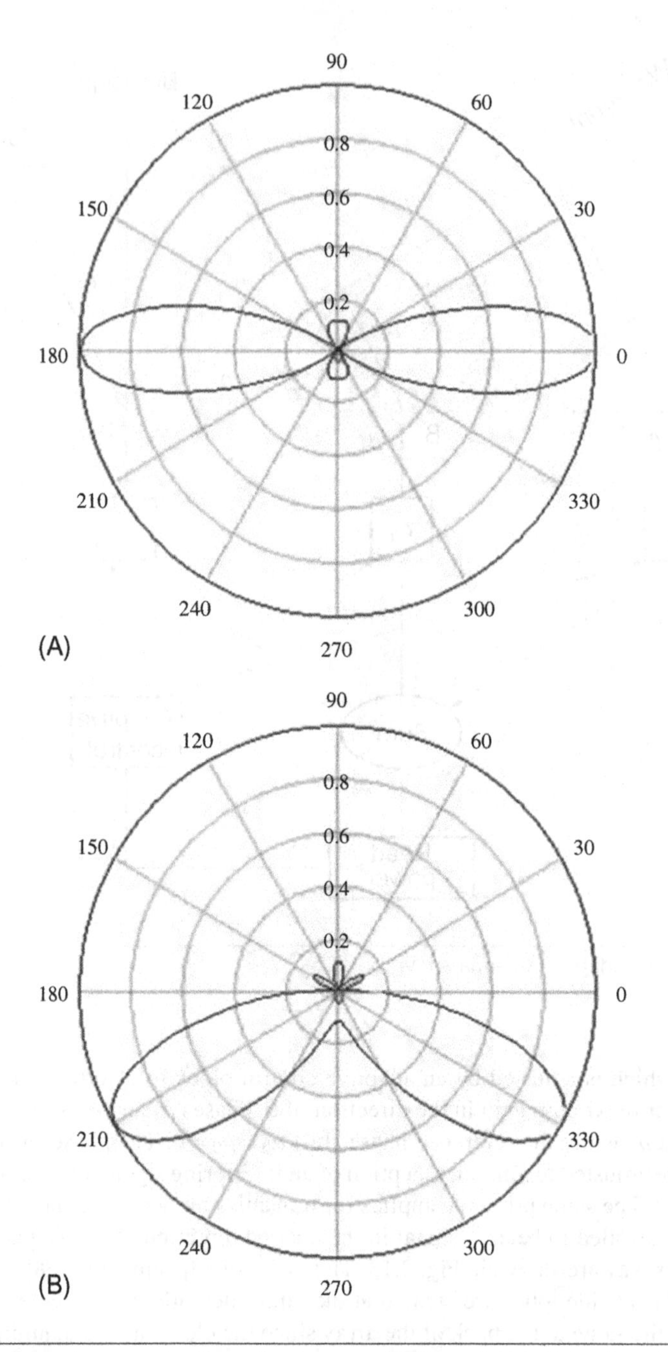


**FIG. 2.13**

Antenna patterns from an adaptive antenna array. (A) Bore sight pattern, $\theta = 0°$. (B) $\theta = -30°$.

## 2.10.2 Spatial diversity

We discussed above spatial diversity with two or more antennas at the receiver and one antenna at the transmitter. As an introduction to a more comprehensive discussion of spatial diversity and the next topic of spatial multiplex, we first give a representation of the signal paths between multiple antennas on one or both sides of the communication link. Fig. 2.14 shows the general situation where $N$ transmitting antennas and $M$ receiving antennas create $M \times N$ signal paths, each subjected to reflections and obstruction from surrounding objects. The relationship between transmitted and received signals over flat fading channels, can be described by

$$\mathbf{Y} = \mathbf{H}\mathbf{x} + \mathbf{n} \tag{2.20}$$

where $\mathbf{Y}$ is the complex vector of the received signals on the $M$ antennas, $\mathbf{H}$ is a matrix of the complex gains over each of the $M \times N$ paths between the transmitter and receiver antennas, $\mathbf{x}$ is a vector of the transmitted signals on $N$ antennas, and $\mathbf{n}$ is a vector of the noise components affecting the $M$ receiving channels. As an example of the channel matrix, for four transmitting antennas and three receiving antennas:

$$\mathbf{H} = \begin{bmatrix} h_{1,1} & h_{1,2} & h_{1,3} & h_{1,4} \\ h_{2,1} & h_{2,2} & h_{2,3} & h_{2,4} \\ h_{3,1} & h_{3,2} & h_{3,3} & h_{3,4} \end{bmatrix} \tag{2.21}$$

The output of any one receiving antenna is the sum of the signals over the paths to it from all transmitting antennas:

$$y_i = (h_{i,1} \cdot x_1 + h_{i,2} \cdot x_2 + \cdots + h_{i,M} \cdot x_N) + n_i \tag{2.22}$$

where $i$ is a subscript designating that particular receiver antenna output.

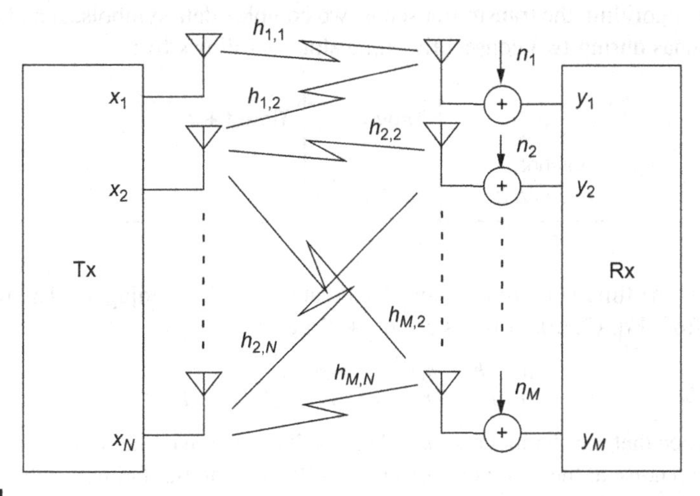


**FIG. 2.14**

MIMO with $N$ transmitting antennas and $M$ receiving antennas.

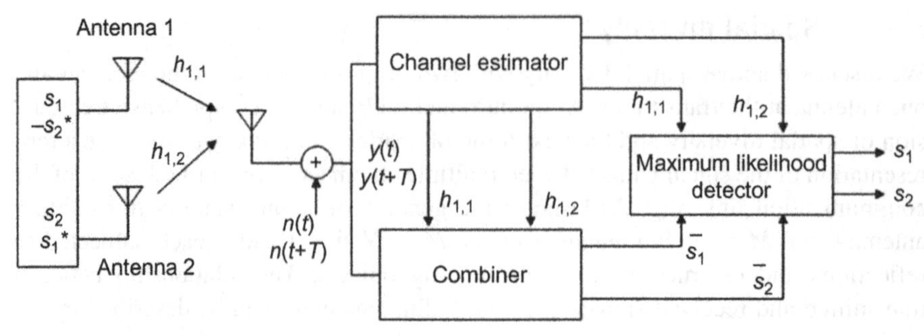


**FIG. 2.15**

Transmit diversity communication link.

*After S.M. Alamouti, A simple transmit diversity technique for wireless communications, IEEE J. Sel. Areas Commun. 16 (8) (1998).*

We now look into the specifics of a communication link with two antennas at the transmitter and one antenna at the receiver and describe spatial diversity based on multiple transmitter antennas, as contrasted to the case of diversity reception based on spaced receiver antennas as described above. Note that it is necessary for the receiver to estimate the channel matrix **H**. Conceptually, this estimate can be carried out by sending pilot symbols known to the receiver sequentially over each transmitter antenna. The receiver can then apply a correction to the distorted received symbols to make them conform to the expected pilot symbols. From the corrections, the channel matrix can be estimated.

The transmit diversity communication link looks like Fig. 2.15. According to the Alamouti algorithm, the transmitter sends two complex data symbols, $s_1$ and $s_2$, over two antennas during two consecutive time slots as follows [6]:

| | Time $t$ | Time $t+T$ |
|---|---|---|
| Antenna 1 | $s_1$ | $-s_2^*$ |
| Antenna 2 | $s_2$ | $s_1^*$ |

where $T$ is the time slot duration and * indicates complex conjugate. The received signals, from Eq. (2.22), at times $t$ and $t+T$ are

$$\begin{aligned} y(t) &= h_{1,1}\cdot s_1 + h_{1,2}\cdot s_2 + n(t) \\ y(t+T) &= -h_{1,1}\cdot s_2{}^* + h_{1,2}\cdot s_1{}^* + n(t+T) \end{aligned} \tag{2.23}$$

It is assumed that the channel response **H** doesn't change over the time interval T but the random noise at the receiver does change. We assume flat fading, so the channel factor is modeled by a complex multiplier, $h = \alpha e^{j\theta}$ where $\alpha$ is the magnitude response and $\theta$ is the phase response over a signal path.

The received signals at times $t$ and $t+T$ are combined in the combiner block with the estimates of elements of the channel matrix and their conjugates to create estimates of the transmitted signals:

$$\begin{aligned}\bar{s}_1 &= h_{1,1}^* \cdot y(t) + h_{1,2} \cdot y(t+T)^* \\ \bar{s}_2 &= h_{1,2}^* \cdot y(t) - h_{1,1} \cdot y(t+T)^*\end{aligned} \tag{2.24}$$

Substituting Eq. (2.23) in Eq. (2.24) gives

$$\begin{aligned}\bar{s}_1 &= \left(|h_{1,1}|^2 + |h_{1,2}|^2\right) s_1 + h_{1,1}^* \cdot n(t) + h_{1,2} \cdot n(t+T)^* \\ \bar{s}_2 &= \left(|h_{1,1}|^2 + |h_{1,2}|^2\right) s_2 + h_{1,2}^* \cdot n(t) - h_{1,1} \cdot n(t+T)^*\end{aligned} \tag{2.25}$$

Now the receiver has an estimate of the pair of transmitted symbols multiplied by scalar constants and distorted by noise. It decides in the maximum likelihood detector block what those symbols are by a *maximum likelihood decision* rule, which chooses the most likely symbol as the one with the smallest vectorial distance between the outputs in Eq. (2.25) and all possible symbols. Mathcad Worksheet **"diversity.mcdx"** gives an example of the transmitter antenna diversity calculations and result.

Diversity gain is improved with additional antennas at the transmitter and receiver [6, 7]. Performance is reduced by correlation between the signals over the different propagation paths, which depends on the separation of the antennas and number and closeness of reflecting objects.

### 2.10.3 Spatial multiplex

Spatial multiplexing is the ultimate way to use multiple antennas on both sides of a wireless communications link to increase capacity [5]. Effectively, multiple parallel and independent SISO channels are created, increasing capacity significantly. The number of independent channels, $n$, is equal to the minimum of the number of uncorrelated array elements on the transmitter or receiver side of the link, that is $n = \min(M, N)$ where $N$ is the number of transmitter elements and $M$ is the number of elements on the receiving side. The channel matrix $\mathbf{H}$ must be known at the receiver and transmitter [7]. Fig. 2.16 is an example of a $3 \times 2$ spatial multiplex system. The system input consists of two independent data flows. The transmitter has two antennas and the receiver three. There is a precoding block at the transmitter input and a shaping block at the receiver output. They create a virtual link where independent data streams can flow without cross-talk, or the rate of a single flow of data can be multiplied by the number of individual streams.

The precoding and shaping matrices, which create the virtual multiple independent SISO channels, are derived from the result of a single value decomposition, SVD, on the channel matrix $\mathbf{H}$, which, as mentioned, must be known to the transmitter and receiver. The mathematical operations are shown on the worksheet **MIMO.mcdx**. The symbols in the following explanation are shown in Fig. 2.16.

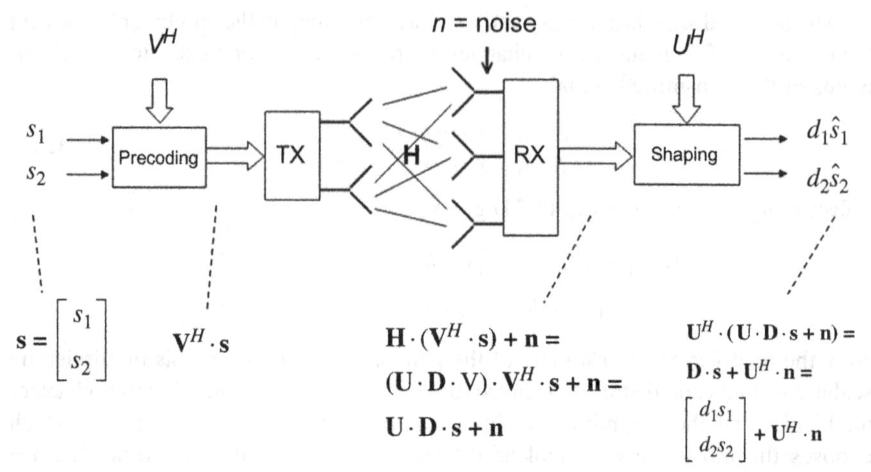


**FIG. 2.16**

MIMO multiplexing.

**(1)** The SVD function in Mathcad yields three components: **d**, **U**, and **V**. After converting the **d** vector to a diagonal matrix **D**, we get

$$\mathbf{H} = \mathbf{U} \cdot \mathbf{D} \cdot \mathbf{V} \tag{2.26}$$

(This definition fits Mathcad, but in other programs, including MATLAB, SVD produces the complex conjugate transpose of **V**, so $\mathbf{H} = \mathbf{U} \times \mathbf{D} \times \mathbf{V}^H$ where the superscript $H$ indicates complex transpose.)

**(2)** A precoder in the form of $\mathbf{V}^H$ transforms the two data streams, and the result goes to the two spatially separated transmitter antennas.

**(3)** The output of the channel, distorted by multipath, is input to the receiver along with receiver noise. Note that where **H** acts on the output of the transmitter in the form $\mathbf{U} \cdot \mathbf{V} \cdot \mathbf{D}$, the factor $\mathbf{V} \cdot \mathbf{V}^H$ is a unity matrix.

**(4)** The shaper function $\mathbf{U}^H$ restores the original data streams, contaminated by receiver noise. $\mathbf{U}^H \cdot \mathbf{U}$ is a unity matrix. Each data stream is multiplied by a factor in the diagonal matrix **D** that affects the signal-to-noise ratio of that stream. A high ratio of maximum to minimum values of the elements in **D**, called the condition number, can reduce the performance of the multiplexing function. That is, one or more data streams will have a low signal-to-noise ratio and a consequentially high error rate [8]. The transmitter can use **D** to adjust the output powers in order to equalize the signal-to-noise ratios of the data streams over the medium.

Maximum MIMO multiplex performance is achieved only when the transmit-receive paths are not correlated, and accurate channel state information is available. Also a high multipath environment with a small line-of-sight component is needed for the

MIMO multiplex algorithm. On the other hand, a good line-of-sight link with minimum fading and optimum path loss should allow increasing S/N which may offset the loss of MIMO capacity [9].

## 2.11 Summary

In this chapter we looked at various factors that affect the range of reliable radio communication. Propagation of electromagnetic waves is influenced by physical objects in and near the line-of-sight path between transmitter and receiver. We can get a first rough approximation of communication distance by considering free space propagation, or only the reflection of the transmitted signal from the earth. If the communication system site can be classified, an empirically determined exponent developed for different environments may be used to estimate the path loss, and thus the range. A link budget is made using known link parameters among transmission power, receiver sensitivity, path loss and antenna gains in order to estimate the value of an unknown parameter that is required to give a stated communication performance. The multipath phenomenon was described. When the transmitter or receiver is in motion, or surrounding objects are not static, the path loss varies causing fading, which may be flat or frequency selective. We described several techniques of diversity reception that can reduce the required power for a given reliability when the communication link is subject to multipath propagation. Finally, we described the use of arrays of multiple antenna elements on both ends of the communication link, MIMO, to take advantage of multipath for improving signal-to-noise ratio through beam forming and diversity, and for providing independent spatial multiplex channels.

# A Appendix

## A.1 Maxwell's equations

Maxwell's equations describe the relationships among five vectors that define an electromagnetic field. These vectors are:

$\mathbf{E}$ = electric intensity, volts/m
$\mathbf{B}$ = magnetic induction, Webers/$m^2$
$\mathbf{D}$ = Electric flux density, coulombs/$m^2$
$\mathbf{H}$ = magnetic intensity, amp/m
$\mathbf{J}$ = current density, amp/$m^2$

Since these quantities are vectors in three-dimensional space, the expressions of their relationships as differential equations or integral equations require the use of three-dimensional operators. Therefore, in order to understand and use Maxwell's equations, you must have knowledge of the subject of mathematics called vector analysis.

The above quantities are also a function of time, and when they are not static, their interrelationship also involves differentiation in respect to time.

The propagation of electromagnetic waves is a result of the interdependence between electric fields and magnetic fields. Electric current flow in a wire causes a magnetic field around the wire. This is the basis of the operation of an electric motor. On the other hand, a moving wire in a magnetic field, or a changing magnetic field in the vicinity of a stationary wire, creates a potential difference in the wire and a flow of current if there is a connection between the wire ends. True, radio waves propagate in space where there are no wires. However, "displacement" currents can exist in a nonconductor, just as alternating current flows through the dielectric material of a capacitor.

We state here the four basic laws of electromagnetism in their integral form. They each also exist in the form of three-dimensional spatial derivative operators. In the equations, bold face symbols indicate three-dimensional vectors. The dot between vector factors means the scalar dot product.

**Gauss's law (electric).**

$$\oiint \mathbf{D} \bullet \mathbf{dS} = Q \tag{A.1}$$

A net electric charge $Q$ gives rise to an electric flux density **D** over a surface totally enclosing $Q$. The value of the integral of **D** over this surface equals the value of $Q$. This equation connects an electric charge with the electric field surrounding it. In space there is no net charge so the electric flux density integrated around a closed surface (such as the surface of a sphere) must be zero.

**Gauss's law (magnetic).**

$$\oiint \mathbf{B} \bullet \mathbf{dS} = 0 \tag{A.2}$$

Here we have the complement of the previous equation. An isolated magnetic pole or magnetic charge doesn't exist, in contrast to the isolated electric charge of Eq. (A.1). Magnetic flux lines are continuous loops, so lines that enter a closed surface must also come out of it. The result is that the integral of magnetic flux density B over any closed surface, expressed on the left side of this equation, is zero. There's no such thing as a magnet having only a north pole from which may emanate magnetic flux, never to return.

**Faraday's law.**

$$\oint \mathbf{E} \bullet \mathbf{ds} = \iint_S \frac{\partial \mathbf{B}}{\partial t} \bullet \mathbf{dS} = -\frac{\partial \phi_m}{\partial t} \tag{A.3}$$

It shows that the electrical intensity integrated on a closed path equals the negative value of the rate of change of magnetic flux flowing through the surface enclosed by that path. (A partial derivative symbol is used on the right since $\phi$ is also a function of its spatial coordinates, which are constant here.) Compare this equation with Faraday's law of magnetic induction

$$V_{out} = -N\frac{d\phi}{dt}$$

which is the basis for operation of an electric generator ($N$ is the number of turns of a coil exposed to changing magnetic flux $\phi$).

**Ampere-Maxwell law.**

$$\oint \mathbf{H} \bullet \mathbf{ds} = I_{enc} + \varepsilon \iint_S \frac{\partial \mathbf{E}}{\partial t} \bullet \mathbf{dS} = I_{enc} + \varepsilon \frac{\partial \phi_e}{\partial t} \quad \text{(A.4)}$$

This law describes the connection between magnetic intensity and electric current. Note the similarity to Eq. (A.3). On the left is a closed line integral of magnetic intensity around a path that encloses a flow of current. This current has two components: a conduction current $I_{enc}$ enclosed within the path such as current through a wire, and a displacement current that is produced by an electric flux that is changing with time. $\varepsilon$ is the permittivity of the medium. In the case of propagation through space, only displacement current exists since conductors are absent.

## References


[1] Radio Communication Handbook, fifth ed., Radio Society of Great Britain (RSGB), Herts, Great Britain, 1982.

[2] T.S. Rappaport, Wireless Communications, Principles and Practice, Prentice Hall, Upper Saddle River, NJ, 1996.

[3] J.D. Gibson (Ed.), (Editorin-Chief) The Mobile Communications Handbook, CRC Press, Inc, 1996.

[4] B. Sklar, Digital Communications, Fundamentals and Applications, Prentice Hall, Upper Saddle River, NJ, 2001.

[5] A. Bensky, Wireless Positioning and Applications, second ed., Artech House, Norwood, MA, 2016.

[6] S.M. Alamouti, A simple transmit diversity technique for wireless communications, IEEE J. Sel. Areas Commun. 16 (8) (1998).

[7] I. Berenguer, X. Wang, Space-time coding and signal processing for MIMO communications, J. Comput. Sci. Technol. 18 (6) (2003) 689–702.

[8] S. Schindler, Assessing a MIMO Channel, Rhode & Schwarz 1SP18, 2011, pp. 1–18.

[9] D. Gesbert, M. Shafi, D. Shiu, From theory to practice: an overview of MIMO space-time coded wireless systems, IEEE J. Sel. Areas Commun. 21 (3) (2003) 281–302.

CHAPTER

# Antennas and transmission lines

3

The antenna is the interface between the transmitter or the receiver and the propagation medium, and it therefore is a deciding factor in the performance of a radio communication system. The principal properties of antennas—directivity, gain, and radiation resistance—are the same whether referred to as transmitting or receiving. The principle of reciprocity states that the power transferred between two antennas is the same, regardless of which is used for transmission or reception, if the generator and load impedances are conjugates of the transmitting and receiving antenna impedances in each case.

First, we define the various terms used to characterize antennas. Then a number of types of antennas that are commonly used in short-range radio systems are examined, including those based on metamaterials. We review methods of matching the impedances of the antenna to the transmitter or receiver RF circuits and conclude the chapter by describing measuring techniques.

## 3.1 Antenna characteristics

Understanding the various characteristics of antennas is a first and most important step before deciding what type of antenna is most appropriate for a particular application. While antennas have several electrical characteristics, often a primary concern in choosing an antenna type is its physical size. Before dealing with the various antenna types and the shapes and sizes they come in, we first must understand the meaning of antenna parameters.

### 3.1.1 Antenna impedance

As stated in the introduction, the antenna is an interface between circuits and space. It facilitates the transfer of power between the communication medium and the transmitter or receiver. The antenna impedance is the load for the transmitter or the input impedance to the receiver. It is composed of three parts—radiation resistance, loss resistance, and reactance. The radiation resistance is a virtual resistance that, when multiplied by the square of the RMS current in the antenna at its feed point, equals the power radiated by the antenna in the case of a transmitter or extracted from the medium in the case of a receiver. It is customary to refer the radiation resistance to a current maximum in the case of an ungrounded antenna, and to the current at



Short-range Wireless Communication. https://doi.org/10.1016/B978-0-12-815405-2.00003-8
© 2019 Elsevier Inc. All rights reserved.

the base of the antenna when the antenna is mounted over a ground plane. Transmitter power delivered to an antenna will always be greater than the power radiated. The difference between the transmitter power and the radiated power is power dissipated in the resistance of the antenna conductor and in other losses. The efficiency of an antenna is the ratio of the radiated power to the total power absorbed by the antenna. It can be expressed in terms of the radiation resistance $R_r$ and loss resistance $R_l$ as

$$\varepsilon_{ant} = \frac{R_r}{R_r + R_l} \tag{3.1}$$

The resistance seen by the transmitter or receiver at the antenna terminals will be equal to the radiation resistance plus the loss resistance only if these terminals are located at the point of maximum current flow in the antenna. The impedance at this point may have a reactive component too.

When there is no reactive component, the antenna is said to be resonant. Maximum power transfer between the antenna and transmitter or receiver will occur only when the impedance seen from the antenna terminals is the complex conjugate of the antenna impedance.

It is important to match the transmitter to the antenna not only to get maximum power transfer. Attenuation of harmonics relative to the fundamental frequency is maximized when the transmitter is matched to the antenna—an important point in meeting the spurious radiation requirements for license-free transmitters. The radiation resistance depends on the proximity of the antenna to conducting and insulating objects. In particular, it depends on the height of the antenna from the ground. Thus, the antenna matching circuit of a transmitter with integral antenna that is intended to be hand-held should be optimized for the antenna impedance in a typical operating situation.

### 3.1.2 Directivity and gain

The directivity of an antenna relates to its radiation pattern. An antenna which radiates uniformly in all directions in three-dimensional space is called an isotropic antenna. Such an antenna doesn't exist, but it is convenient to refer to it when discussing the directional properties of an antenna. All real antennas radiate stronger in some directions than in others. The directivity of an antenna is defined as the power density of the antenna in its direction of maximum radiation in three-dimensional space divided by its average power density. The directivity of the hypothetical isotropic radiator is 1 or 0 dB. The directivity of a half-wave dipole antenna is 1.64 or 2.15 dB.

The radiation pattern of a wire antenna of short length compared to a half wavelength is shown in Fig. 3.1A. The antenna is high enough so as not to be affected by the ground. If the antenna wire direction is parallel to the earth, then the pattern represents the intersection of a horizontal plane with the solid pattern of the antenna shown in Fig. 3.1B [1]. A vertical wire antenna is omnidirectional; that is, it has a circular horizontal radiation pattern and directivity in the vertical plane.

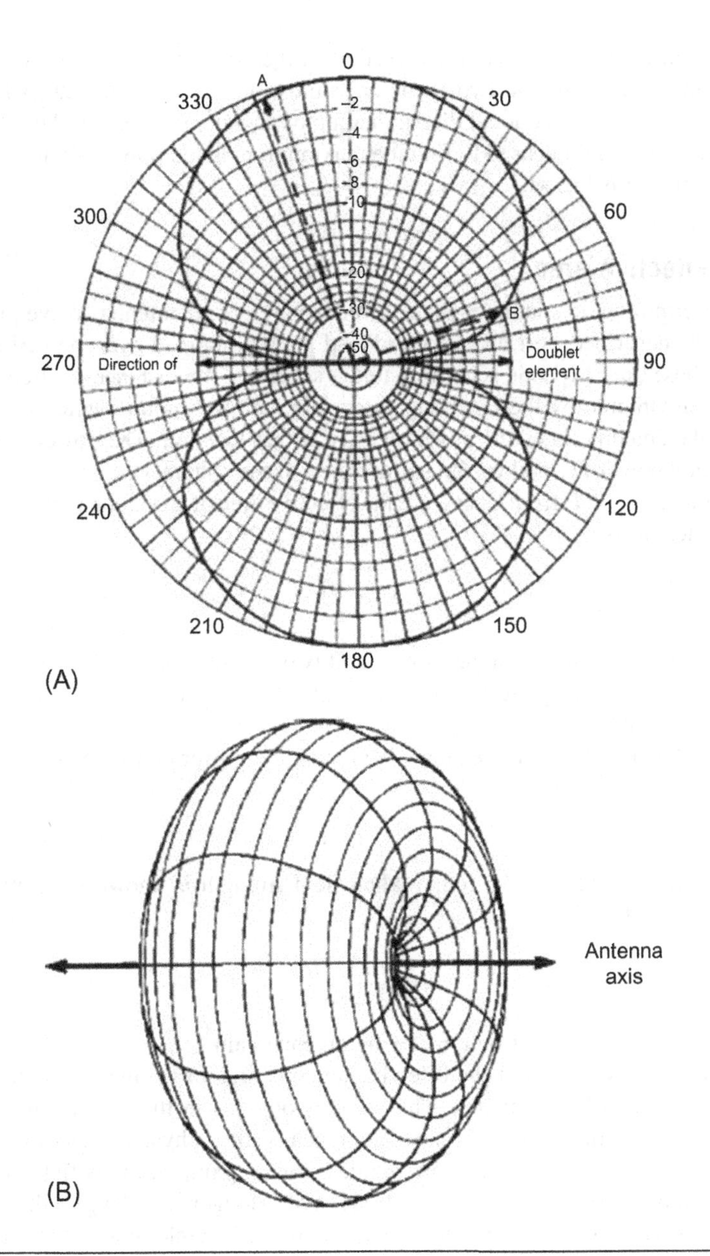


**FIG. 3.1**

Short dipole antenna (ARRL Antenna Book). (A) Directive pattern in plane containing antenna and (B) solid diagram.

*Reprinted with permission, ARRL Antenna Book, 16th edition.*

The gain of an antenna is the directivity times the antenna efficiency. When antenna losses are low, the two terms are almost the same. In general, when you are interested in the directional discrimination of an antenna, you will be interested in its directivity. Gain is used to find the maximum radiated power when the power into the antenna is known.

### 3.1.3 Effective area

Another term often encountered is the effective area of an antenna. Wave propagation can be described as if all of the radiated power is spread over the surface of a sphere whose area expands according to the square of the distance (in free space). The power captured by the receiving antenna is then the capture area, or effective area, of the antenna times the power density at that location. The power density is the radiated power divided by the surface area of the sphere.

Recalling Eq. (2.4) from Chapter 2, the effective area of an antenna related to gain and wavelength is

$$A_e = \frac{\lambda^2 G}{4\pi} \tag{3.2}$$

This expression shows us that the capture ability of an antenna of given gain $G$ grows proportionally as the square of the wavelength. It holds for both transmitting and receiving antennas.

When the electric field strength $E$ is known, the power density $S$ is

$$S = \frac{E^2}{120\pi} \tag{3.3}$$

Thus, received power can be found when field strength is known by multiplying Eq. (3.2) times Eq. (3.3):

$$P_r = \frac{E^2 \lambda^2 G_r}{480\pi^2} \tag{3.4}$$

where $G_r$ expressly refers to the receiving antenna gain.

It's intuitive to note that the effective antenna area has some connection with the physical size of the antenna. This is most obvious in the case of microwave antennas where the effective area approaches the physical aperture. From Eq. (3.4), it is seen that for a given radiated power, gain, and thus field strength, received signals are stronger on lower frequencies (longer wavelength) than higher frequencies. However, short-range devices are often portable or are otherwise limited in size, and their antennas may have roughly the same dimensions, regardless of frequency. The lower frequency antennas whose sizes are small fractions of a wavelength have poor efficiency and low gain and therefore may have effective areas similar to their high-frequency counterparts. Thus, using a low frequency doesn't necessarily mean higher power at the receiver, which Eq. (3.4) may lead us to believe.

### 3.1.4 Polarization

Electromagnetic radiation is composed of a magnetic field and an electric field. In the far field, beyond around a wavelength from the antenna, these fields are at right angles to each other, and both are in a plane normal to the direction of propagation. The direction of polarization refers to the direction of the electric field in relation to the earth. Linear polarization is created by a straight wire antenna. A wire antenna parallel to the earth is horizontally polarized and a wire antenna normal to the earth is vertically polarized.

The electric and magnetic fields may rotate in their plane around the direction of propagation, and this is called elliptical polarization. It may be created by perpendicular antenna elements being fed by RF signals on the communication frequency that are not in the same time phase with each other. Circular polarization results when these elements are fed by equal power RF signals which differ in phase by 90°, which causes the electric (and magnetic) field to make a complete 360° rotation every period of the wave (a time of *1/frequency* seconds). Some antenna types, among them the helical antenna, produce elliptic or circular polarization inherently, without having two feed points. There are two types of elliptical polarization, right hand and left hand, which are distinguished by the direction of rotation of the electric field.

The polarization of a wave, or an antenna, is important for several reasons. A horizontally polarized receiving antenna cannot receive vertically polarized radiation from a vertical transmitting antenna, and vice versa. Similarly, right-hand and left-hand circular antenna systems are not compatible. Sometimes, this quality is used to good advantage. For example, the capacity of a microwave link can be doubled by transmitting two different information channels between two points on the same frequency using oppositely polarized antenna systems.

The degree of reflection of radio signals from the ground is affected by polarization. The phase and amount of reflection of vertically polarized waves from the ground are much more dependent on the angle of incidence than horizontally polarized waves.

Except for directional, line-of-sight microwave systems, the polarity of a signal may change during propagation between transmitter and receiver because of reflections. Thus, in most short-range radio applications, a horizontal antenna will receive transmissions from a vertical antenna, for example, albeit with some attenuation. The term *cross polarization* defines the degree to which a transmission from an antenna of one polarization can be received by an antenna of the opposite polarization. Often, the polarization of a transmitter or receiver antenna is not well defined, such as in the case of a handheld device. A circular polarized antenna can be used when the opposite antenna polarization is not defined, since it does not distinguish between the orientation of the linear antenna. Cross polarization between antennas results in signal attenuation.

### 3.1.5 Bandwidth

Antenna bandwidth is the range of frequencies over which the antenna can operate while some other characteristic remains within a defined range. Very frequently, the bandwidth is related to the antenna impedance expressed as standing wave ratio. Obviously, a device that must operate over a number of frequency channels in a band must have a comparatively wide bandwidth antenna. Less obvious are the bandwidth demands for a single frequency device.

A narrow bandwidth or high $Q$ antenna will discriminate against harmonics and other spurious radiation and thereby will reduce the requirements for a supplementary filter, which may be necessary to allow meeting the radio approval specifications. On the other hand, drifting of antenna physical dimensions or matching components could cause the power output (or sensitivity) to fall with time. Changing proximity of nearby objects or the "hand effect" of portable transmitters can also cause a reduction of power or even a pulling of frequency, particularly in low-power transmitters with a single oscillator stage and no buffer or amplifier stage.

### 3.1.6 Antenna factor

The antenna factor is commonly used with calibrated test antennas to make field strength measurements on a test receiver or spectrum analyzer. It relates the field strength to the voltage across the antenna terminals when the antenna is terminated in its specified impedance (usually 50 or 75 ohms):

$$AF = \frac{E}{V} \tag{3.5}$$

where $AF$ is the antenna factor in m, $E$ is the field strength in V/m, $V$ is the load voltage in V.

Usually the antenna factor is stated in dB/m:

$$AF_{\mathrm{dB/m}} = 20 \log (E/V)$$

The relationship between receiver antenna numerical gain $G_r$ and antenna factor $AF$ is:

$$AF = \frac{4\pi}{\lambda}\sqrt{\frac{30}{R_L G_r}} \tag{3.6}$$

where $R_L$ is the load resistance, usually 50 ohms.

## 3.2 Types of antennas

In this section, we review the characteristics of several types of antennas that are used in short-range radio devices. Variations on these and other configurations of small antennas are described in Refs. [2, 3]. The size of the antenna is related to the wavelength, which in turn is found when frequency is known from *wavelength = (velocity of propagation)/(frequency)*.

The maximum velocity of propagation occurs in a vacuum. It is approximately 300,000,000 m/s, with little difference in air. This figure is less in solid materials, so the wavelength will be shorter for antennas printed on circuit board materials or protected with a plastic coating.

### 3.2.1 Dipole

The dipole is a wire antenna fed at its center. The term often refers to an antenna whose overall length is one-half wavelength. In free space, its radiation resistance is 73 ohms, but that value will vary somewhat in the presence of the ground or other large conducting objects. The dipole is usually mounted horizontally, but if mounted vertically, its transmission line feeder cable should extend from it at a right angle for a distance of at least a quarter wavelength. In free space, the radiation pattern of a horizontal half-wave dipole is similar to that of the small dipole shown in Fig. 3.1 and has a directivity of 1.64 or 2.15 dB in the horizontal plane perpendicular to the wire direction. A three dimensional radiation pattern of a half wave dipole two wavelengths above the ground is shown in Fig. 3.2A. It is seen that the presence of ground,

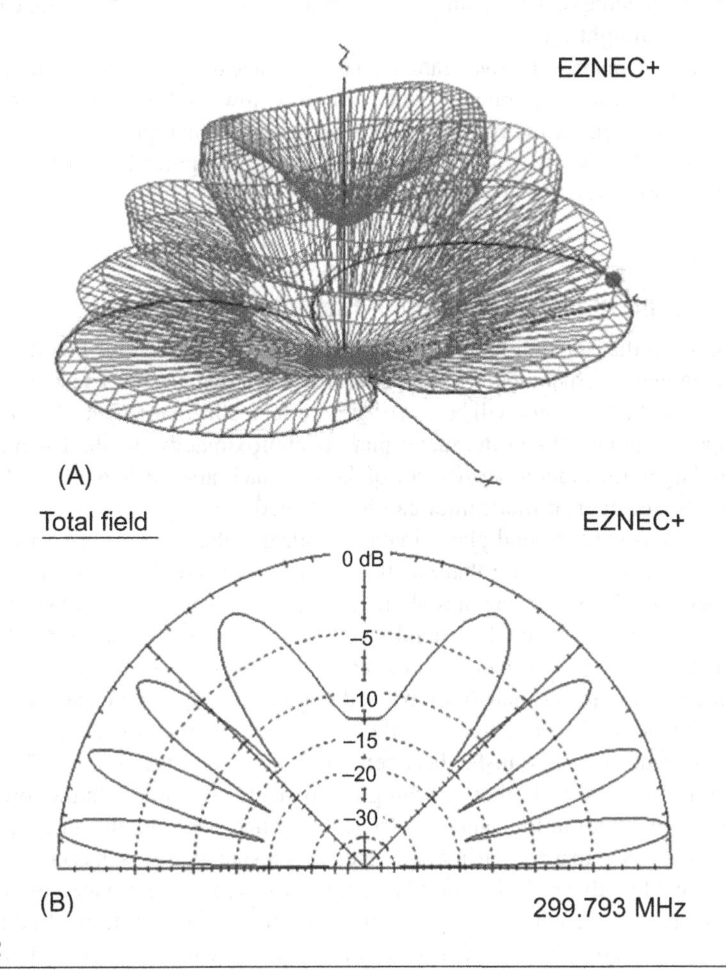


**FIG. 3.2**

Dipole antenna pattern two wavelengths above ground. (A) Three dimension pattern, (B) vertical pattern in plane containing $y$ and $z$ axes.

or any other nearby conducting object, greatly influences the pattern. Maximum gain is 7 dBi on the lower lobe. Fig. 3.2B shows the vertical pattern, on a plane at right angles to the antenna wire direction. When used indoors where there are a multitude of reflections from walls, floor, and ceiling, the horizontal dipole can give good results in all directions.

The half-wave dipole antenna is convenient to use because it is easy to match a transmitter or receiver to its radiation resistance. It has high efficiency, since wire ohmic losses are only a small fraction of the radiation resistance. Also, the antenna characteristics are not much affected by the size or shape of the device it is used with, and it doesn't use a ground plane. Devices whose dimensions are small relative to the antenna size can directly feed the dipole, with little or no transmission line. For increased compactness, the two antenna elements can be extended at an angle instead of being in a straight line.

In spite of its many attractive features, the half-wave dipole is not commonly used with short-range radio equipment. On the common low UHF unlicensed frequency bands, it is too large for many applications, particularly small portable devices. The antenna types below are smaller than the half-wave dipole, and generally give reduced performance.

### 3.2.2 Ground plane or monopole antenna

We mentioned that the dipole can be mounted vertically. If we take one dipole element and mount it perpendicular to a large metal plate, then we don't need the bottom element—a virtual element will be electrically reflected from the plate. This is called a monopole antenna. When the metal plate is approximately one-half wavelength square or larger, the radiation resistance of the antenna is around 36 ohms, and a good match to the receiver or transmitter can be obtained.

The quarter-wave ground plane antenna is ideal if the receiver or transmitter is encased in a metal enclosure that has the required horizontal area for an efficient vertical antenna. However, in many short-range devices, a quarter-wave vertical element is used without a suitable ground plane, or, due to space restrictions, the radiating element may be less than one quarter wave long. In these cases, the radiation resistance is much lower than 36 ohms, and there is considerable capacitance reactance. An inductor is needed to cancel the reactance as well as a matching circuit to assure maximum power transfer between the antenna and the device. The ohmic losses in the inductor and other matching components, together with the low radiation resistance, result in low antenna efficiency. When possible, the antenna length should be increased to a point where the antenna is resonant, that is, has no reactance. The electrical length can be increased and capacitive reactance reduced by winding the bottom part of the antenna element into a coil having several turns. In this way, the loss resistance is reduced and efficiency increased. See the description of the helical antenna below.

### 3.2.3 Inverted F antenna (IFA)

A popular configuration for implementing a more compact antenna while maintaining high efficiency is the inverted F antenna. This is actually a form of the quarter wave monopole antenna where the top part of the radiating element that protrudes from the ground plane is folded ninety degrees, thereby allowing the whole antenna to fit inside the product case. Making part of the antenna parallel to the metal ground plane introduces capacitance and detunes the resistive input impedance, making some sort of matching solution necessary. This matching can be accomplished without discrete components by adding a right-angle stub, connected to the ground, at the antenna bend. This looks like an inverted "F" with the top line connected to ground and the line in the middle being the feed point. An inverted F antenna is shown in Fig. 3.3, along with its radiation pattern, which contains both horizontal and vertical polarization. By adjusting lengths A, B, and C, a good match can be achieved to the feed impedance [4]. For comparison, a quarter wave monopole antenna with elevation radiation patterns is shown in Fig. 3.4. Note that the elevation pattern of the inverted F antenna, in a plane perpendicular to the plane of the antenna, shows both horizontal and vertical polarization, since the antenna has both horizontal and vertical elements. The monopole antenna has only vertical polarization.

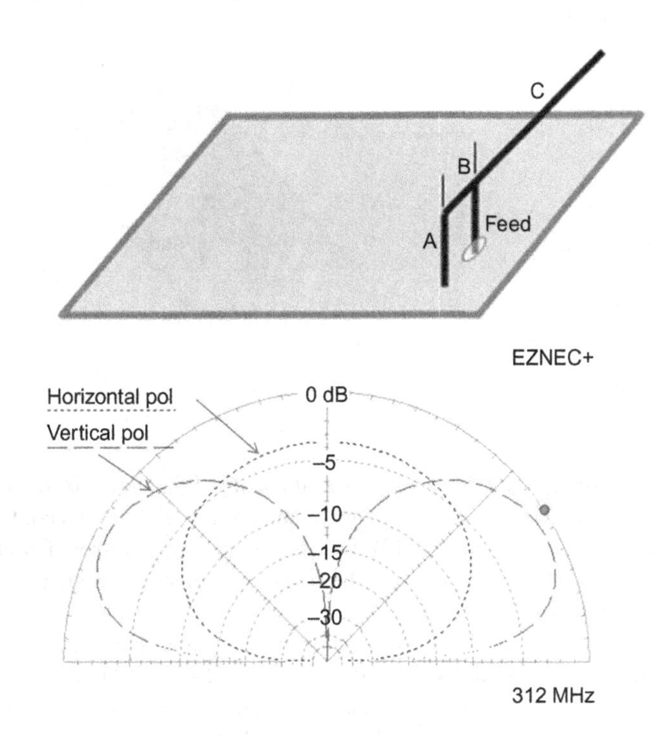


**FIG. 3.3**

Inverted F wire antenna.

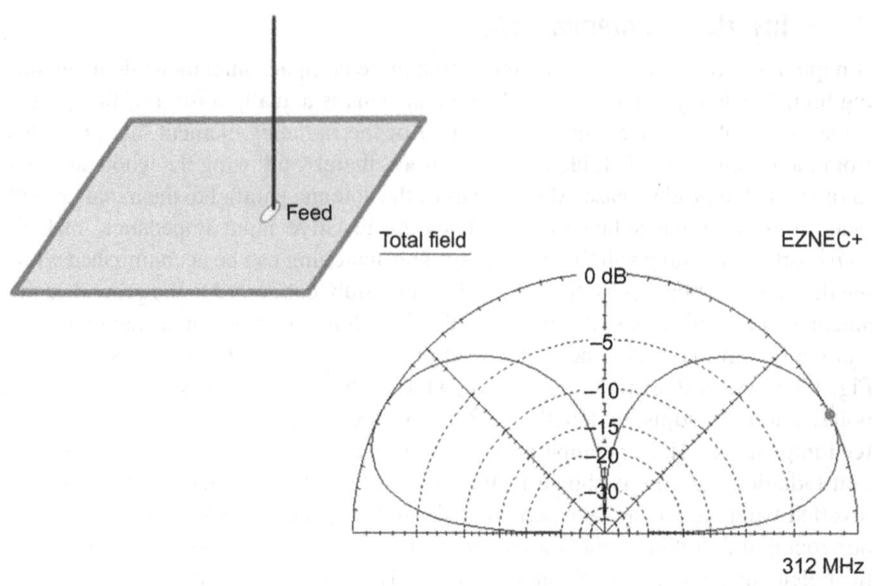


FIG. 3.4

Quarter wave monopole antenna.

FIG. 3.5

Planar inverted F antenna.

A common variant of the essentially two dimensional F antenna is to use a metal plate, or patch, parallel to the ground plane, as shown in Fig. 3.5. This configuration is called planar inverted F antenna (PIFA). Among advantages over IFA are wider bandwidth and availability of a surface (the patch) in which to design a slot or slots for multiband matching [5].

Another configuration of IFA is to make the ground plane and antenna element on the same plane, that is, on the same side of a printed circuit board, as shown in Fig. 3.6. An obvious advantage is that the whole antenna is printed, so there is no component cost other than a small additional board area.

Generally, there is some flexibility in the design parameters of IFA. Design equations are not available except for very specific cases [6]. Referring to Fig. 3.6,

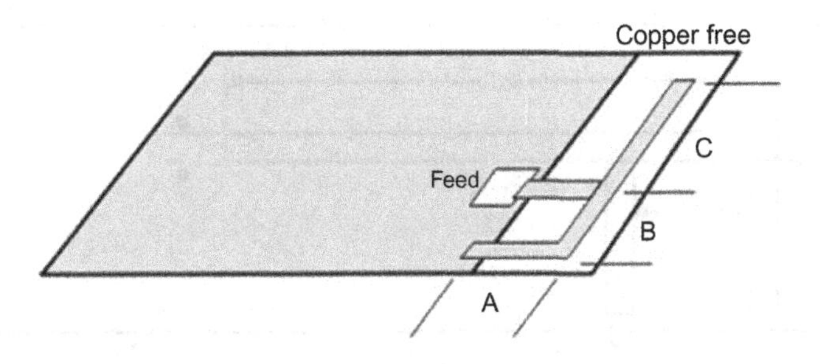


**FIG. 3.6**

Pc board inverted F antenna.

dimension A is chosen to be typically 20% to 30% of one-quarter wavelength. Length A + C will approximately equal one quarter wavelength on the board. The stub length B is adjusted for a good impedance match. Adjust C to attain the exact desired resonant frequency. An example of a design for 2.4 GHz is shown in Fig. 3.7 [7]. The board is 1.6 mm thick, and the relative permittivity, or dielectric constant, is 4.4. In calculating wavelength, note that on the pc board, propagation speed is less than the speed of light $c$. It is inversely proportional to the square root of the effective dielectric constant $\varepsilon_{eff}$ which depends on the thickness and relative permittivity of the board material, the width of the conducting trace and existence or lack of an opposing ground plane. This means that at a given frequency the wavelength in a board with dielectric constant greater than unity is smaller than in air. An estimate of the effective dielectric constant is

$$\varepsilon_{eff} = \frac{\varepsilon_r + 1}{2} + \frac{\varepsilon_r - 1}{2} \cdot \frac{1}{\sqrt{1 + \frac{12h}{w}}} \tag{3.7}$$

where $\varepsilon_r$ is the dielectric constant of the board material, $h$ is the thickness of the board, and $w$ is the trace width [8]. This expression is conceived for propagation on a microstrip line, which runs on a dielectric surface that is backed by a ground plane. The conductors of a pc board inverted- F antenna are not backed by copper, resulting in an effective dielectric constant smaller than that calculated in Eq. (3.7).

In common with other small antennas, load matching and other performance parameters are affected by the surroundings, which may be fixed, like a surrounding plastic case, or variable in the case of hand operated devices. Although the inverted F antenna can provide a good match to the source impedance, many designs have on the pc board near the antenna feed point provision for mounting discrete components for optimizing the match.

### 3.2.4 Loop

The loop antenna is popular for hand-held transmitters particularly because it can be printed on a small circuit board and is less affected by nearby conducting

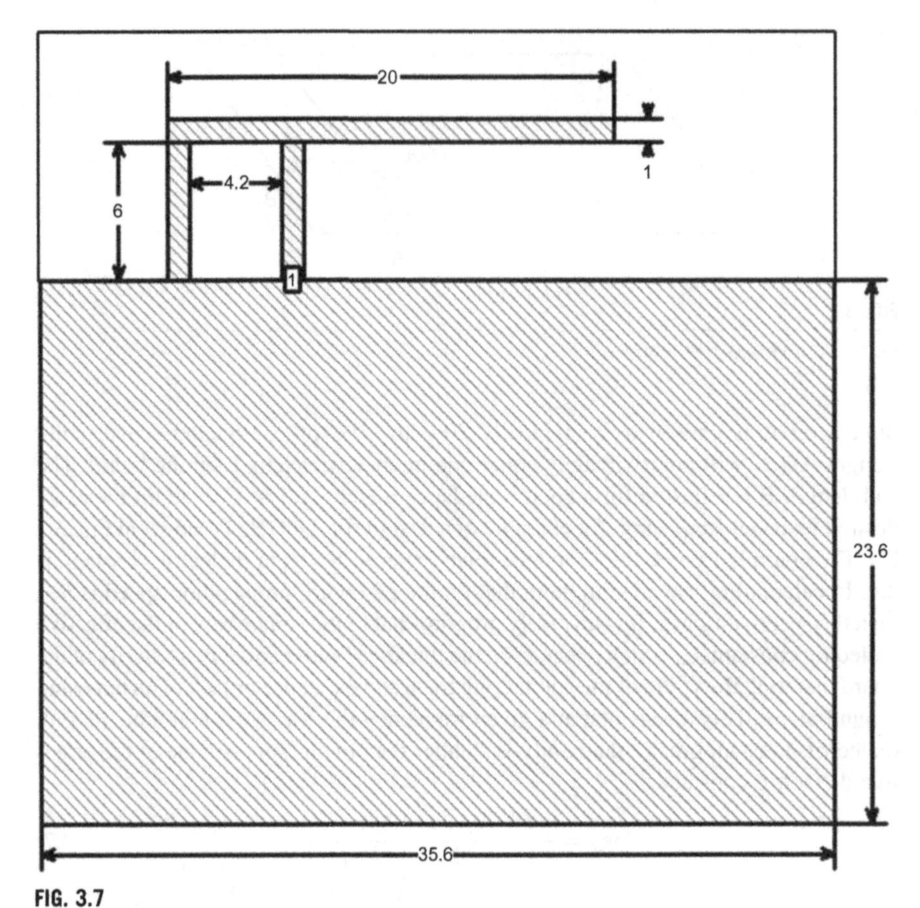


**FIG. 3.7**

Pc board IFA for 2.4 GHz.

objects than other small resonant antennas. Its biggest drawback is that it is very inefficient.

A loop antenna whose dimensions are small compared to a wavelength—less than $0.1\lambda$—has essentially constant current throughout. Its radiation field is expressed as

$$E(\theta) = \frac{120\pi^2 \cdot I \cdot N \cdot A}{r\lambda^2} \cos\theta \tag{3.8}$$

where $I$ is its current, $A$ is the loop area, $r$ is the distance, $\theta$ is the angle from the plane of the loop, $N$ is the number of turns.

From this expression the radiation resistance is derived. It is

$$R_r = 320\pi^2 \cdot \frac{(A \cdot N)^2}{\lambda^4} \tag{3.9}$$

Loop antennas are frequently used in small hand-held remote control transmitters on the low UHF frequencies. The radiation resistance is generally below a tenth of an ohm and the efficiency under 10%. In order to match the transmitter output stage to the low antenna resistance, parallel resonance is created using a capacitor across the loop terminals. While it may appear that the radiation resistance and hence the efficiency could be raised by increasing the number of turns or the area of the loop, the possibilities with this approach are very limited. Increasing turns or area increases the loop inductance, which requires a smaller value of resonating capacitance. The limit on this approach is reached when the resonant capacitance is down to several picofarads.

Because of the low efficiency of the loop antenna, it is rarely used in UHF short-range receivers. An exception is pager receivers, which use low data rates and high sensitivity to help compensate for the low antenna efficiency. One advantage of the loop antenna is that it doesn't require a ground plane.

In low-power unlicensed transmitters, the low efficiency of the loop is not of much concern, since it is the radiated power that is regulated, and at the low powers in question, the power can be boosted enough to make up for the low efficiency. A reasonably high $Q$ is required in the loop circuit, however, in order to keep harmonic radiation low in respect to the fundamental. In many short-range transmitters, it is the harmonic radiation specification that limits the fundamental power output to well below the allowed level.

We will design a loop antenna for a transmitter operating on 315 MHz. The task is easy using the **Mathcad worksheet "Loop.mcdx."**

**Example 3.1**

Given data: $f = 315$ MHz; 1 oz. copper plating; loop sides 25 mm and 40 mm, conductor width 2 mm.

Enter the relevant data in the worksheet.

The results of this example are:

Radiation resistance = 0.038 ohm
Loss resistance = 0.15 ohm
Efficiency = 20.1%
Total resistance for matching = 0.2 ohm
Resonating capacitance 2.9 pF

The results from using the loop antenna worksheet are not particularly accurate, but they do give a starting point for design. Efficiency can be expected to be worse than that calculated because circuit board losses were not accounted for, nor were the effects of surrounding components. There will also be significant losses in the matching circuit because of the difficulty of matching the high output impedance of the low-power transmitter to the very low impedance of the loop. Transmitters designed to operate from low battery voltages can be expected to be better in this respect (see Section 3.3.1).

### 3.2.5 Helical

The helical antenna can give much better results than the loop antenna, when radiation efficiency is important, while still maintaining a relatively small size compared to a dipole or quarter-wave ground plane.

The helical antenna is made by winding stiff wire in the form of a spring, whose diameter and pitch are very much smaller than a wavelength, or by winding wire on a cylindrical form (see Fig. 3.8). This helical winding creates an apparent axial velocity along the spring which is much less than the velocity of propagation along a straight wire which is approximately the speed of light in space. Thus, a quarter wave on the helical spring will be much shorter than on a straight wire. The antenna is resonant for this length, but the radiation resistance will be lower, and consequently the efficiency is less than that obtained from a standard quarter-wave antenna. The helical antenna resonates when the wire length is in the neighborhood of a half wavelength. Impedance matching to a transmitter or receiver is relatively easy.

Radiation from a helical antenna has both vertical and horizontal components, so its polarization is elliptic. However, for the form factors most commonly used, where the antenna length is several times larger than its diameter, polarization is essentially vertical. The helical antenna should have a good ground plane for best and predictable performance. In hand-held devices, the user's arm and body serve as a counterpoise, and the antenna should be designed for this configuration.

The **Mathcad worksheet "helical.mcdx"** helps design a helical antenna. We'll demonstrate by an example.

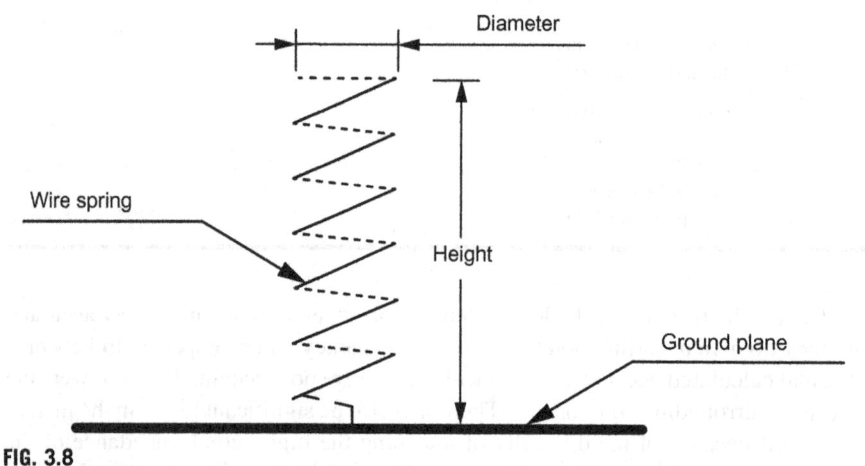


**FIG. 3.8**

Helical antenna.

**Example 3.2**

Our antenna will be designed for 173 MHz. We will wind it on a 10 mm form with AWG 20 wire. We want to find the number of turns to get a resonant antenna 16 cm high. We also want an approximation of the radiation resistance and the antenna efficiency.

Given: The mean diameter of the antenna $D = 10.8$ mm (includes the wire diameter). Wire diameter of AWG 20 is $d = 0.8$ mm. Antenna height $h = 160$ mm. Frequency $f = 173$ MHz.

We insert these values into the helical antenna worksheet and get the following results:

Number of turns = 26
Wire length = 89 cm = 0.514 $\lambda$
Radiation efficiency = 90 percent
Total input resistance = 6.1 ohm

The prototype antenna should have a few more turns than the design value so that the length can be gradually reduced while return loss is measured, until a resonant condition or good match is obtained for the ground plane that results from the physical characteristics of the product. The input resistance of the antenna can be raised by grounding the bottom end of the antenna wire and tapping the wire up at a point where the desired impedance is found.

### 3.2.6 Patch

The patch antenna is convenient for microwave frequencies, specifically on the 2.4-GHz band and higher. It consists of a plated geometric form (the patch) on one side of a printed circuit board, backed up on the opposite board side by a ground plane which extends beyond the dimensions of the radiating patch. Rectangular and circular forms are the most common, but other shapes—for example, a trapezoid—are sometimes used. Maximum radiation is perpendicular to the board. A square half-wave patch antenna has a directivity of 7 to 8 dB.

A rectangular patch antenna is shown in Fig. 3.9. The dimension $L$ is approximately a half wavelength, calculated as half the free space wavelength ($\lambda$) divided by the square root of the effective dielectric constant ($\varepsilon$) of the board material. It must actually be slightly less than a half wavelength because of the fringing effect of the radiation from the two opposite patch edges that are $L$ apart and the ground plane. As long as the feed is on the centerline, the two other edges don't radiate. The figure shows a microstrip feeder, which is convenient because it is etched on the board together with the patch and other component traces on the same side.

The impedance at the feed point depends on the width $W$ of the patch. A microstrip transforms it to the required load (for transmitter) or source (for receiver) impedance. The feed point impedance can be made to match a transmission line directly by moving the feed point from the edge on the centerline toward the center of the board. In this way a 50-ohm coax transmission line can be connected directly to the underside of the patch antenna, with the center conductor going to the feed point through a via and the shield soldered to the ground plane.

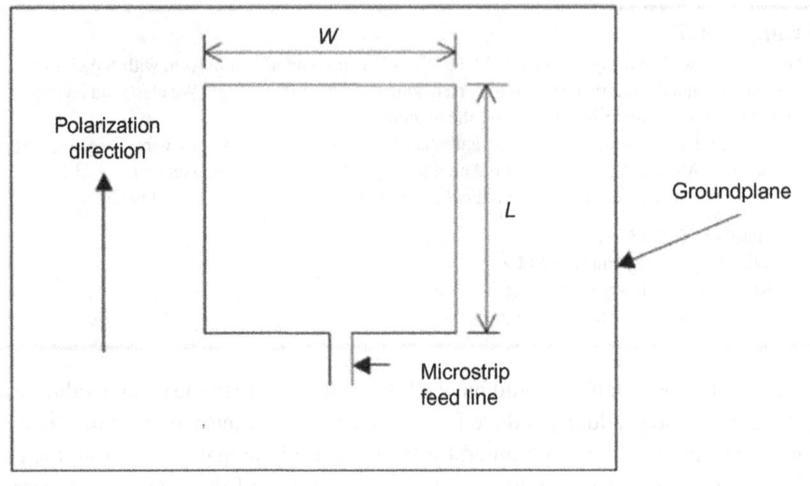


**FIG. 3.9**

Patch antenna.

The **Mathcad "Patch.mcdx" worksheet** helps design a square patch antenna. It includes calculations for estimating the coax cable feed point location.

### 3.2.7 Metamaterial antennas

In general, the antennas used on short-range communication devices are small, in order to fit inside portable devices. When total antenna size, including the finite ground plane when one is used, is shorter than one half wavelength, performance begins to suffer. Efficiency and gain go down, as does bandwidth. For many designers, use of metamaterials may be at least a partial solution to the problem of reducing antenna size without deterioration of performance.

Metamaterials are structures of conductors and dielectrics that emulate material properties not available in nature. They facilitate changing characteristics of the medium through which electromagnetic waves propagate. These characteristics are primarily permittivity and permeability. Specifically, a metamaterial can be configured to have negative values of permeability or permittivity, or both together, in order to react to electromagnetic radiation in ways not obtainable naturally. An example is the patch antenna which was discussed in the previous subsection. Its physical size is determined by the electrical wavelength at the operating frequency in the material it is made of. Its dimensions are on the order of one half or one quarter wavelength, positioned over a larger ground plane. The size of the patch can be reduced by fabricating it on a substrate with a high dielectric constant, or relative permittivity. This is seen from the expression

$$v_p = \frac{1}{\sqrt{\varepsilon_{eff}\varepsilon_0\mu_{eff}\mu_0}} = \frac{1}{\sqrt{\varepsilon_{eff}\mu_{eff}}} \cdot c \quad \varepsilon_{eff}, \mu_{eff} \geq 1 \tag{3.10}$$

where $v_p$ is velocity of propagation in the medium, $\varepsilon_{eff}$ and $\mu_{eff}$ are relative permittivity and permeability in respect to their absolute values $\varepsilon_0$ and $\mu_0$, and $c$ is the speed of light in a vacuum. It follows that the wavelength is reduced in proportion to the reduction of $v_p$:

$$\lambda = \frac{v_p}{f} \tag{3.11}$$

where $f$ is the operating frequency. When $v_p$ is reduced by using a natural substrate with higher dielectric constant, bandwidth decreases, which is often undesirable. Reducing $v_p$ by increasing the permeability in a metamaterial substrate allows decreasing antenna size while maintaining or even increasing the bandwidth.

A consequence of the existence of materials with negative permittivity and permeability is negative index of refraction, expressed as:

$$n = -\sqrt{(-\varepsilon_r)(-\mu_r)} \tag{3.12}$$

where $\varepsilon_r$ and $\mu_r$ are the relative permittivity and permeability of the metamaterial, both equal to or greater than unity. A negative index of refraction causes electromagnetic radiation passing between a material with a positive $n$ (air, for instance) and a material with negative $n$ to change its direction of propagation, in accordance with Snell's Law, but opposed to the deviation of propagation direction in natural materials both of which have positive indices of refraction. This effect is shown in Fig. 3.10. Also, propagation in a material with negative permittivity and negative permeability, called double negative metamaterial, DNG, exhibits a different

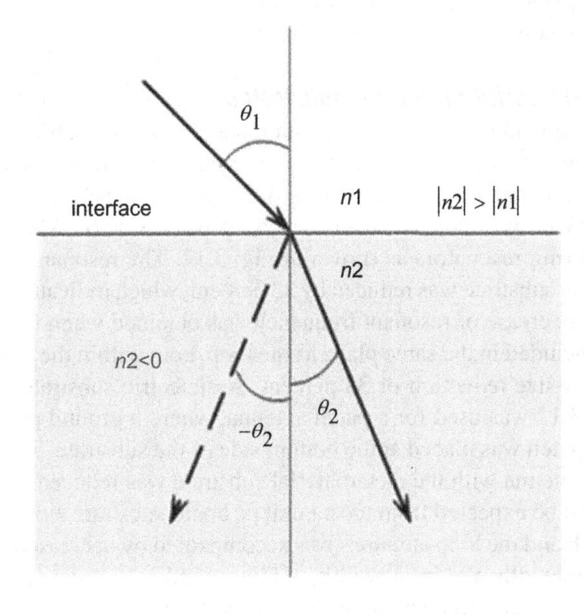


**FIG. 3.10**

Path of wave propagation depends on the sign of the index of refraction.

vectorial relationship between the magnetic field **B**, the electric field **E**, and the phase constant vector **β**. In a natural medium, these vectors follow the right hand rule, and the wavefront proceeds in the direction of the radiated power. However, in a DNG medium, the left hand rule applies, and the wave appears to travel backwards, that is, in the opposite direction of power flow. In such a medium, for example, the Doppler effect is reversed; that is, an observer would see a rise in frequency from a receding emitter and a decrease in frequency from an approaching emitter, just the opposite of the observation in a natural medium.

The characteristics and phenomena described above are responsible for the reduced electrical size (comparing with wavelength) and increased performance relating to efficiency, gain, and bandwidth that have been achieved in antenna designs based on metamaterials. They also are exploited for isolation of multiple antennas in close proximity on a printed circuit board. The examples below demonstrate metamaterial applications for efficient small antennas.

#### 3.2.7.1 Ultra wideband patch antenna

A patch antenna was designed for exceptionally wide bandwidth [9]. Negative permeability and negative permittivity were achieved by etching particular fractal patterns on the patch and on the opposing ground plane. These are shown in Fig. 3.11. Unit cells of top and bottom of the patch are shown in Fig. 3.11A, and the whole patch antenna, top and bottom, is shown in Fig. 3.11B. The substrate is 0.79 mm thick with a dielectric constant of 2.2. The referenced authors demonstrated that the antenna structure supports backward waves, as described above. The antenna bandwidth was simulated and measured at 4 GHz to 18 GHz for a return loss of 10 dB or better. Average gain over this range was 6 dBi.

#### 3.2.7.2 Metamaterial dielectric substrates

Patch and loop antennas were designed with metamaterial dielectric substrates [10]. The aim here was to significantly decrease the physical size at the operating frequency. A loop antenna with approximately one wavelength perimeter that resonated in air at 2.58 GHz was mounted over a stack of pc boards on which were etched groups of split ring resonators as shown in Fig. 3.12. The resonant frequency with the metamaterial substrate was reduced by 23 percent, which indicates the size reduction. A greater decrease of resonant frequency was obtained when the split ring resonators were included in the same plane as the loop, both within the loop and outside it, resulting in a size reduction of 38 percent. A dielectric substrate similar to that shown in Fig. 3.12 was used for a patch antenna, where a ground plane larger than the size of the patch was placed at the bottom side of the substrate. The resonant frequency of the antenna with the metamaterial substrate was reduced somewhat more than what would be expected from the natural pc board substrate alone. Reduction of size of the patch and the loop antennas was accompanied by decreased bandwidth. In both cases, bandwidth was improved by adding nondriven antenna elements with slightly higher resonant frequencies on the opposite side of the dielectric from the driven element.

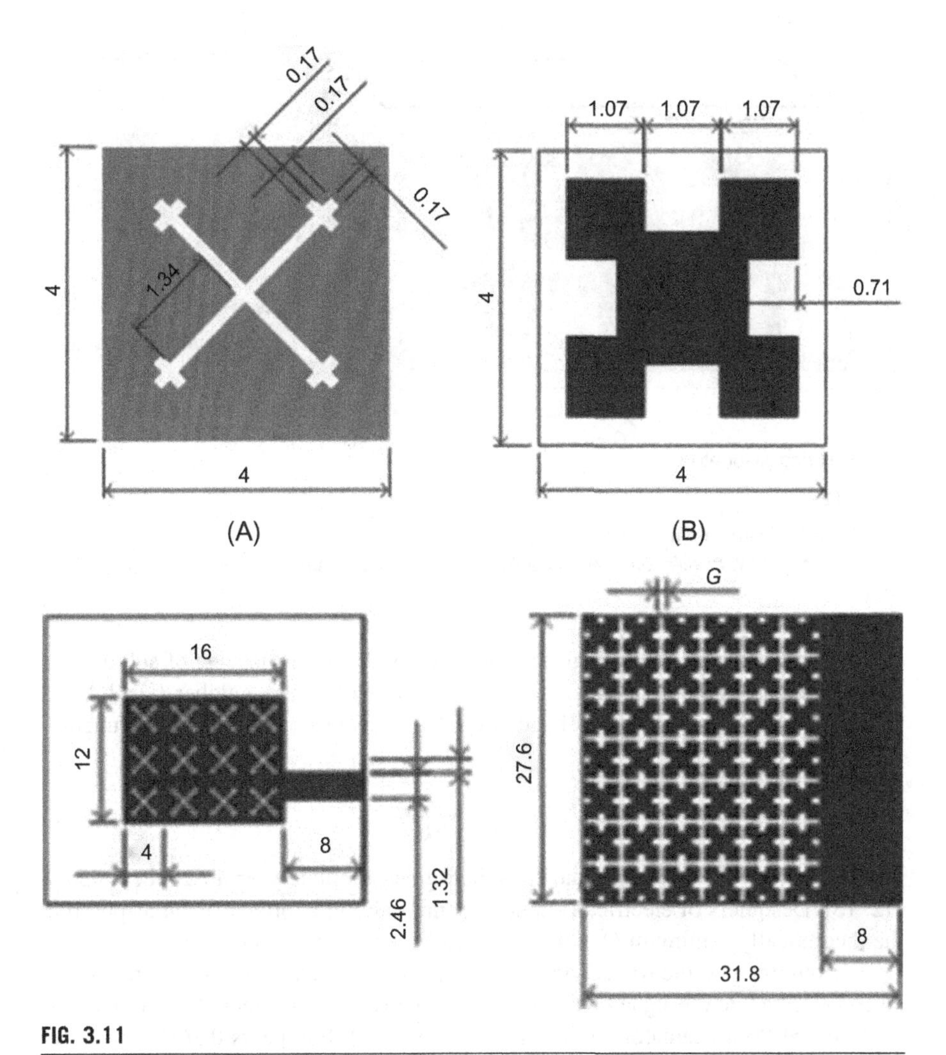


**FIG. 3.11**

Fractal patterns give DNG characteristics to a patch antenna. (A) Left: pattern on top side (patch). Right: pattern on bottom side (ground plane). (B) Repeating patterns on top and bottom side ([9] Open Access).

### *3.2.7.3 Metamaterial inspired small antennas*

Whereas "true" metamaterial is generally considered to be made from repeated unit cells of metallic patterns designed to influence magnetic or electric field properties, as was described in the examples above, characteristics of small antennas can also be modified using a single cell. Antennas designed in this way have been called "metamaterial inspired" small antennas [11]. The basic purpose of such a design appears to

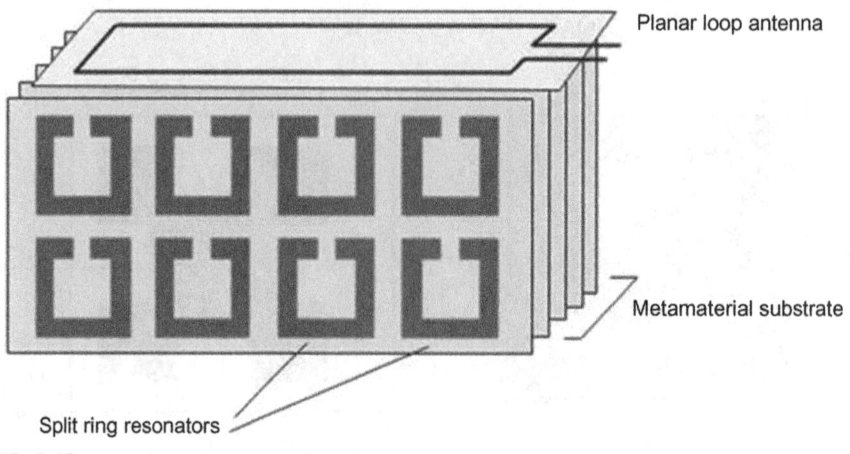


**FIG. 3.12**

Loop antenna with metamaterial substrate.

*After D.W. Prather, Embedded Meta-Material Antennas, U.S. Army Research Office Report, 2009.*

be efficient matching of the high reactance low radiation resistance of sub-quarter wave antennas. As total antenna size goes down below wave number $k = \lambda/2\pi$ its $Q$, and in inverse proportion its bandwidth, has a lower bound which is a function of its size and $k$,

$$Q_{low} = \frac{1}{ka} + \frac{1}{(ka)^3} \tag{3.13}$$

where $a$ is the radius of the smallest sphere that completely encloses the antenna [12, 13]. Designers of electrically small antennas aim to approach as far as possible the theoretically minimum Q and consequently maximum bandwidth.

Various three dimensional and two dimensional metamaterial inspired small antennas have been designed using both electrically based (monopole) and magnetically based (loop) radiators or driving elements [11]. It appears that the most significant feature of these designs is the relative ease of matching as compared with using discrete component matching networks at VHF and UHF frequencies and quarter wave microstripe in microwave bands. Low efficiencies and narrow bandwidths appear to be unavoidable as $ka$ (electrical size metric) becomes much less than unity.

An example of a small antenna based on a split ring resonator (SRR) radiator is shown in Fig. 3.13 [12]. It is fed via a smaller loop and has a good 50 ohm match at the GPS frequency of 1.575 GHz. A second resonant frequency is around 3 GHz, as seen in the return loss plot in Fig. 3.14. The authors of the referenced article assert that the split ring resonator antenna with loop coupling shows improved performance from the point of view of wider bandwidth for a given size than loop and dipole antennas. It can be noticed that the small loop coupling is essentially the same as

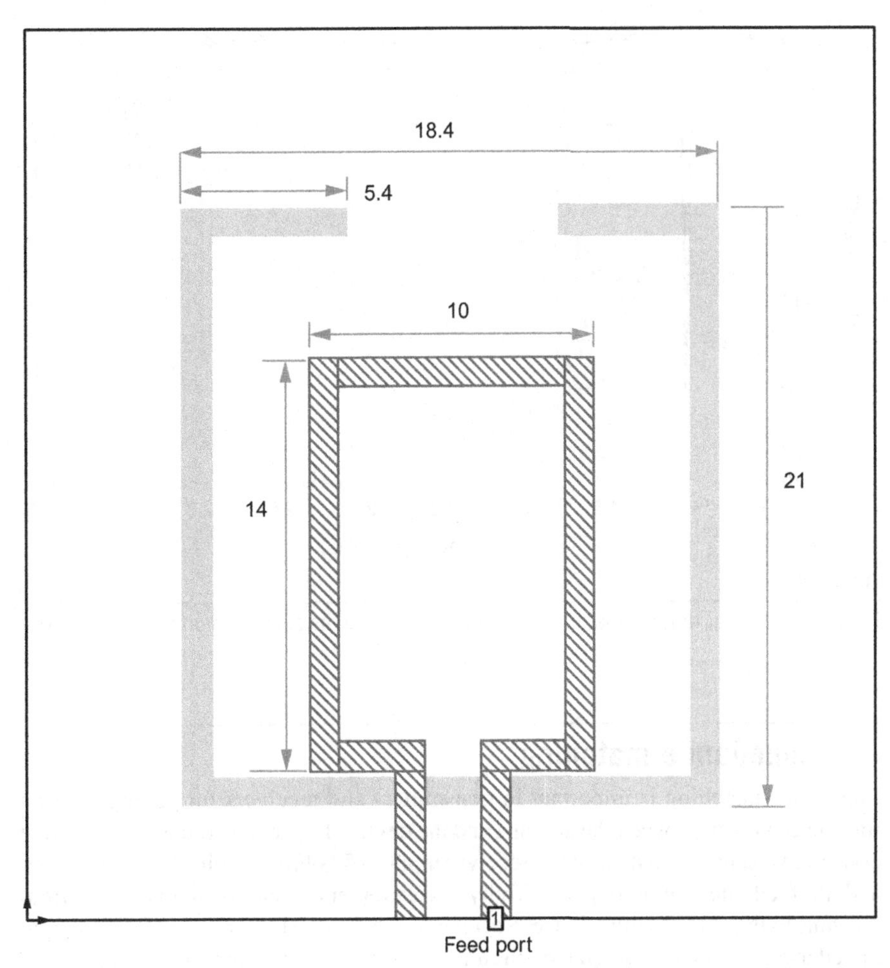


**FIG. 3.13**

Split ring resonator with feed loop coupling. Line widths are 1 mm. Loop is on board top, and shaded SRR is on the bottom. Board is FR4, permittivity 4.4, 1.6 mm thick.

*After S.K. Sharma, D.S. Nagarkoti, Meet the Challenge of Designing Electrically Small Antennas, Microwaves & RF, 2017.*

the mutual inductance coupling described in Section 3.3.5. In that description a small break in the larger loop is bridged by a resonance tuning capacitor. While not apparent explicitly, the SRR in Fig. 3.13 has a parasitic resonating capacitance between the opposing edges of the split making this antenna similar in principle to the mutual inductance matched antenna in Fig. 3.23. In fact, it is debatable whether single cell configuration antennas should be described as having the characteristics of metamaterials [14].

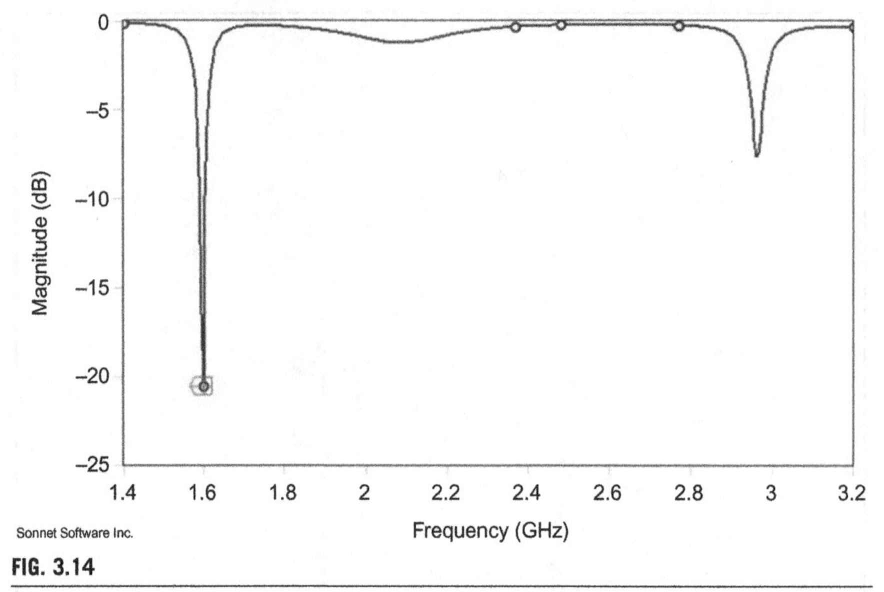


**FIG. 3.14**

Return loss of split ring resonator with feed loop coupling, as simulated on Sonnet Lite [15].

## 3.3 Impedance matching

Impedance matching is important in transmitters and receivers for getting the best transfer of power between the antenna and the device. In a receiver, matching is often done in two stages—matching the receiver input to 50 ohms to suit a band-pass filter and to facilitate laboratory sensitivity measurements, and then matching from 50 ohms to the antenna impedance. Receiver modules most often have 50 ohms input impedance. Receiver integrated circuits or low-noise RF amplifiers may have 50 ohms input, or the input impedance may be specified for various frequencies of operation. Sometimes a particular source impedance is specified that the input RF stage must "see" in order to obtain minimum noise figure.

Impedances to be matched are specified in different ways in component or module datasheets. A complex impedance may be specified directly for the operating frequency or for several possible operating frequencies. Another type of specification is by a resistance with capacitor or inductor in parallel or in series. The degree of matching to a specified impedance, usually 50 ohms, can be expressed by the *reflection coefficient*. This will be discussed later.

There are various circuit configurations that can be used for impedance matching, and we present some simple examples here to match a pure resistance to a complex impedance. First, you should be able to express an impedance or resistance-reactance combination in parallel or serial form, whichever is convenient for the matching topography you wish to use. You can do this using the **Mathcad worksheet**

**"Conversions.mcdx."** Then, use the **worksheet "Matching.mcdx"** to find component values to match a wide range of impedances. The parallel or series source reactance must be separated from the total adjacent derived reactance value to get the value of the component to use in the matching circuit. Example 3.3 demonstrates this for a parallel source capacitor.

Remember that coils and capacitors are never purely reactive. The losses in coils, specified by the quality factor $Q$, are often significant, whereas those in capacitors are usually ignored. In a parallel equivalent circuit (loss resistance in parallel to the reactance), $Q = R/X$ ($X$ is the reactance). In a series equivalent circuit, $Q = X/R$. If the loss resistance is within a factor of up to around five times the resistance to be matched, it should be combined with that resistance before using the impedance matching formula. Example 3.4 shows how to do it.

Use nearest standard component values for the calculated values. Variable components may be needed, particularly in a high $Q$ circuit. Remember also that stray capacitance and inductance will affect the matching and should be considered in selecting the matching circuit components.

## Example 3.3

Fig. 3.15 shows a circuit that can be used for matching a high impedance, such as may be found in a low-power transmitter, to 50 ohms.

Let's use it to match a low-power transmitter output impedance of 1000 ohms (see Section 3.3.1) and 1.5-pF parallel capacitance to a 50-ohm band-pass filter or antenna. The frequency is 315 MHz. We use **Impedance Matching worksheet "Matching.mcdx,"** circuit 3.

Given values are the following:

$f = 315$ MHz, $R_1 = 1000$ ohms, $C_{out} = 1.5$ pF, $R_2 = 50$ ohms

**(a)** At the top of the worksheet, set $f = 315$ MHz. Under circuit (3), set $R_1$ to 1000 and $R_2$ to 50 ohms. Select a value for $Q$.

Let $Q = 10$

**(b)** Rounded off results are:

$C_1 = 5.1$ pF
$C_2 = 20.3$ pF
$L_1 = 60.1$ nH

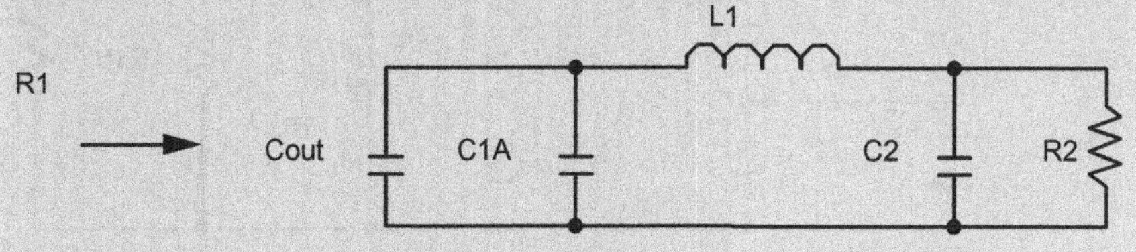


**FIG. 3.15**

Impedance transformation, Example 3.3.

**(c)** $C_1$ of the worksheet is made up of the parallel combination of $C_{out}$ and $C_{1A}$ of Fig. 3.15:

$$C_{1A} = C_1 - C_{out} = 5.1\,\text{pF} - 1.5\,\text{pF} = 3.6\,\text{pF}$$

The closest standard component values for $C_{1A}$, $L_1$, and $C_2$ are used in the impedance transformation matching network of Fig. 3.15.

## Example 3.4

We want to match the input of an RF mixer and IF amplifier integrated circuit (such as Philips NE605) to a 50-ohm antenna at 45 MHz. The equivalent input circuit is 4500 ohms in parallel with 2.5 pF. We choose to use a parallel coil $L_{1A}$ having a value of 220 nH and a $QL_{1A}$ of 50. See Fig. 3.16.

Given:

$f = 45$ MHz, $R_1 = 50$ ohms, $R_{in} = 4.5$K ohms, $C_{in} = 2.5$ pF

Let $L_{1A} = 220$ nH, $Q1A = 50$
Find: $C_1$ and $C_2$

**(a)** Calculate $RL_{1A}$:
$XL_{1A} = 62.2$ ohms (you can use **"Conversions.mcdx" worksheet**)

$$RL_{1A} = Q_{1A} \times XL_{1A} = 3110\,\text{ohms}$$

**(b)** Find equivalent input resistance to be matched, $RL_1$:

$$RL_1 = RL_{1A} \| R_{in} = (3110 \times 4500)/(3110 + 4500) = 1839\,\text{ohms}$$

**(c)** Find equivalent parallel inductance $L_1$:

$XC_{in} = -1415$ ohms (**"Conversions.mcdx" worksheet**)

$$XL_1 = XL_{1A} \| XC_{in} = (62.2 \times (-1415))/(62.2 - 1415) = 65.06\,\text{ohms}$$

**(d)** Find $Q$, which is needed for the calculation of $C_1$ and $C_2$ using the "Impedance Matching" worksheet:

$$Q = RL_1/XL_1 = 1839/65.06 = 28.27$$

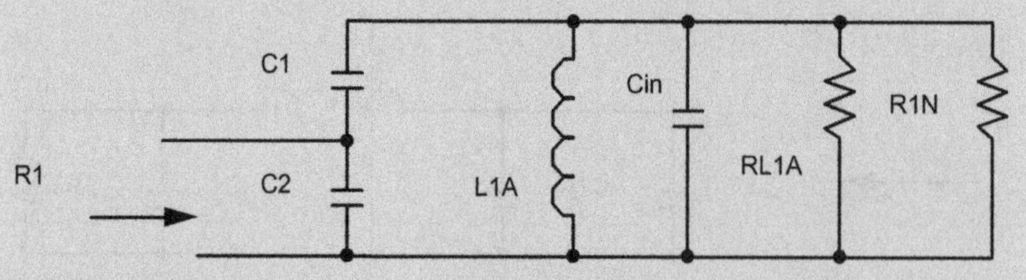


**FIG. 3.16**

Impedance transformation, Example 3.4.

**(e)** Use the **worksheet "Matching.mcdx"** circuit (5) to find $C_1$ and $C_2$, after specifying $f$, $R_1$, $R_2$, and $Q$:

$f = 45$ MHz, $R_1 = 50$ ohms, $R_2 = RL_1 = 1839$ ohms, $Q = 28.27$

Results:

$C_1 = 65$ pF
$C_2 = 322$ pF

It may seem that the choice of the parallel inductor was arbitrary, but that's the designer's prerogative, as long as the resultant $Q$ is greater than the minimum $Q$ given in the worksheet example (in this case approximately 6). The choice of inductance determines the circuit $Q$, and consequently the bandwidth of the matching circuit. The total $Q$ of the circuit includes the loading effect of the source resistance. Its value is one half the $Q$ used in the design procedure, or 28.27/2 = approximately 14 in this example.

## Example 3.5

We have a helical antenna with 15-ohm impedance that will be used with a receiver module having a 50-ohm input. The operating frequency is 173 MHz. The matching network is shown in Fig. 3.17.

Given: $f = 173$ MHz, $R_1 = 50$ ohms, $R_2 = 15$ ohms

Use these values in the **"Matching.mcdx" worksheet**, circuit (1), to get the matching network components:

$$L_1 = 21.1\,\text{nH}$$

$$C_1 = 28.1\,\text{pF}$$

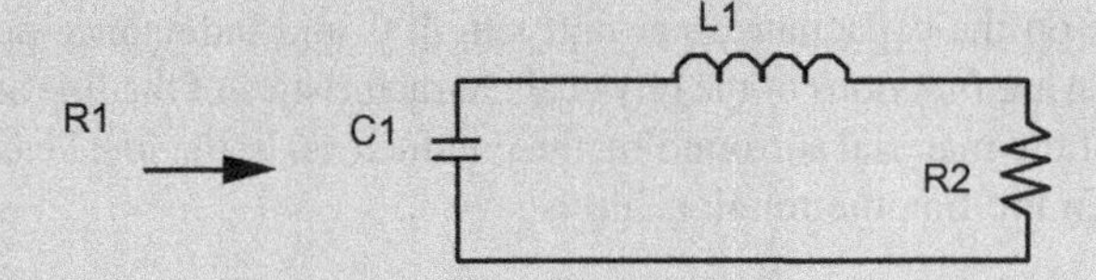


**FIG. 3.17**

Impedance transformation, Example 3.5.

## 3.3.1 Transmitter output impedance

In order to get maximum power transfer from a transmitter to an antenna, the RF amplifier output impedance must be known, as well as the antenna impedance, so that a matching network can be designed as shown in the previous section. For very low-power transmitters with radiated powers of tens of microwatts at the most, close matching is not critical. However, for a radiated power of 10 milliwatts and particularly when low-voltage lithium battery power is used, proper matching can save battery energy due to increased efficiency, and can generally simplify transmitter design. Besides, an output band-pass filter needs reasonably good matching to deliver a predictable frequency response.

The following expression is a simplified estimate of an RF amplifier's output impedance $R_L$:

$$R_L = \frac{\left(V_{CC} - V_{CE(sat)}\right)^2}{2P} \tag{3.14}$$

$V_{CC}$ is the supply voltage to the RF stage, $V_{CE(sat)}$ is the saturation voltage of the RF transistor at the operating frequency, and $P$ is the power output.

### 3.3.2 Transmission lines

In many short-range radio devices, the transmitter or receiver antenna is an integral part of the device circuitry and is coupled directly to the transmitter output or receiver input circuit through discrete components. This is particularly the case with portable equipment. Devices with an external antenna located away from the equipment housing need a transmission line to connect the antenna to the input or output circuit. The transmission line is an example of a distributed circuit, and it affects the coupling or transfer of the RF signal between the device RF circuit and the antenna. At high UHF and microwave frequencies, even the short connection between an internal antenna and the RF circuit is considered a transmission line whose characteristics must be designed to achieve proper impedance matching.

The transmission line can take several forms, among them coaxial cable, balanced two-wire cable, microstrip, and waveguide (a special case not considered below).

A basic characteristic of a transmission line is its *characteristic impedance*. Its value depends on the capacitance per unit length $C$ and inductance per unit length $L$, which in turn are functions of the physical characteristics of the line and the dielectric constant of the material surrounding the conductors. In the ideal case when there are no losses in the line the relationship is

$$Z_0 = \sqrt{\frac{L}{C}} \tag{3.15}$$

where $L$ is the inductance per unit length in henrys and $C$ is the capacitance per unit length in farads. Another important characteristic is the velocity factor, which is the ratio of the propagation velocity, or phase velocity, of the wave in the line to the speed of light. The velocity factor depends on the dielectric constant, $\varepsilon_r$, of the material enclosing the transmission line conductors as

$$VF = \frac{1}{\sqrt{\varepsilon_r}} \tag{3.16}$$

Also important in specifying a transmission line, particularly at VHF and higher frequencies and relatively long lines, is the attenuation or line loss.

For example, the characteristics of a commonly used coaxial cable, RG-58C, are:

| | |
|---|---|
| Characteristic impedance | 50 ohms |
| Inductance per meter | 0.25 microhenry |
| Capacitance per meter | 101 picofarad |
| Velocity factor | 0.66 |
| Attenuation at 50 MHz | 3.5 dB per 30 m |
| Attenuation at 300 MHz | 10 dB per 30 m |

In the previous section, we talked about matching the antenna impedance to the circuit impedance. When the antenna is connected to the circuit through a transmission line, the impedance to be matched, seen at the circuit end of the transmission line, may be different from the impedance of the antenna itself. It depends on the characteristic impedance of the transmission line and the length of the line.

Several terms which define the degree of impedance matching, usually relating to transmission lines and antennas, are presented below.

*Standing wave ratio* is a term commonly used in connection with matching a transmission line to an antenna. When the load impedance differs from the characteristic impedance of the transmission line, the peak voltage on the line will differ from point to point. On a line whose length is greater than a half wavelength, the distance between voltage peaks or between voltage nulls is one-half wavelength. The ratio of the voltage peak to the voltage null is the standing wave ratio, abbreviated *SWR*, or *VSWR* (voltage standing wave ratio). The ratio of the peak current to the minimum current is the same as the *VSWR* or *SWR*.

When the load is a pure resistance, $R$, and is larger than the characteristic impedance of the line (also considered to be a pure resistance), we have

$$SWR = \frac{R}{Z_0} \tag{3.17}$$

If the load resistance is less than the line characteristic impedance, then

$$SWR = \frac{Z_0}{R} \tag{3.18}$$

In the general case where both the line impedance and the characteristic impedance may have reactance components, thus being complex, the *VSWR* is

$$SWR = \frac{1 + \left|\frac{Z_{load} - Z_0}{Z_{load} + Z_0}\right|}{1 - \left|\frac{Z_{load} - Z_0}{Z_{load} + Z_0}\right|} \tag{3.19}$$

*Reflection coefficient* is the ratio of the complex value of the voltage of the reflected wave from a load to the complex voltage value of the forward wave absorbed by the load:

$$\rho = \frac{E_r}{E_f} \tag{3.20}$$

which is complex.

When the load is perfectly matched to the transmission line, the maximum power available from the generator is absorbed by the load, there is no reflected wave, and the reflection coefficient is zero. For any other load impedance, less power is absorbed by the load, and what remains of the available power is reflected back to the generator. When the load is an open or short circuit, or a pure reactance, all of the power is reflected back, the reflected voltage equals the forward voltage, and the reflection coefficient is unity. We can express the reflection coefficient in terms of the load impedance and characteristic impedance as

$$\rho = \frac{Z_{load} - Z_0}{Z_{load} + Z_0} \tag{3.21}$$

The relation between the *SWR* and the reflection coefficient is

$$SWR = \frac{1 + |\rho|}{1 - |\rho|} \tag{3.22}$$

*Return loss* is an expression of the amount of power returned to the source relative to the available power from the generator. It is expressed in decibels as

$$RL = -20\log(|\rho|) \tag{3.23}$$

Note that the return loss is always equal to or greater than zero. Of the three terms relating to transmission line matching, the reflection coefficient gives the most information, since it is a complex number. As for the other two terms, *SWR* may be more accurate for large mismatches, whereas return loss presents values with greater resolution than *SWR* when the load impedance is close to the characteristic impedance of the line.

A plot of forward and reflected powers for a range of *SWRs* is given in Fig. 3.18. This plot is convenient for seeing the effect of an impedance mismatch on the power actually dissipated in the load or accepted by the antenna, which is the forward power minus the reflected power.

Transmission line losses are not represented in the above definitions. Their effect is to reduce the *SWR* and increase the return loss, compared to a lossless line with the same load. This may seem to contradict the expressions given, which are in terms of load impedance, but that is not so. For instance, the load impedance in the expression for *SWR*, Eq. (3.19), is the impedance at a particular point on the line where the SWR is wanted and not necessarily the impedance at the end of the line. Thus, a long line with high losses may have a low *SWR* measured at the generator end, but a high *SWR* at the load. Transmission line loss is specified for a perfectly matched line, but when

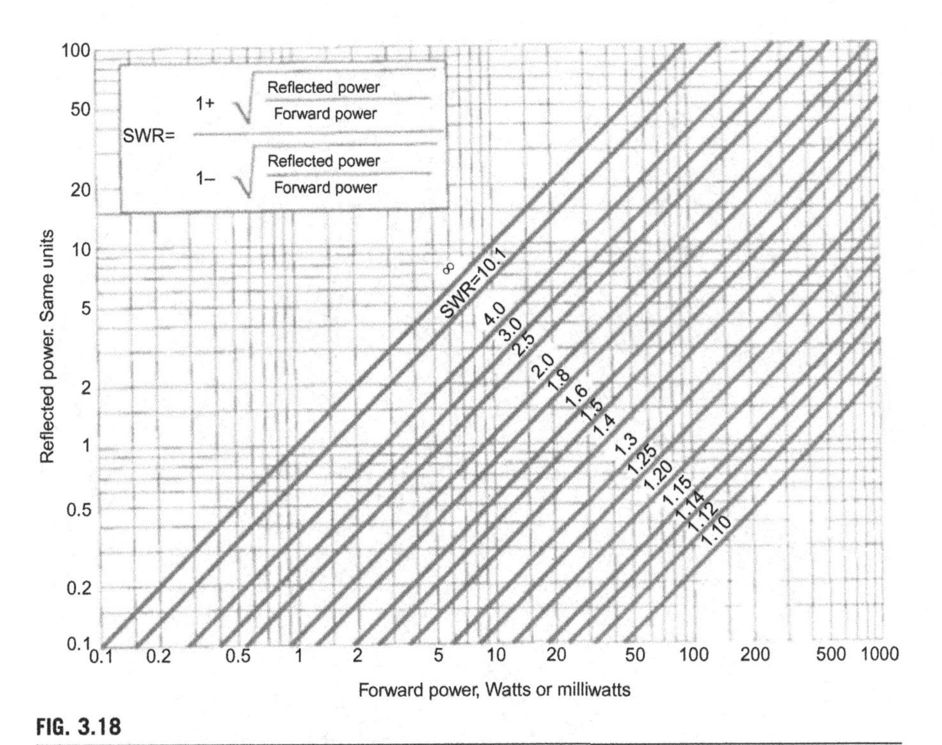


**FIG. 3.18**

SWR, forward and reflected power.

*Reprinted with permission, ARRL Antenna Book, 16th edition.*

a mismatch exists, the loss is higher because of higher peak current and a resulting increased $I^2R$ power dissipation in the line.

Worksheet **Translines.mcdx** facilitates solving transmission line problems.

## 3.3.3 Smith chart

A convenient tool for finding impedances in transmission lines and designing matching networks is the Smith chart, shown in Fig. 3.19. The Smith chart is a graph on which you can plot complex impedances and admittances (admittance is the inverse of impedance). An impedance value on the chart is the intersection of a resistance circle, labeled on the straight horizontal line in the middle, and a reactance arc, labeled along the circumference of the "0" resistance circle. Fig. 3.20 shows an expanded view of the chart with some of the labels. The unique form of the chart was devised for convenient graphical manipulation of impedances and admittances when designing matching networks, particularly when transmission lines are involved. The Smith chart is useful for dealing with distributed parameters which describe the characteristics of circuit board traces at UHF and microwave frequencies.

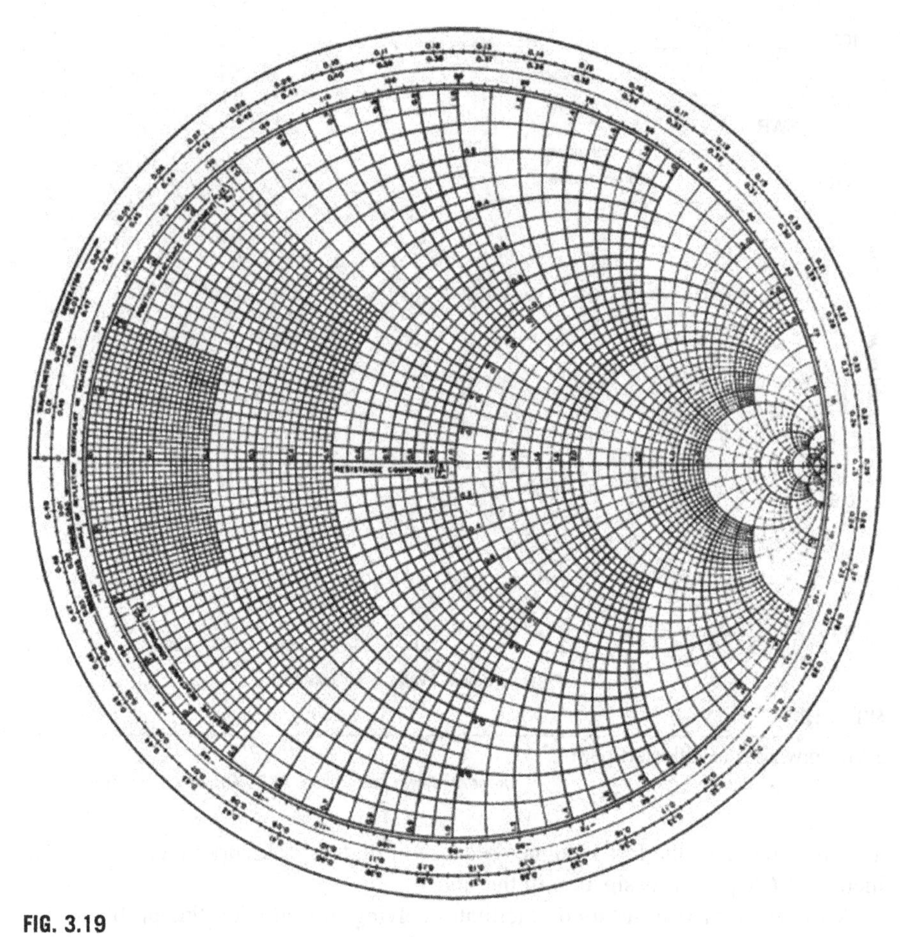

**FIG. 3.19**

Smith chart.

We'll describe some features of the Smith chart by way of an example. Let's say we need to match an antenna having an impedance of 15 ohms resistance in series with a capacitance reactance of 75 ohms to a transmitter with 50 ohms output impedance. The operating frequency is 173 MHz. The antenna is connected to the transmitter through 73 cm of RG-58C coaxial cable. What is the impedance at the transmitter that the matching network must convert to 50 ohms? The example is sketched in Fig. 3.21, and Fig. 3.22 shows the use of the Smith chart.

Step 1. First, we mark the antenna impedance on the chart. Note that the resistance and reactance coordinates are normalized. The center of the chart, labeled 1.0, is the characteristic impedance of the transmission line, which is 50 ohms. We divide the resistance and capacitive reactance of the antenna by 50 and get, in complex form:

$$Z_{load} = 0.3 - j1.5$$

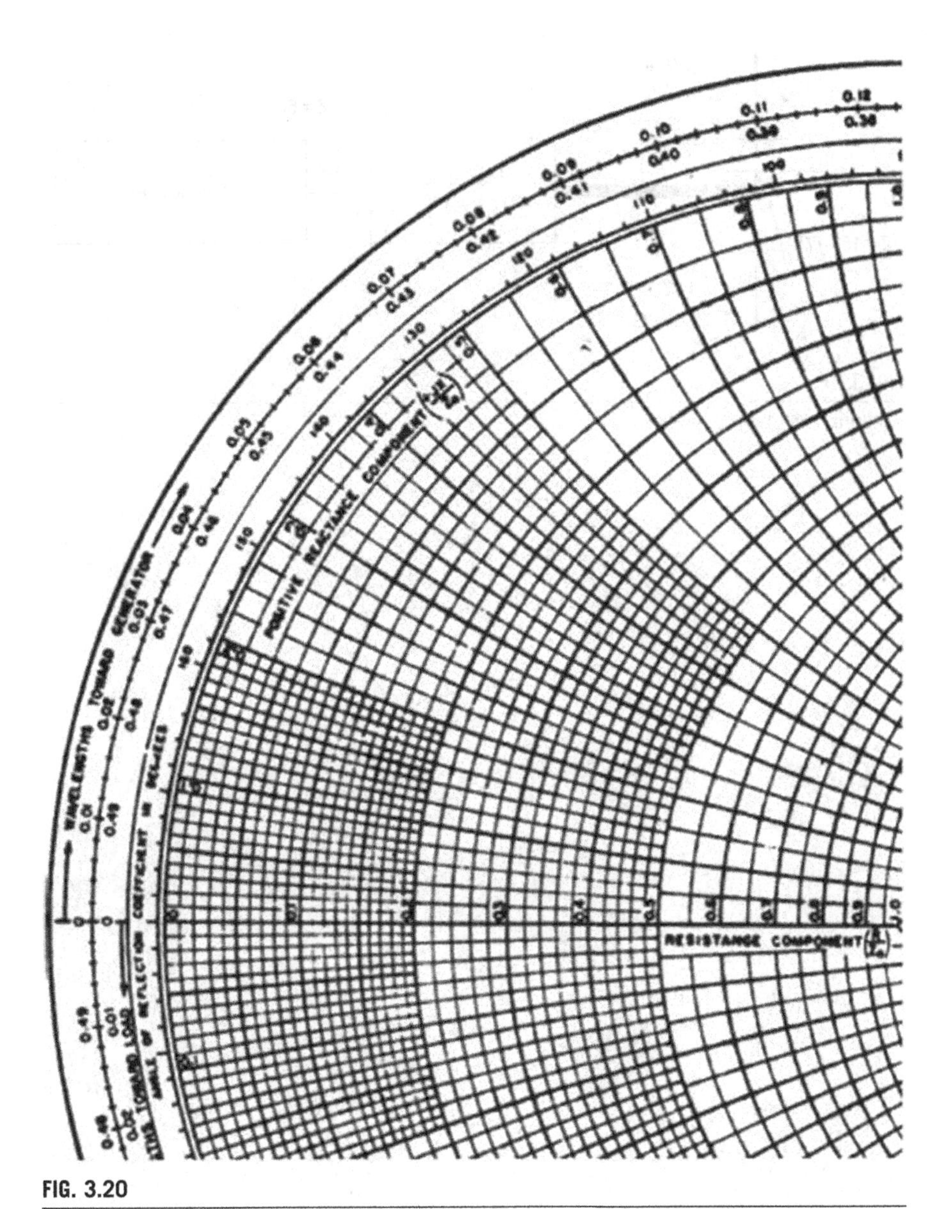


**FIG. 3.20**

Expanded view of Smith chart.

This is marked at the intersection of the 0.3 resistance circle with the 1.5 capacitive reactance coordinate in the bottom half of the chart. This point is marked "A" in Fig. 3.22.

Step 2. The impedance at the transmitter end of the transmission line is located on a circle whose radius is the length of a line from the center of the chart to point "A"

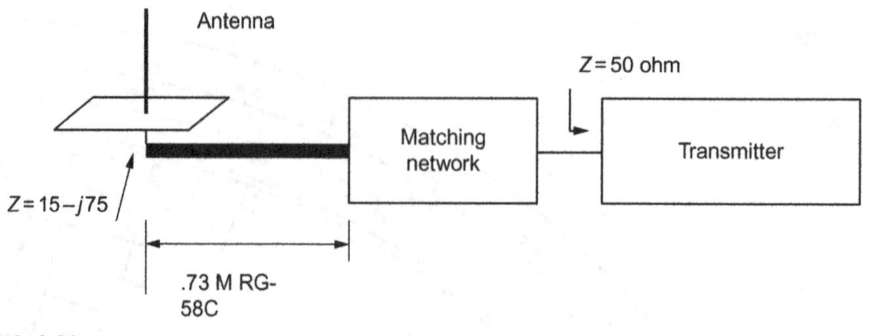


FIG. 3.21

Antenna matching example.

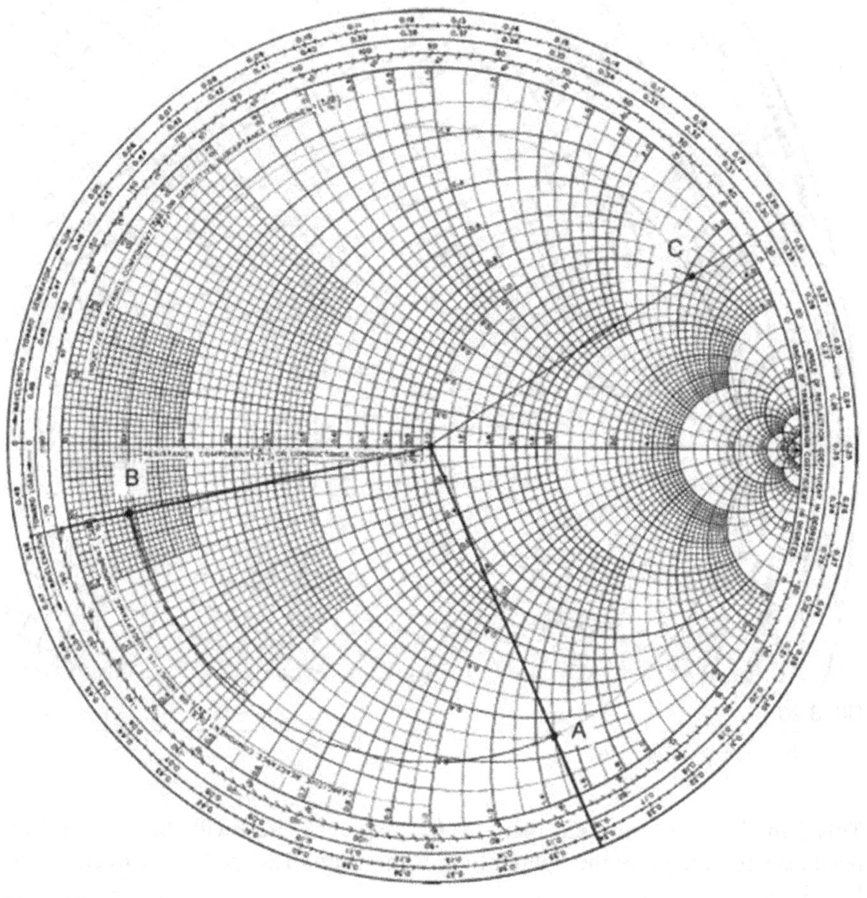


FIG. 3.22

Using the Smith chart.

(assuming no cable losses). In order to find the exact location of the impedance on this circle for the 73-cm coax cable, we must relate the physical cable length, $l$, to the electrical length, $L$, in wavelengths:

$$L = \frac{l}{\eta\lambda} \tag{3.24}$$

where $\eta$ is the velocity factor (0.66) and $\lambda$ is the wavelength in free space (speed of light in free space/frequency). Inserting the values for this example, we find the electrical length of the line is 0.64 wavelengths. The Smith chart instructs us to move toward the generator in a clockwise direction from the load. The wavelength measure is marked on the outmost marked circle. Every 0.5 wavelengths, the load impedance is reflected to the end of the cable with no change, so we subtract a whole number of half wavelengths, in this case one, from the cable length, giving us

$$0.64\lambda_c - 0.5\lambda_c = 0.14\lambda_c$$

where $\lambda_c$ is the wavelength in the cable.

The line drawn in Fig. 3.22 from the center through Zload (point A) intersects the "wavelengths to generator" circle at 0.342. We add 0.14 to get 0.482 and draw a line from the center to this point. Mark the line at point B, which is the same distance from the center as point A. This is done conveniently with a compass.

Step 3. Read off point B. It is 0.095 – $j$0.115. Multiplying by the 50 ohm characteristic impedance, we get the input impedance:

$$Z_{gen} = 4.7 - j5.8 \text{ ohms}$$

Step 4. Now using the procedure of Example 3.5 above, we can design a matching network to match the impedance seen at the coax cable to the 50-ohm impedance of the transmitter. It is necessary to cancel the capacitance at the input to the cable by adding the corresponding inductance. In worksheet **Matching.mcdx** circuit (1), $f =$ 173 MHz, $R_1 = 50$ ohm, and $R_2 = 4.7$ ohm (the real part of $Z_{gen}$). After solving for $L_1$ and $C_1$, add the equivalent inductive reactance of the capacitive reactance of $Z_{gen}$, 5.8 ohms, to the reactance of $L_1$, $XL_1$, and convert the result to inductance. The resulting matching network components are modified $L_1$ = 19 nH and $C_1$ = 57 pF.

By examining the Smith chart in Fig. 3.22, we note that if we use a longer coaxial cable, we can get an impedance at its end that has a real part, or resistance, of 50 ohms, and an inductive reactance of 50 × 3.1 = 155 ohms. This is point C in the figure. We attain this impedance by adding $0.22\lambda_c$ = 25 cm to the original transmission line, for a total coax cable length of 98 cm. Now the only matching component we need is a series capacitor to cancel out the inductive reactance of 155 ohms. Using the **Conversions.mcdx** worksheet, we find its capacitance to be approximately 6 pF. All of the results in this example are approximate and depend on the resolution achieved by plotting on the Smith chart. More precise results are obtained with worksheet **Translines.mcdx**. Using it, we find $Z_{gen}$ = 4.6 – $j$6.3 ohms.

Other transmission-line matching problems can be solved using the Smith chart. Using the chart, you can easily determine *SWR*, reflection coefficient, and return loss. The chart also has provision for accounting for line losses.

The Smith chart is very handy for seeing at a glance the effects on impedance of changing transmission line lengths, and also for using transmission lines as matching networks. Computer programs are available for doing Smith chart plotting. The Mathcad worksheet **Translines.mcdx** solves transmission line problems directly from mathematical formulas.

### 3.3.4 Microstrip

From around 800 MHz and higher, the lengths of printed circuit board conductors are a significant fraction of a wavelength, so they should be considered as transmission lines. Thus, for example, if a 6 cm conductor connects a receiver integrated circuit or low-noise amplifier to an antenna socket, the component will not see 50 ohms when a 50 ohm antenna source is plugged in, unless the conductor is designed to have a characteristic impedance of 50 ohms.

A printed conductor over a ground plane (copper plating on the opposite side of the board) is called microstrip. The transmission line characteristics of conductors on a board are used in UHF and microwave circuits as matching networks between the various components.

Using the attached Mathcad worksheet **Microstrip.mcdx**, you can find the conductor width needed to get a required characteristic line impedance, or you can find the impedance if you know the width. Then, you can use the Smith chart to do impedance transformations and design matching networks using microstrip components. The **Microstrip.mcdx** worksheet also gives you the wavelength on the pc board for a given frequency. In order to use this worksheet, you have to know the dielectric constant of your board material and the board's thickness.

### 3.3.5 Mutual inductance matching

High Q antennas, notably loops, can be matched through inductance by placing a matching coil close to the subject antenna. [16]. This has several advantages: the radiating loop is floating and less susceptible to spurious capacitance, even harmonics are suppressed, component loss is reduced, and there is physical flexibility in some applications when the radiating loop is a small distance from the circuit board, with no connecting wires. A short description of mutual impedance matching is given here. More details are available in the referenced article.

The radiation resistance of a loop antenna increases as its enclosed area, that is, its size, is made larger. The loop must be resonated by a capacitor, whose value decreases as a function of the loop size for a given frequency. Connection of the loop to a matching circuit introduces spurious capacitance which decreases the loop area that can be obtained at resonance. Mutual inductance matching avoids this effect. Fig. 3.23 shows the inductance matching arrangement, with square coils shown as traces on printed circuit material. The two loops form a transformer coupled through their mutual inductance. An equivalent circuit is shown in Fig. 3.24. It can be viewed as a matching circuit whose virtual components must be adjusted to transform the

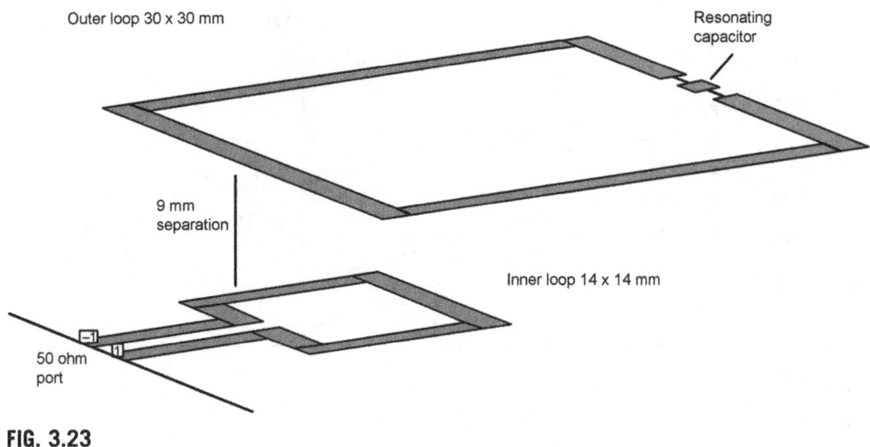


**FIG. 3.23**

Mutual induction matching loops.

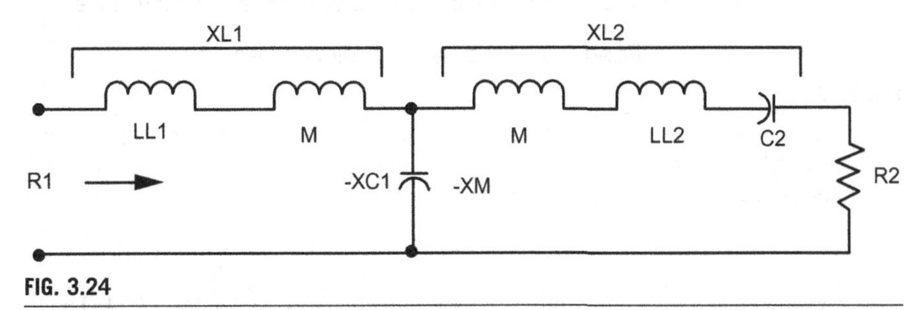


**FIG. 3.24**

An equivalent circuit for mutual inductance matching.

radiation and loss resistances of the antenna coil, $R_2$, to the typically 50 ohm resistance $R_1$ presented to the transmitter or receiver feed. The values of the inductive and capacitive reactances depend on the size and placement of the matching feed coil in respect to the antenna coil. Worksheet **Loop_match.mcdx** gives the expression for the mutual inductance of two parallel circular loops as a function of loop radius, distance between the planes of the two loops, and the skew of the loops—the distance between the axes going through the centers of each loop. Fig. 3.25 is a plot of mutual inductance versus feed loop area, normalized to the area of the antenna loop, with distance between the planes of the loops as a parameter.

These are the steps in designing the mutual inductance matching arrangement at a given frequency:

**(1)** Determine the required radiating loop dimensions and loop resistance—radiation resistance plus loss resistance—with the help of worksheet **loop.mcdx**. The loop resistance estimate to use should be somewhat higher than that calculated to account for additional losses, such as nearby objects, not

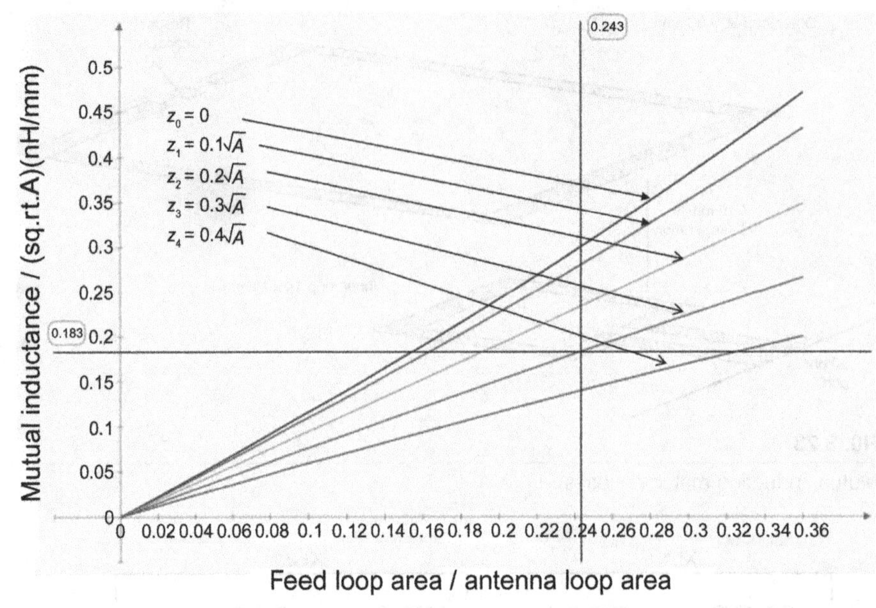


**FIG. 3.25**

Curve set for finding feed loop area for mutual impedance matching network for loop antenna. Parameters "*z*" are the separation between antenna loop and feed loop planes. "*A*" is the area of the antenna loop. Loops are concentric.

considered in the calculation. The loop inductance is $LL_2$ in Fig. 3.24. The worksheet also estimates the resonating capacitance, $C_2$.

**(2)** Using circuit (4) in **worksheet "Matching.mcdx,"** chose a value for $Q$ and solve for the matching components $L_1$, $L_2$, and $C_1$ required to match the loop resistance $R_2$ to the feedline resistance $R_1$, usually 50 ohms. The mutual inductance $M$ is found from

$$M = \frac{-XC1}{2\pi f_0} \tag{3.25}$$

where $f_0$ is the resonating frequency of the circuit.

**(3)** From knowledge of the radiating loop area and the calculated mutual inductance $M$, find the area of the feeding loop on the appropriate curve in Fig. 3.25. Find on the $y$-axis the known mutual impedance $M$ from step 2, normalized by dividing by the square root of the radiating loop area. Then cross horizontally to the right to the curve representing the closest distance between the coils, $z$, divided by the square root of the radiating coil area. The feed loop area is the corresponding abscissa value multiplied by the area of the radiating coil.

**(4)** Determine desirable feed loop dimensions that result in the area found in step 4. From those dimensions, estimate the loop inductance, $LL_1$ in Fig. 3.24, from worksheet **Loop.mcdx**. Add to this inductance the mutual inductance $M$ and compare the result with $L_1$ found in step 2. If the two values are not close, repeat the steps 2 to 4 with another value of $Q$ in step 2 until $L_1$, derived from inductive reactance $XL_1$ of Fig. 3.24, is approximately realized.

**(5)** After a suitable estimating of feed loop area has been found, a prototype of the circuit can be fabricated and tested for adequate return loss. Achieve resonance at the desired frequency by adjusting the value of the resonating capacitor (see Fig. 3.23 and $C_2$ in Fig. 3.24). The circuit design should allow the possibility of adding a series inductor or capacitor at the terminals of the coupling coil to give an additional degree of freedom to the matching circuit.

**Example 3.6**

Design a coupling loop for the 30-mm square loop antenna (outer loop) shown in Fig. 3.23. Frequency is 315 MHz, and the concentric loops are 9 mm apart. Conductor width is 2 mm.

*Solution*

**(1)** From "**Loop.mcdx,**" loop inductance is 81.6 nH, and resonating capacitance is 3.1 pF. Assume total resistance is 1 ohm.

**(2)** In **Matching.mcdx** circuit (4), $f = 315$ MHz, $R_1 = 50$ ohm, $R_2 = 1$ ohm. Let $Q = 12$. From Eq. (3.25), $M = 5.5$ nH.

**(3)** Area of radiating loop is 900 mm$^2$. Normalized $M$ is 5.48 nH/30 mm = 0.18. In Fig. 3.25, use $z_3 = 0.3 \times 30 = 9$ mm, which is the given loop separation. From corresponding curve, (feed loop area)/(antenna loop area) is 0.24. Feed loop area is $0.24 \times 900 = 216$mm$^2$.

**(4)** For a square loop, the side is 14.7 mm. Its inductance is 31.6 nH. Adding $M$ gives 37.1 nH. From **Matching.mcdx,** $L_1 = 34.82$ nH. The two results are reasonably close, and a good match should be obtained by tuning the resonating capacitor.

Worksheet **Loop_match.mcdx** facilitates design of mutual impedance coupling to a loop antenna.

## 3.4 Measuring techniques

Using a vector analyzer, you can measure the impedances you want to match, design a matching network, and check the accuracy of your design. When a matching network is designed and adjusted correctly, the impedance looking into the network where it is connected to the load or source is the complex conjugate of the impedance of the load or of the source impedance. The complex conjugate of an impedance has the same real part minus the imaginary part of the impedance. For example, if $Z_{source} = 30 - j12$ ohms, then the impedance seen at the input to the matching network should be $30 + j12$ ohms when its output port is connected to the load.

Without a vector analyzer, you need considerable cut-and-try to optimize the antenna and matching components. Other instruments, usually available in RF electronics laboratories, can be a big help. Here are some ideas for adjusting antennas and circuits for resonance at the operating frequency using relatively inexpensive equipment (compared to a vector analyzer).

A grid dip meter (still called that, although for years, it has been based on a transistor oscillator, not a vacuum tube) is a simple, inexpensive tool, popular with radio amateurs. It consists of a tunable RF oscillator with external coil, allowing it to be lightly coupled to a resonant circuit, which can be an antenna of almost any type. When the dip meter is tuned across the resonant frequency of the passive circuit under test, its indicating meter shows a current dip due to absorption of energy from the instrument's oscillator. A loop antenna with resonating capacitor is easy to adjust using this method. A dipole, ground plane, or helical antenna can also be checked for resonance by connecting a small one-turn loop to the antenna terminals with matching circuit components disconnected. Set the dip meter coil close to the loop and tune the instrument to find a dip.

The main limitation to the grid dip meter is its frequency range, usually no more than 250 MHz. Higher frequency resonances can be measured with a return loss bridge, also called directional bridge or impedance bridge. This device, which is an integral part of a scalar network analyzer, can be used with a spectrum analyzer and tracking generator or a noise source to give a relative display of return loss versus frequency.

The return loss bridge is a three-port device that indicates power reflected from a mismatched load. Fig. 3.26 shows a diagram of the bridge. Power applied at the source port passes to the device under test at the test port with a nominal attenuation of 6 dB. Power reflected from the device under test appears at the measurement port, attenuated approximately 6 dB. If the tested circuit presents the same impedance as the characteristic impedance of the bridge, there will be no output at the measurement port, except for a leakage output on the order of 50 dB below the output of the source. If the test port sees an open or short circuit, all power will be reflected, and the measurement port output will be around −12 dB. The return loss is the difference between the

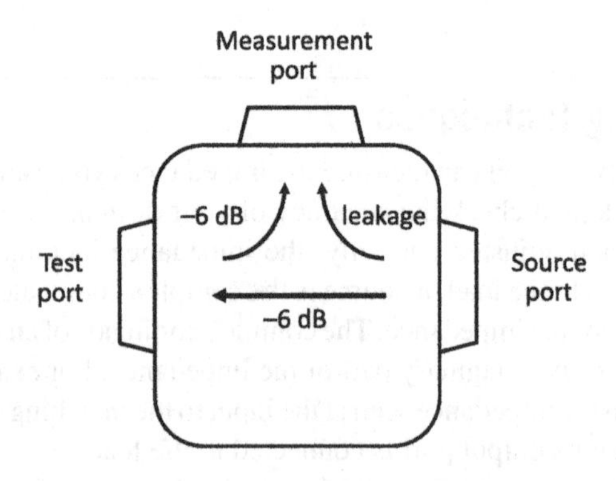


**FIG. 3.26**

Return loss bridge.

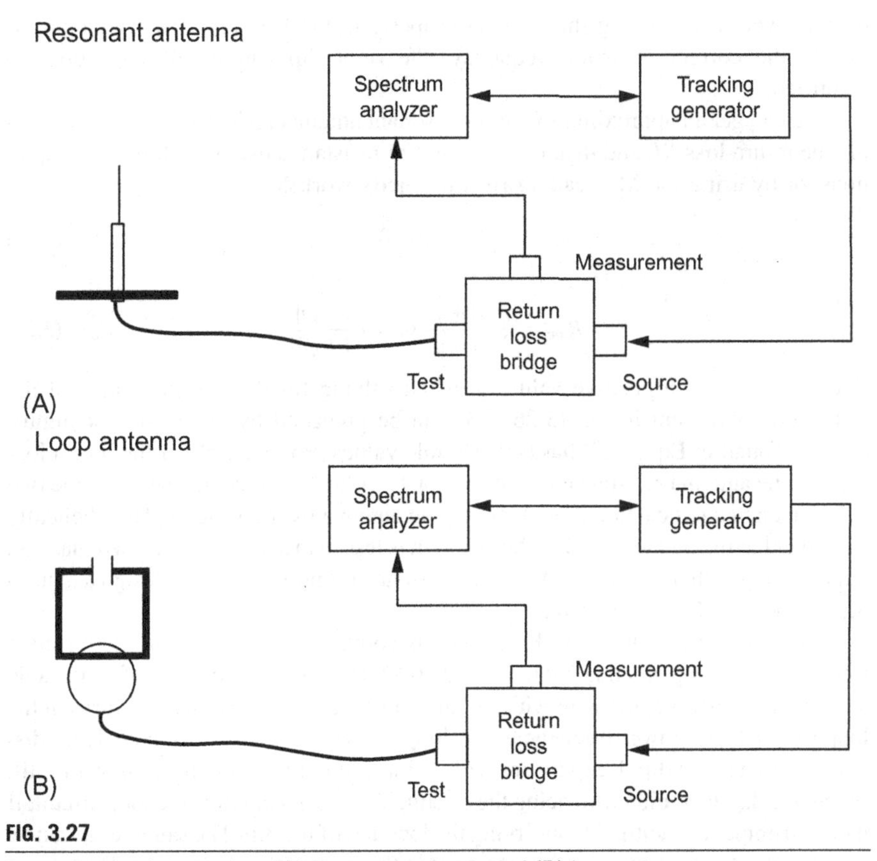


**FIG. 3.27**

Resonant circuit test setup. (A) Resonant antenna and (B) loop antenna.

output measured in dBm at the measurement port when the test port is open or shorted, and the dBm output when the circuit under test is connected to the test port.

A setup to determine the resonant frequency of an antenna is shown in Fig. 3.27. In Fig. 3.27A the antenna is connected to the test terminal of the bridge through a short length of 50-ohm coaxial cable. A spectrum analyzer is connected to the measurement port, and a tracking generator, whose frequency is swept in tandem with the frequency sweep of the spectrum analyzer, drives the source port.

When the swept frequency passes the resonant frequency of the antenna, the analyzer display dips at that frequency. At the resonant frequency, reactance is canceled, and the antenna presents a pure resistance. The closer the antenna impedance is to 50 ohms, the deeper the dip. Antenna parameters may be changed—length, loading coil, or helical coil dimensions, for example—until the dip occurs at the desired operating frequency. Dips are usually observed at several frequencies because of more than one resonance in the system. By noting the effect of changes in the antenna on the various dips, as well as designing the antenna properly in the first place to give approximately the correct resonant frequency, the right dip can usually be correctly identified.

You can get an approximation of the resonant antenna resistance $R_{ant}$ by measuring the return loss $RL$ and then converting it to resistance using the following equations, or by using the Mathcad **Translines.mcdx** worksheet:

$$|\rho| = \pm 10^{-\frac{RL}{20}} \tag{3.26}$$

$$R_{ant} = Z_0 \cdot \frac{1+|\rho|}{1-|\rho|} \text{ or } Z_0 \cdot \frac{1-|\rho|}{1+|\rho|} \tag{3.27}$$

The return loss is a positive value, so when solving for the absolute value of the reflection coefficient in Eq. (3.26), $|\rho|$ can be preceded by either plus or minus, and $R_{ant}$ found in Eq. (3.27) has two possible values. For example, if the return loss is 5 dB, the antenna resistance is either 14 or 178 ohms. You decide between the two values using an educated guess. A monopole antenna over a ground plane, helically wound or having a loading coil, whose length is less than a quarter wave, will have an impedance less than 50 ohms. Once the resistance is known, you can design a matching network as described above.

The arrangement shown in Fig. 3.27B is convenient for checking the resonant frequency of a loop antenna, up to around 500 MHz. Use a short piece of coax cable and a loop of stiff magnet wire with a diameter of 2 cm. Use two or three turns in the loop for VHF and lower frequencies. At loop resonance, the spectrum analyzer display shows a sharp dip. Keep the test coil as far as possible from the loop, while still seeing the dip, to avoid influencing the circuit. You can easily tune the loop circuit, if it has a trimmer capacitor, by observing the location of the dip. The same setup can be used for checking resonance of tuning coils in the transmitter or receiver. There must be no radiation from the circuit when this test is made. If possible, disable the oscillator and apply power to the device being tested. The resonant frequency of a tuned circuit that is coupled to a transistor stage will be different when voltage is applied and when it is not.

## 3.5 Summary

We have covered in this chapter the most important properties of antennas and transmission lines that one needs to know to get the most from a short-range radio system. Antenna characteristics were defined. Then, we discussed some of the types of antennas commonly used in short-range systems including the dipole, monopole over a ground plane, planar and inverted F, and gave examples of design. Metamaterials were described, and we showed how their special features have been applied to produce a wideband patch antenna. Impedance matching is imperative to get the most into, and out of, an antenna, and we presented several matching circuits and gave examples of how to use them. We showed how to design with the Smith chart, which may not be as widely used now as it once was, but understanding it helps visualize the concepts of circuit matching, particularly with distributed components. Impedance matching can be accomplished through the coupling of air coils, and we outlined the steps on how to carry it out. Finally, we showed some simple measurements which help in realizing a design and which shorten the cut-and-try routine that is almost inevitable when perfecting a product.

## References

[1] The ARRL Antenna Book, American Radio Relay League, Newington, CT, USA, 1991.

[2] K. Smith, Antennas for Low Power Wireless Applications, Murata Manufacturing Co. Ltd, 2009.

[3] K. Fujimoto, H. Morishita, Modern Small Antennas, Cambridge University Press, Cambridge, United Kingdom, 2013.

[4] Z.H. Lwin, S.S.Y. Mon, H.M. Tun, Analysis and simulation of low profile planar inverted-F antenna design for WLAN operation in portable devices, Int. J. Sci. Technol. Res. 5 (06) (2016).

[5] H.F. AbuTarboush, R. Nilavalan, H.S. Al-Raweshidy, D. Budimird, Design of planar inverted-F antennas (PIFA) for multiband wireless applications, in: ICEAA09 International Conference on Electromagnetics in Advance Applications, Torino, Italy, 14–18 September, 2009.

[6] P. Hui, Design of Integrated Inverted F Antennas Made of Asymmetrical Coplanar Striplines, Applied Microwave & Wireless, 2002.

[7] Silicon Labs Application Note, AN1088: Designing with an Inverted-F 2.4 GHz PCB Antenna, Rev. 0.1.

[8] F. Mohammed Ali, Estimate Microstrip Substrate Relative Dielectric Constant, Microwaves and RF, 2007.

[9] R. Kubacki, et al., A broadband left-handed metamaterial microstrip antenna with double-fractal layers, Hindawi Int. J. Anten. Propag. (2017).

[10] D.W. Prather, Embedded Meta-Material Antennas, U.S. Army Research Office Report, 2009.

[11] A. Erentok, R.W. Ziolkowski, Metamaterial-inspired efficient electrically small antennas, IEEE Trans. Anten. Propag. 56 (3) (2008).

[12] S.K. Sharma, D.S. Nagarkoti, Meet the Challenge of Designing Electrically Small Antennas, Microwaves & RF, 2017.

[13] J.S. McLean, A re-examination of the fundamental limits on the radiation $Q$ of electrically small antennas, IEEE Trans. Anten. Propag. 44 (5) (1996).

[14] P. Ikonen, Electrically small metamaterial-based antennas—have we seen any real practical benefits? in: 3rd European Conference on Antennas and Propagation, 2009, pp. 866–869.

[15] Sonnet Precision Electromagnetics, https:/www.sonnetsoftware.com/.

[16] A. Bensky, Match Loop Antennas Via Mutual Inductance, Microwaves & RF, 2010.

CHAPTER

# Communication protocols and modulation

4

In this chapter we take an overall view of the characteristics of the communication system. While these characteristics are common to any wireless communication link, for detail we'll address the peculiarities of short-range systems.

A simple block diagram of a digital wireless link is shown in Fig. 4.1. The link transfers information originating at one location, referred to as source data, to another location where it is referred to as reconstructed data. A more concrete implementation of a wireless system, an intrusion detection system, is shown in Fig. 4.2.

## 4.1 Baseband data format and protocol

Let's first take a look at what information we may want to transfer to the other side. This is important in determining what bandwidth the system needs.

### 4.1.1 Change-of-state source data

Many short-range systems only have to relay information about the state of a contact. This is true of the security system of Fig. 4.2 where an infrared motion detector notifies the control panel when motion is detected. Another example is the push-button transmitter, which may be used as a panic button or as a way to activate and deactivate the control system, or a wireless smoke detector, which gives advance warning of an impending fire. There are also what are often referred to as "technical" alarms—gas detectors, water level detectors, and low and high temperature detectors—whose function is to give notice of an abnormal situation.

All these examples are characterized as very low-bandwidth information sources. Change of state occurs relatively rarely, and when it does, we usually don't care if knowledge of the event is signaled tens or even hundreds of milliseconds after it occurs. Thus, required information bandwidth is very low—several hertz.

It would be possible to maintain this very low bandwidth by using the source data to turn on and off the transmitter at the same rate the information occurs, making a very simple communication link. This is not a practical approach, however, since the receiver could easily mistake random noise on the radio channel for a legitimate signal and thereby announce an intrusion, or a fire, when none occurred. Such false alarms are highly undesirable, so the simple on/off information of the transmitter must be coded to be sure it can't be misinterpreted at the receiver.



Short-range Wireless Communication. https://doi.org/10.1016/B978-0-12-815405-2.00004-X
© 2019 Elsevier Inc. All rights reserved.

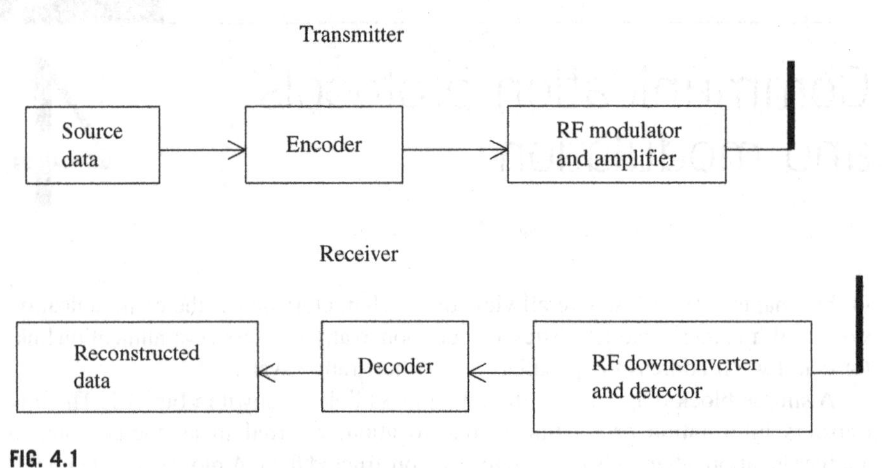


FIG. 4.1

Radio communication link diagram.

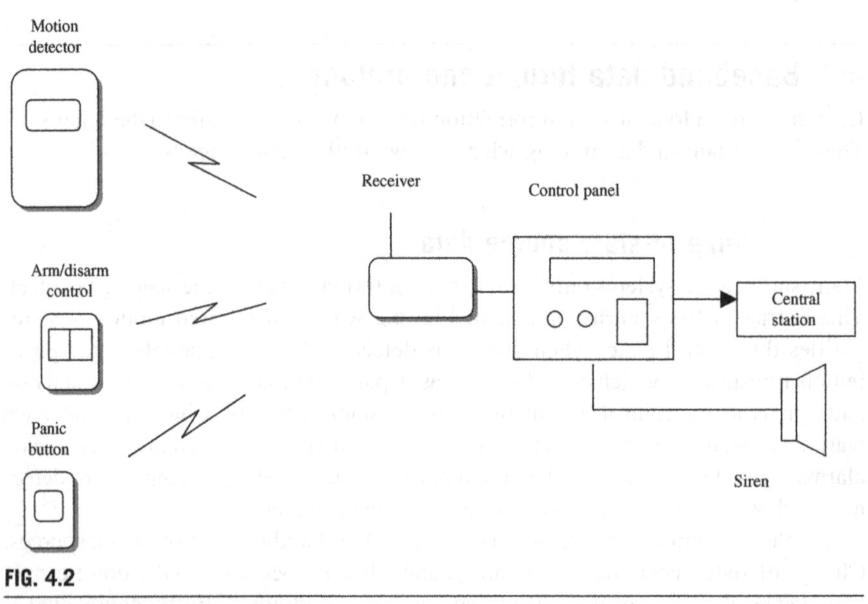


FIG. 4.2

Intrusion detection system.

This is the purpose of the encoder shown in Fig. 4.1. This block creates a group of bits, assembled into a frame, to make sure the receiver will not mistake a false occurrence for a real one. Fig. 4.3 is an example of a message frame. The example has four fields. The first field is a preamble which conditions the receiver for the transfer of information and tells it when the message begins. The next field is an identifying address. This address is unique to the transmitter and its purpose is to notify the

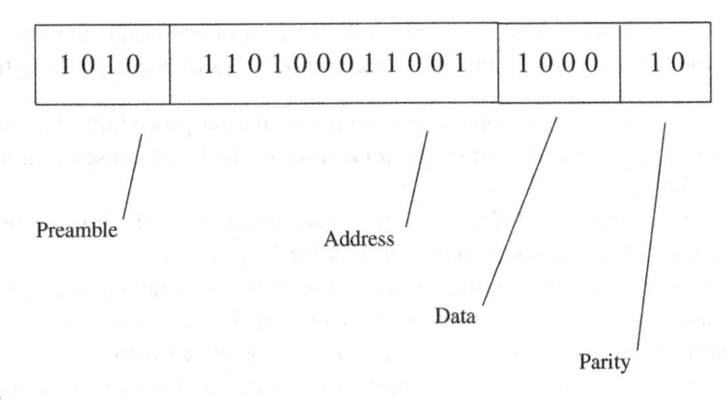


FIG. 4.3

Message frame.

receiver from where, or from what unit, the message is coming. The data field follows, which may indicate what type of event is being signaled, followed, in some protocols, by a parity bit or bits to allow the receiver to determine whether the message was received correctly.

### 4.1.2 Address field

The number of bits in the address field depends on the number of different transmitters there may be in the system. Often the number of possibilities is far greater than this, to prevent confusion with neighboring, independent systems and to reduce the statistically possible chance that random noise will duplicate the address. The number of possible addresses in the code is $2^{L1}$, where $L1$ is the length of the message field. In many simple security systems the address field is determined by dip switches set by the user. Commonly, eight to ten dip switch positions are available, giving 256 to 1024 address possibilities. In other systems, the address field, or device identity number, is a code number set in the unit microcontroller during manufacture. This code number is longer than that produced by dip switches, and may be 16 to 24 bits long, having 65,536 to 16,777,216 different codes. The longer codes greatly reduce the chances that a neighboring system or random event will cause a false alarm. On the other hand, the probability of detection is lower with the longer code because of the higher probability of error. This means that a larger signal-to-noise ratio is required for a given probability of detection.

In all cases, the receiver must be set up to recognize transmitters in its own system. In the case of dip-switch addressing, a dip switch in the receiver is set to the same address as in the transmitter. When several transmitters are used with the same receiver, all transmitters must have the same identification address as that set in the receiver. In order for each individual transmitter to be recognized, a subfield of two to four extra dip switch positions can be used for this differentiation. When a built-in individual fixed identity is used instead of dip switches, the receiver must be taught to recognize the identification numbers of all the transmitters used in the system; this is done at the time of installation. Several common ways of accomplishing this are:

**(a)** *Wireless "learn" mode.* During a special installation procedure, the receiver stores the addresses of each of the transmitters which are caused to transmit during this mode.
**(b)** *Infrared transmission.* Infrared emitters and detectors on the transmitter and receiver serve as the communication link for learn mode.
**(c)** *Direct key-in.* Each transmitter is labeled with its individual address, which is then keyed into the receiver or control panel by the system installer.
**(d)** *Wired learn mode.* A short cable temporarily connected between the receiver and transmitter is used when performing the initial address recognition procedure during installation.

Advantages and disadvantages of dip switch or fixed internal addressing systems

| Advantages | Disadvantages |
|---|---|
| ***Dip switch*** | |
| Large number of transmitters can be used with a receiver | Limited number of bits increases false alarms and interference from adjacent systems |
| Can be used with commercially available data encoders and decoders | Device must be opened for coding during installation |
| Transmitter or receiver can be easily replaced without recoding the opposite terminal | Multiple devices in a system are not distinguishable in most simple systems |
| | Control systems are vulnerable to unauthorized operation since the address code can be duplicated by trial and error |
| ***Internal fixed code identity*** | |
| Large number of code bits reduces possibility of false alarms | Longer code reduces probability of detection |
| System can be set up without opening transmitter | Replacing transmitter or receiver involves redoing the code learning procedure |
| Each transmitter is individually recognized by receiver | Limited number of transmitters can be used with each receiver |
| | Must be used with a dedicated microcontroller. Cannot be used with standard encoders and decoders |

### 4.1.3 Code-hopping addressing

While using a large number of bits in the address field reduces the possibility of false identification of a signal, there is still a chance of purposeful duplication of a transmitter code to gain access to a controlled entry. Keyless entry devices are used widely for access control to vehicles and buildings. A replay attack, called code grabbing, involves recording a wireless remote control signal and retransmitting it at a convenient time to carry out a threat. Such attacks can be thwarted by incorporating in a keyless entry system an algorithm for insuring that each transmission, instigated by pressing a button on the remote control, differs from all previous ones such that it is very difficult to predict new control messages. This may be done by including a counter field in the transmitted signal, encrypting the message and using a counter synchronization procedure for transmitter and receiver. This method is variously called code rotation, code hopping, or rolling code addressing. Here is a simplified description [1].

The remote control transmitter (encoder) employs a counter that is incremented on each activation, generally by a push button. The state of the counter is transmitted, along with the transmitter address and the control command. The transmitted message frame is shown in Fig. 4.4. Each remote control transmitter has a unique serial number assigned by the manufacturer, stored in nonvolatile memory in the device. This serves as the device address, transmitted with each activation of the unit. The address, and control data which is determined by the button pressed, is sent in the clear, not encrypted. Counter state along with a copy of the address and the data is encrypted using an encryption key unique to that device. After transmission, the counter is incremented. On incrementation beyond the highest counter state, the counter is reset to zero.

The receiver detects the radio transmission and checks the device address to see if it is part of the network. If so, it retrieves the corresponding secret symmetrical encryption key from a table and decodes the message. Comparing the decrypted address and control data with the corresponding plain text fields verifies the decryption. The receiver has a stored copy for each transmitter in the network of the last received synchronization counter state. See Fig. 4.5. If the presently received counter state is greater than the previous one up to a fixed number of increments above the last one received, the acceptance zone, the message is considered valid and the command is carried out. Then the receiver counter for that device is set to the counter value in the last received message, to maintain synchronization with the transmitter. If the received counter is not within the acceptance zone, no action is taken. The acceptance zone may bridge the counter overflow and zero reset. Allowance of counter values within a range above the last received value is necessary, since the transmitter device may have been activated one or more times and its counter

**FIG. 4.4**

Code hopping address frame.

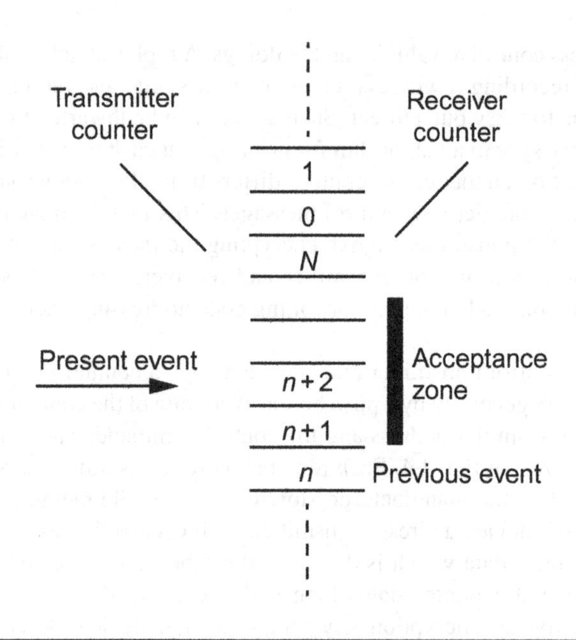


**FIG. 4.5**

Code hopping synchronization counter. *n* is the count of the last received event, and *N* is the highest state of the counter, after which the count starts from zero.

incremented, without detection at the receiver, because of excessive distance or interference on the communication channel.

In order to add a remote control transmitter to the system, a special procedure has to be followed. The receiver is put into a training mode. Upon transmission from the new device, the receiver includes the received address in a device table. The receiver has a built in shared code key that is common for similar devices from the same manufacturer. Using this key, and the transmitter's address, the receiver can compute the secret encryption code of the remote control device and include it in the new transmitter's entry in the data table. Note that the secret encryption code is not sent over the air. The receiver will also record the state of the transmitter's counter, to use to determine validity of subsequent signals received from that transmitter.

Although a rolling code, or code hopping, goes a long way to provide wireless security for keyless entry devices, protection can be compromised [2]. A replay attach may succeed as follows. A rogue device planted near the receiver in a protected property (vehicle or gate, for example) detects a remote control transmission and simultaneously records the message while blocking it from being received by the intended receiver. A repeat transmission attempt, now with incremented counter, is also blocked from the intended receiver, but saved in the rogue device. Now the rogue terminal can retransmit the first message, which will allow access of the authorized person to the protected property since it wasn't detected earlier. The second message can later be used to steal the vehicle or enter the protected property.

A solution to this ruse is to time stamp messages to insure nonrepetition instead of using the incremented counter. The receiver can then compare this time stamp with the current time to know that it wasn't recorded earlier. Since the clocks in transmitter and receiver are not perfectly synchronized, the receiver will allow for some clock drift in making the decision to accept the transmission, and if the message is approved will resynchronize the clocks. The time stamp is encrypted, as was the synchronizing counter, to prevent tampering with the message and to provide authentication of the transmitter.

### 4.1.4 Data field

The next part of the message frame is the data field. Its number of bits depends on how many pieces of information the transmitter may send to the receiver. For example, the motion detector may transmit three types of information: motion detection, tamper detection, and low battery indication.

### 4.1.5 Parity bit field

The last field is for error detection bits, or parity bits. Some protocols have inherent error detection features so the last field is not always needed.

### 4.1.6 Baseband data rate

Once the data frame is defined, the appropriate baseband data rate can be determined. For the security system example, this rate will usually be several hundred hertz up to several kilohertz. Since a rapid response is not needed, a frame can be repeated several times to be more certain it will get through. Frame repetition is needed in systems where space diversity is used in the receiver. In these systems, two separate antennas are periodically switched to improve the probability of reception. If signal nulling occurs at one antenna because of the multipath phenomena, the other antenna will produce a stronger signal which can be correctly decoded. Thus, a message frame must be sent more often to give it a chance to be received after unsuccessful reception by one of the antennas.

### 4.1.7 Supervision

Another characteristic of digital event systems is the need for link supervision. Security systems and other event systems, including medical emergency systems, are one-way links. They consist of several transmitters and one receiver. As mentioned above, these systems transmit relatively rarely, only when there is an alarm or possibly a low-battery condition. If a transmitter ceases to operate, due to a component failure for example, or if there is an abnormal continuing interference on the radio channel, the fact that the link has been broken will go undetected. In the case of a security system, the installation will be unprotected, possibly until a routine system inspection is carried out. In a wired system, such a possibility is usually covered by a normally energized relay connected through closed contacts to a control panel. If a fault occurs in the device, the relay becomes unenergized and the panel detects the opening of the contacts. Similarly, cutting the connecting wires will also be detected by the panel. Thus, the advantages of a wireless system are compromised by the lower confidence level accompanying its operation.

Many security systems minimize the risk of undetected transmitter failure by sending a supervisory signal to the receiver at a regular interval. The receiver expects to receive a signal during this interval and can emit a supervisory alarm if the signal is not received. The supervisory signal must be identified as such by the receiver so as not to be mistaken for an alarm.

The duration of the supervisory interval is determined by several factors:

- Devices certified under FCC Part 15 paragraph 15.231, which applies to most wireless security devices in North America, are constrained to limit supervision transmissions to a total of 2 s/h for each transmitter.
- The more frequently regular supervision transmissions are made, the shorter the battery life of the device.
- Frequent supervisory transmissions when there are many transmitters in the system raise the probability of a collision with an alarm signal, which may cause the alarm not to get through to the receiver.
- The more frequent the supervisory transmissions, the higher the confidence level of the system.

While it is advantageous to notify the system operator at the earliest sign of transmitter malfunction, frequent supervision raises the possibility that a fault might be reported when it doesn't exist. Thus, most security systems determine that a number of consecutive missing supervisory transmissions must be detected before an alarm is given. A system which specifies security emissions once every hour, for example, may wait for eight missing supervisory transmissions, or eight hours, before a supervisory alarm is announced. Clearly, the greater the consequences of lack of alarm detection due to a transmitter failure, the shorter the supervision interval must be.

### 4.1.8 Continuous digital data

In other systems flowing digital data must be transmitted in real time and the original source data rate will determine the baseband data rate. This is the case in wireless LANs and wireless peripheral-connecting devices. The data is arranged in message frames which contain fields needed for correct transportation of the data from one side to the other, in addition to the data itself.

An example of a frame used in *synchronous data link control* (SDLC) is shown in Fig. 4.6. It consists of beginning and ending bytes that delimit the frame in the message, address and control fields, a data field of undefined length, and check bits or parity bits for letting the receiver check whether the frame was correctly received. If it is, the receiver sends a short acknowledgment and the transmitter can continue with the next frame. If no acknowledgment is received, the transmitter repeats the

| Beginning flag - 8 bits | Address - 8 bits | Control - 8 bits | Information - any no. of bits | Error detection - 16 bits | Ending flag - 8 bits |
|---|---|---|---|---|---|

**FIG. 4.6**

Synchronous data link control frame.

message again and again until it is received. This is called an ARQ (automatic repeat query) protocol. In high-noise environments, such as encountered on radio channels, the repeated transmissions can significantly slow down the message throughput.

Instead of just detecting errors, a *forward error control* (FEC) protocol can correct them. In this case, there is enough information in the parity bits to allow the receiver to correct a small number of errors in the message so that it will not have to request retransmission. Although more parity bits are needed for error correction than for error detection alone, the throughput may be increased when using FEC on noisy channels.

In all cases, we see that extra bits must be included in a message to insure reliable transmission, and the consequently longer frames require a higher transmission rate than what would be needed for the source data alone. This message overhead must be considered in determining the required bit rate on the channel, the type of digital modulation, and consequently the bandwidth.

### 4.1.9 Analog transmission

Analog transmission devices, such as wireless microphones, also have a baseband bandwidth determined by the data source. A high-quality wireless microphone may be required to pass 50 to 15,000 Hz, whereas an analog wireless telephone needs only 100 to 3000 Hz. In this case determining the channel bandwidth is more straightforward than in the digital case, although the bandwidth depends on whether AM or FM modulation is used. In most short-range radio applications, FM is preferred—narrowband FM for voice communications and wide-band FM for quality voice and music transmission.

## 4.2 Baseband coding

The form of the information signal that is modulated onto the RF carrier we call here baseband coding. We refer below to both digital and analog systems, although strictly speaking the analog signal is not coded but is modified to obtain desired system characteristics.

### 4.2.1 Digital systems

Once we have a message frame, composed as we have shown by address and data fields, the information must be formed into signal levels that can be effectively transmitted, received, and decoded. Since we're concerned here with binary data transmission, the baseband coding selected has to give the signal the best chance to be decoded after it has been modified by noise and the response of the circuit and channel elements. This coding consists essentially of the way that zeros and ones are represented in the signal sent to the modulator of the transmitter.

There are many different recognized systems of baseband coding. We will examine only a few common examples.

These are the dominant criteria for choosing or judging a baseband code:

**(a)** *Timing*. The receiver must be able to take a data stream polluted by noise and recognize transitions between each bit. The bit transitions must be independent of the message content—that is, they must be identifiable even for long strings of zeros or ones.

**(b)** *DC content*. It is desirable that the average level of the message— that is, its DC level—remains constant throughout the message frame, regardless of the content of the message. If this is not the case, the receiver detection circuits must have a frequency response down to DC so that the levels of the message bits won't tend to wander throughout the frame. In circuits where coupling capacitors are used, such a response is impossible.

**(c)** *Power spectrum*. Baseband coding systems have different frequency responses. A system with a narrow frequency response can be filtered more effectively to reduce noise before detection.

**(d)** *Inherent error detection*. Codes that allow the receiver to recognize an error on a bit-by-bit basis have a lower possibility of reporting a false alarm when error-detecting bits are not used.

**(e)** *Probability of error*. Codes differ in their ability to properly decode a signal, given a fixed transmitter power. This quality can also be stated as having a lower probability of error for a given signal-to-noise ratio.

**(f)** *Polarity independence*. There is sometimes an advantage in using a code that retains its characteristics and decoding capabilities when inverted. Certain types of modulation and demodulation do not retain polarity information. Phase modulation is an example.

Now let's look at some common codes (Fig. 4.7) and rate them according to the criteria above [3]. It should be noted that coding considerations for event-type reporting are far different from those for flowing real-time data, since bit or symbol times are not so critical, and message frames can be repeated for redundancy to improve the probability of detection and reduce false alarms. In data flow messages, the data rate is important and sophisticated error detection and correction techniques are used to improve system sensitivity and reliability.

#### 4.2.1.1 Non-return to zero (NRZ)—*Fig. 4.7A*

This is the most familiar code, since it is used in digital circuitry and serial wired short-distance communication links, like RS-232. However, it is rarely used directly for wireless communication. Strings of ones or zeros leave it without defined bit boundaries, and its DC level is very dependent on the message content. There is no inherent error detection. If NRZ coding is used, an error detection or correction

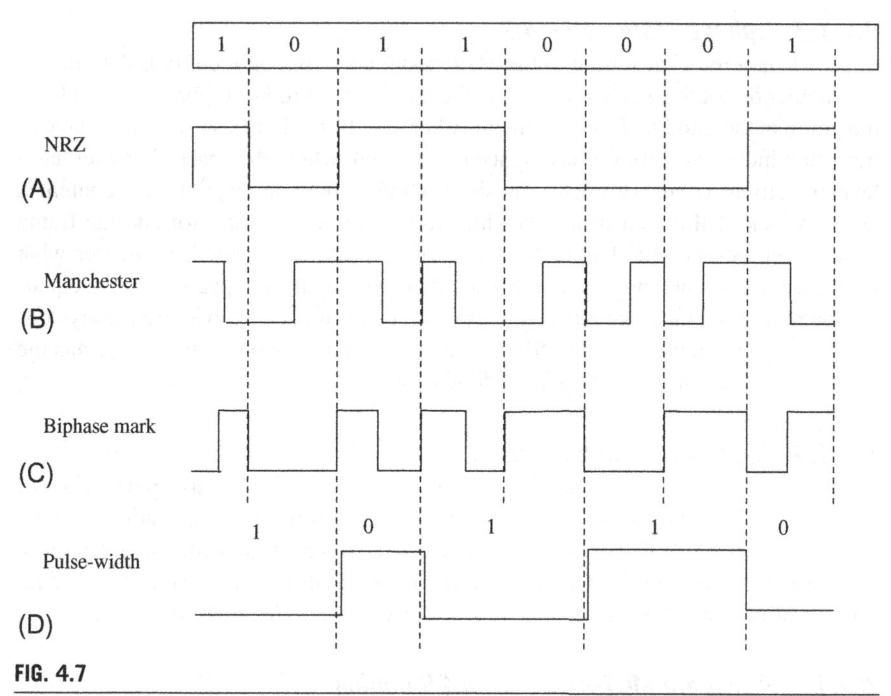


**FIG. 4.7**

Baseband bit formats. (A) Non-return-to-zero (NRZ), (B) Manchester, (C) Biphase Mark, and (D) Pulse-width.

field is imperative. If amplitude shift keying (ASK) modulation is used, a string of zeros means an extended period of low power or no transmission at all. In any case, if NRZ signaling is used, it should only be for very short frames of no more than eight bits.

#### *4.2.1.2 Manchester code—Fig. 4.7B*

A primary advantage of this code is its relatively low probability of error compared to other codes. It is the code used in Ethernet local area networks. It gives good timing information since there is always a transition in the middle of a bit, which is decoded as zero if this is a positive transition and a one otherwise. The Manchester code has a constant DC component and its waveform doesn't change if it passes through a capacitor or transformer. However, a "training" code sequence should be inserted before the message information as a preamble to allow capacitors in the receiver detection circuit to reach charge equilibrium before the actual message bits appear. Inverting the Manchester code turns zeros to ones and ones to zeros. The frequency response of the Manchester code has components twice as high as NRZ code, so a low-pass filter in the receiver must have a cut-off frequency twice as high as for NRZ code with the same bit rate.

#### 4.2.1.3 Biphase Mark—*Fig. 4.7C*

This code is somewhat similar to the Manchester code, but bit identity is determined by whether or not there is a transition in the middle of a bit. For biphase mark, a level transition in the middle of a bit (going in either direction) signifies a one, and a lack of transition indicates zero. Biphase space is also used, where the space character has a level transition. There is always a transition at the bit boundaries, so timing content is good. A lack of this transition gives immediate notice of a bit error and the frame should then be aborted. The biphase code has constant DC level, no matter what the message content, and a preamble should be sent to allow capacitor charge equalization before the message bits arrive. As with the Manchester code, frequency content is twice as much as for the NRZ code. The biphase mark or space code has the added advantage of being polarity independent.

#### 4.2.1.4 Pulse width modulation—*Fig. 4.7D*

As shown in the figure, a one has two timing durations and a zero has a pulse width of one duration. The signal level inverts with each bit so timing information for synchronization is good. There is a constant average DC level. Since the average pulse width varies with the message content, in contrast with the other examples, the bit rate is not constant. This code has inherent error detection capability.

#### 4.2.1.5 Freescale MC145026-145028 coding

Knowledge of the various baseband codes is particularly important if the designer creates his own protocol and implements it on a microcontroller. However, there are several off-the-shelf integrated circuits that are popular for simple event transmission transmitters where a microcontroller is not needed for other circuit functions, such as in panic buttons or door-opening controllers.

The Freescale Semiconductor (originally Motorola) chips are an example of nonstandard coding developed especially for event transmission where three-state addressing is determined as high, low, or open connections to device pins [4]. See Fig. 4.8. A bit symbol can be one of three different types. The receiver, MC145028, must recognize two consecutive identical frames to signal a valid message. The Motorola protocol gives high reliability and freedom from false alarms. Its signal does have a broad frequency spectrum relative to the data rate and the receiver filter passband must be designed accordingly. The DC level is dependent on the message content.

### 4.2.2 Analog baseband conditioning

Wireless microphones and headsets are examples of short-range systems that must maintain high audio quality over the vagaries of changing path lengths and indoor environments, while having small size and low cost. To help them achieve this, they have a signal conditioning element in their baseband path before modulation. Two features used to achieve high signal-to-noise ratio over a wide dynamic range are

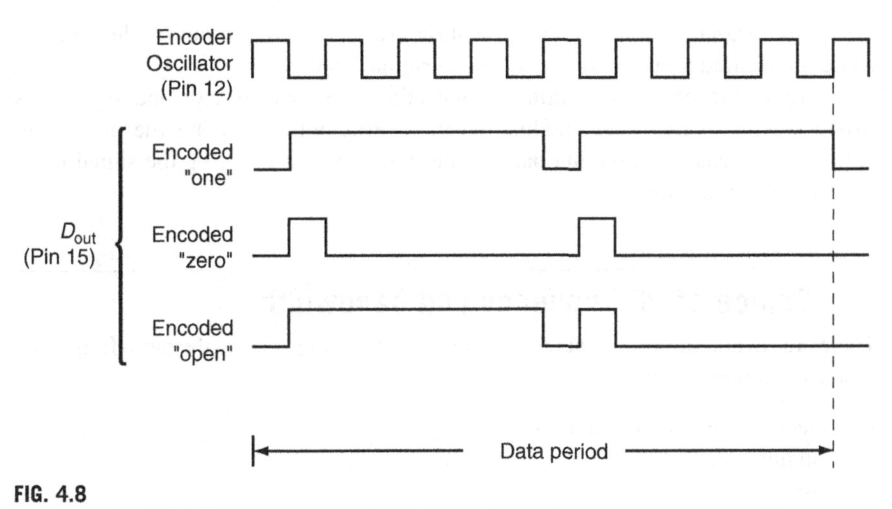


**FIG. 4.8**

Freescale semiconductor MC145026.

pre-emphasis/de-emphasis and compression/expansion. Their positions in the transmitter/receiver chain are shown in Fig. 4.9.

The transmitter audio signal is applied to a high-pass filter (pre-emphasis) which increases the high frequency content of the signal. In the receiver, the detected audio goes through a complementary low-pass filter (de-emphasis), restoring the signal to its original spectrum composition. However, in so doing, high frequency noise that

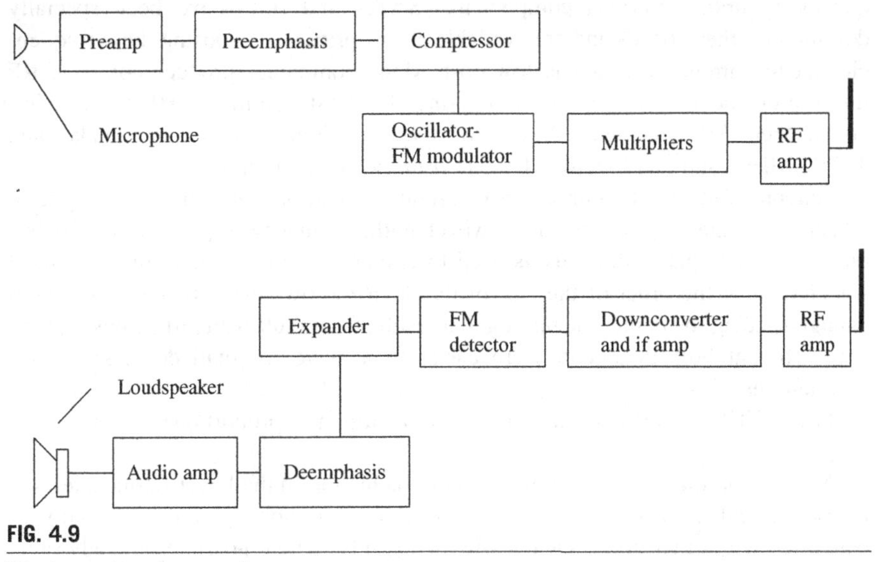


**FIG. 4.9**

Wireless microphone system.

entered the signal path after the modulation process is filtered out by the receiver while the desired signal is returned to its original quality.

Compression of the transmitted signal raises the weak sounds and suppresses strong sounds to make the modulation more efficient. Reversing the process in the receiver weakens annoying background noises while restoring the signal to its original dynamic range.

## 4.3 Choice of RF frequency and bandwidth

There are several important factors to consider when determining the radio frequency of a short-range system:

- Telecommunication regulations
- Antenna size
- Cost
- Interference
- Propagation characteristics

When a device is to be marketed in several countries and regions of the world, it is advantageous to choose a frequency that can be used in the different regions, or at least frequencies that don't differ very much so that the basic design won't be changed by changing frequencies. We'll discuss the matter of frequency choice and regulations in detail in Chapter 10.

The UHF frequency bands are usually the choice for wireless alarm, medical, and control systems. The bands that allow unlicensed operation don't require rigid frequency accuracy. Various components—SAWs and ICs—have been specially designed for these bands and are available at low prices, so choosing these frequencies means simple designs and low cost. Many companies produce complete RF transmitter and receiver modules covering the most common UHF frequencies, among them 315 MHz and 902 to 928 MHz (the United States and Canada), and 433.92 MHz band and 868 to 870 MHz (European Community).

Antenna size may be important in certain applications, and for a given type of antenna, its size is proportional to wavelength, or inversely proportional to frequency. When spatial diversity is used to counter multipath interference, a short wavelength of the order of the size of the device allows using two antennas with enough spacing to counter the nulling that results from multipath reflections. In general, efficient built-in antennas are easier to achieve in small devices at short wavelengths.

From VHF frequencies and up, cost is directly proportional to increased frequency.

Natural and man-made background noise is higher on the lower frequencies. On the other hand, certain frequency bands available for short-range use may be very congested with other users, such as the ISM bands. Where possible, it is advisable to choose a band set aside for a particular use, such as the 868–870 MHz band available in Europe.

Propagation characteristics also must be considered in choosing the operating frequency. High frequencies reflect easily from surfaces but penetrate insulated barriers less readily than lower frequencies.

The radio frequency bandwidth is a function of the baseband bandwidth and the type of modulation employed. For security event transmitters, the required bandwidth is small, of the order of several kilohertz. If the complete communication system were designed to take advantage of this narrow bandwidth, there would be significant performance advantages over the most commonly used systems having a bandwidth of hundreds of kilohertz. For given communication link, the range is inversely dependent on the receiver bandwidth. Also, narrow-band unlicensed frequency allotments can be used where available in the different regions, reducing interference from other users. However, cost and complexity considerations tend to outweigh communication reliability for these systems, and manufacturers decide to make do with the necessary performance compromises. The bandwidth of the mass production security devices is thus determined by the frequency stability of the transmitter and receiver frequency determining elements, and not by the required signaling bandwidth. The commonly used SAW devices dictate a bandwidth of at least 200 kHz, whereas the signaling bandwidth may be only 2 kHz. Designing the receiver with a passband of 20 kHz instead of 200 kHz would increase sensitivity by 10 dB, roughly doubling the range. This entails using highly stable crystal oscillators in the transmitter and in the receiver local oscillator.

## 4.4 Modulation

Amplitude modulation (AM) and frequency modulation (FM), known from commercial broadcasting, have their counterparts in modulation of digital signals, but you must be careful before drawing similar conclusions about the merits of each. The third class of modulation is phase modulation, not used in broadcasting, but its digital counterpart is commonly used in high-end, high-data-rate digital wireless communication.

Digital AM is referred to as ASK—amplitude shift keying—and sometimes as OOK—on/off keying. FSK is frequency shift keying, the parallel to FM. Commercial AM has a bandwidth of 10 kHz whereas an FM broadcasting signal occupies 180 kHz. The high post-detection signal-to-noise ratio of FM is due to this wide bandwidth. However, on the negative side, FM has what is called a threshold effect, also due to wide bandwidth. Weak FM signals are unintelligible at a level that would still be usable for AM signals. When FM is used for two-way analog communication, narrow-band FM, which occupies a similar bandwidth to AM, also has comparable sensitivity for a given *S/N*.

### 4.4.1 Modulation for digital event communication

For short-range digital communication we're not interested in high fidelity, but rather high sensitivity. Other factors for consideration are simplicity and cost of modulation and demodulation. Let's now look into the reasons for choosing one form of modulation or the other.

An analysis of error rates versus bit energy to noise density shows that there is no inherent advantage of one system, ASK or FSK, over the other. This conclusion is based on certain theoretical assumptions concerning bandwidth and method of detection. While practical implementation methods may favor one system over the other, we shouldn't jump to conclusions that FSK is necessarily the best, based on a false analogy to FM and AM broadcasting.

In low-cost security systems, ASK is the simplest and cheapest method to use. For this type of modulation we must just turn on and turn off the radio frequency output in accordance with the digital modulating signal. The output of a microcontroller or dedicated coding device biases on and off a single SAW-controlled transistor RF oscillator. Detection in the receiver is also simple. It may be accomplished by a diode detector in several receiver architectures, to be discussed later, or by the RSSI (received signal strength indicator) output of many superheterodyne receiver ICs employed today. Also, ASK must be used in simple superregenerative receivers.

For FSK, on the other hand, it's necessary to shift the transmitting frequency between two different values in response to the digital code. More elaborate means is needed for this than in the simple ASK transmitter, particularly when crystal or SAW devices are used to keep the frequency stable. In the receiver, also, additional components are required for FSK demodulation as compared to ASK. We have to decide whether the additional cost and complexity is worthwhile for FSK.

In judging two systems of modulation, we must base our results on a common parameter that is a basis for comparison. This may be peak or average power. For FSK, the peak and average powers are the same. For ASK, average power for a given peak power depends on the duty cycle of the modulating signal. Let's assume first that both methods, ASK and FSK, give the same performance—that is, the same sensitivity—if the average power in both cases are equal. It turns out that in this case, our preference depends on whether we are primarily marketing our system in North America or in Europe. This is because of the difference in the definition of the power output limits between the telecommunication regulations in force in the United States and Canada as compared to the common European regulations.

The US FCC Part 15 and similar Canadian regulations specify an *average* field strength limit. Thus, if the transmitter is capable of using a peak power proportional to the inverse of its modulation duty cycle, while maintaining the allowed average power, then under our presumption of equal performance for equal average power, there would be no reason to prefer FSK, with its additional complexity and cost, over ASK.

In Western Europe, on the other hand, the low-power radio specification, ETSI 300 220 limits the *peak* power of the transmitter. This means that if we take advantage of the maximum allowed peak power, FSK is the proper choice, since for a given peak power, the average power of the ASK transmitter will always be less, in proportion to the modulating signal duty cycle, than that of the FSK transmitter.

However, is our presumption of equal performance for equal average power correct? Under conditions of added white Gaussian noise (AWGN) it seems that it is. This type of noise is usually used in performance calculations since it represents the noise present in all electrical circuits as well as cosmic background noise on the radio channel. But in real life, other forms of interference are present in the receiver passband that have very different, and usually unknown, statistical characteristics from AWGN. On the UHF frequencies normally used for short-range radio, this interference is primarily from other transmitters using the same or nearby frequencies. To compare performance, we must examine how the ASK and FSK receivers handle this type of interference. This examination is pertinent in the United States and Canada where we must choose between ASK and FSK when considering that the average power, or signal-to-noise ratio, remains constant. Some designers believe that a small duty cycle resulting in high peak power per bit is advantageous since the presence of the bit, or high peak signal, will get through a background of interfering signals better than another signal with the same average power but a lower peak. To check this out, we must assume a fair and equal basis of comparison. For a given data rate, the low duty cycle ASK signal will have shorter pulses than for the FSK case. Shorter pulses means higher baseband bandwidth and a higher cutoff frequency for the post detection bandpass filter, resulting in more broadband noise for the same data rate. Thus, the decision depends on the assumptions of the type of interference to be encountered and even then the answer is not clear cut.

An analysis of the effect of different types of interference is given by Anthes [5]. He concludes that ASK, which does not completely shut off the carrier on a "0" bit, is marginally better than FSK.

### 4.4.2 Continuous digital communication

For efficient transmission of continuous digital data, the modulation choices are much more varied than in the case of event transmission. We can see this in the three second generation cellular digital radio systems, all of which have the same use and basic requirements. The system referred to as DAMPS or TDMA (time division multiple access) uses a type of modulation called Pi/4 DPSK (differential phase shift keying). The GSM network is based on GMSK (Gaussian minimum shift keying). The third major system is CDMA (code division multiple access). Each system claims that its choice is best, but it is clear that there is no simple cut-and-dried answer. We aren't going into the details of the cellular systems here, so we'll look at the relevant trade-offs for modulation methods in what we have defined as short-range radio applications. At the end of this chapter, we review the basic principles of digital modulation and spread-spectrum modulation.

For the most part, license-free applications specify ISM bands where signals are not confined to narrow bandwidth channels. However, noise power is directly proportional to bandwidth, so the receiver bandwidth should be no more than is required for the data rate used. Given a data rate and an average or peak power limitation, there are several reasons for preferring one type of modulation over another. They involve error rate, implementation complexity, and cost. A common way to compare performance of the different systems is by curves of bit error rate (BER) versus the signal-to-noise ratio, expressed as energy per bit divided by the noise density

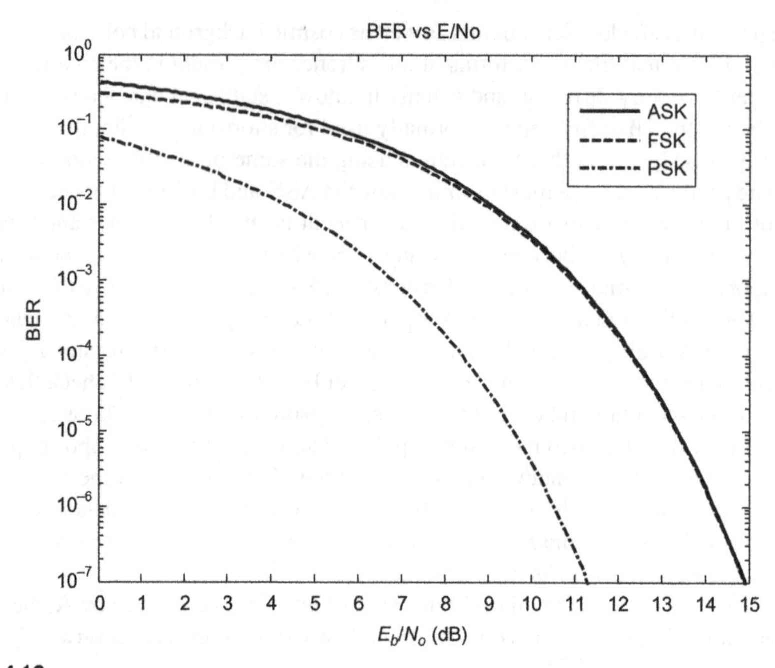


FIG. 4.10

Bit error rates.

(defined below). The three most common types of modulation system are compared in Fig. 4.10. Two of the modulation types were mentioned above. The third, phase shift keying (PSK), is described below.

### 4.4.3 Phase shift keying

Whereas in ASK and FSK the amplitude and frequency are varied according to the digital source data, in PSK it is the phase of the RF carrier that is varied. In its simplest form, the waveform looks like Fig. 4.11. Note that the phase of the carrier wave shifts 180 degrees according to the data signal bits. Similar to FSK, the carrier remains constant, thus giving the same advantage that we mentioned for FSK—maximum signal-to-noise ratio when there is a peak power limitation.

### 4.4.4 Comparing digital modulation methods

In comparing the different digital modulation methods, we need a common reference parameter that reflects the signal power and noise at the input of the receiver, and the bit rate of the data. This common parameter of signal-to-noise ratio for digital

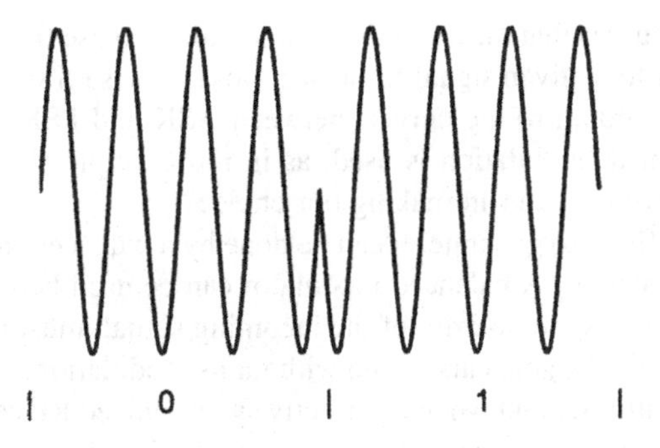


FIG. 4.11

Phase shift keying.

systems is expressed as the signal energy per bit divided by the noise density, *E/No*. We can relate this parameter to the more familiar signal-to-noise ratio, *S/N*, and the data rate *R*, as follows:

**(1)** *S/N* = signal power/noise power. The noise power is the noise density $N_o$ in watts/Hz times the transmitted signal bandwidth $B_T$ in Hz:

$$S/N = S/(N_o B_T)$$

**(2)** The signal energy in joules is the signal power in watts, *S*, times the bit period in seconds, which is 1/(data rate) = 1/*R*, thus

$$E = S(1/R)$$

**(3)** The minimum transmitted bandwidth (Nyquist bandwidth) to pass a bit stream of *R* bits per second is $B_T = R$ Hz.

**(4)** In sum:

$$E/N_o = S/N_o R = S/N_o B_T = S/N$$

The signal power for this expression is the power at the input of the receiver, which is derived from the radiated transmitted power, the receiver antenna gain, and the path loss. The noise density $N_o$, or more precisely the one sided noise power spectral density, in units of watts/Hz, can be calculated from the expression:

$$N_o = kT = 1.38 \times 10^{-23} \times T\,(\mathrm{K})\ \mathrm{W/Hz}$$

The factor *k* is Boltzmann's constant and *T* is the equivalent noise temperature that relates the receiver input noise to the thermal noise that is present in a resistance at the same temperature *T*. Thus, at standard room temperature of 290 degrees Kelvin, the noise power density is $4 \times 10^{-21}$ W/Hz, or −174 dBm/Hz.

The modulation types making up the curves in Fig. 4.10 are

- PSK
- Noncoherent FSK
- Noncoherent ASK

We see from the curves that the best type of modulation to use from the point of view of lowest bit error for a given signal-to-noise ratio ($E/N_o$) is PSK. There is essentially no difference, according to the curves, between ASK and FSK. (This is true only when noncoherent demodulation is used, as in most simple short range systems.) What then must we consider in making our choice?

PSK is not difficult to generate. It can be done by a balanced modulator. The difficulty is in the receiver. A balanced modulator can be used here too but one of its inputs, which switches the polarity of the incoming signal, must be perfectly correlated with the received signal carrier and without its modulation. The balanced modulator acts as a multiplier and when a perfectly synchronized RF carrier is multiplied by the received signal, the output, after low-pass filtering, is the original bit stream. PSK demodulation is shown in Fig. 4.12. Some ripple, at two times the carrier frequency, may remain at the output of the filter.

There are several ways of generating the required reference carrier signal that are synchronized with the received signal. Two examples are the Costas loop and the squaring loop, which include three multiplier blocks and a variable frequency oscillator (VFO) [6]. Because of the complexity and cost, PSK is not commonly used in inexpensive short-range equipment, but it is the most efficient type of modulation for high-performance data communication systems.

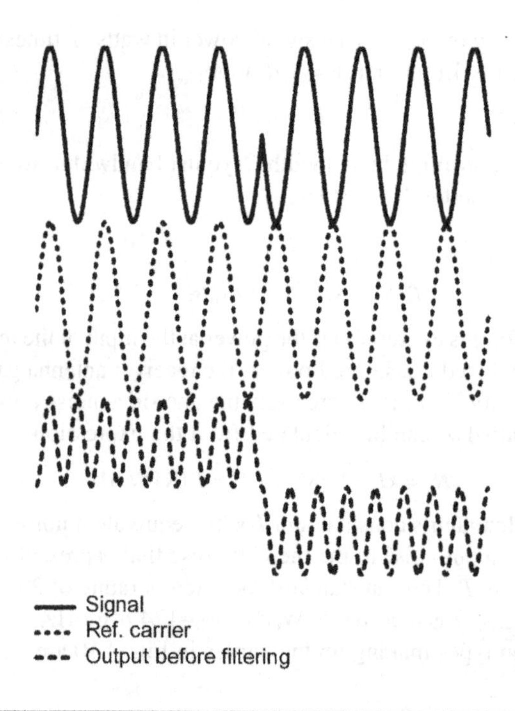


**FIG. 4.12**

PSK demodulation.

ASK is easy to generate and detect, and as we see from Fig. 4.10, its bit error rate performance is essentially the same as for FSK. However, FSK is usually the modulation of choice for many systems. The primary reason is that peak power is usually the limitation, which gives FSK a 3-dB advantage, since it has constant power for both bit states, whereas ASK has only half the average power, assuming a 50% duty cycle and equal probability of marks and spaces. FSK has slightly more complexity than ASK, and that's probably why it isn't used in all short-range digital systems.

A modulation system which incorporates the methods discussed above but provides a high degree of interference immunity is spread spectrum, which we'll discuss later on in this chapter.

### 4.4.5 Analog communication

For short-range analog communication—wireless microphones, wireless earphones, auditive assistance devices—FM is almost exclusively used. When transmitting high-quality audio, FM gives an enhanced post-detection signal-to-noise ratio, at the expense of greater bandwidth. Even for narrow band FM, which doesn't have post-detection signal-to-noise enhancement, its noise performance is better than that of AM, because a limiting IF amplifier can be used to reduce the noise. AM, being a linear modulation process, requires linear amplifiers after modulation in the transmitter, which are less efficient than the class C amplifiers used for FM. Higher power conversion efficiency gives FM an advantage in battery-operated equipment.

### 4.4.6 Advanced digital modulation

Two leading demands of wireless communication are the need for higher data rates and better utilization of the radio spectrum. This translates to higher speeds on narrower bandwidths. At the same time, much of the radio equipment is portable and operated by batteries. So what is needed is:

- high data transmission rates
- narrow bandwidth
- low error rates at low signal-to-noise ratios
- low power consumption

Breakthroughs have occurred with the advancement of digital modulation and coding systems. We deal here with digital modulation principles. Coding will be discussed in Chapter 10.

The types of modulation that we discussed previously, ASK, FSK, and PSK, involve modifying a radio frequency carrier one symbol at a time. The Nyquist bandwidth is the narrowest bandwidth of an ideal filter that permits passing a symbol stream without intersymbol interference. As the bandwidth of the digital symbol stream is further reduced, the symbols are lengthened and interfere with the detection of subsequent symbols. This minimum, or Nyquist, bandwidth equals one-half of the symbol rate at baseband, but twice as much for the modulated signal. We can see this

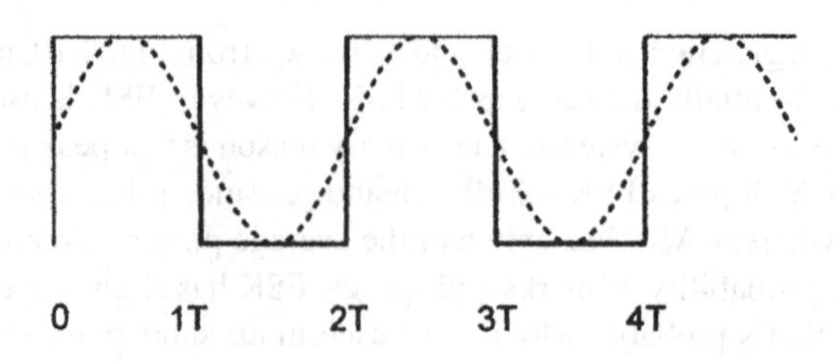


**FIG. 4.13**

Nyquist bandwidth.

result in Fig. 4.13. An alternating series of marks and spaces are represented by a sine wave whose frequency is one-half the bit rate: $f_{\text{sin}} = (1/2)T$. An ideal filter with a lower cutoff frequency will not pass this fundamental frequency component and the data will not get through.

Any other combination of bits will create other frequencies, all of which are lower than the Nyquist frequency. It turns out then that the maximum number of bits per hertz of filter cutoff frequency that can be passed at baseband is two. Therefore, if the bandwidth of a telephone line is 3.4 kHz, the maximum binary bit rate that it can pass without intersymbol interference is 6.8k bits per second. We know that telephone line modems pass several times this rate. They do it by incorporating several bits in each symbol transmitted, for it is actually the symbol rate that is limited by the Nyquist bandwidth, and the problem that remains is to put a sequence of several bits on each symbol in such a manner that they can be effectively interpreted from the symbol at the receiving end with as small as possible chance of error, given a particular $S/N$.

In Fig. 4.14 we see three ways of combining bits with individual symbols, each of them based on one of the basic types of modulation—ASK, FSK, and PSK. Each symbol duration $T$ carries one of four different states of two bits. Using any one of the modulation types shown, the telephone line, or wireless link, can pass a bit rate twice as high as before over the same bandwidth. Combinations of these types are also employed, particularly of ASK and PSK, to make it easier to recognize the individual states in noise. Quadrature amplitude modulation, QAM, modulates in-phase and quadrature versions of the carrier signal with multiple levels to give a high bandwidth efficiency—a high bit-rate relative to the signal bandwidth. It seems then that there is essentially no limit to the number of bits that could be compressed into a given bandwidth. If that were true, the 3.4-kHz telephone line could carry millions of bits per second, and the internet bottleneck to our homes would no longer exist. However, there is a very definite limit to the rate of information transfer over a transmission medium where noise is present, expressed in the Hartley-Shannon law:

$$C = W\log\left(1+\frac{S}{N}\right) \tag{4.1}$$

This expression tells us that the maximum rate of information (the capacity $C$) that can be sent without errors on a communication link is a function of the bandwidth, $W$, and the signal-to-noise power ratio, $S/N$. We examine this equation in greater detail in Chapter 9.

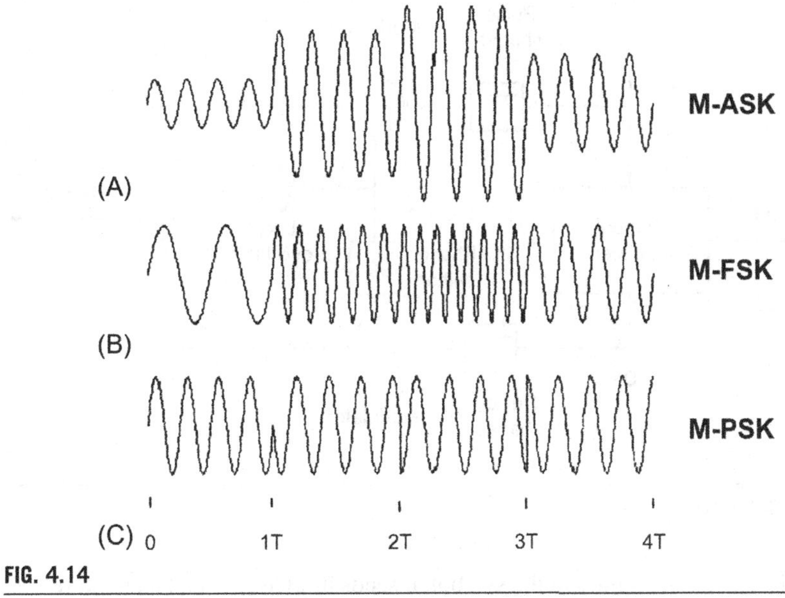


**FIG. 4.14**

M-ary modulation. (A) M-ASK; (B) M-FSK; and (C) M-PSK.

In investigating the ways of modulating and demodulating multiple bits per symbol, we'll first briefly discuss M-ary FSK (Fig. 4.14B). In contrast to the aim we mentioned above of increasing the number of bits per hertz, it increases the required bandwidth as the number of bits per symbol is increased. "M" in "M-FSK" is the number of different frequencies that are transmitted in each symbol period. The benefit of this method is that the required *S/N* per bit for a given bit error rate decreases as the number of bits per symbol increases. This is analogous to analog FM modulation, which uses a wideband radio channel, well in excess of the bandwidth of the source audio signal, to increase the resultant *S/N*. M-ary FSK is commonly used in point-to-point microwave transmission links for high-speed data communication where bandwidth limitation is no problem but the power limitation is.

Most of the high-data-rate bandwidth-limited channels use multiphase PSK or QAM. While there are various modulation schemes in use, the essentials of most of them can be described by the block diagram in Fig. 4.15. This diagram is the basis of what is called vector modulation, IQ modulation, or quadrature modulation. "*I*" stands for "in phase" and "*Q*" stands for "quadrature."

The basis for quadrature modulation is the fact that two independent data streams are simultaneously modulated on a carrier wave of a given frequency. This is possible if each data stream modulates coherent carriers whose phases are 90 degrees apart. We see in Fig. 4.15 that each of these carriers is created from the same source by passing one of them through a 90-degree phase shifter. The symbol encoder block is a serial to parallel converter and symbol mapper. It takes *n* data bits and for each of

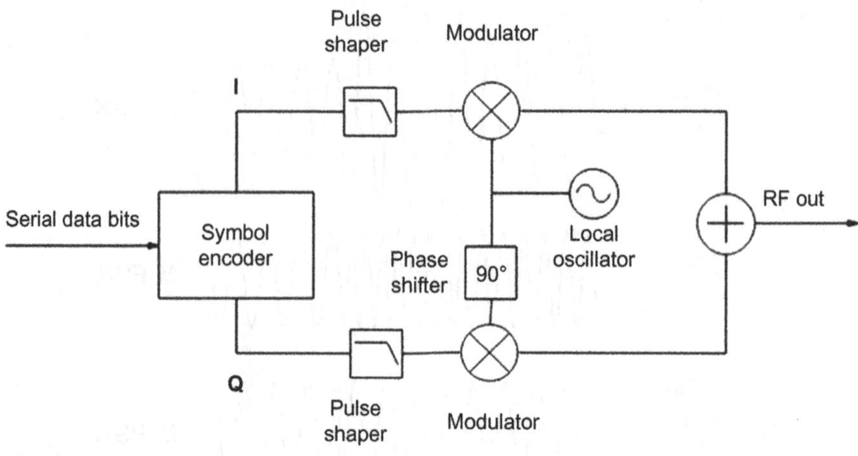


**FIG. 4.15**

Quadrature modulation.

the possible $2^n$ binary values of the symbol, it sends in-phase "*I*" and quadrature "*Q*" levels to the modulators whose outputs are summed to give a distinctive magnitude and phase to the symbol RF output. The pulse shaper filters smooth the transitions between symbols to restrain the bandwidth.

The receiver, by reversing the process used in the transmitter (see Fig. 4.16), can reproduce the bit sequence of the transmitted symbol. Down conversion mixers switched by quadrature versions of the local oscillator output produce in-phase and quadrature baseband signals after lowpass filtering. The symbol decoder performs the opposite function of the transmitter symbol encoder and outputs the binary data corresponding to the values of *I* and *Q* of the symbol.

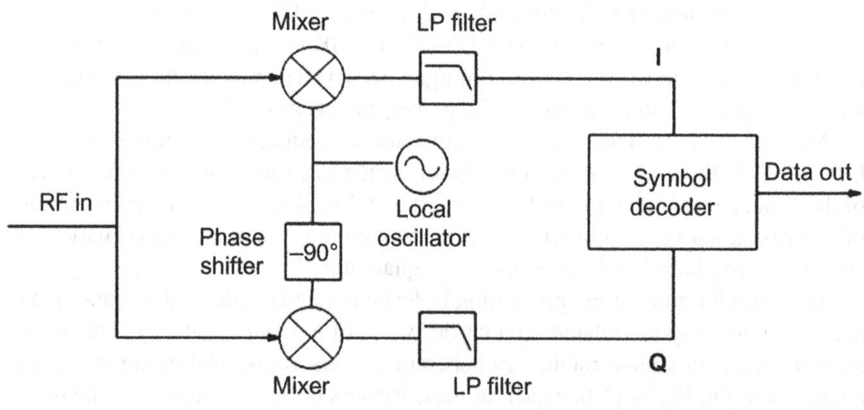


**FIG. 4.16**

Quadrature demodulation.

The magnitude and phase of each possible state of a carrier symbol in response to the modulation is shown on a vector or constellation diagram. The constellation diagram for QPSK, where two data bits form a symbol, is shown in Fig. 4.17. The Xs on the plot are tips of vectors that represent the magnitude (distance from the origin) and phase of the signal that is the sum of the "*I*" and the "*Q*" carriers shown in Fig. 4.15, where each is multiplied by $-1$ or $+1$, corresponding to bit values of 0 and 1. The signal magnitudes of all four possible combinations of the two bits is the same — $\sqrt{2}$ when the *I* and *Q* carriers have a magnitude of 1. *I* and *Q* bit value combinations corresponding to each vector are shown on the plot.

We have shown that two data streams, derived from a data stream of rate *R*2, can be sent in parallel on the same RF channel and at the same bandwidth as a single data stream having half the bit rate, *R*1. At the receiver end, the demodulation process of each of the split bit streams is exactly the same as it would be for the binary phase shift modulation shown in Fig. 4.12, and its error rate performance is the same as is shown for the BPSK curve in Fig. 4.10. Thus, we've doubled the data rate on the same bandwidth channel while maintaining the same error rate as before. In short, we've doubled the efficiency of the communication.

However, there are some complications in adopting quadrature modulation. Similar to the basic modulation methods previously discussed, the use of square waves to modulate RF carriers causes unwanted sidebands that may exceed the allowed channel bandwidth. Thus, pulse shaping low-pass filters shown in Fig. 4.15 are inserted in the signal paths before modulation. Even with these filters, the change in the phase of the RF signal at the change of data state cause increased sidebands. We see in

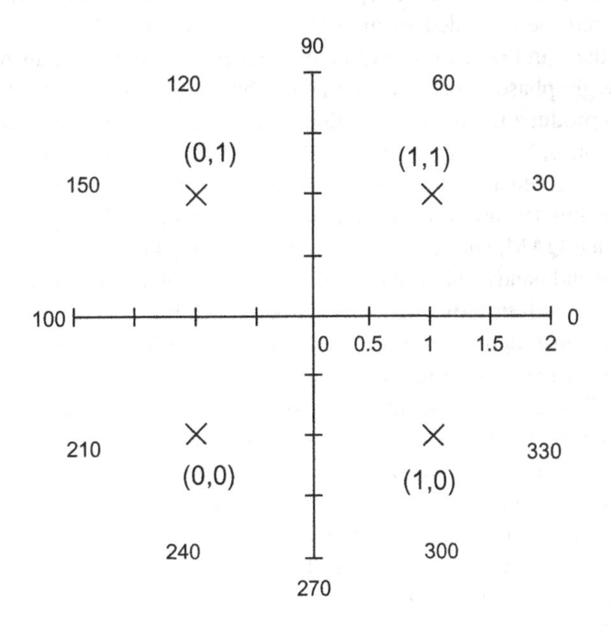


**FIG. 4.17**

QPSK constellation diagram.

Fig. 4.14C that changes of data states may cause the RF carrier to pass through zero when changing phase. Variations of carrier amplitude make the signal resemble amplitude modulation, which requires inefficient linear amplifiers, as compared to nonlinear amplifiers that can be used for frequency modulation, for example.

Two variations of quadrature phase shift keying (QPSK) have been devised to reduce phase changes between successive symbols and to prevent the carrier from going through zero amplitude during phase transitions. One of these is offset PSK. In this method, the *I* and *Q* data streams in the transmitter are offset in time by one-half of the bit duration, causing the carrier phase to change more often but more gradually. In other words, the new "*I*" bit will modulate the cosine carrier at half a bit time earlier (or later) than the time that the "*Q*" bit modulates the sine carrier.

The other variant of QPSK is called pi/4 DPSK. It was used in the US second generation TDMA (time division multiple access) digital cellular system. In it, the constellation diagram is rotated pi/4 radians (45 degrees) at every bit time such that the carrier phase angle can change by either 45 or 135 degrees. This system reduces variations of carrier amplitude so that more efficient nonlinear power amplifiers may be used in the transmitter.

Another problem with quadrature modulation as described above is the need for a coherent local oscillator in the receiver in order to separate the in-phase and quadrature data streams. As for bipolar PSK, this problem may be ameliorated by using differential modulation and by multiplying a delayed replica of the received signal by itself to extract the phase differences from symbol to symbol.

The principle of transmitting separate data streams on in-phase and quadrature RF carriers may be extended to multiple bits per symbol. If there are three bits per symbol, the *I* and *Q* streams will be modulated such that the transmitted carrier can assume eight phase states, each with equal amplitude, that can be detected at the receiver to reproduce the transmitted three bit pattern. An even number of bits per symbol, ($b = 4, 6, 8, \ldots$) are applied to the *I* and *Q* paths to create a carrier signal with $2^b$ states of given amplitude and phase. For example, 4 bits per each carrier vector symbol creates 16 combinations of amplitude levels and phase. This type of modulation is called QAM, and in this example, 16-QAM, 4 data bits are transmitted in the same time and bandwidth as one bit sent by binary phase shift keying (BPSK), so the data rate is one fourth of what it would be if each bit modulates the carrier individually. The constellation of 16-QAM (4 bits per symbol) is shown in Fig. 4.18.

The process of concentrating more and more bits per symbol cannot go on without limit. While the bit rate per hertz goes up, the required *S/N* of the channel for a given bit error rate (BER) also increases—that is, more power must be transmitted, or sophisticated coding algorithms have to be incorporated in the data protocol. This can be realized by comparing Figs. 4.17 and 4.18. For equal average power, or average length of the vectors to the constellation points in each figure, those points are closer together for 16-QAM than for QPSK, and less noise around each point (higher *S/N*) is needed for 16-QAM so that the receiver can distinguish between neighboring points, which represent symbols.

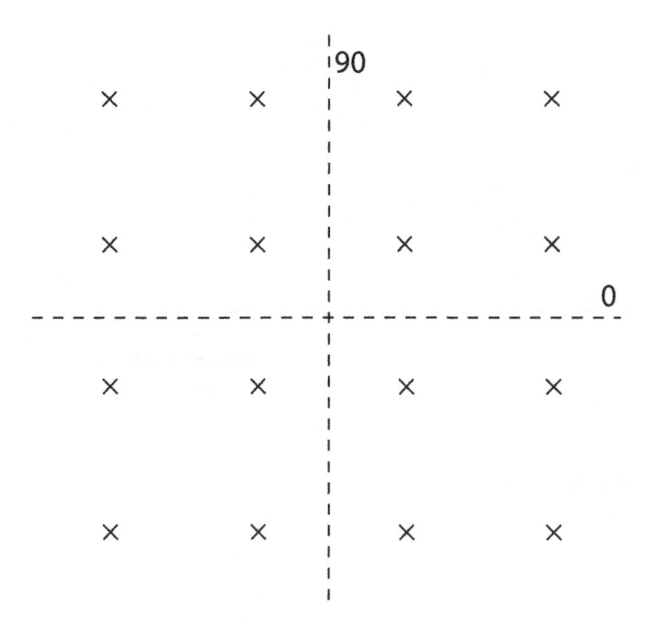


**FIG. 4.18**

16-QAM constellation diagram.

We now turn to examine another aspect of digital modulation that is used in a number of short-range wireless applications—spread-spectrum modulation.

### 4.4.7 Spread spectrum

The regulations for unlicensed communication using unspecified modulation schemes, both in the United States and in Europe, determine maximum power outputs ranging from tens of microwatts up to 10 mW in most countries. This limitation greatly reduces the possibilities for wireless devices to replace wires and to obtain equivalent communication reliability. However, the availability of frequency bands where up to 1 W may be transmitted in the United States and 100 mW in Europe, under the condition of using a specified modulation system, greatly enlarges the possible scope of use and reliability of unlicensed short-range communication.

Spread spectrum has allowed the telecommunication authorities to permit higher transmitter powers because spread-spectrum signals can coexist on the same frequency bands as other types of authorized transmissions without causing undue interference or being unreasonably interfered with [6]. The reason for this is evident from Fig. 4.19. Fig. 4.19A shows the spread-spectrum signal spread out over a bandwidth much larger than the narrow-band signals. Although its total power if transmitted as a narrow band signal could completely overwhelm another narrow-band signal on the same or adjacent frequency, the part of it occupying the narrow-band signal bandwidth is small related to the total, so it doesn't interfere with it. In other words,

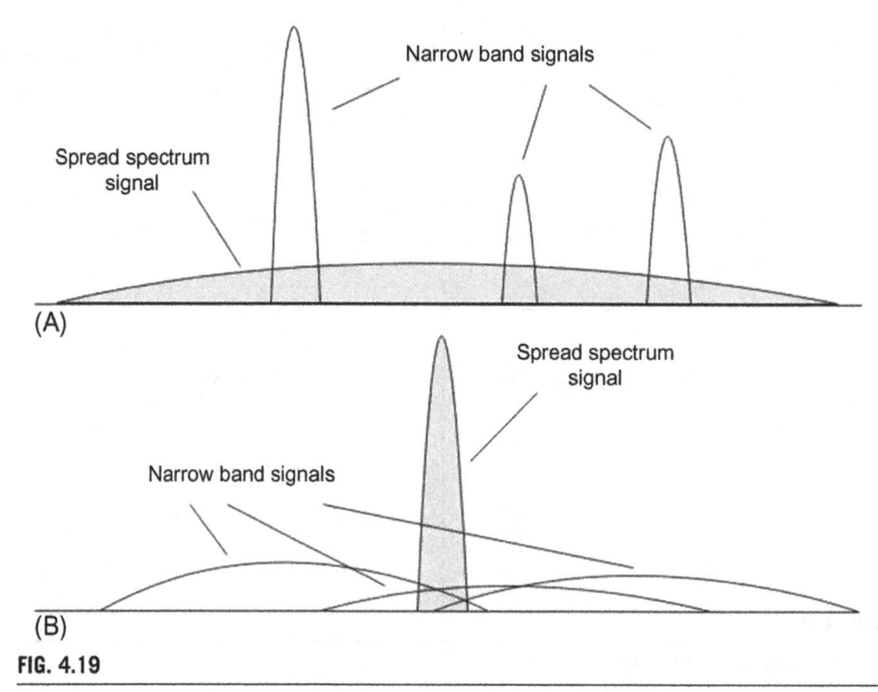


**FIG. 4.19**

Spread spectrum and narrow-band signals. (A) Signals at receiver input and (B) signals after despreading.

spreading the transmitted power over a wide frequency band greatly reduces the signal power in a narrow bandwidth and thus the potential for interference.

Fig. 4.19B shows how spread-spectrum processing reduces interference from adjacent signals. The despreading process concentrates the total power of the spread-spectrum signal into a narrow-band high peak power signal, whereas the potentially interfering narrow-band signals are spread out so that their power in the bandwidth of the desired signal is relatively low.

These are some advantages of spread spectrum modulation:

- Allows permitting higher power for nonlicensed devices
- Reduces co-channel interference—good for congested ISM bands
- Reduces multipath interference
- Resists intentional and unintentional jamming
- Reduces the potential for eavesdropping
- Permits code division multiplexing of multiple users on a common channel.

There is sometimes a tendency to compare spread spectrum with wideband frequency modulation, such as used in broadcasting, since both processes achieve performance advantages by occupying a channel bandwidth much larger than the bandwidth of the information being transmitted. However, it's important to understand that there are principal differences in the two systems.

In wide band FM (WBFM), the bandwidth is spread out directly by the amplitude of the modulating signal and at a rate determined by its frequency content. The result achieved is a signal-to-noise ratio that is higher than that obtainable by sending the same signal over baseband (without modulation) and with the same noise density as on the RF channel. The post detection signal-to-noise ratio (*S/N*) is a multiple of the *S/N* at the input to the receiver, but the input *S/N* must be higher than a threshold value, which depends on the deviation factor of the modulation.

In contrast, spread spectrum has no advantage over baseband transmission from the point of view of signal-to-noise ratio. Usual comparisons of modulation methods are based on a channel having only additive wideband Gaussian noise. With such a basis for comparison, there would be no advantage at all in using spread spectrum compared to sending the same data over a narrow-band link. The advantages of spread spectrum are related to its relative immunity to interfering signals, to the difficulty of message interception by a chance eavesdropper, and to its ability to use code selective signal differentiation. Another often-stated advantage to spread spectrum—reduction of multipath interference—is not particularly relevant to short-range communication because the pulse widths involved are much longer than the delay times encountered indoors.

The basic difference between WBFM and spread spectrum is that the spreading process of the latter is independent of the baseband signal itself. The transmitted signal is spread by one of several different spreading methods, and then despread in a receiver that knows the spreading code of the transmitter.

The methods for spreading the bandwidth of the spread-spectrum transmission are (1) frequency-hopping spread spectrum (FHSS), (2) direct sequence spread spectrum (DSSS), (3) pulsed-frequency modulation or chirp modulation, and (4) time-hopping spread spectrum. The last two types are not allowed in the FCC rules for unlicensed operation and after giving a brief definition of them we will not consider them further.

**(1)** In FHSS, the RF carrier frequency is changed relatively rapidly at a rate of the same order of magnitude as the bandwidth of the source information (analog or digital), but not dependent on it in any way. At least several tens of different frequencies are used, and they are changed according to a pseudo-random pattern known also at the receiver. The spectrum bandwidth is roughly the number of the different carrier frequencies times the bandwidth occupied by the modulation information on one hopping frequency.

**(2)** The direct-sequence spread-spectrum signal is modulated by a pseudo-random digital code sequence known to the receiver. The bit rate of this code is much higher than the bit rate of the information data, so the bandwidth of the RF signal is consequently higher than the bandwidth of the data.

**(3)** In chirp modulation, the transmitted frequency is swept for a given duration from one value to another. The receiver knows the starting frequency and duration so it can despread the signal.

**(4)** A time-hopping spread-spectrum transmitter sends low duty cycle pulses with pseudo-random intervals between them. The receiver despreads the signal by gating its reception path according to the same random code as used in the transmitter.

Actually, all of the above methods, and their combinations that are sometimes employed, are similar in that a pseudo-random or arbitrary (in the case of chirp) modulation process used in the transmitter is duplicated in reverse in the receiver to unravel the wide-band transmission and bring it to a form where the desired signal can be demodulated like any narrowband transmission.

The performance of all types of spread-spectrum signals is strongly related to a property called process gain. It is this process gain that quantifies the degree of selection of the desired signal over interfering narrow-band and other wide-band signals in the same passband. Process gain is the difference in dB between the output *S/N* after despreading and the input *S/N* to the receiver:

$$PG_{dB} = (S/N)_{out} - (S/N)_{in} \tag{4.2}$$

The process gain factor may be approximated by the ratio

$$PG_f = (\text{RF bandwidth})/(\text{rate of information}) \tag{4.3}$$

A possibly more useful indication of the effectiveness of a spread-spectrum system is the jamming margin:

$$\text{Jamming Margin} = PG_{dB} - \left(L_{sys} + S/J\right) \tag{4.4}$$

where *S/J* is the minimum signal power to jamming power in dB for acceptable receiver performance, and $L_{sys}$ is system implementation losses, which may be of the order of 2 dB. The jamming margin is the amount by which a potentially interfering signal in the receiver's passband may be stronger than the desired signal without impairing the desired signal's ability to get through.

**Example 4.1**

Process gain $PG_{dB} = 30$ dB and system losses $L_{sys} = 2$ dB. The received output signal power at the input to the detector must be at least 9dB over jammer power to have an acceptable bit rate to error rate ratio, that is, *S/J* = 9dB. The jamming margin is 30dB − (2 + 9) dB = 19 dB, which is the amount of jammer power over input signal power that the system can tolerate.

Let's now look at the details of frequency hopping and direct-sequence spread spectrum.

#### *4.4.7.1 Frequency hopping*

FHSS can be divided into two classes—fast hopping and slow hopping. A fast-hopping transmission changes frequency one or more times per data bit (or symbol if there are multiple data bits per symbol). In slow hopping, several bits are sent per hopping frequency. Slow hopping is used for fast data rates since the frequency synthesizers in the transmitter and receiver are not able to switch and settle to new frequencies fast enough to keep up with the data rate if one or fewer bits per hop are

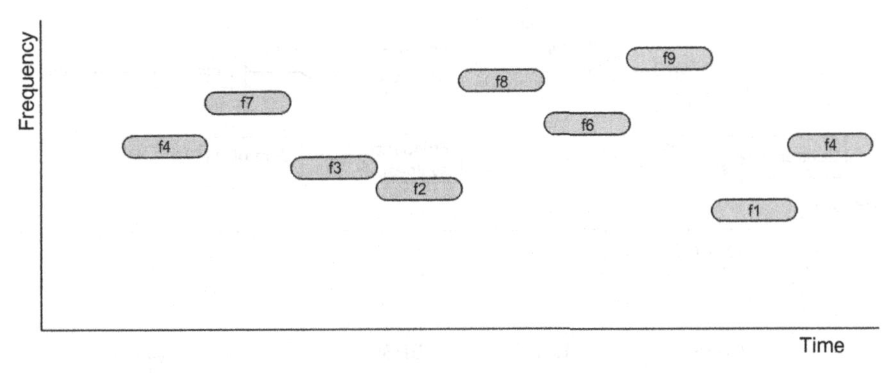


**FIG. 4.20**

Frequency hopping spread spectrum. Frequency hops vs. time.

transmitted. Frequency hopping as a function of time is shown in Fig. 4.20. The spectrum of a frequency-hopping signal looks like Fig. 4.21.

Fig. 4.22 is a block diagram of FHSS transmitter and receiver. Both transmitter and receiver local oscillator frequencies are controlled by frequency synthesizers. The receiver must detect the beginning of a transmission and synchronize its synthesizer to that of the transmitter. When the receiver knows the pseudo-random pattern of the transmitter, it can lock onto the incoming signal and must then remain in synchronization by changing frequencies at the same time as the transmitter. Once exact synchronization has been obtained, the IF frequency will be constant and the signal can be demodulated just as in a normal narrow-band superheterodyne receiver. If one or more of the frequencies that the transmission occupies momentarily is also occupied by an interfering signal, the bit or bits that were transmitted at that time may be lost. Thus, the transmitted message must contain redundancy or error correction coding so that the lost bits can be reconstructed.

### 4.4.7.2 Direct sequence

Fig. 4.23 is a diagram of a DSSS transmitter. Two configurations are shown. In Fig. 4.23A, the carrier frequency modulated by data is spread by a high rate pseudo-random spreading code. The direct sequence spread spectrum signal is expressed as [3]:

$$s(t) = A \cdot \cos(2\pi f_c t + \theta_d(t) + \theta_c(t)) \tag{4.5}$$

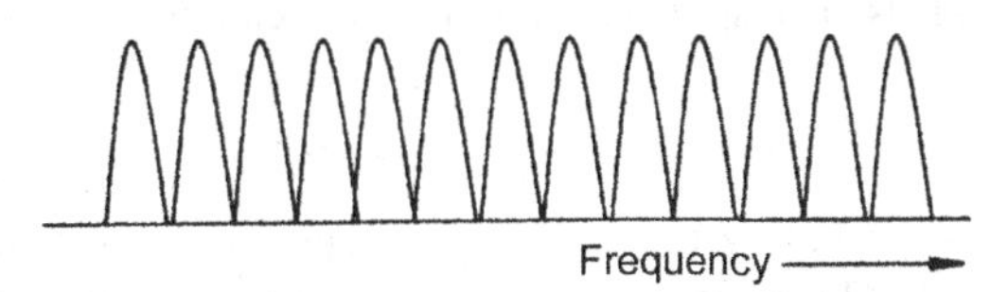


**FIG. 4.21**

Spectrum of frequency-hopping spread spectrum signal.

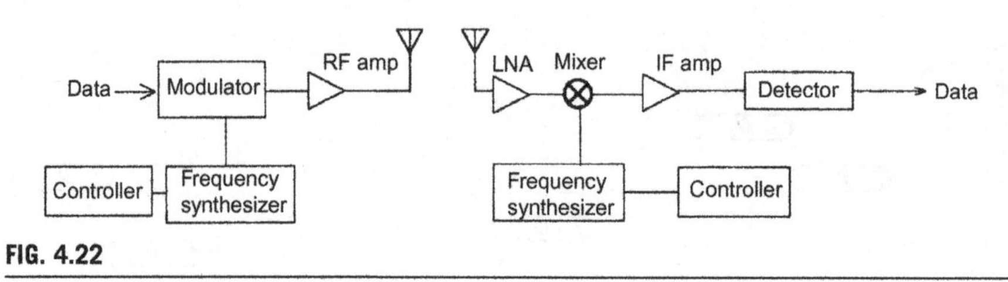


FIG. 4.22

FHSS transmitter and receiver.

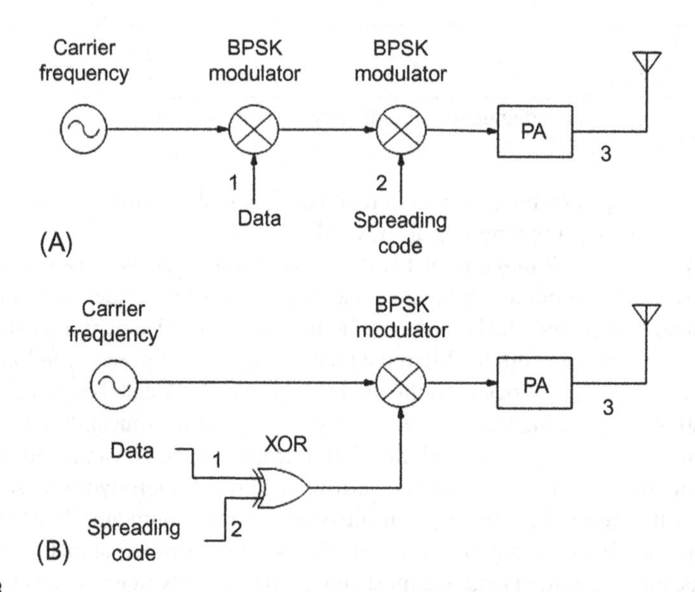


FIG. 4.23

DSSS Transmitter. (A) Data modulates the local oscillator and the result is spread by a pseudo random spreading code. (B) The spreading code is modulated at base band by binary data and the result modulates the local oscillator output.

where $A$ is peak amplitude and $\theta_d(t)$ and $\theta_c(t)$ are the phase modulation functions of the data and the spreading code on the carrier with frequency $f_c$. Since BPSK is used, each phase component in Eq. (4.5) can be 0 or 180 degrees, which is equivalent to multiplying the carrier signal by +1 or –1, in accordance with the logic states of the two modulating signals. The spread spectrum signal can now be expressed alternatively as

$$s(t) = A \cdot d(t) \cdot c(t) \cdot \cos(2\pi f_c t) \qquad (4.6)$$

where $d(t)$ and $c(t)$ are binary bipolar data and spreading code streams scaled to values of +1 and –1. The form of $s(t)$ in Eq. (4.6) implies that the DSSS signal can be produced as shown in Fig. 4.23B where the logic level data and spreading code are XOR'd at baseband—the equivalent of multiplication of bipolar signals—with the logic gate output applied to a single BPSK modulator. Data bit transitions are typically timed to coincide with the transitions of the chips of the spreading code. When the data bit is a logic "0," the spreading code is passed to the modulator without change. When the data bit is "1," the spreading code is inverted for the duration of the bit. The elements of the spreading code are called chips.

The unfiltered direct sequence spread spectrum around the carrier is shown in Fig. 4.24. The width of the major lobe is $2 \times R_c$, which is twice the chip rate. It contains approximately 90% of the signal power. Due to the wide bandwidth, the signal-to-noise ratio at the receiver input is very low, often below 0 dB.

Two alternative receiver architectures are shown in Fig. 4.25. In Fig. 4.25A, a replica of the transmitter's pseudo-random spreading code is multiplied with the incoming signal. The timing of the local code replica is varied by the phase control block according to feedback from the correlator output and when the transmitter and receiver spreading codes are the same and are in phase, that is, synchronized, a narrow-band IF signal results that can be demodulated in a conventional fashion. A common way to obtain this synchronization is for the receiver to adjust its replica code to a rate slightly different from that of the received signal. Then the phase difference between the two codes will change steadily until they are within one bit of

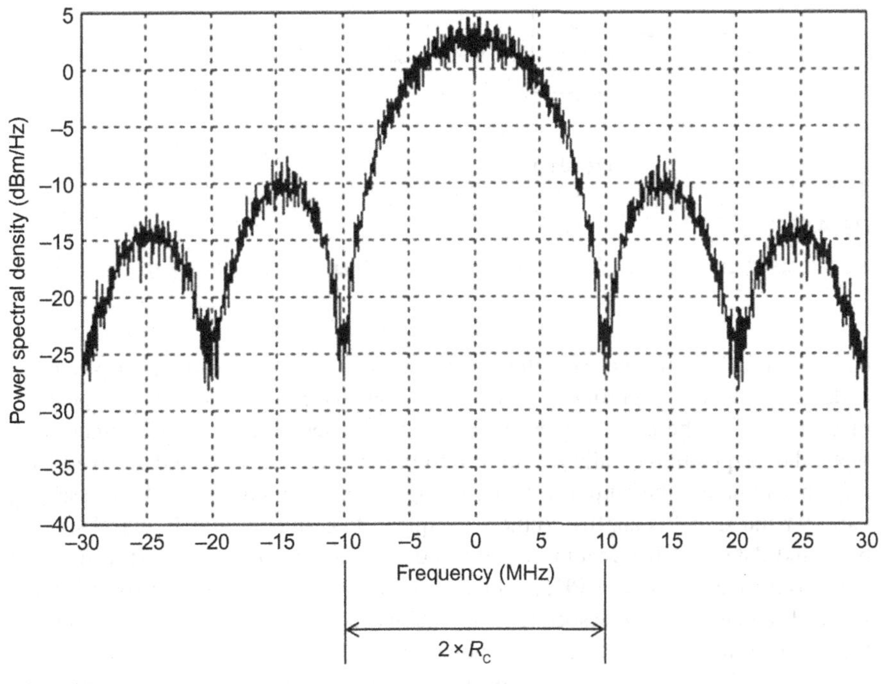


**FIG. 4.24**

DSSS frequency spectrum.

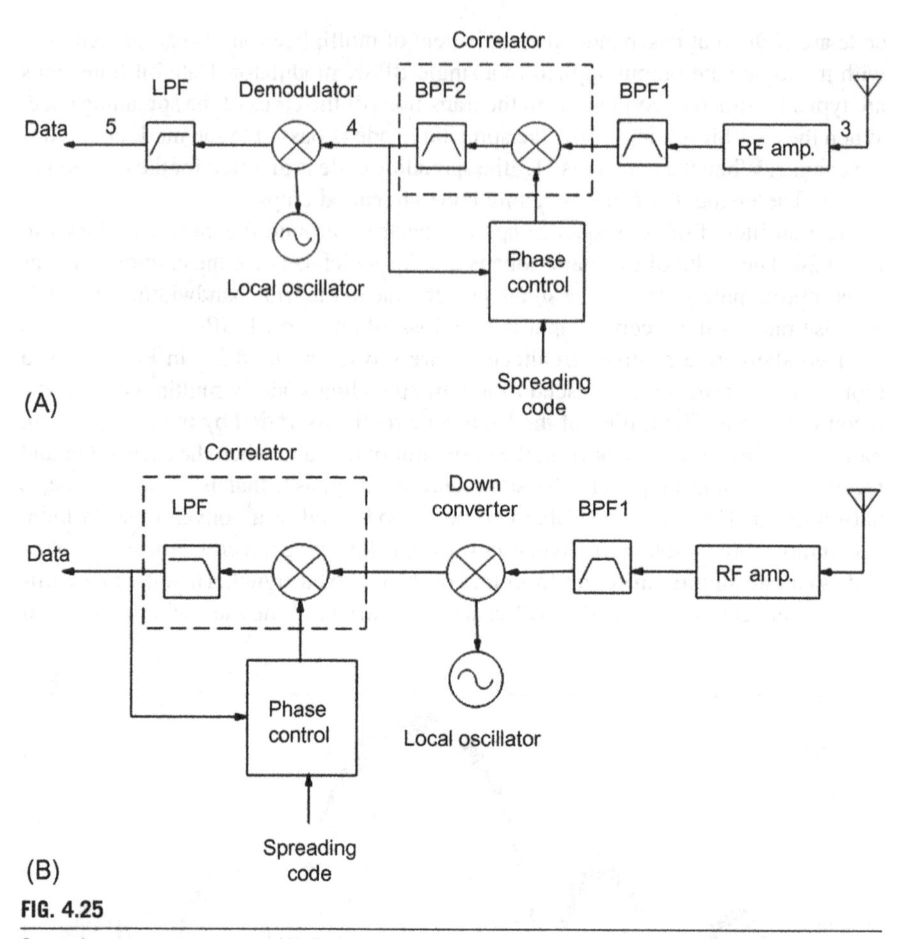


**FIG. 4.25**

Spread spectrum receiver. (A) Despreading at RF. (B) Despreading at baseband.

each other, at which time the output of the correlator increases to a peak when the codes are perfectly synchronized. The meaning of phase in this context is the position of each chip in the code sequence relative to the sequence's starting point. This method of acquisition is called a sliding correlator. When the local and received code sequences line up, the output of BPF2 in Fig. 4.25A is maximum and the resulting narrow band signal is applied to the subsequent demodulator which reproduces the baseband data. The bandwidth of BPF1 must fit the spectrum of the spread signal whereas the bandwidth of BPF2 is of the order of the spectrum of the dispread signal and wide enough to include spectral components due to the synchronizing process. In the alternate receiver configuration shown in Fig. 4.25B the received spread spectrum signal is down shifted to baseband before despreading by the correlation function.

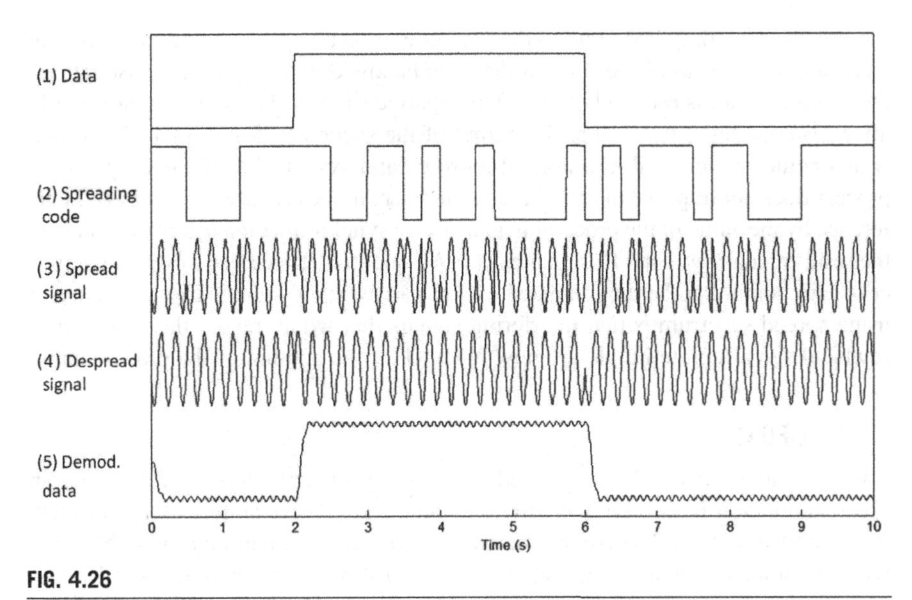


**FIG. 4.26**

DSSS demodulation waveforms.

The waveforms of Fig. 4.26 demonstrate DSSS modulation and demodulation. Wave numbers refer to the numbered locations in the block diagrams of Figs. 4.23 and 4.25A. Consider the signal levels +1 and −1. Wave 1 and wave 2 are the data and the spreading code, and wave 3 is the modulated spread spectrum signal at the transmitter and receiver antennas, the result of multiplying wave 1 times wave 2, then times the unmodulated carrier. Phase reversals of the carrier, equivalent to a phase shift of 180 degrees, are evident in the transmitted signal of wave 3 at the times of transitions of the spreading code times the data. When the data changes from "1" to "−1," there is no change in carrier phase, since the change of polarity of the code signal as seen by the modulator is cancelled by the inversion of the data. Multiplication of the RF signal by a bipolar locally generated replica of the spreading code that is perfectly aligned with the code imbedded in the received wave despreads the signal—collapses the bandwidth to that required by the data alone—giving the signal labeled wave 4. At the transition of dissimilar data bits the RF carrier is inverted, just as in a normal narrow band BPSK signal. This BPSK wave is coherently demodulated to reproduce the data, wave 5. The data stream at this point has a second harmonic ripple that isn't completely eliminated by the lowpass filter in Fig. 4.25A. The despreading and detection operation in the alternative receiver configuration of Fig. 4.25B is similar except that the correlation function is at base band.

BPF1 following the RF amplifier in Fig. 4.25A must have a bandwidth sufficient to pass the spread spectrum signal, usually at least ten times greater than the bandwidth required to pass a signal modulated by data alone. The ratio of the spread signal bandwidth to the data bandpass bandwidth is called the processing gain. It is also the ratio of the spreading code rate to the data rate, or symbol rate if multidimensional data modulation is used. Because of the wide bandwidth, the signal-to-noise ratio at the receiver input is relatively low, often negative (in dB). The signal-to-noise ratio after despreading is restored at the output of the second BPF in Fig. 4.25A to that which could be obtained in a normal narrow band system. While the despreading process does not improve the *S/N* in random noise, it does reduce narrow band interference by the value of the processing gain. Once synchronization has been achieved, the resulting narrow-band IF signal has a *S/N* equal to the received *S/N* plus the process gain, and it can be demodulated as any narrow-band signal. The difference in using spread spectrum is that interfering signals that would render the normal narrowband signal unintelligible are now reduced by the amount of the process gain.

### 4.4.8 OFDM

The continuing quest for improved spectral efficiency, good performance in urban and indoor high multipath environments, and relative simplicity has brought to the forefront of wireless communication, as well as some particularly problematic wired communication uses, the technique of OFDM—orthogonal frequency division multiplex. Based on the use of mathematical Fourier transform manipulations, OFDM forms the basis of diverse wideband communications technologies, among them:

- IEEE 802.11 (Wi-Fi)
- IEEE 802.16 (WiMAX)
- Long Term Evolution Cellular (LTE)
- Terrestrial Digital Video Broadcasting (DVB-T)
- Digital Audio Broadcast (DAB)
- Asymmetric Digital Subscriber Line (ADSL)
- Power Line Networking (HomePlug)

OFDM is advantageous in dealing with high bit rate communication in a multipath environment. It avoids frequency selective fading by converting a high rate data stream to multiple low rate signals, which are subjected to flat fading which can more easily be equalized and corrected by coding and interleaving. Intersymbol interference, often a problem in indoor systems, is essentially eliminated by posing a guard interval in front of each OFDM symbol.

The relative spectral efficiency advantage of OFDM is seen by comparing the OFDM and DSSS power spectra as shown in Fig. 4.27. The OFDM spectrum is flat-topped, with transmission power distributed equally over the bandwidth, and shows sharp attenuation outside of the passband. The DSSS spectrum, on the other hand, has a rounded main lobe with highest power density around the carrier frequency and reduced power density approaching the passband edges. It also has minor lobes which must be filtered out.

Fig. 4.28 demonstrates creation of low rate bit streams from a high rate data flow. This is similar to a demultiplexing procedure. The symbol period is increased in each subchannel by the number of subchannels, so the overall symbol rate doesn't change.

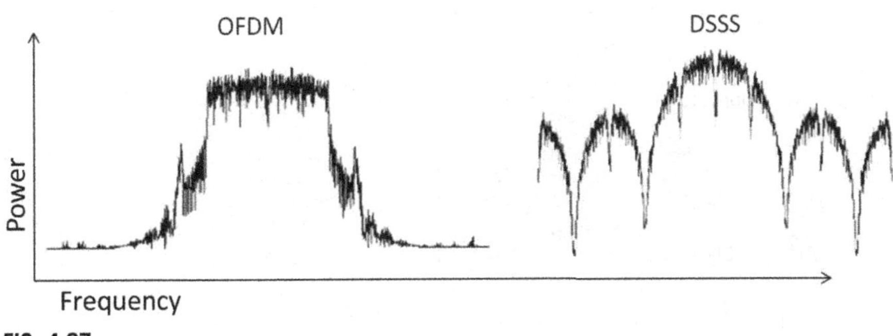


FIG. 4.27

Comparison of OFDM and DSSS power spectra.

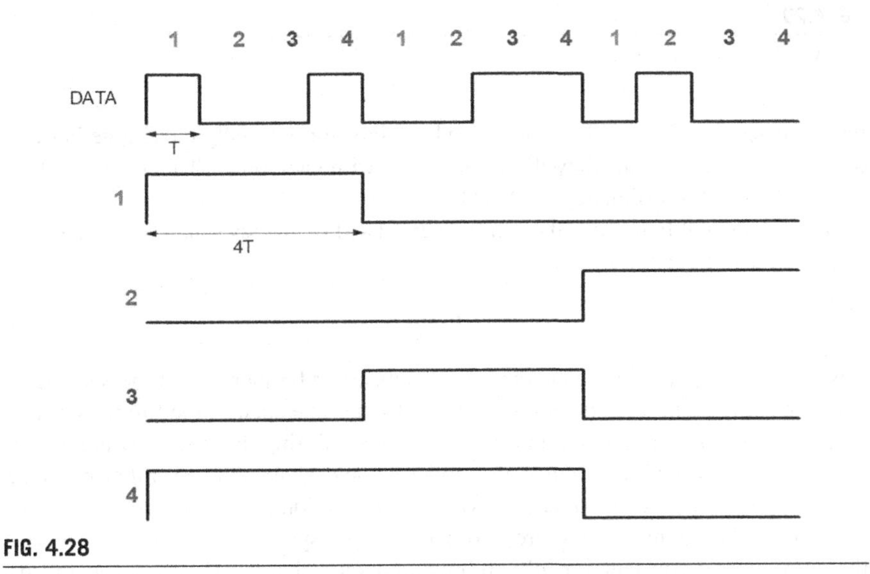


FIG. 4.28

Create multiple data channels from high bit rate stream.

While four subchannels are shown in the figure, in an actual system there are tens or hundreds of subchannels. It is the increase of the symbol period in each subchannel that changes potential frequency selective fading to flat fading. The low rate subchannels each modulate a subcarrier, as seen in Fig. 4.29. The result is frequency division multiplex (FDM), where individual data streams are transmitted simultaneously on adjacent carrier frequencies. In classical FDM, each subcarrier is generated by a separate oscillator, and the distance between the subcarriers has to be large enough, after bandpass filtering, so that there is no interference between them. However, in OFDM, there are no separate oscillators and no bandpass filters; the subcarriers are created through an inverse discrete Fourrier transform, implemented as a

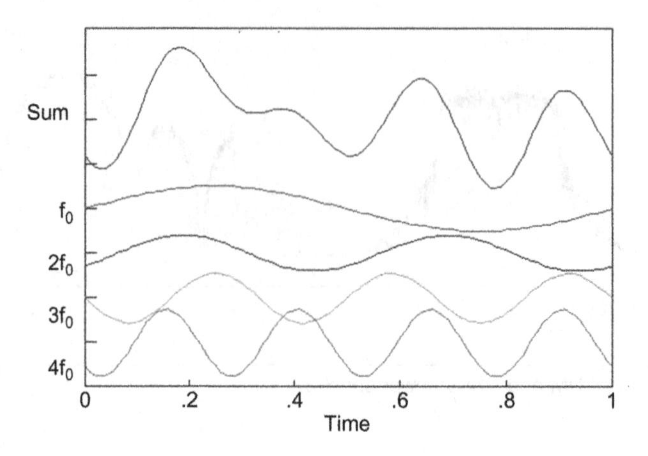


**FIG. 4.29**

OFDM orthogonal subcarriers.

fast Fourrier transform, IFFT, and the subcarriers are optimally close together and orthogonal, which means they do not interfere with each other. This gives the best obtainable spectral efficiency for FDM.

For orthogonality to exist, the symbol period and frequency separation of the subcarriers must be related by

$$T = \frac{1}{f_0} \tag{4.7}$$

$T$ is the symbol period and $f_0$ is the lowest subcarrier frequency and the separation between subcarriers. This is seen in Fig. 4.29. The time span of the figure is $T$, and each subcarrier has a constant amplitude and phase during that period which represents the modulation. The range of amplitudes and phase that a symbol can take depends on the type of modulation, which is a function of the number of bits per symbol. If, for example, there are two data bits per symbol, QPSK is used, where the amplitude is constant for all symbols, and four phases, 90 degrees apart, are defined. See Fig. 4.17. With four bits per symbol, 16 QAM may be employed, with sixteen combinations as shown in Fig. 4.18. The sum of $N$ carriers, shown in the upper curve of Fig. 4.29 where $N = 4$, is

$$s(t) = \text{Re}\left\{ \sum_{k=1}^{N} x_k A_k e^{j2\pi \cdot k \cdot f_0 \cdot t} \right\} \tag{4.8}$$

where $k$ identifies each individual subchannel, $x_k$ is the complex symbol which is constant over each period $T$, $A_k$ is the maximum amplitude of the subcarrier, $f_0$ is the channel separation, and $t$ is time. Fig. 4.30 is a frequency domain plot showing the power spectra of four modulated subchannels over many periods. It appears that the main and minor lobes of the spectrum of each subchannel overlap but there is no interference on the center frequencies. Because the subchannels are orthogonal, there is no interaction between the data streams. The figure illustrates how close the subchannels can be spaced without interference between each other.

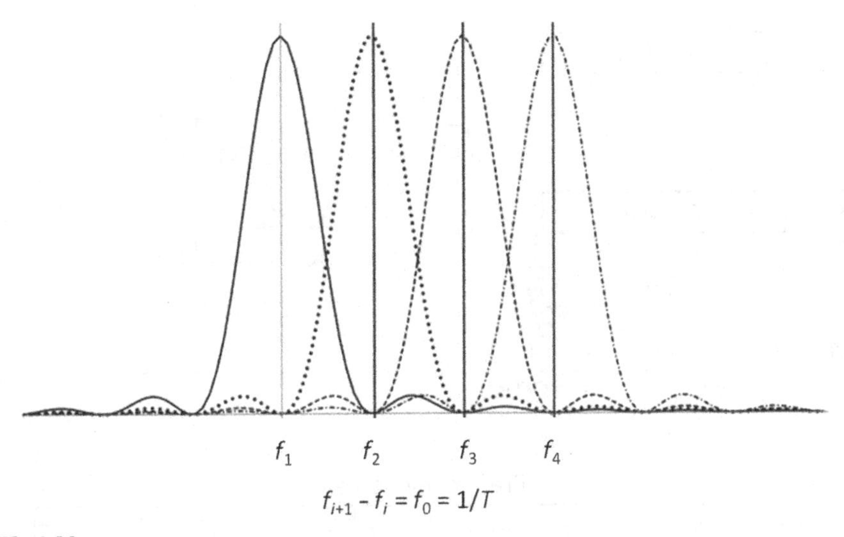


**FIG. 4.30**

Frequency domain power spectra of OFDM orthogonal subcarriers.

Fig. 4.31 shows the inputs and outputs of the inverse fast Fourier transform (IFFT) that creates the orthogonal subchannels illustrated in Fig. 4.30. For each period $T$, the complex symbol values of $N$ subchannels form the elements of a vector that is the input to a IFFT calculation. The position of each subchannel in the vector determines its baseband subcarrier frequency. The output vector of the IFFT contains complex sample values that make up a signal flow during a time period $T$. The frequency and time domains are illustrated in Fig. 4.32. In the frequency domain on the left the samples taken at the subcarrier frequencies are the complex values, magnitude $M$ and phase $\theta$, of each subchannel during the symbol period $T_{\text{FFT}}$. Time domain samples, also complex values, are shown as a jagged continuously changing signal. A cyclic prefix, *CP*, is appended, making a symbol time $T_{\text{SYM}}$. The cyclic prefix is a copy of a sequence of time samples taken near the end of the symbol period. Simulation of signals in the time and frequency domains is shown in Fig. 4.33. The upper

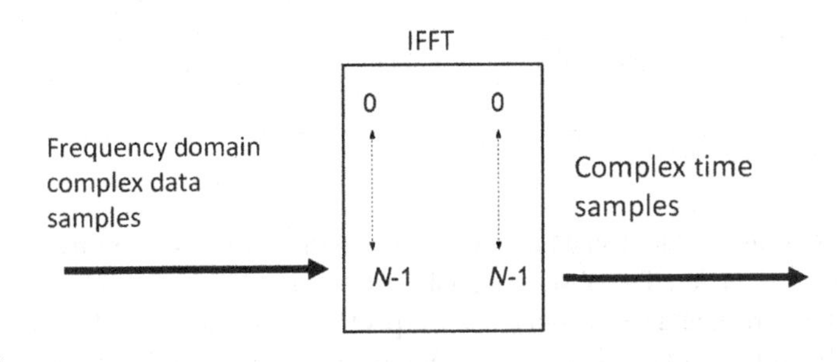


**FIG. 4.31**

OFDM implementation by inverse Fourier transform.

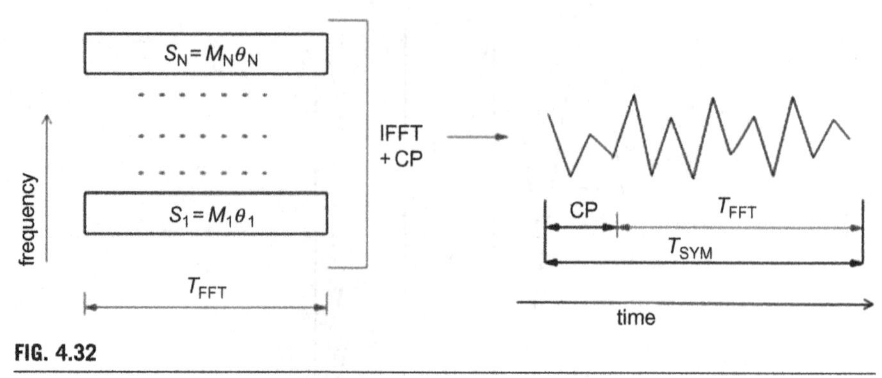


FIG. 4.32

Frequency and time domains in OFDM.

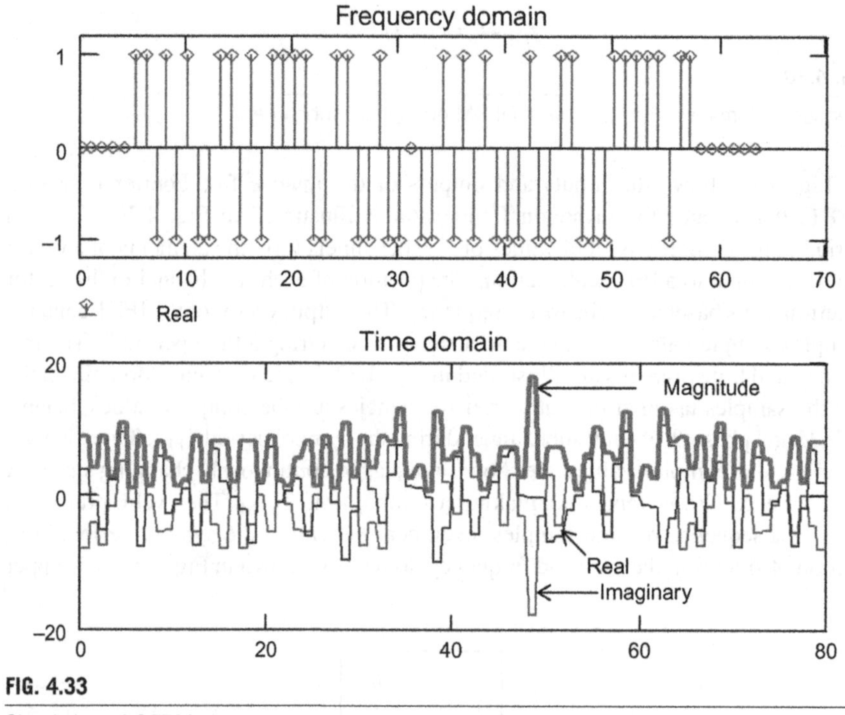


FIG. 4.33

Simulation of OFDM signals.

plot is a BPSK signal which can be viewed as a bandpass representation centered on a high frequency carrier. The abscissa is labeled by sample number. Note that in this example, the carrier and subcarriers at both spectrum edges are null. The corresponding complex time domain signal, which includes the cyclic prefix, is shown at the bottom of Fig. 4.33.

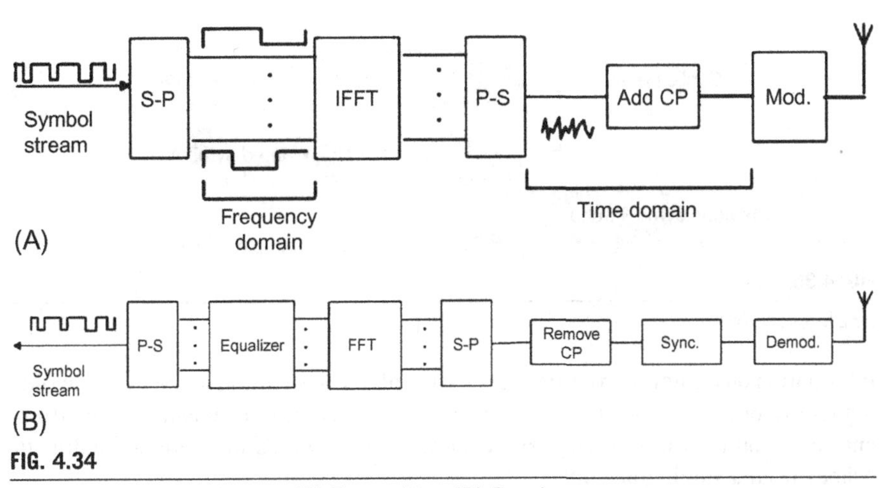


**FIG. 4.34**

OFDM device block diagram. (A) Transmitter. (B) Receiver.

Signal flow in an OFDM transmitter and receiver is shown in Fig. 4.34. In the transmitter, the high rate symbol stream is demultiplexed in a serial to parallel converter (S-P) and the symbols in each of the low rate parallel streams are applied at the start of each symbol time to the inverse fast Fourier transform block IFFT. The parallel output of the IFFT becomes a baseband time domain signal at the output of the parallel to serial converter block (P-S). The baseband symbol length is increased by the cyclic prefix (CP) and upconverted in a balanced modulator to the transmitted pass band. In the receiver, Fig. 4.34B, the OFDM signal is down converted in the demodulator, and synchronized to symbol and frame timing. The cyclic prefix is removed and after serial to parallel conversion (S-P), the time samples within an OFDM period are transformed by the FFT block to frequency domain symbols. An equalizer compensates for the flat fading over the communication channel and then a parallel to serial converter restores the original symbol stream. Not shown in Fig. 4.34 are the processes of FEC coding and interleaving that are essential to achieving a low bit error rate.

The OFDM cyclic prefix, shown in Fig. 4.35, has several objectives. First of all, echoes of the previous symbol, which are the result of multipath delays, fall in the cyclic prefix and intersymbol interference is avoided. The length of the cyclic prefix is chosen to be long enough to absorb all multipath reflections that would be expected in the environments where the communication system is deployed. Echoes of the present symbol, however, will actually improve the signal-to-noise ratio since, as shown in the figure, the sampling period can include multiple complete symbols due to the way the cyclic prefix is formed from a copy of a given number of the last to be received time domain samples in the symbol period. Addition of the symbol echoes does cause distortion, but this distortion is removed by equalization in the same manner as the flat fading. Another effect of the cyclic prefix is to facilitate a single frequency network (SFN) where multiple broadcast stations transmitting the same

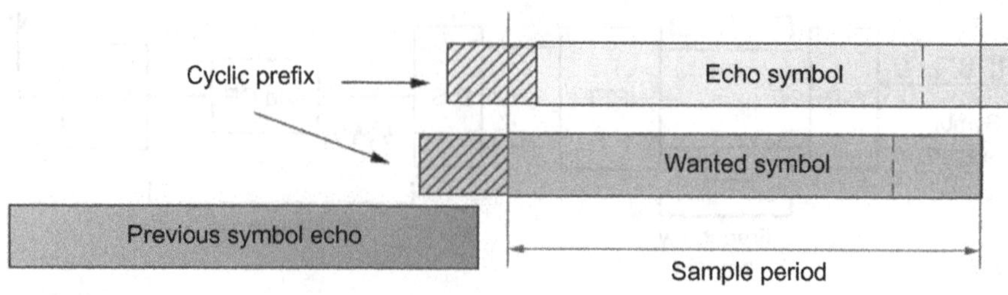


FIG. 4.35

OFDM cyclic prefix.

information on a single carrier frequency can enlarge a reception region while avoiding interference with each other. Transmissions from different antennas have different delays on arrival at a receiver, so they can be handled the same as multipath echoes from a single transmitter.

The key to transmitter-receiver synchronization and equalization is the transmission of symbols, known to the receiver, on specific subchannels and time period slots. Fig. 4.36 shows a spectrum of 52 subchannels where four of them, numbered −21, −7, 7, and 21, are dedicated to pilot subcarriers. The receiver knows what symbols are transmitted by these pilots, so it can apply the proper correction to restore them. Through interpellation between the pilots, it can also restore symbols on all of the other subchannels.

While OFDM has many benefits for wideband communication, it does have some disadvantages:

- It is particularly sensitive to phase noise and changing carrier offsets between transmitter and receiver due to the Doppler effect between moving terminals.
- The signal has a large peak to average power ratio which reduces transmitter power amplifier efficiency.
- The cyclic prefix and pilot carriers reduce spectrum efficiency.

On the other hand, a big advantage of OFDM is its scalability feature. A network can be designed for multiple bandwidths to satisfy different throughput needs, and bandwidth, along with enhanced data rate, can be extended through aggregation of channels on adjacent or nonadjacent frequency bands.

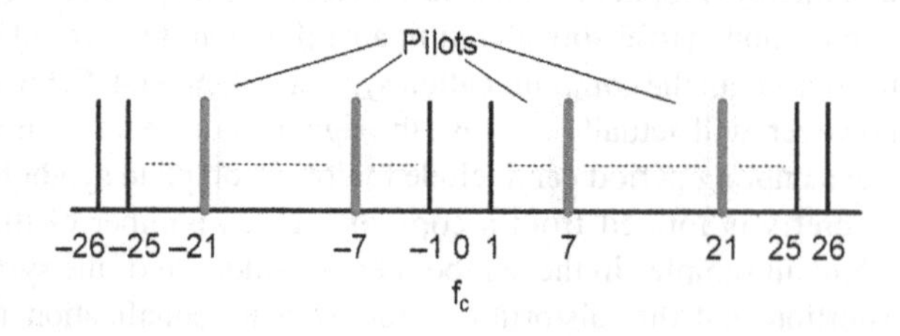


FIG. 4.36

Pilot subcarriers.

## 4.5 Summary

This chapter has examined various characteristics of short-range systems. It started with the ways in which data are formatted into different information fields for transmission over a wireless link. We looked at several methods of encoding the one's and zero's of the baseband information before modulation in order to meet certain performance requirements in the receiver, such as constant DC level and minimum bit error probability. Analog systems were also mentioned, and we saw that pre-emphasis/de-emphasis and compression/expansion circuits in voice communication devices improve the signal-to-noise ratio and increase the dynamic range.

Reasons for preferring frequency or amplitude digital modulation were presented from the points of view of equipment complexity and of the different regulatory requirements in the United States and in Europe. Similarly, there are several considerations in choosing a frequency band for a wireless system, among them background noise, antenna size, and cost.

The three basic modulation types involve impressing the baseband data on the amplitude, frequency, or phase of an RF carrier signal. Modern digital communication uses combinations and variations in the basic methods to achieve high bandwidth efficiency and low error rate. Quadrature modulation and demodulation were explained.

We described the principles of spread spectrum, with details of direct sequence spread spectrum and frequency hopping spread spectrum. Finally, OFDM was explained.

## References

[1] C. Toma, Introduction to Ultimate Keeloq® Technology, Application Note AN1683, Microchip Technology Inc., 2014.
[2] A. Greenberg, This Hacker's Tiny Device Unlocks Cars and Opens Garages, Wired Magazine, Security, 2015.
[3] B. Sklar, Digital Communications, Prentice Hall, Upper Saddle River, New Jersey, 2001.
[4] Freescale Semiconductor Technical Data, MC145026/D, Rev. 4, 2005.
[5] J. Anthes, OOK, ASK and FSK Modulation in the Presence of an Interfering Signal, Application Note, Murata Manufacturing Co. Ltd., 2007.
[6] R.C. Dixon, Spread Spectrum Systems, John Wiley & Sons, New York, 1984.

CHAPTER

# Signal generation and transmitters

5

In this chapter we examine the details of transmitter design. The basic constituents of a radio transmitter are shown in Fig. 5.1. Source data, which may be analog or digital, is imposed on a radio-frequency carrier wave by a modulator, then amplified, filtered, and applied to an antenna. The transmitter must also have a source of power. First we'll look at the definition of each block, and then see various ways in which the whole transmitter can be implemented.

## 5.1 RF source

Frequency control in short-range transmitters is based on LC, SAW, crystal, synthesizer and direct digital synthesis. In comparing the attributes of the several frequency-controlling methods, we refer to their accuracy and stability. Accuracy is the degree of deviation from a nominal frequency at a given temperature. Stability expresses how much the frequency may change over a temperature range and under other external conditions such as proximity to surrounding objects.

### 5.1.1 LC control

Only the simplest and least expensive portable transmitters may be expected to employ an inductor-capacitor (*LC*) resonator frequency control source. It has largely been supplanted by the SAW resonator. Like the SAW device, the *LC* circuit, together with a transistor, directly generates the RF carrier, almost always in the UHF range. Its stability is in the order of hundreds of kilohertz, and it suffers from hand effect and other proximity disturbances, which can move the frequency. Because of its poor stability and accuracy, it is almost always used with a superregenerative receiver, which has a broad bandwidth response. The *LC* oscillator circuit was used in garage door openers and automobile wireless keys, but has been virtually supplanted by SAW devices.

### 5.1.2 SAW resonators

SAW stands for *surface acoustic wave*. Many inexpensive short-range wireless control and alarm devices use it as the frequency determining element. It's more stable than the *LC* resonator, and simple circuits based on it have a stability of around $\pm 30$ kHz.



Short-range Wireless Communication. https://doi.org/10.1016/B978-0-12-815405-2.00005-1
© 2019 Elsevier Inc. All rights reserved.

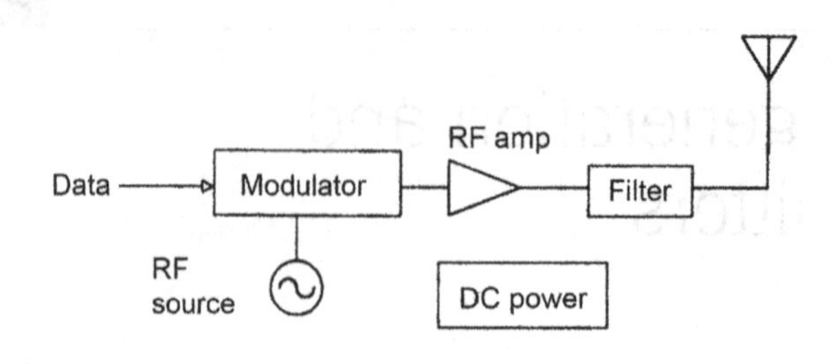


**FIG. 5.1**

Basic transmitter block diagram.

Its popularity is based on its reasonable price and the fact that it generates relatively stable UHF frequencies directly from a simple circuit.

Fig. 5.2 shows the construction of a SAW resonator. An RF voltage applied to metal electrodes generates an acoustic wave on the surface of the quartz substrate. The frequency response, quality factor $Q$, and insertion loss depend on the pattern and dimensions of the deposited metal electrodes. The frequency vs. temperature characteristic of the SAW device is shown in Fig. 5.3. We see that it has the form of a parabola peaking at room temperature, so that temperature stability of the device is very good at normal indoor temperatures, being within—50 ppm or 0.005% between −40 and +40°C.

SAW resonators are made in one-port and two-port configurations. The one-port devices are most commonly used for simple transmitters, whose oscillator circuits are of the Colpitts type, shown in Fig. 5.4A. A two-port SAW is convenient for designing an oscillator from a monolithic amplifier block, as shown in Fig. 5.4B. Since positive feedback is needed for oscillation, a device with 0- or 180-degree phase shift must be chosen according to the phase shift of the amplifier. The oscillator designer must use impedance matching networks to the SAW component and take into account their accompanying phase shifts.

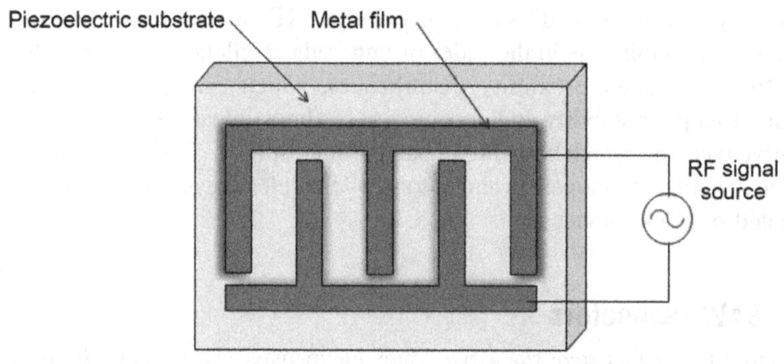


**FIG. 5.2**

SAW resonator construction.

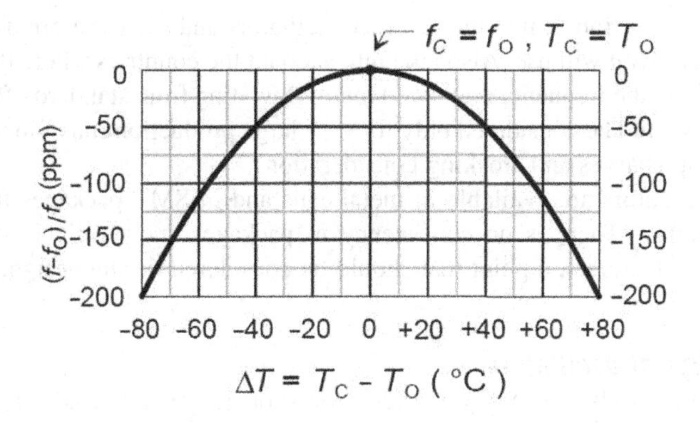


**FIG. 5.3**

SAW frequency vs. temperature.

*With permission from Murata Electronics N.A. Inc.*

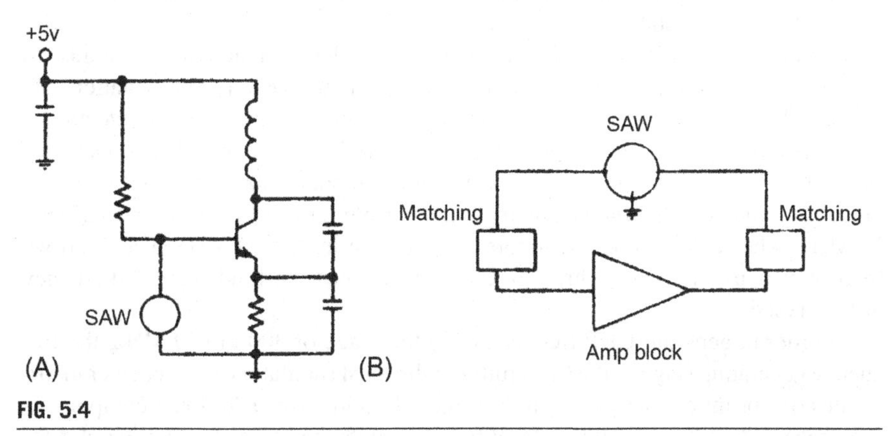


**FIG. 5.4**

SAW oscillators. (A) 1-port SAW and (B) 2-port SAW.

The frequency accuracy for SAW resonators is between $\pm 100$ and $\pm 200$ kHz and it's usually necessary to include a tuning element in the circuit for production frequency adjustment. This may be a trimmer capacitor or slug-tuned coil in the resonant circuit, or an air coil which can be tuned by distorting the turns.

In addition to resonators, SAW UHF filters are also available for short-range transmitters and receivers.

SAW components, both resonators and filters, are available from around 200 to 950 MHz. They cover frequencies used for unlicensed devices both in the United States and Canada, the United Kingdom and continental Europe and selected other countries. You can obtain resonators at offset frequencies for local oscillators in superheterodyne receivers. When designing a short-range radio system based on a SAW device, you should check the frequencies available from the various manufacturers for transmitter and receiver oscillators and filters before deciding on the frequencies you will use. Also take into account the countries where the product will be sold and the frequencies allowed there. Deviating from standard off-the-shelf frequencies should be considered only for very large production runs, since it entails special set-up charges and stocking considerations.

SAW resonators are available in metal cans and in SMT packages for surface mount assembly. There is no consistency in packages and terminal connections between manufacturers, a point that should be considered during design.

#### 5.1.2.1 Crystal oscillators

While the availability of SAW devices for short-range unlicensed applications resulted in a significant upgrading of performance compared to *LC* circuits, they still are not suitable for truly high performance uses. Sensitive receivers need narrow bandwidth, and this requires an accurate, stable local oscillator and correspondingly accurate transmitter oscillator. Stability in the range of several kilohertz is obtainable only with quartz crystals.

Crystals, like SAWs, are made of quartz piezoelectric material. In contrast to SAWs, whose operation we have seen is based on surface wave propagation, the crystal vibration is a bulk effect caused by high-frequency excitation to electrodes positioned on opposite sides of the bulk material. Crystal oscillators used for short-range transmitters are of two basic forms: fundamental mode and overtone mode. The fundamental mode usually extends from tens of kilohertz up to around 25 MHz, whereas overtone oscillators may operate up to 200 MHz. The overtone frequency is usually about three, five or seven times the fundamental frequency of the crystal.

In order to generate UHF frequencies in the range of 300 to 950 MHz, the frequency generating stage of the transmitter or the local oscillator of the receiver incorporates one or more doubler or tripler stages. This involves additional components and space and explains why the SAW based circuits are so much more popular for low-cost, short-range equipment, particularly hand-held and keychain transmitters. However, considering the frequency multiplier stages of a crystal oscillator as a disadvantage compared to the SAW-based oscillator isn't always justified. The single-stage SAW oscillator, when coupled directly to an antenna, is subject to frequency pulling due to the proximity effect or hand effect. Conducting surfaces, including the human body, within centimeters of the antenna and resonant circuits can displace the frequency of a SAW oscillator by 20 kHz or more. To counter this effect, a buffer circuit is often needed to isolate the antenna from the oscillator. With this buffer, there is hardly a cost or board space difference between the SAW transmitter and crystal transmitter with one doubler or tripler stage. Another factor in favor of the crystal oscillator transmitter is that it can be designed more easily without need for production tuning. Production frequency accuracy without tuning of up to 10 kHz is acceptable for many short-range applications.

A circuit for a crystal oscillator transmitter is shown in Fig. 5.5.

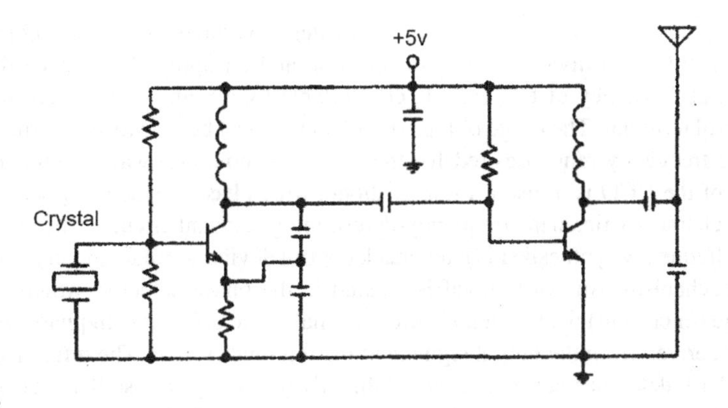


FIG. 5.5

Crystal oscillator transmitter.

## 5.1.3 Synthesizer control

The basic role of a frequency synthesizer is to generate more than one frequency using only one frequency-determining element, usually a crystal. Other reasons for using a synthesizer are

- The output frequency is derived directly from an oscillator without using multipliers.
- It allows higher deviation in frequency modulation than when directly modulating a crystal oscillator.

A block diagram of a frequency synthesizer, essentially a phase locked loop (PLL), is shown in Fig. 5.6. Its basic components are the reference crystal oscillator, a phase

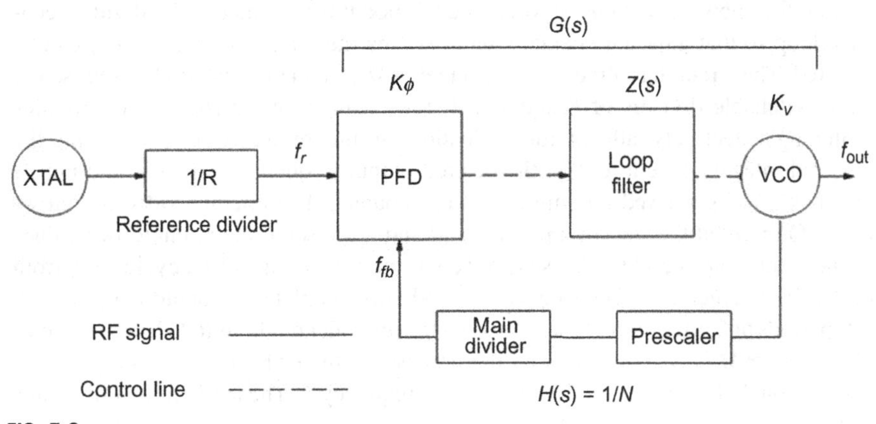


FIG. 5.6

Frequency synthesizer.

frequency detector (PFD), a low-pass loop filter, a voltage (or current) controlled oscillator (VCO), a divider in a feedback loop and an optional reference divider. The output frequency of the VCO is controllable by a voltage or current input at the control terminal. The range of the control signal must be able to sweep the output over the frequency range desired for the device. A common way to vary the frequency of the VCO is to use a varactor diode, which has a capacitance vs. voltage transfer characteristic, in the frequency-determining resonant circuit of the oscillator.

The frequency synthesizer is a feedback control device, similar in many ways to a servo mechanism. An error signal is created in the phase detector, which outputs a voltage (or current) whose value is proportional to the difference in phase between a signal derived from the output signal and the reference signal. The reference input to the phase detector may be connected directly to the crystal oscillator or derived from it through a frequency divider. The low-pass filter (LPF) following the phase detector determines the stability of the PLL, its speed of response, and the rejection of spurious output frequencies and noise. The division ratio N of the main frequency divider determines the output frequency, since the control voltage will adjust itself through the feedback such that $f_{out}/N$ will equal the crystal oscillator derived reference frequency. Increasing or decreasing $N$ by 1 increases or decreases the output frequency by the amount of the reference frequency. The reference divider is inserted in the circuit when frequency resolution smaller than the value of the crystal is needed. A prescaler, included in the main divider, is a low value high frequency divider, which is required when the maximum input frequency specified for the variable ratio divider is lower than the output frequency from the VCO. The use of a prescaler having a single frequency division ratio increases the incremental frequency of the synthesizer (the difference between adjacent channels) by a factor equal to the prescaler division ratio. In order not to lose the channel frequency resolution desired, or to be forced to divide the reference frequency further because of a single divisor prescaler, a dual modulus prescaler can be used.

Fig. 5.7 shows how to incorporate a dual modulus prescaler in the divider feedback loop so that adjacent channels separated by the reference frequency $f_r$ can be selected. The circuit has three counters, labeled $M$, $A$, and the dual modulus prescaler with switchable divisors of $P$ and $P + 1$, for example 8/9 or 16/17. The modulus switching effectively allows multiplication of the reference frequency by the required integer value to obtain the desired output frequency, which would not be obtainable using a fixed modulus prescaler counter. The circuit works as follows [1, 2]. On reset at the start of a $f_r$ period, $M$ and $A$ are set to their countdown values by the control block and $P+1$ is selected on the prescaler. Each $P+1$ cycles of $f_v$ from the VCO decrements a count from $A$ and $M$. When $A$ completes its countdown after $A(P+1)$ periods of $f_v$, the control block switches the prescaler modulus to $P$. Every $P$ cycles of $f_v$ counter $M$ is decremented and the process continues for the remaining $M - A$ counts, concluding a period of the reference frequency $f_r$. The total number of counts $N$ in one $f_r$ cycle is thus

$$N=(P+1)A+P(M-A)=A+PM \tag{5.1}$$

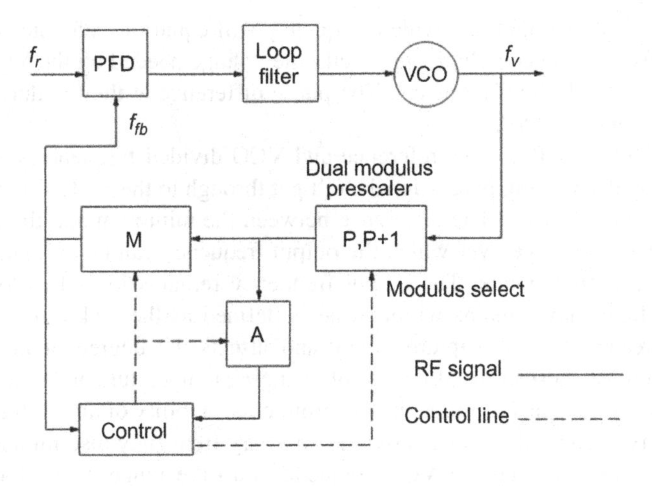


**FIG. 5.7**

Divider with dual modulus prescaler.

The choice of $P$, $A$, and $M$ can be shown by an example.

**Example 5.1**

**(1)** A transmitter lower edge frequency is 915 MHz and consecutive frequency channels are 1 MHz apart. The reference frequency is 1 MHz. The smallest $N$ is 915 MHz/1 MHz = 915.

**(2)** Choose $P$ under the constraint that $P^2 - P \leq N$. Let $P = 16$. For output frequency 915 MHz $N = P \times M_1$, $M_1 = N/P = 57.1875$. Set $M = 57$ and $A = 915 \bmod P = 16 \times 0.1875 = 3$. Remember that the prescaler and counters have to be chosen such that their maximum input frequency specifications are adhered to.

**(3)** Check result using Eq. (5.1). $A + PM = 3 + 57 \times 16 = 915$.

**(4)** Consecutive channels are obtained by incrementing $A$. Maximum $A$ for given $M$ and $P$ is $A \leq M$.

Essentially, switching the prescaler divisor gives an output that is a mixed number, or improper fraction, which, when multiplied by divider $M$ equals the total integer divider $N$. From the example above, the average $P$, $P_{av} = N/M = 915/57 = 16 + 3/57$. So, we may call the dual modulus prescaler a non-integer prescaler.

Many PLL integrated circuits have an output which is designed to be connected to the switching terminal on the dual modulus prescaler to facilitate its use. Complete VHF and UHF synthesizer integrated circuits, which include phase detector and VCO and reference dividers, also include a dual or even higher modulus prescaler.

There are two basic types of phase detectors (PD) in use. The analog multiplier (also a digital PD using simple combinatorial logic) outputs a voltage (or current) proportional to the phase difference between its two input signals. When these signals are not on the same frequency, the phase difference varies at a rate equal to the difference of the two frequencies. The negative feedback of the loop acts to lock the

VCO frequency such that its divided frequency will equal the reference frequency and the phase difference will create exactly the voltage needed on the VCO control input to give the desired frequency. The phase difference at the PD during lock is equal to or very near 90°.

Because of the LPF, if the reference and VCO divided frequencies are too far apart, the rapidly varying phase signal can't get through to the VCO control and the synthesizer will not lock. The difference between the minimum and the maximum frequencies of the range over which the output frequency can be brought into lock is called the capture range. The output frequency remains locked as long as the open loop frequency remains within what is defined as the lock range. The lock range is greater than the capture range, and affects the degree of immunity of the frequency synthesizer to influences of changing temperature and other environmental and component variations that determine the stability of an oscillator. When the analog type phase detector is used, an auxiliary frequency discriminator circuit may be required to bring the VCO frequency into the range where lock can be obtained.

The other type of phase detector is called a charge-pump PLL. It's a digital device made from sequential logic circuitry (flip-flops). It works by sending voltage (or current) pulses to the LPF, which smoothes the pulses before application to the VCO. The polarity of these pulses and their width depend on the relationship between the digital signal edge transitions of the reference oscillator and the divided down VCO signals at its input. In contrast to the analog phase detector, the charge-pump PLL always locks both of its input signals to zero phase difference between them. This device acts as both frequency discriminator and phase detector so an auxiliary frequency discriminator is not required.

The characteristics of the LPF determine the transient response of the synthesizer, the capture range, capture time, and rejection of noise originating in the VCO and the reference frequency source. Its design requires some compromises according to system requirements. For example, a relatively low cut-off frequency reduces spurious outputs due to modulation of the carrier by the reference source, and allows lower baseband frequencies with frequency modulation applied to the VFO. However, the average time to obtain lock is relatively long, and the output signal passes more noise originating in the VCO and its control circuit. A higher cut-off frequency allows the loop to lock in quicker but reference oscillator noise and spurious frequencies will be more prominent in the output. Filter design affects the use of frequency modulation, specifically analog. Changing the output frequency is exactly what the PLL is supposed to prevent, so to allow the modulation to get through, the spectrum of the modulating signal, applied to the frequency control input of the VCO, must be above the cut-off frequency of the LPF.

Although the output signal from a frequency synthesizer comes directly from an oscillator operating on the output frequency, it does contain spurious frequencies attributable to the PLL and in particular to the reference oscillator and its harmonics. A clean, noise-free signal is particularly important for the local oscillator of a superheterodyne receiver, since the spurious signals on the oscillator output can mix with undesired signals in the receiver passband and cause spurious responses and interference. Careful design and layout of the PLL is necessary to reduce them [3].

Worksheet **PLL.mcdx** facilitates calculating the component values of the loop filter for a charge pump PLL. Fig. 5.6 shows some variables used in Laplace transform analysis of phase lock loop behavior and the blocks with which they are associated [4].

### 5.1.4 Fractional-*N* synthesizer

As seen above, use of an integer divider in the PLL feedback loop obligates a reference frequency equal to (or a submultiple of) the channel width. This has some disadvantages. The necessity to have a reference frequency equal to, or lower than (in the case of the use of a prescaler), the channel frequency increment causes a high noise on the output carrier. This is because the reference signal and phase detector noise are amplified by the total division ratio in the feedback loop. Take, for example, 100 kHz channels in the 900 MHz band. At 915 MHz the divider counter is set to 915 MHz/100 kHz = 9150. Amplification of the phase detector noise floor is 20 log(9150) = 79 dB. In addition, a low reference frequency entails a low loop filter bandwidth, and consequently a relatively long lock time. The fractional-*N* synthesizer can use a reference frequency that is higher than the channel increment while accommodating fast switching between channels. This technique permits switching the output frequency in increments that are smaller than the reference frequency. It is the higher reference frequency, compared to a normal synthesizer with the same frequency increments that gives the fractional-*N* synthesizer its improved performance from the point of view of phase noise and switching speed.

A diagram of a fractional-*N* synthesizer is shown in Fig. 5.8. Note the similarity to the dual modulus prescaler of Fig. 5.7. One important difference is the addition of an accumulator circuit whose input comes from the output of the divider chain connected to the input of the PFD and its output intermittently increments the feedback divider. In order to select consecutive narrow channels, the fractional-*N* synthesizer effectively multiplies the reference frequency by a mixed number—a whole number plus fraction—which is the average of the integer divisors in the feedback loop, just as the dual modulus prescaler outputs an average mixed divisor to the input of an integer divisor. In the latter case, the PLL remains locked all the time, seeing a feedback frequency, $f_{fb}$, equal to the reference frequency $f_r$, whereas in the fractional-*N* synthesizer $f_{fb}$ dithers between two frequencies whose average is $f_r$. Let $\Delta f$ stand for the channel spacing, $F = f_r/\Delta f$ (must be an integer), $N = \lfloor f_v/f_r \rfloor$ (greatest integer less than or equal to $f_v/f_r$), $K = (f_v \bmod f_b)/f_r$. Let $N' = f_v/f_r$, the average divisor. Then

$$N' = \frac{K(N+1)+(F-K)N}{F} = N + \frac{K}{F} \tag{5.2}$$

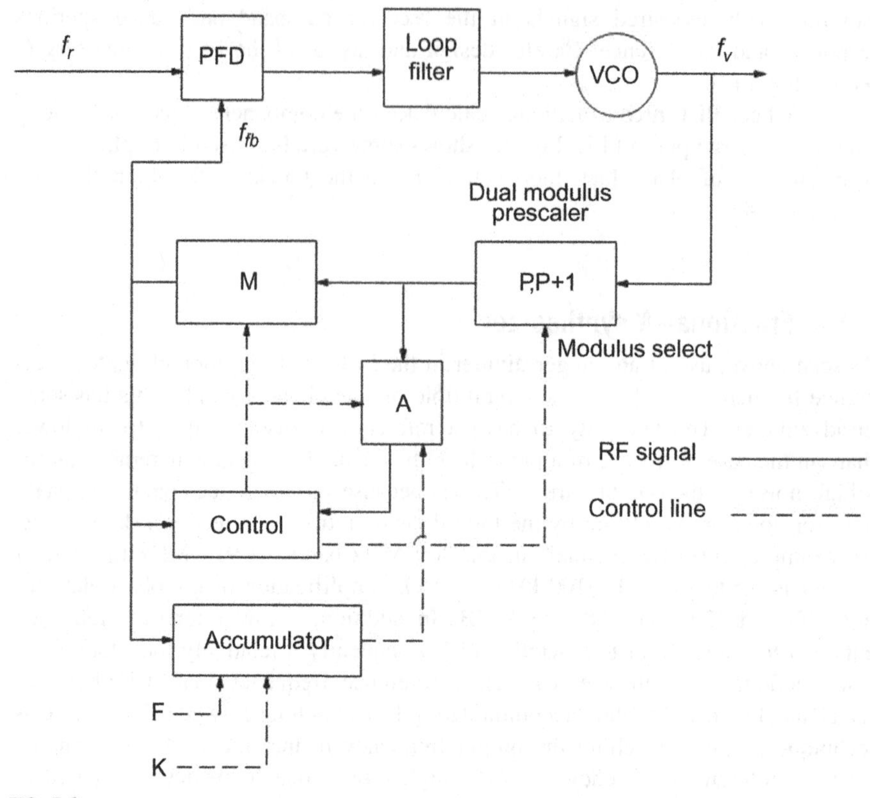


**FIG. 5.8**

Fractional-*N* synthesizer.

A simplified explanation of how this is carried out in Fig. 5.8 is as follows. A counter in the accumulator block is loaded with *F* and is advanced on each cycle of $f_{fb}$. The accumulator circuitry increments *A*, which effectively increments *N*, *K* times per period of $\Delta f$, which is *F* periods of $f_{fb}$. During the remainder *F*-*K* periods of $\Delta f$, the divisor is *N* [2, 5]. Here is an example of choosing parameters and operation.

**Example 5.2**

A transmitter operates on 100 kHz channels on the 900 MHz band. $\Delta f = 100$ kHz. Reference frequency $f_r = 1.6$ MHz, so $F = 16$. At 915 MHz, $N' = 915/1.6 = 571.875$, $N = 571$, $K = (915 \bmod 1.6)/0.1 = 14$. During 14 periods of 1.6 MHz, the loop feedback divisor is 572 and during two periods of 1.6 MHz it is 571. Average $f_{fb}$ is

$$(f_{fb})_{average} = \frac{14}{16} \cdot \frac{915\text{MHz}}{572} + \frac{2}{16} \cdot \frac{915\text{MHz}}{571} = 1.6\text{MHz} = f_r$$

During the division process, the jumps in the feedback loop divider cause spurious disturbances at the output of the PFD, resulting in spurs in the spectrum of the output of the VCO. Details of the spurious signals and methods of alleviation and compensation are described in Refs. [2, 3, 5].

### 5.1.5 Direct digital synthesis

Direct digital synthesis is a method for radio frequency wave generation that allows precise digital control over frequency, phase and amplitude is. A direct digital synthesizer (DDS) generates waveforms digitally instead of being based on an analog oscillator as in the methods described above. Digital words representing samples of a sine wave are applied to a digital-to-analog converter, whose output is LPF.

Fig. 5.9 is a numerically controlled oscillator (NCO), the basis of the DDS. It shows both cosine and sine digital outputs, but when quadrature signals are not required, the sine output need not be included. The inclusion of digital-to-analog converters followed by anti-alias filters make up the DDS.

The NCO operates as follows. The phase accumulator is a binary counter with $N$ bits that is incremented periodically by a master clock of frequency $f_S$. An example of the output of the phase accumulator is plotted in Fig. 5.10A. On each clock pulse, the contents of the frequency register, $\Delta\phi$, are added to the accumulator. The period of the generated frequency $f_O$ is determined by the accumulator overflows. The output frequency is

$$f_O = \frac{\Delta\phi \cdot f_S}{2^N} \tag{5.3}$$

The phase of the output is determined by the contents of the phase register. The linearly stepped digital accumulator output can be changed to a digital sine or cosine output using look-up tables in read-only memory (ROM). These outputs, in turn, are input to digital-to-analog converters, followed by anti-alias filters, to produce analog signals. Fig. 5.10B shows the sine output before filtering. The output frequency is generally limited to 40% of the sampling frequency. The minimum frequency and frequency increments are $fs/2N$ and possible phase increments are $2\pi/2^N$.

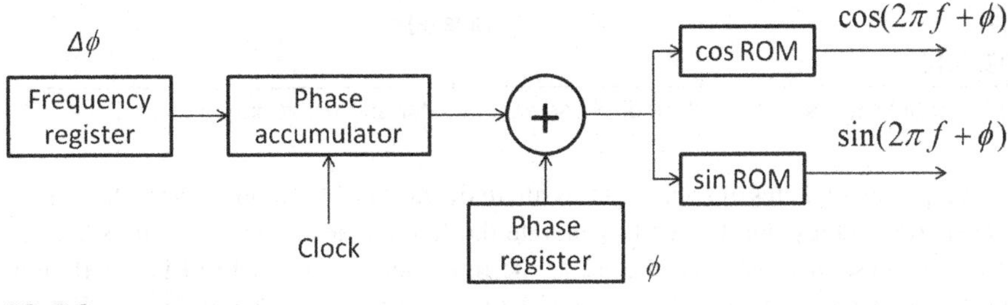


**FIG. 5.9**

Numerically controlled oscillator (NCO).

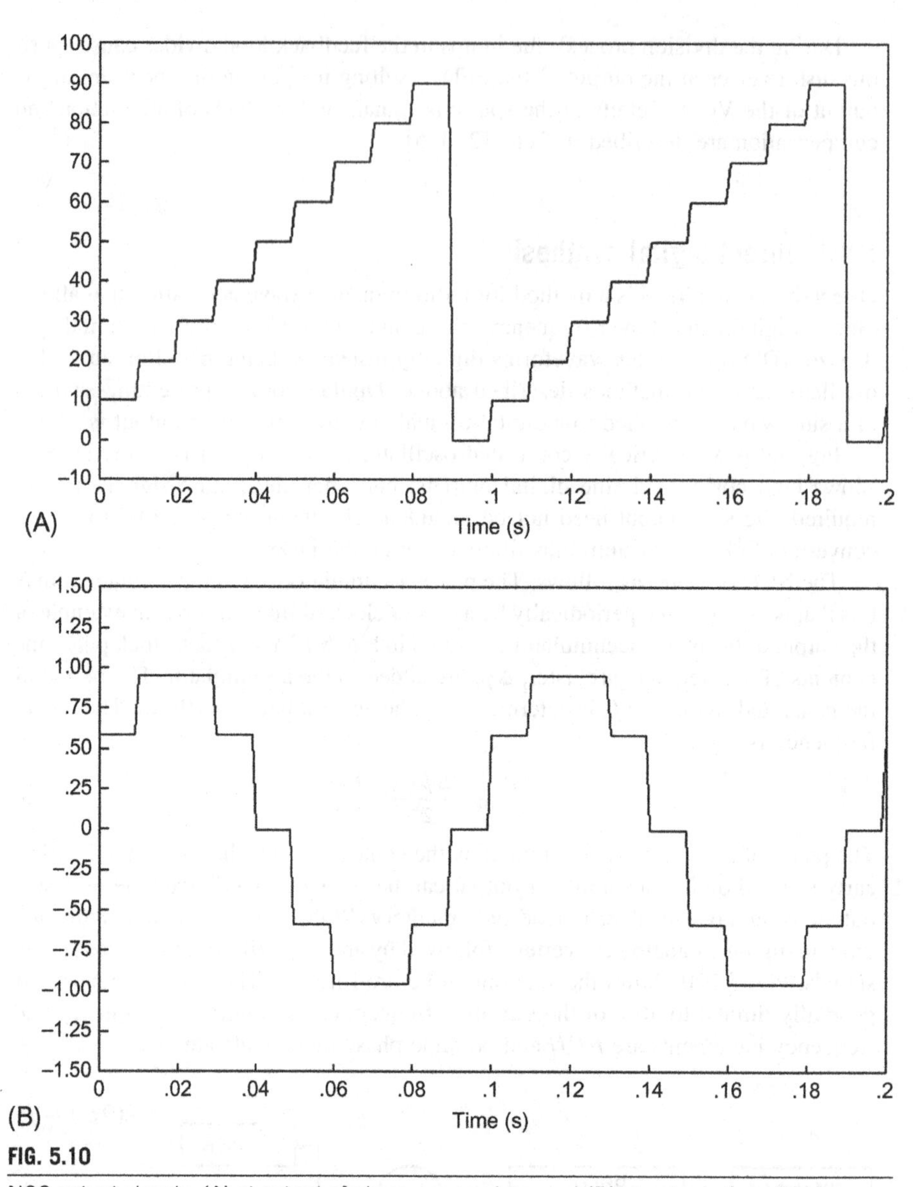


**FIG. 5.10**

NCO output signals: (A) at output of phase accumulator and (B) at output of sin ROM.

Output frequencies of DDS systems are in the range of fractions of hertz to tens of megahertz, so they don't directly generate the frequencies required for most short-range devices. However, when used as the reference frequency in PLLs, UHF and microwave signals can be established with high accuracy and resolution. Amplitude, frequency and phase modulation can be applied directly by software to the generated signals, which gives great flexibility where complicated and precise waveforms are required.

## 5.2 Modulation

In this section we look at the most common methods of implementing amplitude shift keying (ASK) and frequency shift keying (FSK) in short-range transmitters. Details of PSK were discussed in Chapter 4.

### 5.2.1 ASK

ASK can be applied to *LC* and SAW oscillators by keying the oscillator stage bias, as shown in Fig. 5.11. Resistor *Rb* is chosen to provide the proper bias current from the code generator, and by varying its value the power from the stage can be regulated in order to meet the requirements of the applicable specifications.

Switching the oscillator on and off works satisfactorily for low rates of modulation—up to around 2 kbits/s. At higher rates, the delay time needed for the oscillations to turn on will distort the transmitted signal and may increase the error rate at the receiver. In this case a buffer or amplifier stage following the oscillator must be keyed while the oscillator stage runs continuously. The problem in this situation with low-power transmitters is that the continuous signal from the oscillator may leak through the buffer stage when it is turned off, or may be radiated directly, reducing the difference between the ON and OFF periods at the receiver. To reduce this effect, the oscillator layout must be as compact as possible and its coil in particular must be small. In many cases the oscillator circuit will have to be shielded.

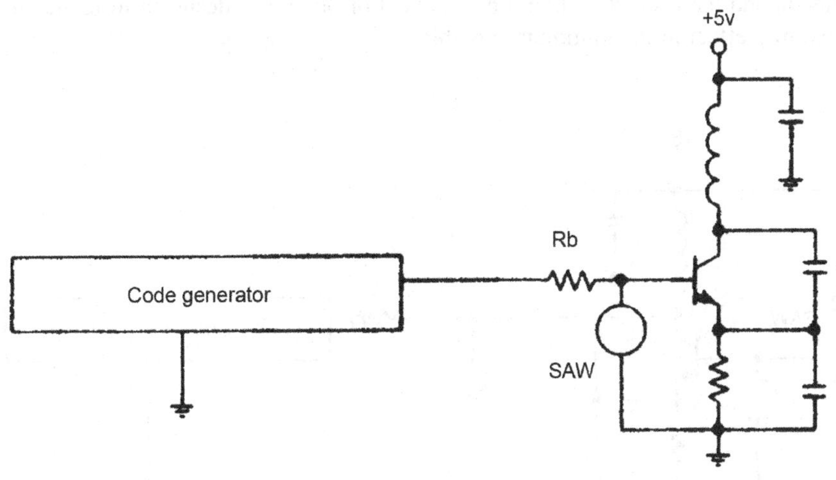


**FIG. 5.11**

Amplitude shift keying of SAW oscillator.

## 5.2.2 FSK

FSK of a SAW oscillator can be achieved by switching a capacitance in and out of the tuned circuit in step with the modulation. A simple and inexpensive way to do it is shown in Fig. 5.12. You can also use a transistor instead of a diode to do the switching, or use a varicap diode, whose capacitance changes with the applied voltage. The effect of changing the capacitance in the tuned circuit is to pull the oscillation frequency away from its static value, which is held by the SAW resonator. Deviation of several thousand hertz can be obtained by this circuit but the actual value depends on the insertion loss of the individual resonator. As we have seen in Section 5.1, SAW-controlled devices can vary as much as 30 kHz due to component aging and proximity effects. This forces compromises in the design of the receiver frequency discriminator, which must accept variable deviation of say 5 to 10 kHz while the center frequency may change 30 kHz. Thus, the performance of FSK SAW devices is not likely to realize the advantages over ASK that are expected, and indeed FSK is not often used with SAWs.

An improved SAW oscillator FSK developed by RF Monolithics [6] is shown in Fig. 5.13. This circuit is not based on indefinite pulling, as is the design in Fig. 5.8, but on varying the resonance of the SAW as seen by the circuit, because of the insertion of a small-valued capacitance in series with the SAW. In the RFM circuit, this capacitance is that of the reverse connected transistor (emitter and collector terminal are reversed) when it is biased off. Reversing the transistor connection decreases its storage time, allowing a higher switching rate. RFM (now Murata) states that the circuit is patented and that licensing is required for using it.

If FSK is used with a SAW resonator, the oscillator must be buffered from the antenna and the oscillator should be shielded or otherwise designed to reduce the proximity effect to the minimum possible.

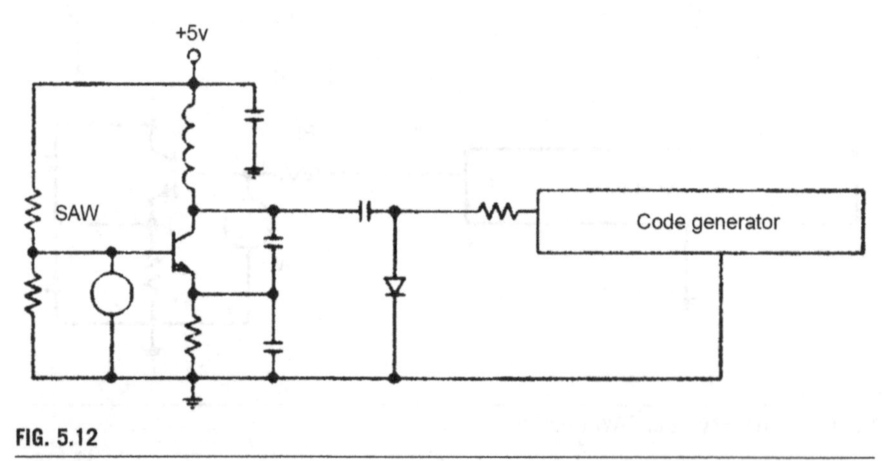


**FIG. 5.12**

Frequency shift keying of SAW oscillator.

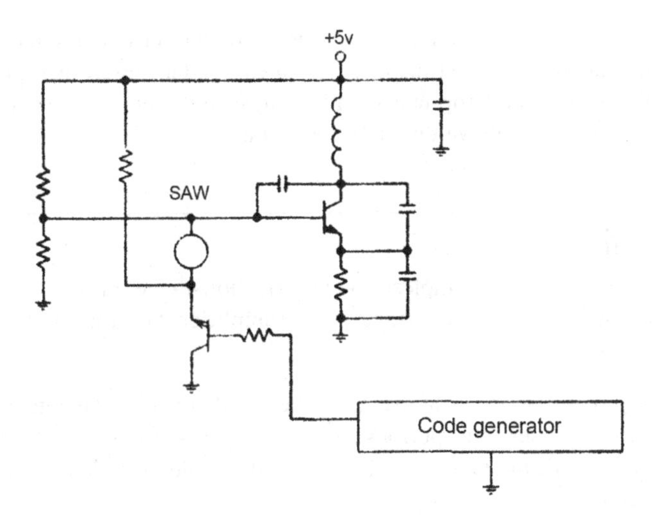


**FIG. 5.13**

Improved FSK modulator.

A crystal oscillator's frequency can be similarly pulled by detuning its tuned circuit or by switching a capacitor or inductor across or in series with the crystal terminals. A frequency deviation of several hundred hertz is possible. The frequency multiplier stages needed to step up the frequency to UHF also multiply the deviation. The required frequency deviation for effective demodulation is at least half the bit rate, so higher deviation and consequently more multiplication stages are needed for high data rates.

The high deviation required for high data rates is more readily obtainable when a PLL is used in a frequency synthesizer. Voltage pulses representing the modulation are applied to the VCO control input, together with the output of the LPF following the phase detector, as shown in Fig. 5.14. The modulation spectrum must be higher than the cut-off frequency of the filter to prevent the PLL from correcting the output

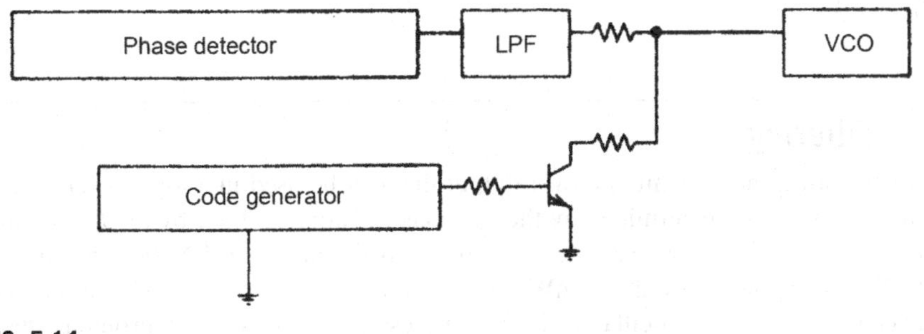


**FIG. 5.14**

FSK with synthesizer.

frequency and canceling the modulation. NRZ modulation is not suitable since a string of ones cannot be held at the offset frequency. Manchester or Biphase Mark or Space modulation is appropriate for FSK since its frequency spectrum is well defined and is displaced above one half the bit rate.

## 5.3 Amplifiers

As we have seen earlier, the simplest low-power, short-range transmitters may consist of a single oscillator transistor stage. An RF amplifier stage may be added for the following reasons:

- to reduce proximity effects by isolating the oscillator from the antenna,
- to make up for losses in a spurious response rejection filter,
- to provide for ASK modulation when it is not desirable to modulate the oscillator directly, and
- to increase power output for greater range.

Proper input and output matching is very important for achieving maximum power gain from an amplifier. In order to design the matching circuits, you have to know the transistor input and output impedances. Manufacturer's data sheets give the linear circuit parameters for finding these impedances. However, in many cases, the amplifier stage does not operate in a linear mode so these parameters are not accurate. Trial and error may be used to find good matching, or a non-linear SPICE simulator may be used to determine the matching components. Non-linear amplification is most efficient but linear amplification is necessary for analog AM systems and digital phase-modulated signals for preventing bandwidth spread due to signal amplitude distortion.

In order to design matching networks, the output impedance of the previous stage, which may be the oscillator, must be known, and also the input impedance of the following stage, which may be a filter or an antenna. Low-power transistor amplifiers typically have low-impedance inputs and high-impedance outputs. RF amplifier output impedance and impedance-matching circuits were discussed in Chapter 3.

## 5.4 Filtering

In many cases, the maximum power output that can be used in a low-power transmitter is essentially determined by the spurious radiation and not by the maximum power specification. For example, the European specification EN300 220 allows 10 mW output power but only 250 nW harmonics, a difference of 46 dB. The harmonics radiated directly from an oscillator will rarely be less than 20 dB down so an output filter, before the antenna, must be used in order to achieve the full allowed power.

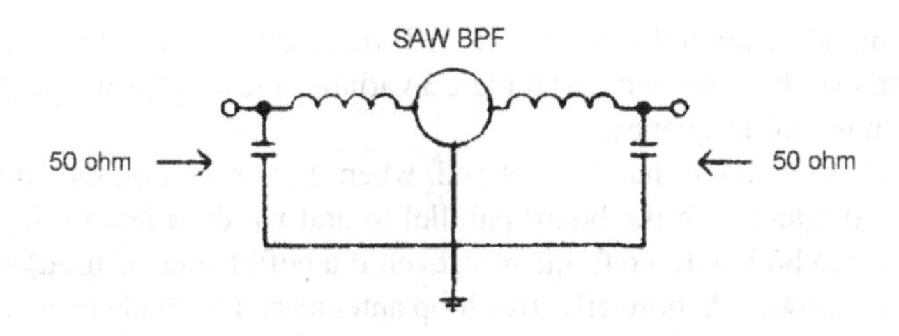


FIG. 5.15

SAW bandpass filter matching.

An effective filter for a transmitter is a SAW-coupled filter. It has a wide enough pass-band and very steep sides. However, to be effective, the oscillator must be shielded to prevent its radiation from bypassing the filter. The filter has an insertion loss of several dB. It has relatively high input and output impedances and can be matched to a low impedance, usually 50 Ω, by the circuit shown in Fig. 5.15. Values for the matching circuit components are given by the device manufacturer.

## 5.5 Antennas

We reviewed the theory of various antennas used in short-range wireless operations in Chapter 3. Here we discuss practical implications of their use.

Probably the most popular antenna for small remote control and sensor transmitters used in security and medical monitoring systems is the loop antenna, discussed in Chapter 3. It's compact, cheap and relatively non-directional. A typical matching circuit for a loop antenna is shown in Fig. 5.16. Resonant capacitance for a given size loop is calculated in the Loop Antenna worksheet **Loop.mcdx**. C1, C2, and C3 contribute to this capacitance. Some trial and error is needed to determine their

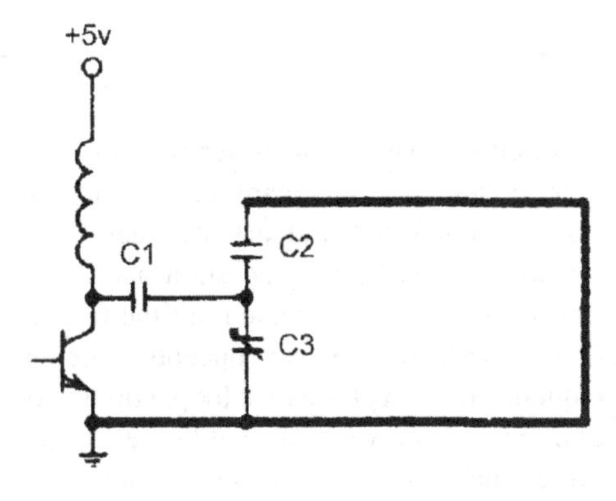


FIG. 5.16

Loop antenna matching circuit.

values. C1 should be around 1 or 2 pF at UHF frequencies if the output transistor is the oscillator, to reduce detuning. Making C3 variable allows adjustment of the transmitter to its nominal frequency.

The loop is sensitive to hand effect and, when it is printed on the circuit board, installing the product with the board parallel to and within a few millimeters of a metal surface can both pull the frequency (even if a buffer stage is used) and greatly reduce radiated power. A more effective loop antenna can be made from a rigid wire loop supported above the circuit board.

Helical patch and inverted F antennas are described in Chapter 3. The helical antenna has higher efficiency than the loop and doesn't take up printed circuit board space, but it's also affected more by the hand effect and therefore may not be suitable for handheld devices. The patch antenna is practical for the 900-MHz ISM band and higher, although variations such as the quarter-wave patch, which has one edge grounded, and the trapezoid patch, resonate with smaller dimensions and may even be considered for 433 MHz.

Other antenna forms are variations on monopole and dipole antennas and involve printing the antenna element or elements in irregular shapes around the circuit board. Inductive matching components are needed to resonate the antenna. Antenna performance is affected by proximity of circuit printed conductors and components, including batteries.

While most of the reduced-size antennas are very inefficient, that fact alone doesn't affect the range of the product, since the specifications limit radiated power, not generated power. Thus, the transmitter can be designed to have excess power to make up for the loss of power in the antenna. However, antenna directivity in three dimensions is a limiting factor for products whose position in respect to the receiver cannot be controlled. Another problem is spurious radiation, which is more difficult to control when an inefficient antenna is used.

Detailed descriptions of compact antennas for low-power transmitters are given in application notes by RF Murata [7] and from Micrel [8].

## 5.6 Summary

In this chapter we described the basic transmitter building blocks—the oscillator, modulator, RF amplifier and antenna. There are many options for oscillator configurations. We discussed advantages and tradeoffs between LC, SAW resonator and crystal control, then examined frequency generation using frequency synthesizers and direct digital synthesis. Synthesizers create multiple frequencies using a single stable oscillator. First we described the basic concepts of generating integer multiples of an oscillator frequency using a phase lock loop, voltage (or current) controlled oscillator and a digital divider. Then we looked at the refinements of the dual modulus prescaler and the fractional-$N$ synthesizer. We saw how another approach, direct digital synthesis, creates source frequencies directly with high precision and can combine modulation with signal generation through software.

ASK and FSK modulation methods depend on data rate and oscillator stability, and amplifier stages may be needed for buffering. Frequency drift due to proximity effects is a particular problem with inexpensive low power devices. The necessity of output filtering to reduce spurious radiation and consequences of using reduced-size antennas were discussed in this chapter.

## References

[1] Determining Valid Divide Ratios Using Peregrine PLL Frequency Synthesizers, Peregrine Semiconductor Application Note AN38, 2015.

[2] Fractional/Integer-N PLL Basics, Texas Instruments Technical Brief SWRA029, 1999.

[3] B. Goldberg, Digital Frequency Synthesis Demystified, LLH Technology Publishing, Eagle Rock, VA, 1999.

[4] LMX1501A/LMX1511 PLLatinumTM 1.1 GHz Frequency Synthesizer for RF Personal Communications, National Semiconductor Specification, 1995.

[5] AN-1879, Fractional N Frequency Synthesis, Texas Instruments Application Report SNAA062A, 2008. revised April 2013.

[6] Murata, A SAW-Stabilized FSK Oscillator, https:/wireless.murata.com/media/products/apnotes/fskfinal.pdf, 2017. Accessed 18 November 2017.

[7] K. Smith, Antennas for Low Power Wireless Applications, Murata Manufacturing Co., Ltd, 2009.

[8] C. Ribeiro, Small PCB Antennas for Micrel RF Products, Application Note 52, Micrel Inc, 2007.

CHAPTER

# Receiver and digital radio architectures

# 6

It is the receiver that ultimately determines the performance of the wireless link. Given a particular transmitter power, which is limited by the regulations, the range of the link will depend on the sensitivity of the receiver, which is not legally constrained. Of course, not all applications require optimization of range, as some are meant to operate at short distances of several meters or even centimeters. In these cases simplicity, size, and cost are the primary considerations. By far most short-range receiver designs today use the superheterodyne topology, which was invented in the 1920s. However, anyone who starts designing a short-range wireless system should be aware of the other possibilities so that an optimum choice can be made for a particular application.

## 6.1 Tuned radio frequency (TRF)

The tuned radio frequency (TRF) receiver is the simplest type conceptually. A diagram of it is shown in Fig. 6.1. The antenna is followed by a band-pass filter (BPF) or turning circuit for the input RF frequency. The signal is amplified by high-gain RF amplifier stages and then demodulated, typically by a diode detector. The digital baseband signal is reconstructed by a comparator circuit.

The gain of the RF amplifier is restricted because very high gain in a UHF amplifier would be subject to positive feedback and oscillation due to the relatively low impedance of parasitic capacitances at RF frequencies. An advantage of this type of receiver is that it doesn't have a local oscillator or any other radiating source so it causes no interference and doesn't require FCC or European-type approval.

A clever variation of the TRF receiver is the amplifier-sequenced hybrid (ASH) receiver developed by RFM. A basic block diagram is shown in Fig. 6.2 [1].

In the ASH receiver, two separate RF gain stages are used, each of them switched on at a different time (ASH). The gain of each stage is restricted to prevent positive feedback as mentioned above, but since the incoming signal goes through both stages, the total effective gain is the sum of the gains (in decibels) of each stage. While one stage is on the other is off, and the signal that was amplified in the first stage has to be retained in a delay or temporary storage element until the first stage is turned off and the second stage is turned on. The delay element is a SAW filter with a delay of approximately 0.5 μs.



Short-range Wireless Communication. https://doi.org/10.1016/B978-0-12-815405-2.00006-3
© 2019 Elsevier Inc. All rights reserved.

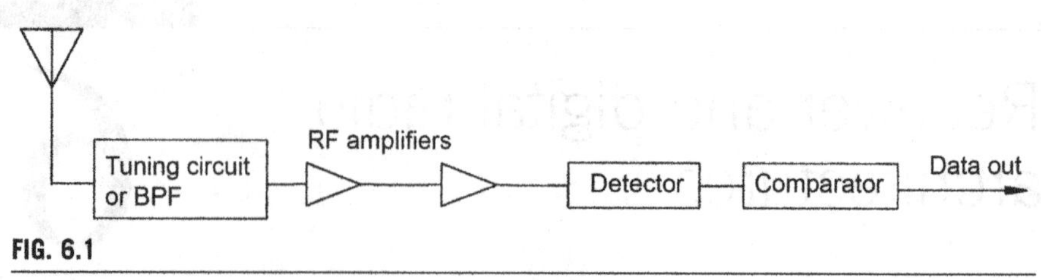


**FIG. 6.1**

Tuned radio frequency amplifier.

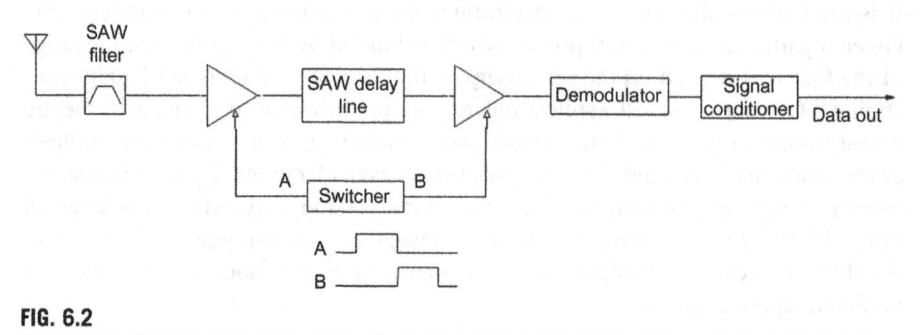


**FIG. 6.2**

ASH receiver.

*After D.L. Ash, SAW-Based Hybrid Transceivers in SLAM Packaging With Frequency Range From 200 to 1000 MHz, AN41 11/21/16, Murata Manufacturing Co. Ltd., 2016.*

The selectivity of the ASH receiver is determined principally by the SAW BPF that precedes the first RF amplifier and to a lesser degree by the SAW delay line.

The TRF receiver is a good choice for wireless communication using ASK at distances of several meters, which is adequate for a wireless computer mouse or other very short-range control device. Longer ranges are achieved by the ASH design, whose sensitivity is as good as the best superregenerative receivers and approaches that of superheterodyne receivers using SAW-controlled local oscillators. The TRF and ASH receivers have low current consumption, on the order of 3 to 5 mA, and are used in battery operated transceivers where the average supply current can be reduced even more using a very low-current sleep mode and periodic wake up to check if a signal is being received.

## 6.2 Superregenerative receiver

For many years the most widely used receiver type for garage door openers and security systems was the superregenerative receiver. It has relatively high sensitivity, is very inexpensive and has a minimum number of components. Fig. 6.3 is a schematic diagram of such a receiver.

The high sensitivity of the superregenerative receiver is obtained by creating a negative resistance to cancel the losses in the input tuned circuit and thereby

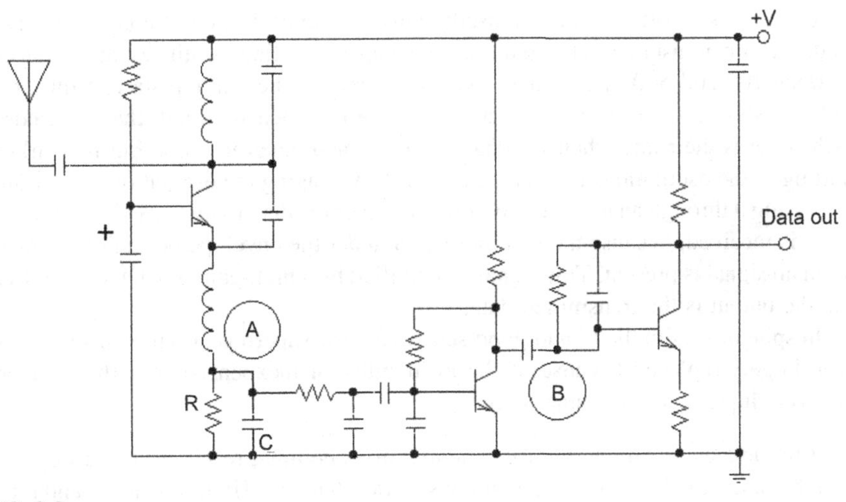


FIG. 6.3

Superregenerative receiver.

increasing its $Q$ factor. This is done by introducing positive feedback around the input stage and bringing it to the verge of oscillation. Then the oscillation is quenched, and the cycle of oscillation buildup and quenching starts again. The process of introducing positive feedback and then quenching it is controlled by periodically altering the bias voltage on the transistor input stage, either by using a separate low-frequency oscillator or by dynamic effects within the input stage itself (self-quenched).

Fig. 6.4 shows the principle of operation of the superregenerative receiver. The two curves refer to points A and B on the schematic in Fig. 6.3. Like any oscillator when it is turned on, random noise starts the buildup of oscillations when there is no signal, shown as "0" in the diagram. After the oscillations have been established, the

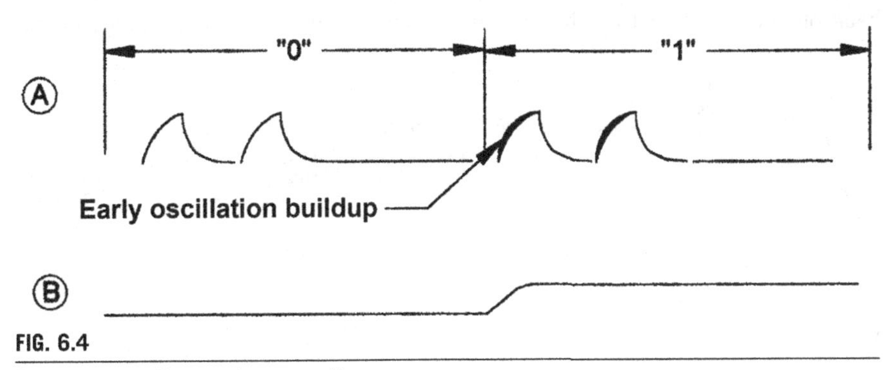


FIG. 6.4

Superregenerative receiver operation.

circuit reaches cutoff bias and the oscillations stop, until the conditions for positive feedback are reinstated. This buildup and cut-off cycling continues at a rate of between 100 and 500 kHz. When a signal is present, the buildup starts a little bit early, as shown as "1" in the diagram. In the self-quenched circuit, the area under each pulse is the same when a signal is present and when it is not, but the earlier starting of the oscillations raises the pulse rate. Averaging the oscillation pulse train by passing it through an integrator or low-pass filter results in a higher DC level when a "1" is received, because there is more area under the envelope per unit time than when no signal is present. This signal is amplified by a high-gain baseband amplifier and the output is the transmitted data.

In spite of its simplicity and good sensitivity, the superregenerative receiver has been largely replaced because of the availability of inexpensive superheterodyne receiver chips. Its disadvantages are

- it re-radiates broad-band noise centered on its nominal receiving frequency;
- it has a relatively broad bandwidth of several MHz on UHF frequencies, and thus is sensitive to interference;
- dynamic range is limited; and
- it is usable only with ASK modulation.

## 6.3 Superheterodyne receiver

The superheterodyne receiver is the most common configuration for radio communication. Its basic principle of operation is the translation of all received channels to an intermediate frequency (IF) band where the weak input signal is amplified before being applied to a detector. The high performance of the receiver is due to the fact that amplification and bandpass filtering of input signals are done at one or more frequencies that do not change with the input tuning of the receiver, and at the lower IF, greater amplification can be used without causing instability.

Fig. 6.5 shows the basic architecture of a superheterodyne receiver. The antenna is followed by a low-noise amplifier (LNA) that assures a low noise figure. Its wide band output goes to a BPF that passes all signals within the tuning range of the

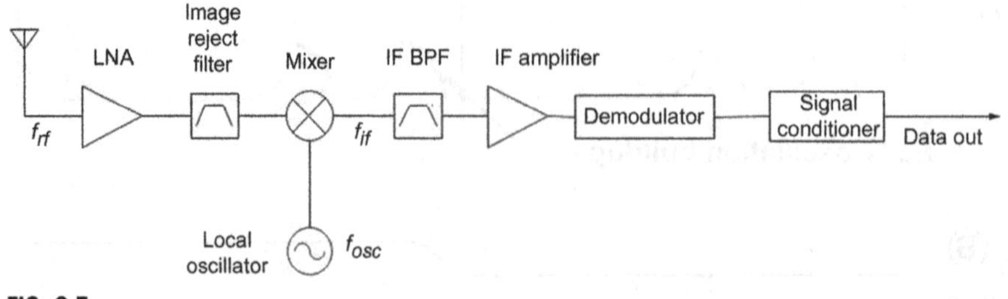


**FIG. 6.5**

Superheterodyne receiver.

receiver and rejects image frequencies that could interfere with the desired signal. The mixer multiplies the RF signals by a tunable signal from a local oscillator and outputs the sum and the difference of $f_{rf}$ and $f_{osc}$. The IF BPF at its output determines receiver selectivity and sensitivity. In VHF and higher frequency receivers it is the lower, difference frequency signal ($f_{rf} - f_{osc}$) that is retained. Receivers in the HF (high frequency 3 to 30 MHz) and LF (low frequency 0.3 to 3 MHz) bands may use an IF that is higher than the received frequency. In both cases the local oscillator frequency is tuned to keep the IF constant over changes in the received frequency. The result is an IF signal that has all of the characteristics of the input RF signal except for the shift in frequency. Most of the amplification in the receiver is provided by the IF amplifier. Its output must be high enough for demodulation in the detector. Any type of modulation may be used. A low-pass filter (not shown in Fig. 6.5) that gives additional noise reduction follows the detector. In a digital data receiver, a signal-conditioning stage converts the baseband received signal to binary levels for digital signals. The signal conditioner in an analog receiver is often a signal expander for audio analog signals that were compressed in the transmitter to improve their dynamic range.

Image frequency rejection is essential in a superheterodyne receiver. The IF BPF should be as narrow as possible in order to reduce the noise without affecting the bandwidth required by the modulation components of the signal. The lower the IF frequency, the narrower the BPF can be. However, using a low IF frequency means that the oscillator frequency must be close to the received RF frequency. Since the IF frequency is the absolute value of the difference between the received frequency and the oscillator frequency, two received frequencies can give the same IF frequency—one at $f_{lo} + f_{if}$ and the other at $f_{lo} - f_{if}$. Fig. 6.6 shows the relationship between $f_{rf}$, $f_{lo}$, $f_{image}$, and $f_{if}$. In the diagram, the local oscillator frequency is higher than the desired input signal, so the span of the image reject BPF must include all frequencies that are above the range of desired signal frequencies by twice the IF.

It is the function of the input BPF to reject the undesired image frequency. Even if there is no interfering signal on the image frequency, the noise at this frequency will get through and reduce the signal-to-noise ratio (SNR). To reduce input noise at the

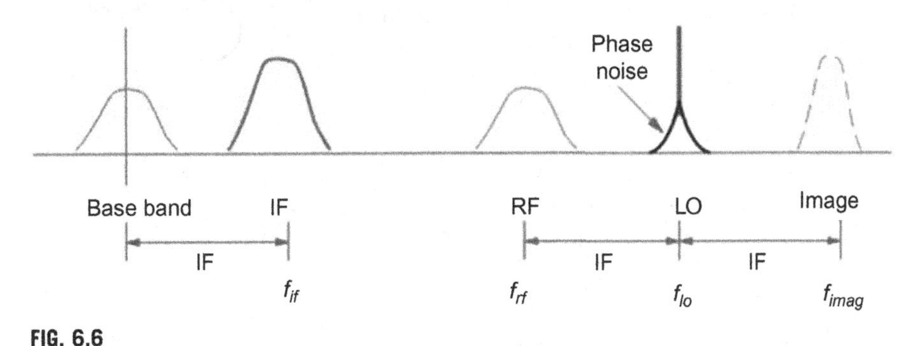


**FIG. 6.6**

Frequency translation and image frequency in a superheterodyne receiver.

image frequency, including the circuit noise of the LNA, the image frequency BPF should be included between the RF amplifier and the mixer (as shown in Fig. 6.5). When a low IF is used in order to obtain a narrow passband, the input BPF may not reject the image frequency, thereby increasing the possibility of interference. In this case, an image reject mixer can be used to cancel out the image frequency signal. An image reject mixer architecture is shown in Fig. 6.7. Note the 90° phase lag introduced in the quadrature mixer output branch, after the low-pass filter. When the desired RF signal is below the local oscillator frequency, as depicted in Fig. 6.6, the image frequency signals at IF in the inphase (upper) and quadrature (lower) branches cancel out, leaving the desired RF signal at the image reject mixer IF out. If the desired signal is above the local oscillator frequency, the −90° phase shifter should be inserted in the upper inphase branch, or the adder should be changed to a subtractor. The configuration of Fig. 6.7, and an alternate architecture which may give better image rejection, are described in Ref. [2].

Multiple frequency conversion is another way to reduce the response to image frequencies and have a low IF for effective filtering. UHF superheterodyne receivers, and HF and LF receivers whose IF is above the received signal frequency, may employ dual or triple conversion. A dual conversion superheterodyne receiver has a first mixer and an IF high enough to reject the image frequency using a simple BPF, followed by another mixer and IF at a low frequency for effective filtering.

The IF frequency must be chosen according to the image rejection and filtering considerations discussed above, but the final choice often is a frequency for which

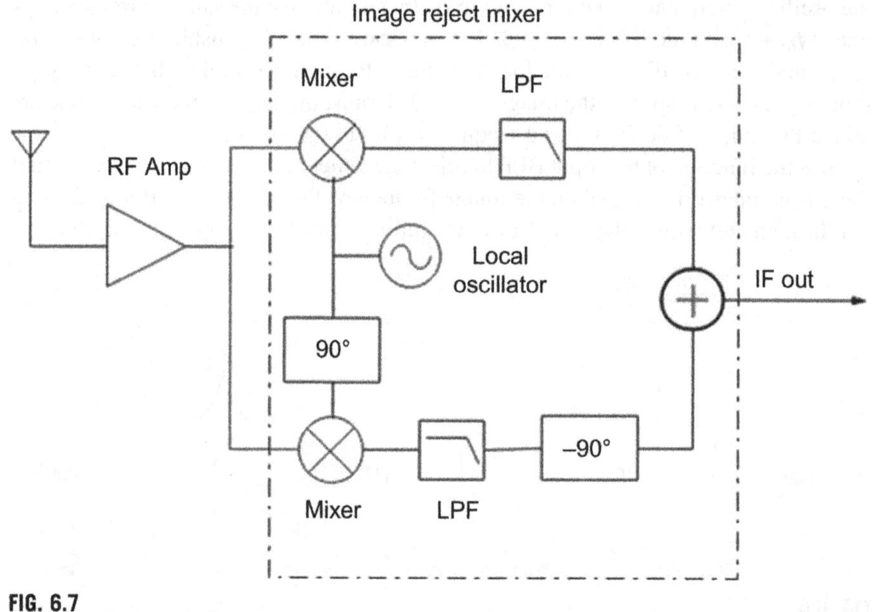


FIG. 6.7

Image frequency reject mixer.

standard BPF and frequency determining components are readily available. The two IF frequencies 455 kHz and 10.7 MHz, for which ceramic IF filters are standard catalog items, are the most commonly used for AM and FM broadcast bands. Pairs of SAW devices exist for unlicensed ISM bands, one for the receiver local oscillator and one for the transmitter, whose frequency difference is on those IF frequencies,

## 6.4 Direct conversion receiver

The direct conversion receiver, also called zero-IF receiver, is similar to the superheterodyne in that a local oscillator and mixer are used, but in this case the IF frequency is zero. The image frequency, a potential problem in the superheterodyne, coincides with the desired signal, so it is no issue in this topography. Very high-gain baseband amplification is used, and a baseband low-pass filter achieves high sensitivity with high noise and adjacent channel interference rejection. On the negative side, the local oscillator is at the same frequency as the received signal, so there is a potential for self-interference, and for interference with close-by receivers tuned to the same frequency. Design and layout are very important to limit radiation from the local oscillator and prevent leakage back through the mixer and RF amplifier to the antenna. Also, because of the very narrow bandwidth, the crystal-controlled local oscillator frequency must be accurate and stable.

A block diagram of an frequency-shift keying (FSK) direct conversion data receiver is shown in Fig. 6.8. The output of an RF amplifier is applied to two mixers. A local oscillator output at the same frequency as the RF signal is applied directly to one mixer (Inline). The other mixer receives the local oscillator signal after being shifted inphase by 90° (quadrature). The "*I*" and "*Q*" outputs of the mixers are each passed through low-pass filters and limiters and then applied to a phase detector for demodulation. A delay, optimally of ¼ the period of the frequency deviation, is inserted in the *Q* path before the phase detector. The three low-pass filters LPF1, LPF2, and LPF3 attenuate the double frequency terms that are included in the mixer

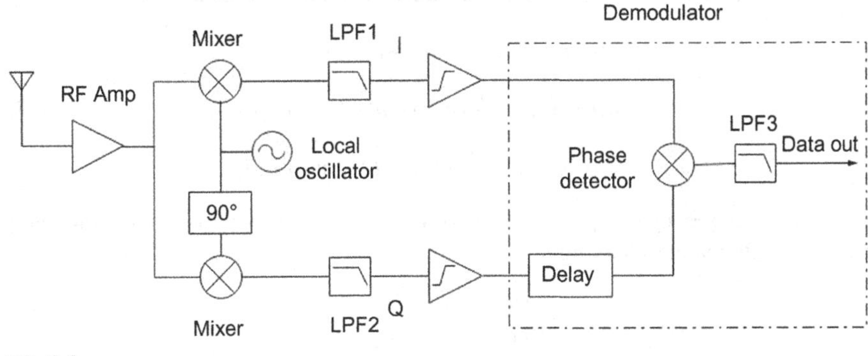


**FIG. 6.8**

Direct conversion receiver.

outputs. The frequency shift keyed signals appear with opposed relative phase at the phase detector, giving a binary MARK or SPACE output according to whether the input signal is higher or lower than the local oscillator frequency. This can be seen as follows:

Let input MARK and SPACE signals be

$$\begin{aligned} S_M &= \cos\left[2\pi(f+d)t\right] \\ S_S &= \cos\left[2\pi(f-d)t+\theta\right] \end{aligned} \tag{6.1}$$

where $f$ is the nominal receiver frequency and $d$ is the frequency deviation of the FSK signal. $\theta$ is a random phase shift. The peak amplitudes are not shown since they are not relevant to the results.

The quadrature oscillator signals to the mixers are

$$\begin{aligned} LO_I &= \cos(2\pi ft) \\ LO_Q &= \sin(2\pi ft) \end{aligned} \tag{6.2}$$

The mixer outputs when a MARK is sent are:

$$\begin{aligned} I_M &= S_M \cdot LO_I \\ &= \cos\left[2\pi(f+d)t\right] \cdot \cos(2\pi ft) \end{aligned} \tag{6.3}$$

$$\begin{aligned} Q_M &= S_M \cdot LO_Q \\ &= \cos\left[2\pi(f+d)t\right] \cdot \sin(2\pi ft) \end{aligned} \tag{6.4}$$

Similarly when a SPACE is sent

$$\begin{aligned} I_S &= S_S \cdot LO_I \\ &= \cos\left[2\pi(f-d)t+\theta\right] \cdot \cos(2\pi ft) \end{aligned} \tag{6.5}$$

$$\begin{aligned} Q_S &= S_S \cdot LO_Q \\ &= \cos\left[2\pi(f-d)t+\theta\right] \cdot \sin(2\pi ft) \end{aligned} \tag{6.6}$$

Use the following trigonometric identities as needed to write the results in Eqs. (6.3) through (6.6) write the product terms as individual sine or cosine terms.

$$\begin{aligned} \cos(\alpha)\cos(\beta) &\equiv \frac{1}{2}\left[\cos(\alpha-\beta)+\cos(\alpha+\beta)\right] \\ \sin(\alpha)\sin(\beta) &\equiv \frac{1}{2}\left[\cos(\alpha-\beta)-\cos(\alpha+\beta)\right] \\ \sin(\alpha)\cos(\beta) &\equiv \frac{1}{2}\left[\sin(\alpha-\beta)+\sin(\alpha+\beta)\right] \end{aligned} \tag{6.7}$$

The double frequency components of $I_M$, $Q_M$, $I_S$, and $Q_S$ are removed in the low-pass filter of each channel, leaving

$$\begin{aligned} I'_M &= \cos(2\pi dt) \\ Q'_M &= -\sin(2\pi dt) \\ I'_S &= \cos(2\pi dt-\theta) \\ Q'_S &= \sin(2\pi dt-\theta) \end{aligned} \tag{6.8}$$

where the multiplying constants, 1/2, have been left out. Now we insert a 90° delay in the $Q$ path to the phase detector. The $I$ and $Q$ inputs to the phase detector when receiving a MARK or SPACE are:

*MARK*

$$\begin{aligned} I'_M &= \cos(2\pi dt) \\ Q''_M &= -\sin(2\pi dt - 90°) \end{aligned} \quad (6.9)$$

*SPACE*

$$\begin{aligned} I'_S &= \cos(2\pi dt - \theta) \\ Q''_S &= \sin(2\pi dt - \theta - 90°) \end{aligned} \quad (6.10)$$

$I'_M$ and $Q''_M$ are multiplied in the phase detector when a MARK is received, and similarly $I'_S$ and $Q''_S$ are multiplied when a SPACE is received. Using the sin($\alpha$)cos($\beta$) identity from Eq. (6.7), and removing the double frequency terms that are filtered out in LPF3, we see that the data out when MARK is received is +1/2, and data out when SPACE is received is −1/2.

The limiter amplifiers square up the filter outputs so that they can be applied to a digital phase detector, which outputs MARK or SPACE according to the phase difference in each pair of inphase ($I$) and quadrature ($Q$) signals, similarly to the explanation above based on sine waves.

Although in the above explanation, the local oscillator frequency is set exactly to the nominal transmitter frequency, small differences are tolerated as long as they are less than the modulation deviation minus the data bit rate. The problem of local oscillator drift and how it is dealt with using other techniques for FSK detection in zero-IF receivers are described in Ref. [3]. The demodulator block in Fig. 6.8 can indicate other types of demodulation. For example, without the "Delay" sub-block, it detects phase shift keying.

## 6.5 Software defined radios

A software defined radio (SDR) is defined as a radio in which some or all of the *physical layer* functions are *Software Defined,* that is, they use software processing within the radio system or device to implement operating (but not control) functions [4]. In practical terms, digital signal processing components extensively replace analog functions, such as mixers, filters and oscillators, and much of the signal flow is digital words instead of continuous analog signals. As a consequence, many performance characteristics are determined by software and can be modified as required for multiple tasks, without hardware changes. The basic architecture is superheterodyne, but down-conversion, up-conversion, and baseband processing are done by digital components after analog-to-digital conversion. SDR has become more prevalent as costs come down and demands for uncompromising wireless performance and compliance with multiple standards in a single radio force a break from the traditional topologies.

Fig. 6.9 shows a diagram of a digital receiver and transmitter. As in a conventional superheterodyne receiver (Fig. 6.9A), the signals from the antenna are amplified by a LNA, then down-converted by mixer and local oscillator to an IF. At that point, the signal flow becomes digital at the output of an analog-to-digital converter (ADC). The ADC is probably the most critical component of a digital radio design. Two basic parameters are the number of converter bits and the sampling frequency (sample rate). The number of bits determines the SNR which, along with the sample rate and the full scale signal input voltage to the ADC, determine the noise spectral density (NSD). The NSD is a factor in the required minimum conversion gain of the receiver up to the ADC [5]. The ideal relationship between number of bits N and SNR is

$$\mathrm{SNR} = 6.02N + 1.8 \tag{6.11}$$

although clock jitter and thermal noise primarily reduce this theoretical maximum. The required sample rate is determined according to Nyquist's theorem: "Any signal can be represented by discrete samples if the sampling frequency is at least twice the bandwidth of the signal" [6]. At baseband, this means that the sampling frequency is above two times the highest frequency component of the signal. The response of a low-pass filter at the input of the ADC must be such that there are no signals higher than one-half the sample rate that is greater than the converter's noise floor. In the case of bandpass signals at IF, the sampling frequency can be below IF, but a BPF

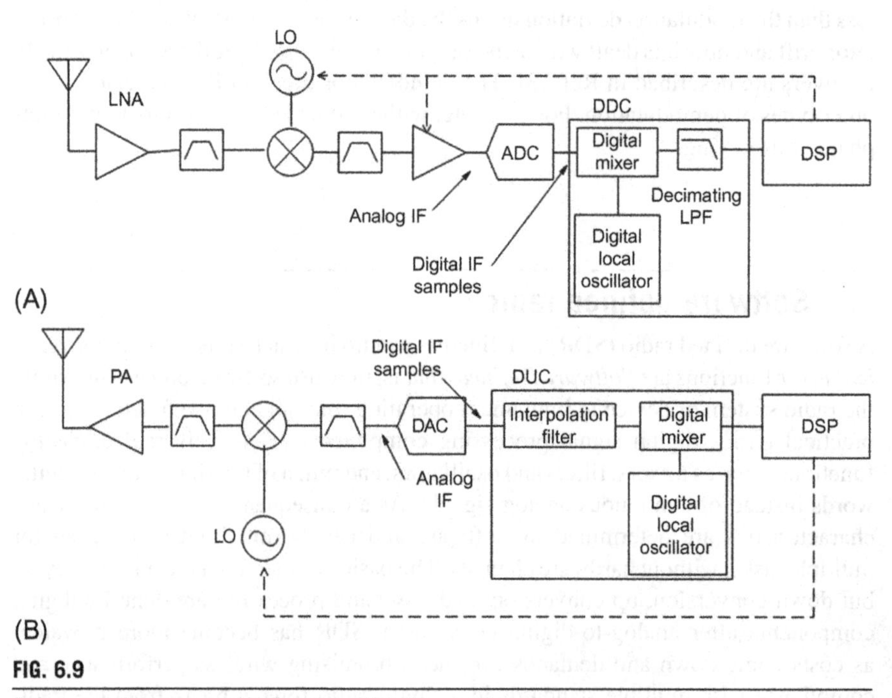


**FIG. 6.9**

Software defined radio. (A) Receiver. (B) Transmitter.

must remove the signals outside of the desired passband. Signals outside the limits determined by half the sampling frequency will *alias* with the desired signals, causing interference that cannot be removed. A broad multichannel IF bandwidth may be prescribed for an SDR, and the sampling frequency has to be chosen accordingly. Oversampling, that is, using a sampling frequency higher than the Nyquist rate, can be advantageous as it lessens the demands on the anti-alias filter. It also provides processing gain, which is the increase in SNR over that of the converter alone, since noise outside the desired signal band can be removed by the digital filter after the analog-to-digital conversion, which improves the in-band SNR. The processing gain is

$$PG_{\mathrm{dB}} = 10\log\left(\frac{\frac{1}{2}f_s}{BW_{filter}}\right) \tag{6.12}$$

where $f_s$ is the sampling frequency and BW is the digital filter bandwidth [7].

At the output of the ADC a digital down converter, DDC, translates the desired received signal to base band. A numerically controlled oscillator (NCO) is set to the center frequency of the desired channel in the IF passband, which may contain multiple channels, and the digital low-pass filter limits response to the required signal bandwidth for that channel. The filter also performs decimation of the sample rate, since the required sampling frequency at baseband is lower than what is required at IF. The output of the mixer consists of digital inphase ($I$) and quadrature ($Q$) components (not shown explicitly in Fig. 6.9) which are fed to a digital signal processor. The DSP performs demodulation, signal conditioning and output filtering. These are some of the functions that are performed by the DSP block [5]:

- Envelope detection (AM)
- Phase/frequency detection (PM, FM)
- Equalization of time division multiplex (TDM) bursts
- Spreading, dispreading direct sequence spread spectrum (DSSS)
- Voice decoding
- Decryption

These functions can be carried out simultaneously on multiple channels in a single device, or in separate devices designed for different applications but using common hardware loaded with individual software. An example in the cellular realm is building mobile telephones, as well as base station sub systems, for use in third generation CDMA networks based on CDMA2000 (prevalent in North America) or UMTS WCDMA (common in Europe and other regions) using common hardware and cellular standard-specific software [5]. Because of the wide range of signal types that may be encountered in a SDR, characteristics of analog RF blocks are also subject to digital control. For example, the DSP can set local oscillator frequency and amplifier gain as shown by the dashed lines in Fig. 6.9.

An SDR transmitter diagram is shown in Fig. 6.9B. It is based on the superheterodyne principle. Baseband data is generated and encrypted in the DSP. The digital transmitter may use a NCO to generate the modulated transmitter frequency. This device outputs digital words which represent the waveform of the signal to be transmitted. Phase, frequency and amplitude variations of the carrier as functions of the baseband data are implemented by the software, and many modulation formats can be used as required with no change in hardware. The digital baseband spectrum is translated to IF through the digital up converter (DUC). The interpolation filter upscales the baseband sampling frequency to the sampling frequency required to maintain signal integrity through the digital mixer and for digital to analog conversion in the DAC (digital-to-analog converter). From the output of the DAC the analog signal is up-converted to the output frequency according to superheterodyne principles and amplified to the required power output.

By performing modulation and demodulation in software, and also filtering, great flexibility is achieved, as well as high performance and stability. Elimination of many hardware components reduce size and cost, while giving high communication efficiency in bandwidth utilization and error correction.

## 6.6 Cognitive radio

Another digital radio technology, related to SDR, is cognitive radio. The SDR forum defines cognitive radio as "Radio in which communication systems are aware of their environment and internal state and can make decisions about their radio operating behavior based on that information and predefined objectives…" Also, "… utilizes *Software Defined Radio*, *Adaptive Radio*, and other technologies to automatically adjust its behavior or operations to achieve desired objectives". Adaptive radio is "Radio in which communications systems have a means of monitoring their own performance and a means of varying their own parameters by closed-loop action to improve their performance." [4]. A cognitive radio must have the ability to convey its operating parameters, which are based on local observations, to the remote terminals with which they are communicating [8].

Two features of cognitive radio are smart antennas and spectrum management. A smart antenna has beam control, so that transmissions can be focused in the direction of the receiver, and transmitter power can be reduced in consideration of the antenna gain. In a receiver, the smart antenna can selectively increase sensitivity in the direction of the transmitter, or create a null to suppress interfering signals from other directions. To do so, the cognitive radio has location awareness, and knows the location of the opposite terminal. A cognitive radio has spectrum sensing, and can use a frequency band that is occupied in one geographical region but not used in another. Licensed spectrum can be used by secondary users while not interfering with a primary licensee. The cognitive radio may access a data base to get information about the radio parameters that may be used according to geographical area and the radio environment. The cognitive radio configures power level, frequency range, modulation type, access technology and protocols in a non interfering manner that results in maximum spectrum utilization [9].

## 6.7 Repeaters

While range and communication link reliability are limited in unlicensed devices by the low power allowed by telecommunication standards, the use of repeaters can overcome these limitations. In the repeater, a weak signal is received and demodulated. The re-created baseband signal then modulates a transmitter whose signal can be received at a distance where the original signal could not be heard. In digital systems, the relaying of a signal through one or more receivers is done without errors. The repeater receiver and transmitter operate at the same frequency as used in the original link, and its operation is transparent in the sense that the receiver doesn't have to know whether the signal is to be received direct or through an intermediary. The repeater does create a time delay, however, since its transmitter must wait until the original transmitter has completed transmitting, in order to avoid interference. Repeaters may be chained, but each link creates an additional delay.

A potential problem when more than one repeater is deployed is that a repeater closer to the transmitter may repeat the transmission from a repeater further along the link, thereby causing a ping-pong effect. One way to avoid this is to include in the message protocol identification of the repeaters so that a repeater will ignore messages received from a device further down the link. Another way is to force a time delay after transmission of a frame during which an identical frame (received from another repeater) will not be retransmitted.

## 6.8 Summary

Most often, short-range radio link performance is determined primarily by the receiver. This chapter has reviewed various topologies that have been developed for different performance levels and applications. While the superheterodyne receiver is dominant, the simplicity and low cost, as well as low power consumption, of TRF and superregenerative receivers can be taken advantage of for very short-range applications—up to several meters. Superheterodyne radios have three dominant architectures—standard where a conventional IF frequency is employed, low IF where the IF is just above the baseband spectrum, and the direct conversion or zero IF receiver which is free of an image frequency. Double and triple conversion in some superheterodyne receivers facilitate image frequency filtering for a high IF while gaining high selectivity and narrow bandwidth at a final low IF. Digital or software radios were introduced here, considering their increasing adoption for short-range applications. Finally, we described how repeaters can be used to extend the range of low-power license-free communication links.

## References

[1] D.L. Ash, SAW-Based Hybrid Transceivers in SLAM Packaging With Frequency Range From 200 to 1000 MHz, AN41 11/21/16, Murata Manufacturing Co. Ltd, 2016.

[2] J.W.M. Rogers, C. Plett, I. Marsland, Radio Frequency System Architecture and Design, Artech House, 2013.

[3] E. Lopelli, J.D. Van Der Tang, A.H.M. Van Roermund, A FSK demodulator comparison for ultra-low power, low data-rate wireless links in ISM bands, in: Proc. 2005 Eur. Conf. Circuit Theory Des., vol. 2, 2005, pp. 259–262.

[4] SDRF Cognitive Radio Definitions, SDR Forum Working Document SDRF-06-R-0011-V1.0.0, 2007.

[5] B. Brannon, Software-defined radio, (Chapter 18) in: RF & Wireless Technologies Know it All, Newnes, 2008.

[6] R.H. Hosking, Software-Defined Radio Handbook, 12th ed., Pentek, Inc, 2016.

[7] Smart Selection of ADC/DAC Enables Better Design of Software-Defined Radio, Texas Instruments Application Report, SLAA407. 2009.

[8] B.A. Fette, History and background of cognitive radio technology, (Chapter 24) in: RF & Wireless Technologies Know it All, Newnes, 2008.

[9] V.T. Nguyen, F. Villain, Y. Le Guillou, Cognitive Radio RF: Overview and Challenges, Hindawi Publishing Corporation, VLSI Design, Volume. 2012.

CHAPTER

# Radio system design

# 7

## 7.1 Introduction

Short-wave radio systems have suffered a bad reputation because of inferior performance and low reliability as a result of poor design. The fact that some of the applications seem deceptively simple, along with the pressures for very low cost, are partially responsible for this. Even in applications where wireless seems obviously more appropriate than wired connections, the market potential has not been realized. While the availability of improved components, such as SAW resonators and advanced integrated RF circuits, has raised the quality and reliability of short-range systems, they do not obviate the need for good design in order to achieve the best performance.

The key to proper design of a short-range wireless system is to match the design to the application. SAW oscillators improved reliability of wireless keys and remote control devices, but in many security systems their frequency accuracy is not sufficient to attain the narrow bandwidth needed for maximum range, given the low power allowed by unlicensed regulations. Current consumption is also an important consideration for many portable battery-operated applications, and compromises in range and spurious response may be necessary to achieve maximum life from small batteries.

The ultimate limitation in radio communication is not the path loss or the fading. Weak signals can be amplified to practically any extent, but it is the noise that bounds the range we can get or the communication reliability that we can expect from our radio system. There are two sources of receiver noise—interfering radiation that the antenna captures along with the desired signal, and the electrical noise that originates in the receiver circuits. In either case, the best signal-to-noise ratio will be obtained by limiting the bandwidth to what is necessary to pass the information contained in the signal. A further improvement can be had by reducing the receiver noise figure, which decreases the internal receiver noise, but this measure is effective only as far as the noise received through the antenna is no more than about the same level as the receiver noise. Finally, if the noise can be reduced no further, performance of digital receivers can be improved by using error correction coding up to a point, which is designated as *channel capacity*. The capacity is the maximum information rate that the specific channel can support without errors, and above this rate error free communication is impossible. Channel capacity is explained in Chapter 9.



**Short-range Wireless Communication. https://doi.org/10.1016/B978-0-12-815405-2.00007-5**
© 2019 Elsevier Inc. All rights reserved.

The aim of this chapter is to describe the main parameters used to define the performance of wireless receivers and transmitters. We give some examples of needed calculations and discuss tradeoffs that may be considered in certain applications. Most of the discussion is centered on digital systems but analog applications are also referred to.

## 7.2 Range

Communication range is probably the most obvious characteristic of a wireless link, and of most interest to the system user. However, it is also the most difficult parameter to specify. We learned in Chapter 2 about the vagaries of radio propagation, about multipath and intersymbol interference. There is also the effect of interfering stations and man-made noise, which we can't predict in advance, particularly on the unlicensed frequencies. The communications path between wireless terminals designed for indoor use can be affected by an infinite variety of obstructions and building materials, radiating devices and mounting heights, and antenna orientations. So to specify a minimum or "typical" range on a data sheet is close to meaningless.

One common way to try to get around the dilemma of specifying the operating range of a wireless system is to specify the "open field" range. However, for this specification to have meaning, say for comparing the capabilities of different products for a similar application, the open field range specification must define at least the following testing conditions:

- Height above ground of transmitter and receiver
- Orientation of transmitter and receiver towards each other
- Type of antenna used, if there is a choice
- Whether one or both devices are handheld
- Whether the definition is "best case" or "worse case"—that is, whether nulls of no reception up to the stated range are allowed, or whether the range means reliable reception at *any* distance up to the stated range (see Fig. 2.2 in Chapter 2)
- Criteria for successful communication—for example, number of correct messages received per number of messages sent.

Rarely are these minimum conditions given when range is stated on a data sheet, so often an inferior device has a superior data sheet range, even "open field," than better equipment that was tested under more tightly controlled and repeatable conditions.

Even though adherence to the conditions described above in specifying range can give some basis for comparison, there are other factors which affect radio range that are almost impossible to control outside of an electromagnetic radiation testing laboratory: ground conductivity, reflecting objects in near vicinity, incidental transmissions, and noise on the communication channel.

So, while the desired criterion for our wireless equipment is elusive indeed, we should make use of well-defined and measurable indications of performance when designing the system in order to achieve optimum results, given known and thought-out compromises.

## 7.3 Sensitivity

Sensitivity is the signal power at the input of a receiver that results in a stated signal-to-noise ratio or bit error rate at the output. It depends on the thermal noise power from the antenna, the internal noise generated in the receiver expressed as noise figure, the noise bandwidth of the receiver, and the required output signal-to-noise ratio. We express this sensitivity as follows:

$$(P_{\min})_{\mathrm{dBm}} = (N_0)_{\mathrm{dBm}} + NF + 10\log B + (SNR_{out})_{\mathrm{dB}} \tag{7.1}$$

Let's look at this expression, term by term. All terms are in dB, or dBm. The place of each term of Eq. (7.1) in a receiver chain is shown in Fig. 7.1

### 7.3.1 Sensitivity $P_{min}$

This is the minimum signal power of $P_{\min}$ applied to the receiver input terminals that gives the required output signal-to-noise ratio, $SNR_{out}$ at the detector input.

### 7.3.2 Input noise power density $N_0$

The input noise power density is the noise originating on the source resistance feeding the receiver, which can be considered as the equivalent resistance of the antenna, or the input resistance of a signal generator used for laboratory testing. The *available* input noise power density, $N_0$, is the maximum power in a bandwidth of one Hertz that this noise source can transfer to a load, achieved when the load resistance equals the source resistance. (When the source has a complex internal impedance, maximum power is delivered when the load is the complex conjugate of the source impedance.)

$$N_0 = kT_0 \ \mathrm{W/Hz}$$

where $k$ is Boltzman's constant $= 1.38 \times 10^{-23}$ J/K, and $T_0 = 290$ K, the standard room temperature in Kelvin.

Note that the available noise power in a resistor depends only on temperature and bandwidth, and not on the resistance.

$N_0$ is in W/Hz, and expressing it in dBm for Eq. (7.1)

$$(N_0)_{\mathrm{dBm}} = -174 \ \mathrm{dBm}$$

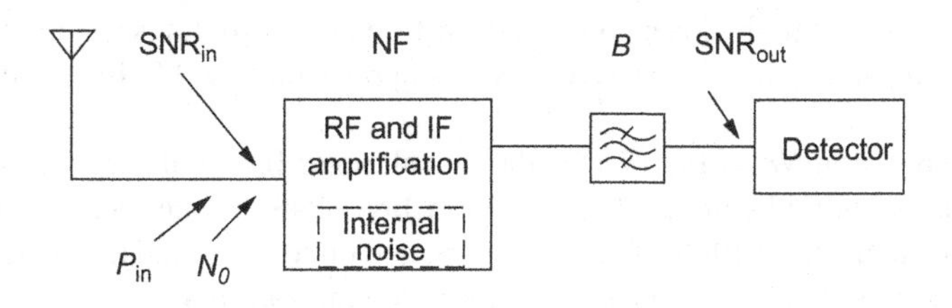


**FIG. 7.1**

Receiver sensitivity terms.

This is the ultimate noise floor of a perfect receiver at room temperature having a bandwidth of 1 Hz, or the sensitivity of a receiver with no internal noise that can operate with a signal having the same power as the input noise.

### 7.3.3 Noise figure *NF*

The noise figure in dB is the difference between the input *SNR* and output *SNR* (at room temperature), expressed in dB as

$$NF = (SNR_{in})_{\mathrm{dB}} - (SNR_{out})_{\mathrm{dB}} \tag{7.2}$$

It's a measure of the deterioration of the signal-to-noise ratio as a signal passes through a two-port network due to the contribution of internal noise [1]. The notation *NF* denotes noise figure in dB. When expressed as a ratio instead of in dB, the noise figure is called noise factor, *F*. One is found from the other by the expression: $NF = 10 \log F$. Pay particular attention in the following discussion to whether reference is to *F*, noise factor, or *NF*, noise figure.

In reference to an amplifier, the gain and the internal noise power can be seen expressly as

$$F = (N_a + G_a N_0)/G_a N_i \tag{7.3}$$

where $N_a$ is the excess noise power of the amplifier (internal noise) referred to the amplifier output, $G_a$ is the gain of the stage, and $N_0$ is the input noise power at standard room temperature.

The noise figure of RF active devices—transistors, mixers, amplifiers—is usually given on the data sheet. Noise figure is dependent on bias conditions and for a bipolar transistor can be roughly approximated by

$$NF = \log(5 \times I_c \times V_c)$$

where $I_c$ and $V_c$ are the collector current in mA and collector-emitter voltage, respectively.

Noise figure is also dependent on input impedance matching, and lowest noise figure usually does not correspond to the optimum matching required for maximum power gain. Thus, the design process involves tradeoffs between noise performance, gain, stability and requirements for matching other components such as filters. The noise figure of passive two-port devices—filters and attenuators for example—is equal to the insertion loss. Insertion loss of filters is stated on the data sheet. The noise figure of a passive mixer is approximately 0.5 dB higher than its conversion loss.

The expression *NF* of Eq. (7.1) is the overall noise figure of the receiver, calculated from the individual noise figures of function blocks in the receiver amplification chain—amplifiers, filters and attenuators, and mixers. To find the noise figure of several cascaded two port devices, use the following formula:

$$F_{total} = F1 + \frac{F2-1}{g1} + \frac{F3-1}{g1 \cdot g2} \cdots$$
$$NF_{total} = 10\log F_{total} \qquad (7.4)$$

where $F1$ is the noise factor of the first stage, $F2$ is the noise factor of the second stage, and so on, and $g1$ is the numerical power gain of the first stage, $g2$ is the power gain of the second stage, and so on. Remember that

$$F = 10^{\frac{NF}{10}}$$

When using the noise factor of a mixer in the expression for overall noise factor in Eq. (7.4), you should note whether the noise figure or noise factor is specified as single sideband or double sideband [2]. Generally in VHF and higher frequency superheterodyne communications receivers, the output intermediate frequency (IF) is the absolute value of the difference between the input frequency and the local oscillator. Thus, when using a wide-band noise source to measure the noise figure, the mixer will convert the noise source power, which takes place of the signal power, in the bands both below and above the local oscillator frequency to IF. That is, the "signal power," which is the noise source, is applied to the image frequency band as well as the desired frequency band. Superheterodyne receivers pass the signal through only one of the two frequency bands that the mixer is sensitive to, so it is the single sideband noise figure that is needed to calculate the cascaded noise figure $(NF)_{dB}$. The mixer single sideband noise figure is 3 dB higher than the double sideband figure (because the mixer sees half the signal power). Thus, the relationship between the single sideband noise factor, $F_{SSB}$, and the double sideband noise factor, $F_{DSB}$ is

$$F_{SSB} = 2F_{DSB} \qquad (7.5)$$

(This is not always the case. For a detailed explanation of mixer noise, see Ref. [2].)

If the receiver front end does not have image suppression the double sideband noise factor of the mixer is used. When an image suppressing bandpass filter is not placed at the input to the mixer, but precedes the low noise amplifier (LNA), the LNA noise factor is modified as $F' = 2F_{LNA} - 1$ and then used to find the total receiver noise figure. Sometimes more conveniently, the rules which follow can be used to find overall receiver noise figure for the case where an image rejection filter does not directly precede the mixer, and the case where there is no image rejection filter at all, but one or more stages of amplification or attenuation precedes the mixer. Worksheet **Sensitivity.mcdx** incorporates these rules. An example is given later in this section.

Case 1. Image rejection filter two or more stages before mixer:

- Modify noise factor of each stage between filter and mixer, including filter attenuation, by: $F_i' = 2F_i - 1$.
- Use $F_{SSB}$ for mixer noise factor.

Case 2. No image rejection filter before mixer as for zero IF receiver:

- Use $F_{DBS} = F_{SSB}/2$ for mixer noise factor.
- Double the denominator for each stage after mixer in Eq. (7.4).

It can be seen from Eq. (7.4) that when a preamplifier is used, the noise figure of a chain of devices is determined predominantly by the first high gain stage, since the influence of noise in subsequent stages is reduced by the gains of previous stages. The preamplifier gain should only be enough to reduce the effect of the noise figure of subsequent stages, and no more— excess gain can be detrimental, as it may increase intermodulation distortion (IMD).

#### 7.3.3.1 Noise temperature

Sometimes the internal noise of an amplifier or receiver is expressed in terms of noise temperature instead of noise factor or noise figure. This is convenient when antenna noise is given in terms of temperature, $T_A$, particularly when it is not at room temperature, and when the noise factor is low, close to unity. Wide band noise power is related to temperature by

$$P = kTB \tag{7.6}$$

where $P$ is power, $k$ is the Boltzman constant given above, $T$ is absolute temperature in Kelvin and $B$ is the bandwidth. Dividing both sides of Eq. (7.6) by $B$ gives noise power density in terms of absolute temperature: $P_d = kT$. Noise factor is converted to an *effective noise temperature* $T_e$ by

$$T_e = T_0(F - 1) \tag{7.7}$$

The *system temperature*, $T_s$, is

$$T_s = T_A + T_e \tag{7.8}$$

which lets us define an *effective noise factor*, $F_e$:

$$F_e = 1 + \frac{T_0}{T_A}(F - 1) \tag{7.9}$$

$F_e$ can be used in Eq. (7.1) (after converting it to $NF$ dB) when the input noise density is not $N_0$, that is, the equivalent noise of the antenna is not at room temperature. Alternatively, the sensitivity can be expressed in terms of $T_s$ in dB:

$$(P_{\min})_{\text{dBm}} = (T_S)_{\text{dB}} + k_{\text{dBm/Hz}} + (10\log B)_{\text{dBHz}} + (SNR_{out})_{\text{dB}} \tag{7.10}$$

where the Boltzman constant for dBm is 10 log$(1.38 \times 10^{-20})_{\text{dBm}}$ and $(T_s)_{dB} = 10$ log $(T_s/1\text{ K})$.

### 7.3.4 Bandwidth *B*

Eq. (7.1) shows the importance of narrow bandwidth in attaining high sensitivity. The receiver bandwidth should match the bandwidth of the modulated signal for maximum sensitivity. It cannot be less, since a narrower receiver bandwidth will distort the demodulated signal causing intersymbol interference and increasing the bit error rate in digital communications and reducing fidelity and intelligibility of analog signals.

In digital systems, the minimum theoretical bandwidth is the Nyquist bandwidth, which is one-half of the symbol rate. In a superheterodyne receiver the channel bandwidth is primarily determined by the IF bandpass filter, whose minimum bandwidth in a double sideband receiver is twice the Nyquist bandwidth. Approaching this lower limit of bandwidth is possible only using high cutoff filters that have a constant delay factor in the passband to prevent distortion. Receivers using these filters must have accurate timing for sampling the data in order to reduce intersymbol interference, which is the blurring of a data symbol due to residual effects from earlier symbols. Such filters are not normally used in simple short-range devices. Instead, discrete component filters make the IF bandwidth much higher than the symbol rate to prevent distortion of the signal.

The signal rates in low-cost short-range devices used for control and security may range from several hundred hertz up to around 2 kHz. The bandwidth of an appropriate IF filter could be 10 kHz. However, a great number of these products use SAW resonators for the transmitter and receiver oscillators and standard 10.7 MHz ceramic IF filters with a bandwidth of 300 kHz. Let's assume for the SAW resonators a stability of 20 kHz and an accuracy, after production tuning, of 50 kHz. Then the worst-case frequency divergence of the transmitter and the receiver would be $2 \times (50\text{ kHz} + 20\text{ kHz}) = 140\text{ kHz}$, considering that the divergence directions are opposite in transmitter and receiver. Adding 10 kHz for the modulated signal passband, we see that the receiver IF bandwidth should be at least $(2 \times 140\text{ kHz}) + 10\text{ kHz} = 290\text{ kHz}$. Thus, the 300 kHz IF filter is justifiable. Referring to Eq. (7.1), the reduction in receiver sensitivity because of the worst-case frequency variation of the SAW resonators is $10 \log (290\text{ kHz}/10\text{ kHz}) = 14.6\text{ dB}$.

The effective total receiver bandwidth can be less than that determined by the IF bandwidth when considering baseband low-pass filtering after the detector, as is done in most receivers. However, baseband filtering is not at all as effective as IF filtering unless synchronous detection is used, or the receiver is a direct conversion type as described in Chapter 6. (In synchronous detection a local oscillator signal having the same frequency and phase-locked to the received signal is multiplied with the received signal recreating, after low-pass filtering, the transmitted baseband data.) The baseband filter will reduce noise, but it will not filter out interfering signals appearing in the IF passband, particularly when the desired signal is relatively weak. The nonsynchronous detector is a nonlinear device, and the signal-to-noise ratio at its output is not a linear function of the signal strength. Thus, the bandwidth of the post-detection filter should not be used for ($10 \log B$) in Eq. (7.1) and its effect should better be expressed in predetection signal-to-noise ratio, as shown below.

In discussing bandwidth for use in Eq. (7.1), we should define exactly what it means. Commonly, the bandwidth of a bandpass filter is defined as the difference between the upper and lower frequencies where the insertion loss increases by 3 dB as compared to the center frequency. The bandwidth of a low-pass filter is

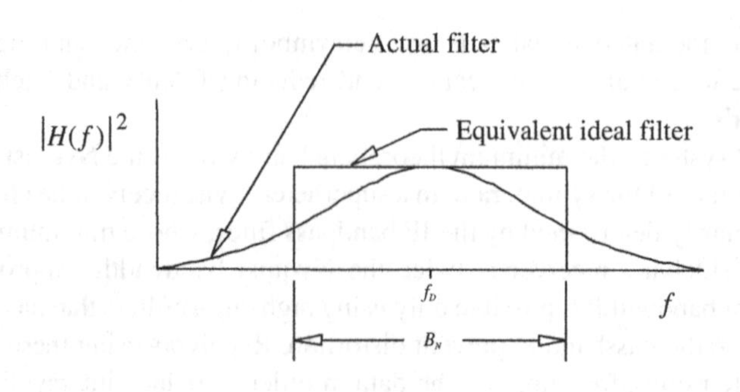


FIG. 7.2
Noise equivalent bandwidth.

the frequency where the response is reduced by 3 dB compared to the DC value. The bandwidth to be used in Eq. (7.1) is the noise equivalent bandwidth ($B_N$), which is not necessarily the same as the 3-dB bandwidth as it depends on type and order of the filter. The noise equivalent bandwidth is the bandwidth of an ideal rectangular filter that passes the same noise power from a white noise source as the filter whose bandwidth is being determined. White noise has a constant power density over the frequency band of interest. This is exemplified in Fig. 7.2, which shows power density spectrums of the two filters. The power spectrum curve of the filter we are interested in is $|H(f)|^2$ and its noise equivalent bandwidth is

$$B_N = \frac{1}{H_0^2} \cdot \int_0^\infty |H(f)|^2 df \tag{7.11}$$

where $H_0^2$ is the maximum power gain of a bandpass filter or DC power gain of a low-pass filter.

For example, the noise equivalent bandwidth of a 3-pole Butterworth type filter is 1.047 times, or around 5% higher, than its 3-dB bandwidth. The noise bandwidth of a single-pole low-pass filter (consisting of a series resistor and a bypass capacitor) is 57% greater than its 3-dB bandwidth. In most cases, except perhaps for the single pole filter, using the known 3-dB bandwidth in Eq. (7.1) gives acceptable accuracy.

### 7.3.5 Predetection signal-to-noise ratio

The last term in the sensitivity Eq. (7.1) is the signal-to-noise ratio required at the detector input to achieve a defined performance criterion. It may be the signal-to-noise ratio needed for a specified BER, or the specified probability of successfully decoding a message frame. For narrow band FM analog detection, the required (signal + noise + distortion)/(noise + distortion) in dB is commonly specified as the performance criterion. The improvement in sensitivity due to the post-detection filter can be absorbed into the predetection *S/N*.

### 7.3.6 Sensitivity calculations: Example

Let's now look at an example using the sensitivity equation, Eq. (7.1). Fig. 7.3 is a diagram of a superheterodyne receiver. Two configurations are shown, differing in the placement of the image rejection bandpass filter. Noise figure and gain are given in Tables 7.1 and 7.2, corresponding to blocks on the diagram marked by an index number.

$(N_O)_{dBm}$ is constant and equals −174 dBm/Hz.

The noise figure, calculated up to the IF amplifier, is

$$F = F1 + \frac{F2-1}{g1} + \frac{F3-1}{g1 \cdot g2} + \frac{F4-1}{g1 \cdot g2 \cdot g3} + \frac{F5-1}{g1 \cdot g2 \cdot g3 \cdot g4} \tag{7.12}$$

where the $F$'s are noise factors and the $g$'s are numerical gains.

The noise figures must be converted to noise factors and gains to ratios, using the relationships:

$$F = 10^{\frac{NF}{10}}, \quad g = 10^{\frac{G}{10}}$$

First, we'll find the sensitivity for the configuration shown in Fig. 7.3A, where the bandpass (image rejection) filter directly precedes the mixer. The noise factor of the

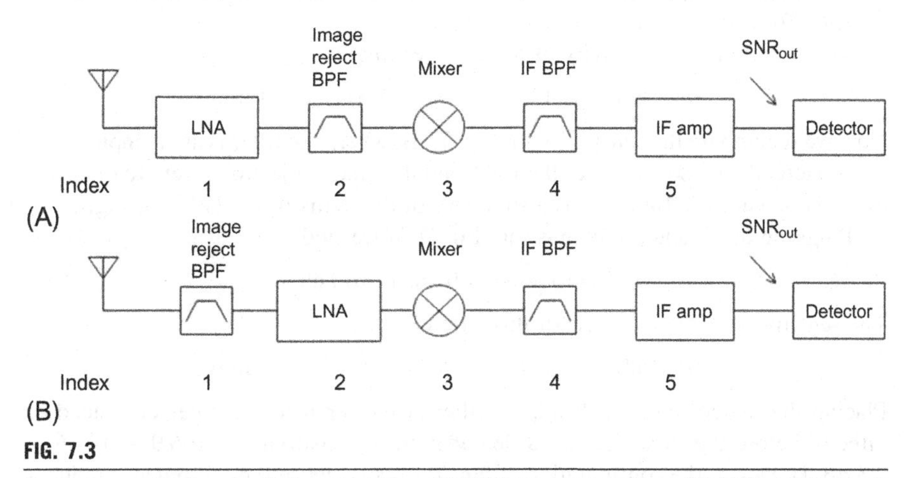


**FIG. 7.3**

Example receiver for sensitivity calculation. (A) BPF at mixer input and (B) BPF at LNA input.

**Table 7.1** Terms for finding cascaded noise figure for Fig. 7.3A

| Index | Noise figure dB (*NF*) | Noise factor (*F*) | Gain dB (*G*) | Gain ratio (*g*) |
|---|---|---|---|---|
| 1 | 2.5 | 1.78 | 15 | 31.62 |
| 2 | 1 | 1.26, $F' = 1.52$ | −1 | 0.79 |
| 3 | 10 | 10 | 8 | 6.31 |
| 4 | 6 | 4 | −6 | 0.25 |
| 5 | 10 | 10 | – | – |

**Table 7.2** Terms for finding cascaded noise figure for Fig. 7.3B

| Index | Noise figure dB (*NF*) | Noise factor (*F*) | Gain dB (*G*) | Gain ratio (*g*) |
|---|---|---|---|---|
| 1 | 1 | 1.26, $F' = 1.52$ | −1 | 0.79 |
| 2 | 2.5 | 1.78, $F' = 2.56$ | 15 | 31.62 |
| 3 | 10 | 10 | 8 | 6.31 |
| 4 | 6 | 4 | −6 | 0.25 |
| 5 | 10 | 10 | – | – |

filter has to be modified (see discussion on noise figure above) because its attenuation is effectively after a lossless filter. Table 7.1 is a summary of the terms in Eq. (7.12) for Fig. 7.3A.

Plugging the $F$ and $g$ columns into Eq. (7.12) we find:

$$F = 2.4 \rightarrow NF = 10 \log F = 3.8\,\text{dB}$$

The bandwidth term in Eq. (71) is 10 log(300,000) = 54.8 dB Hz and we will assume that the detector needs an 8.5 dB $S/N$ to output four correct frames out of five, our criterion for minimum sensitivity.

We now find the sensitivity of the receiver to be:

$$(P_{\min})\text{dBm} = -174 + 3.8 + 54.8 + 8.5 = -106.9\,\text{dBm}$$

Now we recompute the sensitivity for Fig. 7.3B, where the filter is at the input to the LNA. Here, the noise factors of the LNA and the image rejection filter are modified as explained above. Table 7.1 is a summary of the terms in Eq. (7.7) for Fig. 7.3B.

Plugging the $F$ and $g$ columns into Eq. (7.4) we find:

$$F = 4.1 \rightarrow NF = 10 \log F = 6.1\,\text{dB}$$

The sensitivity of the receiver for this case is

$$(P_{\min})\text{dBm} = -174 + 4.9 + 54.8 + 8.5 = -104.6\,\text{dBm}$$

Placing the image rejection bandpass filter at the input to the receiver instead of directly before the mixer causes a degradation of sensitivity of 106.9 − 104.6 = 2.3 dB. However, the configuration of Fig. 7.3B is sometimes preferred as it reduces the strength of out-of-band signals that could block the LNA or cause IMD. Use of an image reject mixer in this case makes putting the bandpass filter directly in front of the mixer unnecessary. Such a mixer subdues the image frequency, and noise around the image frequency through phase cancellation instead of filtering [3].

Calculation of noise figure for cascaded stages, and receiver sensitivity, is conveniently carried out using the Mathcad worksheet **Sensitivity.mcdx**.

The sensitivity derived from Eq. (7.1) is what would be measured in the laboratory using a signal generator source instead of the antenna. It is useful for comparing different receivers, but it doesn't give the whole picture regarding real life performance, as shown in Section 7.4.

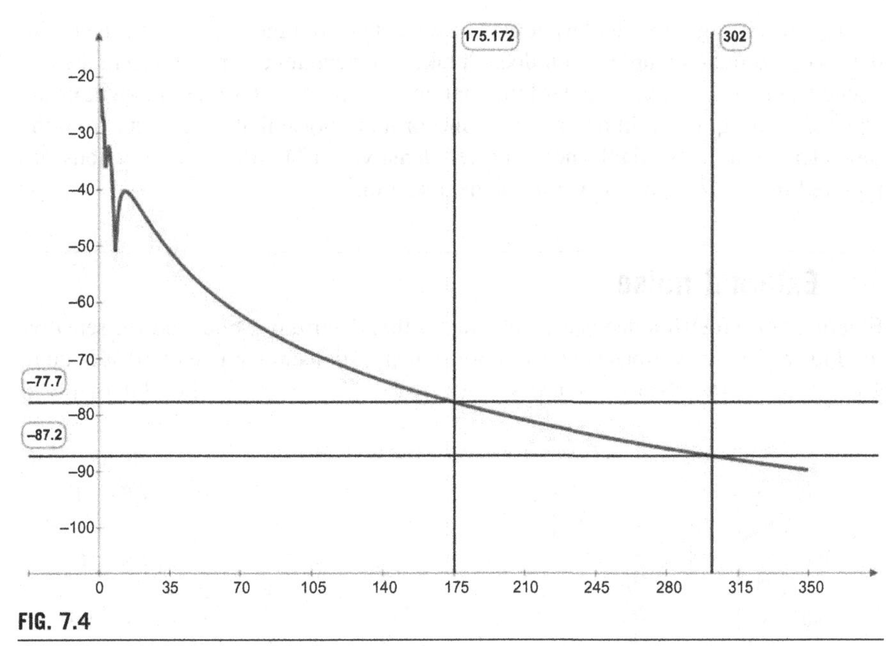


**FIG. 7.4**

Finding range from path gain.

### 7.3.7 Finding range from sensitivity

We can tie sensitivity to range with the aid of a plot of path gain vs. distance in an open field. Fig. 7.4 is drawn for Example 7.1. The frequency of 315 MHz with transmitting and receiving antennas 2 m above ground, horizontal polarization. The curve shows the ratio, in decibels, of the power at the receiver to the radiated power at the transmitter.

**Example 7.1**

Assume that the transmitter is radiating the maximum allowed power. At 315 MHz, a common frequency used in North America, the maximum allowed average field strength per FCC regulations is 6000 μV/m at a distance of 3 m. The corresponding radiated transmitted power $P_t$ can be found from:

$$P_t = \frac{E^2 \cdot d^2}{30} \tag{7.13}$$

Solving for $E = 6000$ μV/m and $d = 3$ m, we find $P_t = .0108$ mW, which is 10 log (0.0108) = −19.7 dBm.

Worksheet **Radiate.mcdx** is useful for RF conversion calculations

We'll assume a receiver antenna gain of −2 dB compared to a dipole, or approximately zero dB isotropic gain. We also assume that the antenna is matched to the receiver input impedance.

Now we can find the path gain (see Section 2.6) corresponding to the receiver sensitivity of −106.9 dBm when the receiving antenna has 0 dB gain:

Pathgain (dB) = −106.9 dBm − (−19.7 dBm) = −87.2 dB

From the plot in Fig. 7.4 we see that the maximum range of the receiver is 302 m.

This result is greater than what may be expected from the example receiver with the specified field strength, but it doesn't take into account external noise in the frequency band, beyond the standard thermal noise of the first term on the right of the equal sign in Eq. (7.1). In order to account for input noise that is different from the room temperature available noise power density, −174 dBm, modifications are required in Eq. (7.1), described in the next section.

## 7.4 External noise

External noise interference is a critical factor that determines true receiver sensitivity. Fig. 7.5 shows various sources of noise over a frequency range of 10 kHz up to 100 GHz [4]. Noise sources that are documented in International

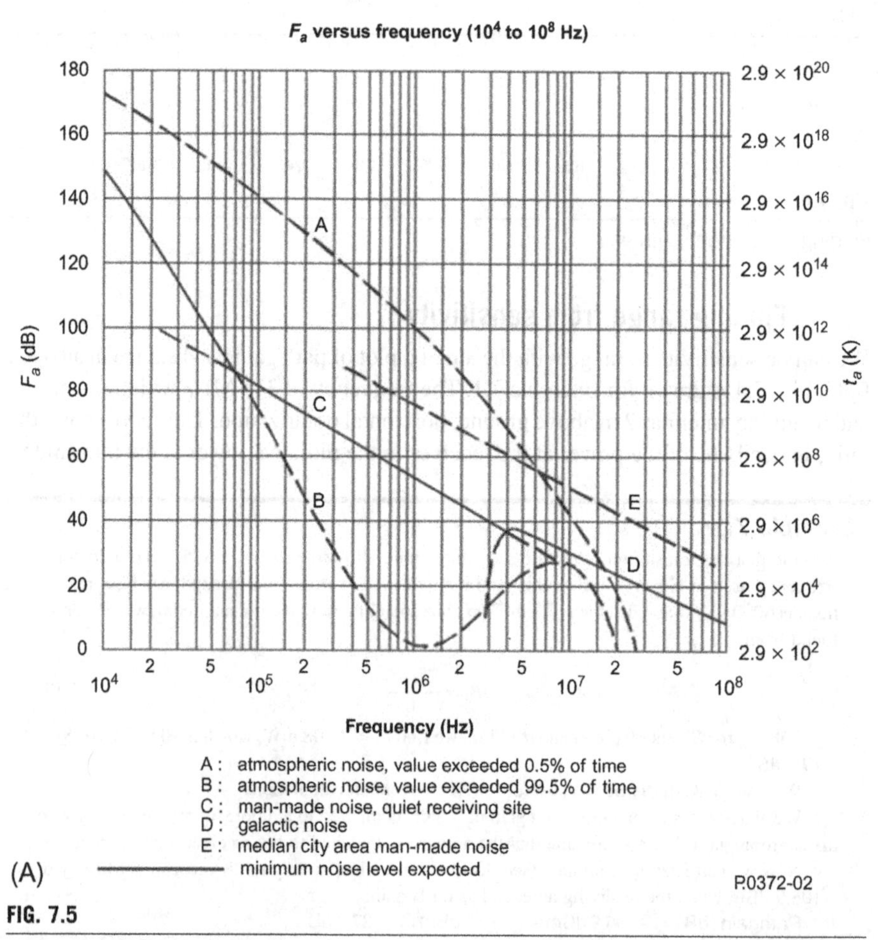


**FIG. 7.5**

External noise sources. (A) $F_a$ vs. frequency, 10 kHz to 100 MHz and

*(Continued)*

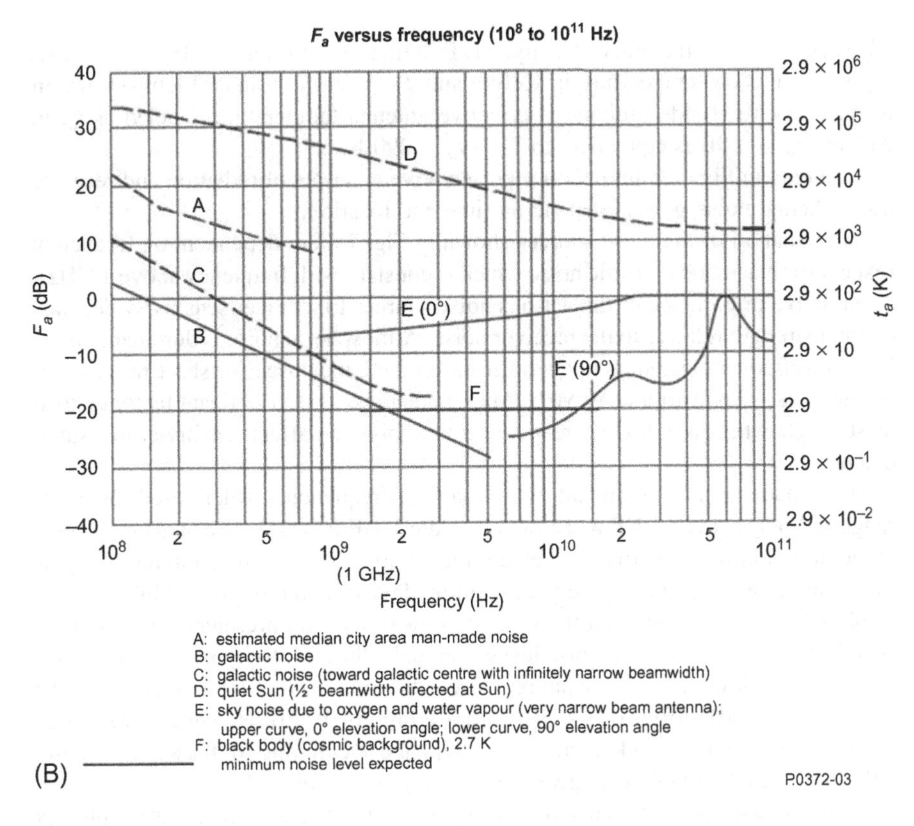


**FIG. 7.5, CONT'D**

(B) $F_a$ vs. frequency, 100 MHz to 100 GHz.

*Reference Radio Noise, Recommendation ITU-R P.372-11 (09/2013) with permission.*

Telecommunications Union document ITU-R P.372-11 are atmospheric noise, including noise due to lightning, galactic noise and man-made noise. Noise levels are for background noise and do not include co-channel transmissions or spurious emissions from individual transmitting or receiving systems. Input noise to the receiver can be lower than equivalent standard room temperature when a highly directional antenna is pointed at a "cool" region of the sky and is not exposed to earthly radiation. This is the case for satellite communication and radio astronomy. For the short-range wireless communication we're considering, external noise received through an antenna will virtually always reduce sensitivity compared to laboratory measurements. In Fig. 7.5, relative noise power $F_a$ is given, on the left side ordinate, in dB. It indicates the noise power density from the source relative to standard room temperature noise power density:

$$F_a = 10\log\left(\frac{p_n}{kT_0}\right) = 10\log\left(\frac{T_A}{T_0}\right) \tag{7.14}$$

where $p_n$ is the external noise density, $k$ is Boltzman's constant = $1.38 \times 10^{-23}$ J/K, $T_A$ is the antenna temperature in Kelvin and $T_0 = 290$ is standard temperature in Kelvin. The right side ordinate is effective antenna temperature, labeled as $t_a$, in Kelvin. $F_a = 0$ dB is equivalent to $T_0 = t_a = 290$ K.

The data in Fig. 7.5 are only a representative example of radiation and receiver noise. Actual noise powers depend on time and location.

Note that all of the noise sources shown in Fig. 7.5 are dependent on frequency (except for black body cosmic noise which is constant with frequency above 1 GHz). The relative importance of the various noise sources to receiver sensitivity depends on their strength relative to the receiver noise. Atmospheric noise is dominant on the low radio frequencies but is not significant on the bands used for short-range communication—above around 40 MHz. Man-made noise is the dominant factor up to at least 1 GHz whereas at higher frequencies sky noise can affect sensitivity to a small extent.

Man-made noise is dominant in the range of frequencies widely used for short-range radio systems—VHF and low to middle UHF bands. It is caused by a wide range of ubiquitous electrical and electronic equipment, including automobile ignition systems, electrical machinery, computing devices and monitors. While we tend to place much importance on the receiver sensitivity data presented in equipment specifications, high ambient noise levels can make the sensitivity irrelevant in comparing different devices. For example, a receiver may have a laboratory measured sensitivity of −105 dBm for a signal-to-noise ratio of 10 dB. However, when measured with its antenna in a known electric field and accounting for the antenna noise, −95 dBm may be required to give the same signal-to-noise ratio.

From around 800 MHz and above, receiver sensitivity is essentially determined by the noise figure. Improving the noise figure must not be at the expense of other characteristics—IMD, for example, which can be degraded by using a very high-gain amplifier in front of a mixer to improve the noise figure. IMD causes the production of in-band interfering signals from strong signals on frequencies outside of the receiving bandwidth. There's more on this subject in Section 7.7. External noise will often be reduced when a directional antenna is used, unless it's pointed at the sun or directly at the source of man-made interference. Regulations on unlicensed transmitters limit the peak radiated power. When possible, it is better to use a high-gain antenna and lower transmitter power to achieve the same peak radiated power as with a lower gain antenna. The result is higher sensitivity through reduction of external noise. Man-made noise is usually less with a horizontal antenna than with a vertical antenna.

Antenna noise above the standard noise density which results from external sources can be included in sensitivity calculations through this equation:

$$(P_{\min})_{dBm} = N_0 + F_a + 10\log\left[1 + (F-1)\left(10^{\frac{-F_a}{10}}\right)\right] + 10\log B + (SNR_{out})_{dB} \tag{7.15}$$

which is equivalent to Eq. (7.10) but possibly a more convenient way to express and use it.

The effect of external noise on sensitivity and consequently on range is demonstrated in Example 7.2.

**Example 7.2**

Link specifications are the same as in Example 7.1 where radiated power = −19.7 dBm. From Fig. 7.5B, relative noise density due to man-made noise at 315 MHz is $K_a = 13$ dB. Using Eq. (7.15) receiver sensitivity is −97.4 dBm. Find the estimated range of the link.

$$\text{Path gain(dB)} = -97.4\,\text{dBm} - (-19.7\,\text{dBm}) = -77.7\,\text{dB}$$

From the path gain vs. range curve in Fig. 7.4, the range corresponding to the path gain of -77.7 dB is 175 m.

Similar problems can be worked out on Worksheet **range.mcdx**.

## 7.5 Internal receiver noise

After all measures have been taken to properly design the receiver and prototypes are made and tested, you may discover that the performance falls short of your best intentions. The reason may be interference sources that weren't eliminated completely in the design stage, and trial and error may be necessary to be sure your efforts haven't been in vain.

**(a)** *Oscillator noise*

A noisy local oscillator can add noise to the mixer output that cannot be eliminated by the IF filter, since most of it occurs in the passband. This noise could be phase noise from the oscillator or spurs in a frequency synthesizer. Changing the oscillator level to the mixer may help to improve the situation. For a detailed explanation of phase noise see [5].

**(b)** *Oscillation in the high gain IF amplifier*

Positive feedback can occur around the high gain IF amplifier. Some IF chip makers specify attenuation between two separate sections of this amplifier—the linear stages and the limiter. Check the data sheet and manufacturer's application notes. Bypass capacitors should be very close to the IC terminals and must have very low impedance at IF. Electrolytics are not recommended for bypassing—use tantalum or ceramic capacitors—preferably SMT.

**(c)** *Radiation from logic circuits*

The microcomputer and accompanying logic circuits operate with switching frequencies in the neighborhood of IF, and harmonics can even reach the received frequency band. Logic circuits and grounds should be isolated from the radio circuits, and shielding may be necessary to prevent reduction of sensitivity because of digital radiation.

**(d)** *Power supply noise*

Hum or ripple on the power bus can cause noise in the system. Also, the outputs of a switching voltage regulator must be well filtered, and radiation from the supply must be prevented.

## 7.6 Superheterodyne image and spurious response

The superheterodyne receiver uses one or more mixers and local oscillators to convert the received signal channel to another frequency band for more convenient filtering and amplification. A detrimental by-product of this frequency transfer process is the susceptibility of the receiver to unwanted signals on other frequencies. We can see this with the help of Fig. 7.6, which shows a dual conversion superheterodyne receiver that has two mixers and two local oscillators.

The desired signal is on 433 MHz which, when combined with the local oscillator at 422.3 MHz in the first mixer, results in a first IF of 433 − 422.3 = 10.7 MHz. After filtering and amplification the signal is applied to a second mixer and a local oscillator at 10.245 MHz, which gives a second IF frequency of 10.7 MHz − 10.245 MHz = 455 kHz. This second IF signal is band pass filtered to the desired narrow passband of ±5 kHz, then amplified and detected.

Why not down-convert directly to 455 kHz and save one local oscillator and mixer? We could do this with a local oscillator of 432.545 MHz or 433.455 MHz, either of which converts 433 MHz to 455 kHz. However, mixing with 432.545 MHz will also convert the undesired frequency of 432.09 MHz to 455 kHz and similarly a 433.455 MHz local oscillator would receive an undesired frequency of 433.455 MHz + 455 kHz = 433.91 MHz. In order to prevent the undesired frequency from reaching the IF filter, it must be filtered out before reaching the mixer. When the undesired frequency is so close to the desired frequency, filtering is particularly difficult.

Returning to Fig. 7.6, we see that when using a 10.7 IF to precede the 455 kHz IF, the input image frequency is 422.3 − 10.7 = 411.6 MHz, which is practical to separate from the desired 433-MHz signal.

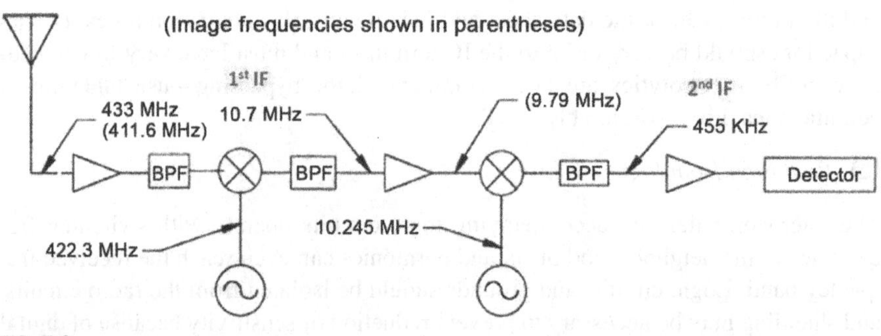


**FIG. 7.6**

Receiver spurious response.

A second IF could be eliminated by using a 10-kHz bandpass filter at 10.7 MHz. However, such filters are relatively expensive, and a high-gain amplifier at 10.7 MHz with such a narrow bandwidth could easily turn into an oscillator because of stray positive feedback. So, the dual conversion approach is more practical and is widely used.

It is important in designing a superheterodyne receiver to calculate the image responses in order to determine the characteristics of the bandpass filters. In the case of Fig. 7.6, a simple LC filter or matching network is not sufficient to eliminate the image frequency, and a helical filter or SAW filter is required at the input.

Spurious response can also result from harmonics of strong interfering input signals and of the local oscillator. The following table shows image frequencies and some spurious responses of the receiver of Fig. 7.6. Not only must the bandpass filter have high attenuation at the image frequencies and spurious frequencies, but the receiver front-end circuit has to be well laid out and may have to be shielded to prevent very strong interfering signals from getting in directly, not via the antenna, and bypassing the input filter. The curve of a SAW bandpass filter for 433 MHz is shown in Fig. 7.7.

The responses shown in Table 7.3 are particular to the mixing process in superheterodyne receivers. Harmonics of the local oscillators and of the input signal are the result of nonlinear effects in the amplifier and mixers, discussed in detail in the next section.

The input RF bandpass filter is very important in reducing spurious responses of superheterodyne receivers. At the least, it must attenuate the image frequency significantly. Other measures must be taken to reduce spurious responses closer to the desired signal frequency.

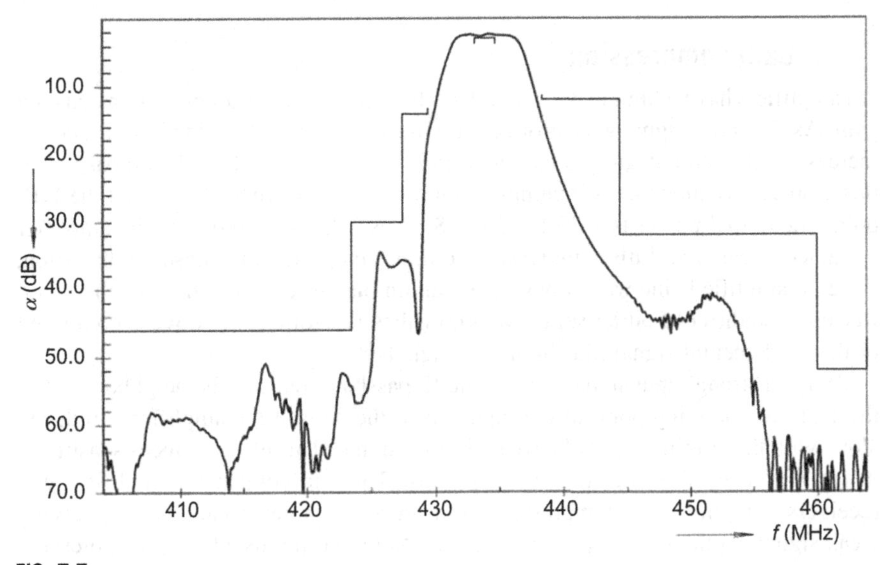


**FIG. 7.7**

SAW filter frequency response.

*Copied from TDK B3721 with permission from RF 360 Europe GmbH.*

**Table 7.3** Spurious responses $f_{in}$

| $f_{in}$ MHz | At 1st IF MHz | At 2nd IF kHz | Comments |
|---|---|---|---|
| 433 | 10.7 | 455 | Wanted signal |
| 411.6 | 10.7 | 455 | 1st IF image |
| 432.09 | 9.79 | 455 | 2nd IF Image |
| 427.65 | 10.7 | 455 | Half IF spurious response (2nd order) |
| 855.3 | 10.7 | 455 | 1st local osc. harmonic mixes with input |
| 833.9 | 10.7 | 455 | 1st local osc. harmonic mixes with input |
| 216.5 | 10.7 | 455 | Sub harmonic of input |

*Note:* $f_{LO1} = 422.3$ MHz, $f_{LO2} = 10.245$ MHz.

## 7.7 Intermodulation distortion and dynamic range

In the previous section we saw that the superheterodyne receiver is susceptible to interference from signals on frequencies other than the one we are trying to receive. In this section we see that new interfering signals can be created in the receiver passband because of the inherent nonlinearity of amplifiers and mixers. Also, strong signals outside the IF band of the receiver can reduce sensitivity to desired signals.

### 7.7.1 Gain compression

All amplifiers have a limit to the strength of the signal they can amplify with constant gain. As the input signal gets stronger, a point is reached where the output doesn't increase to the same degree as the input; that is, the gain is reduced. A measure for this property is called the 1 dB compression point. It is shown in Fig. 7.8. At the 1 dB point, the output power is 1 dB less, or 0.8 times, what it would be if the amplifier remained linear. The 1 dB compression point is a measure of the onset of distortion, since an amplified sine wave does not maintain the perfect sine shape of the input. Recalling studies of Fourier series, we know that a distorted sine wave is composed of the fundamental signal plus harmonic signals.

When a strong input signal outside the IF passband reaches the neighborhood of the 1 dB compression point of the input LNA, the gain of the amplifier is reduced along with the amplification of a weak desired signal. This effect reduces sensitivity in the presence of strong interfering signals. Thus, in comparing amplifiers and receivers, a higher 1 dB compression point means a larger capability of receiving weak signals in the presence of strong ones. Note that the listed 1 dB compression point may refer to the input or the output. The output 1 dB point is most commonly given, and it is the input 1 dB point plus the gain in dB.

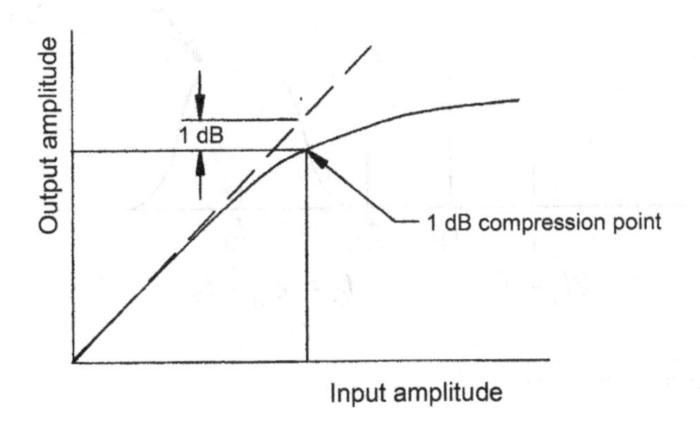


**FIG. 7.8**

1 dB compression point.

### 7.7.2 Third-order intermodulation distortion

A more sensitive indication of signal-handling capability is IMD, represented by the third-order intercept. This property is related to the dynamic range of a receiver.

When there are two or more input signals to an amplifier or mixer, the existence of harmonics due to distortion creates spurious signals whose frequencies are the sum and difference of the fundamental signals and their harmonics. If there are two unwanted input signals, $f_1$ and $f_2$, and the distortion creates second harmonics, these are the frequencies of the signals at the output of the amplifier (input frequencies plus spurious frequencies attributed to distortion up to third-order):

$$f_1, f_2, 2f_1, 2f_2, |f_1 - f_2|, 2f_1 - f_2, 2f_2 - f_1$$

If $f_1$ and $f_2$ are fairly close to a received frequency $f_0$ but not in the IF passband, either of the two last terms, $2f_1 - f_2$ and $2f_2 - f_1$, could produce interfering spurious signals closer to $f_0$ depending on whether $f_1$ and $f_2$ are above or below $f_0$. This is demonstrated in Fig. 7.9. Both $f_1$ and $f_2$ are strong signals outside the IF passband, but they create $f_3$, which interferes with the desired signal $f_0$.

The strength of the spurious interfering signal, $f_3$ in Fig. 7.9 increases in proportion to the cube, or third power, of the amplitudes of $f_1$ and $f_2$. The intercept point, IP3, is the point on a plot of amplifier power output vs. input, expressed in decibels, where the power of $f_3$, the third-order distortion product $2f_1 - f_2$ or $2f_2 - f_1$ at the output of the amplifier, equals the output power that equal amplitude signals $f_1$ and $f_2$ would have if the amplification remained linear. Fig. 7.10 shows the amplification curves of $f_1$ or $f_2$, and $f_3$, for a 15-dB gain amplifier with relative output IP3 (OIP3) = 10 dB and input *IP3 (IIP3)* = −5 dB. The slope of the $f_3$ curve is three whereas that of the normal amplification curve is one. The actual output power in a real amplifier will never reach the intercept point because of the amplitude

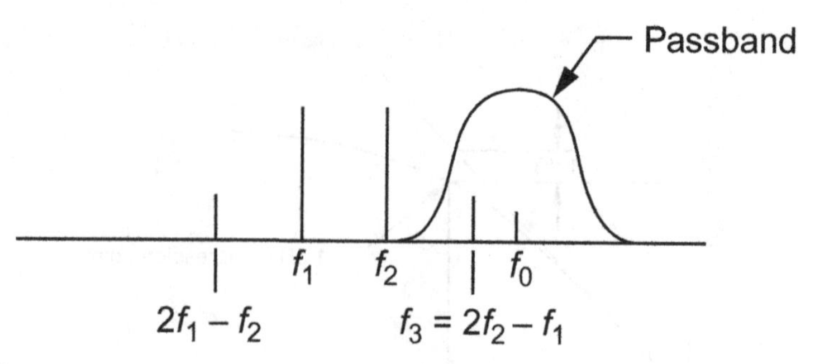


**FIG. 7.9**

Intermodulation interference.

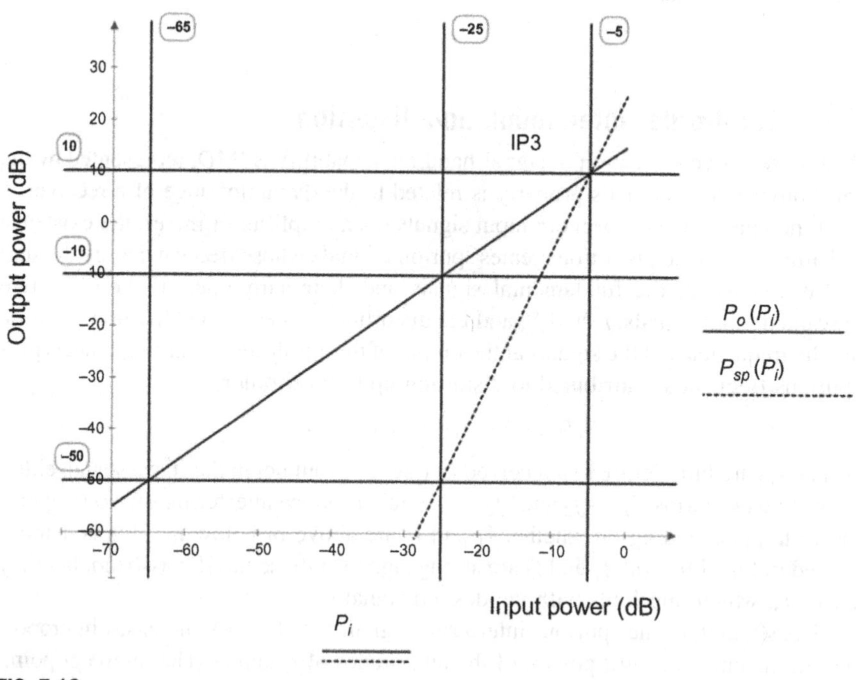


**FIG. 7.10**

Third-order intercept.

compression that starts to flatten out the output power at around the 1-dB compression point, which is below the intercept.

The equation of the amplifier ouput power $P_o$ vs. input power $P_i$ in dB is

$$P_o = G + P_i \tag{7.16}$$

and the equation of the spurious output power vs. interfering input power in dB, indicated by a dotted line is

$$P_{sp} = 3 \cdot P_i - 2 \cdot IIP3 + G \quad (7.17)$$

where $P_{sp}$ is output power, $P_i$ is input power, and $G$ is the gain, all terms in dB.
An example illustrates the use of the curves.

### Example 7.3

An amplifier has an output third-order intercept $OIP3 = 10$ dBm. Gain $G = 15$ dB. Find the spurious output power resulting from two interfering signals of −25 dBm at the amplifier input, the virtual spurious input power, and the output power of the interfering signals.

*Solution*

Using the dotted $P_{sp}$ curve in Fig. 7.10, the spurious output corresponding to interfering inputs of −25 dBm is **−50 dBm**. From the solid $P_o$ curve, the virtual spurious input corresponding to an output of −50 dBm is **−65 dBm**. The power of the interfering signals at the amplifier output, read off the $P_o$ curve at −25 dBm input, is **−10 dBm**.

The answers can be found directly from Eqs. (7.16), (7.17).

**(1)** Spurious output power: in Eq. (7.17), $P_i = -25$, $IIP3 = OIP3 - G = -5$. Then

$$P_{sp} = 3(-25) - 2(-5) + 15 = -50$$

**(2)** Spurious input power: in Eq. (7.16), $P_o = -50$. Then, subtracting gain:

$$P_i = (-50) - 15 = -65$$

**(3)** Power output of interfering signals: in Eq. (7.16), $P_i = -25$. Then, adding gain:

$$P_o = -25 + 15 = -10$$

Note from the dotted curve in Fig. 7.10 that as the interference increases, the resulting spurious signal increases three times as fast, and when the interfering inputs increase by 20 dB from −25 dBm, the spurious output becomes equal to the output of the interfering signals, at the third-order intercept point. As mentioned above, this point won't be reached in practice because of nonlinearity at high input signal levels. In any case, the larger the OIP3 relative to the rated output power, the greater the linearity of the amplifier over its operating range and the less susceptible it is to spurious responses.

Problems relating to third order distortion can be easily solved using Worksheet **Intermodulation.mcdx**.

If we don't have the intercept point explicitly from a data sheet, we can use the following rules of thumb to find it and the 1-dB compression point of a bipolar transistor amplifier stage.

Output third-order intercept:

$$OIP3 \approx 10\log(V_{ce} \cdot I_c \cdot 5)\ \text{dBm} \quad (7.18)$$

1 dB compression:

$$P1dB \approx OIP3 - 10\,\text{dBm} \tag{7.19}$$

where $V_{ce}$ is the collector to emitter voltage and $I_c$ is the collector current.

### 7.7.3 Minimum discernible signal (MDS) and dynamic range

We saw that there is a minimum signal strength that a given receiver can detect, and there is also some upper limit to the strength of signals that the receiver can handle without affecting the sensitivity. The range of the signal-handling capability of the receiver is its dynamic range.

The lowest signal level of interest is not the sensitivity, but rather the noise floor, or minimum discernible signal, MDS, as it is often called. This is the signal power that equals the noise power at the entrance to the demodulator. It can be found through Eq. (7.1) or (7.15), minus the last term, *S/N*.

The reason for using the MDS and not the sensitivity for defining the lower limit of the dynamic range can be appreciated by realizing that interfering signals smaller than the sensitivity but above the noise floor or MDS, such as those that arise through IMD, will prevent the receiver from achieving its best sensitivity.

The upper limit of the dynamic range is commonly taken to be the level of interfering signals that create a third-order spurious signal with an equivalent input power equal to the MDS. This is called two-tone dynamic range, TTDR. It is determined from the input intercept *IIP*3 with the following relation:

$$\text{TTDR} = \frac{2}{3} \cdot (IIP3 - \text{MDS}) \tag{7.20}$$

While Eq. (7.20) is the definition preferred by high-level technical publications, you may also see articles where dynamic range is used as the difference between the largest wanted signal that can be demodulated correctly and the receiver sensitivity. This definition doesn't account for the effect of spurious responses and is less useful than Eq. (7.20). Another way to define dynamic range is to take the upper limit as the 1-dB compression point and the lower level as the MDS. This dynamic-range definition emphasizes the onset of desensitization. It may be called single-tone dynamic range.

The TTDR of the receiver in Fig. 7.3A using an *IIP*3 of −5 dBm and MDS of (−106.9 − 8.5) = −115.4 is, from Eq. (7.20)

$$\text{TTDR} = \frac{2}{3} \cdot (-5 - (-115.4)) = 73.6\,\text{dB}$$

We note here that the intercept point *IIP3* to use in finding dynamic range is not necessarily that of the LNA, since the determining intercept point may be in the following mixer and not in the LNA. The formula for finding the intercept point of three cascaded stages is

$$IIP3 = \frac{1}{\dfrac{1}{IIP3_1} + \dfrac{G_1}{IIP3_2} + \dfrac{G_1 \cdot G_2}{IIP3_3}} \tag{7.21}$$

where $G_1$ and $G_2$ are the *numerical gains* in of the first two stages and $IIP3_1$, $IIP3_2$, and $IIP3_3$ are the *numerical input intercepts* expressed in milliwatts of the three stages. Then the total input third-order intercept in dBm will be 10 log($IIP3$).

Note that in calculating IMD of a receiver from its components in the RF chain, you must base the IMD on interfering signals outside of the IF passband and take account of the resulting strength of those interfering signals as they pass through bandpass filters.

It should be evident from this description that adding a preamplifier to improve (reduce) noise figure to increase sensitivity may adversely affect the dynamic range because its gain reduces overall IP3, and using an input attenuator to control IMD and compression will raise the noise figure and reduce sensitivity. The design of a receiver front end must take the conflicting consequences of different measures into account in order to arrive at the optimum solution for a particular application.

### 7.7.4 Second-order intermodulation distortion

In Table 7.3 we saw that our receiver has a spurious response at 427.65 MHz, which is the desired frequency of 433 MHz less one half the first IF—10.7/2. This response is the difference between the second harmonic of the interfering signal—2 × 427.65—and the second harmonic of the first local oscillator—2 × 422.3—which results in the first IF of 10.7 MHz. These second harmonics are the result of second-order IMD. This spurious response can be reduced by the image rejection filter between the LNA (Fig. 7.6) and the first mixer, which should also reject the harmonics produced in the LNA.

### 7.7.5 Automatic gain control

Dynamic range is so important in some short-range radio systems that sometimes special measures are taken to improve it. One of them is the incorporation of automatic gain control, AGC. Fig. 7.11 shows its principle of operation.

The incoming signal strength is proportional to the amplitude of the IF signal, which is taken off to a level detector. A low-pass filter averages the level over a time constant considerably larger than the bit time if amplitude shift keying (ASK) is used, and longer than short-term radio channel fluctuations if the signal is frequency shift keying (FSK) or phase shift keying (PSK). The DC amplifier has a current or voltage output that will control the gain of the LNA and other stages as appropriate to reduce the gain proportionally to the strength of the input signal.

There are some variations to the basic system of Fig. 7.11. In one, level detection is taken from the baseband signal after demodulation. For an ASK system, a peak detector with a long time constant compared to the bit time samples the amplitude of demodulated pulses. Also in digital data systems, a keyed AGC implementation is used. Relatively weak signals are unaffected by the AGC, which starts reducing the gain only when the signal strength surpasses a fixed threshold point. Instead of

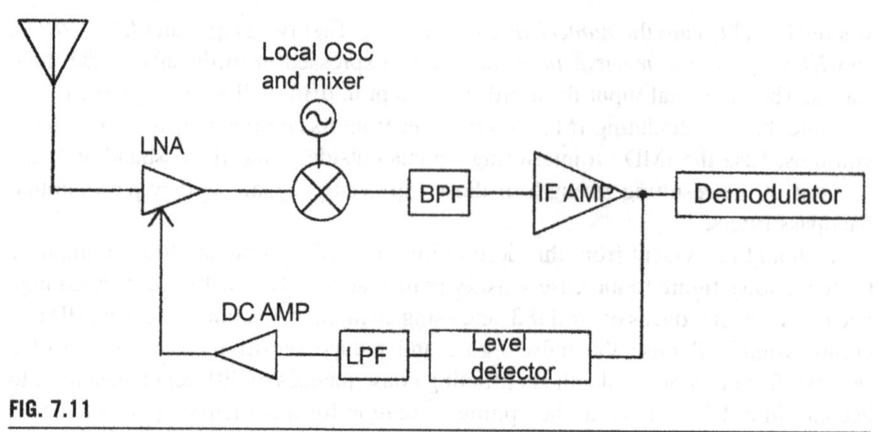


**FIG. 7.11**

Automatic gain control.

regulating the gain continuously, the gain may be switched to a lower value after the threshold is reached, instead of being adjusted continuously.

While AGC is effective in increasing the single-tone dynamic range, it will not improve TTDR when the IMD is mainly contributed by the LNA.

## 7.8 Demodulation

The RF and IF sections of the receiver are common to a wide range of system types, but the demodulator is specific to the type of modulation used. We will consider the three basic digital data modulation methods—ASK, FSK, and PSK.

### 7.8.1 ASK

ASK is the general term referring to modulation by changing the level of the RF carrier. It is sometimes referred to as OOK (on-off keying) although ASK can be understood as having multiple levels. ASK can be demodulated very simply using a diode detector at the output of the IF amplifier. Another way is to use the received signal strength indicator (RSSI) provided in many integrated circuit receiver ICs. This output is a DC current or voltage that is a logarithmic function of the signal strength. The RSSI output facilitates adding AGC, and when used in cellular radios can be part of the transmitter power control function. An RSSI current output must be shunted by a resistor and capacitor for current-to-voltage conversion and filtering before going to the data comparator. The time constant of the resistor-capacitor combination should be less than 10% of the width of the narrowest data pulse to avoid unreasonable distortion.

### 7.8.2 FSK

On receiver and IF circuit ICs, a common FM or FSK demodulator is a quadrature detector. This is essentially a multiplier that acts as a phase comparator. The amplitude limited IF signal is applied to both inputs of the detector, but a phase difference dependent on frequency is created on one of its inputs by a capacitance $C_S$ in series with a tuned circuit, as shown in Fig. 7.12. Fig. 7.13 shows how the shape of the output curve depends on the quality factor $Q$ of the tuned circuit. The demodulated output voltage is proportional to the phase difference of the inputs to the multiplier, which is the ordinate of the curve in Fig. 7.13. The following example shows how to find the required $Q$.

**Example 7.4**

Given: Maximum frequency deviation = 60 kHz. Worst-case frequency drift = ±15 kHz. $f_{IF}$ = 10.7 MHz.

**(a)** Set $\Delta f$ to the sum of the maximum frequency deviation and worst-case drift:

$$\Delta f = 60 + 15 = 75\,\text{kHz}$$

**(b)** Get maximum normalized deviation from $f_{IF}$:

$$\text{Max } \Delta f / f_{IF} = 75\,\text{kHz}/10.7\,\text{MHz} = 0.007$$

**(c)** Choose the highest $Q$ with a reasonably straight curve in Fig. 7.13 between normalized deviation limits:

$$Q = 60$$

The resonant circuit components shown in Fig. 7.12 can be found using the formulas

$$f_{IF} = \frac{1}{2\pi\sqrt{L(C_p + C_s)}} \tag{7.22}$$

$$Q = \frac{R}{L \cdot 2\pi \cdot f_{IF}} \tag{7.23}$$

Consult the data sheet of the particular IC used to find the recommended $Cs$ for the IF used (it may even be included on chip). $R$ may also be built in and its value must be used in the expression for $Q$ above. For the NE605A IC $R$ = 40 kΩ and the recommended value of $C_s$ is 1 pF. First find $L$ using Eq. (7.23) then solve Eq. (7.22) for $C_p$ after inserting the calculated $L$ and the recommended $C_s$. Using the above equations for our example with $Q$ = 60, $L$ = 9.9 μH, and $Cp$ = 21.3 pF.

Note that the actual circuit $Q$ will be less than that calculated because of the losses in the inductor $L$. Both ceramic and $LC$ modular components are available for quadrature detectors and using them will make tuning unnecessary and may save board space.

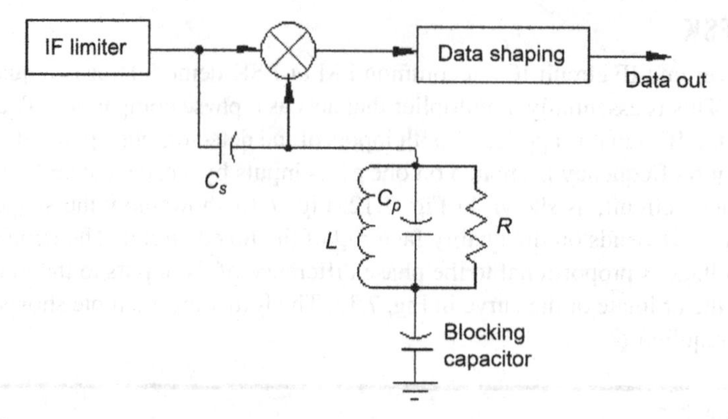


FIG. 7.12

FSK quadrature detector.

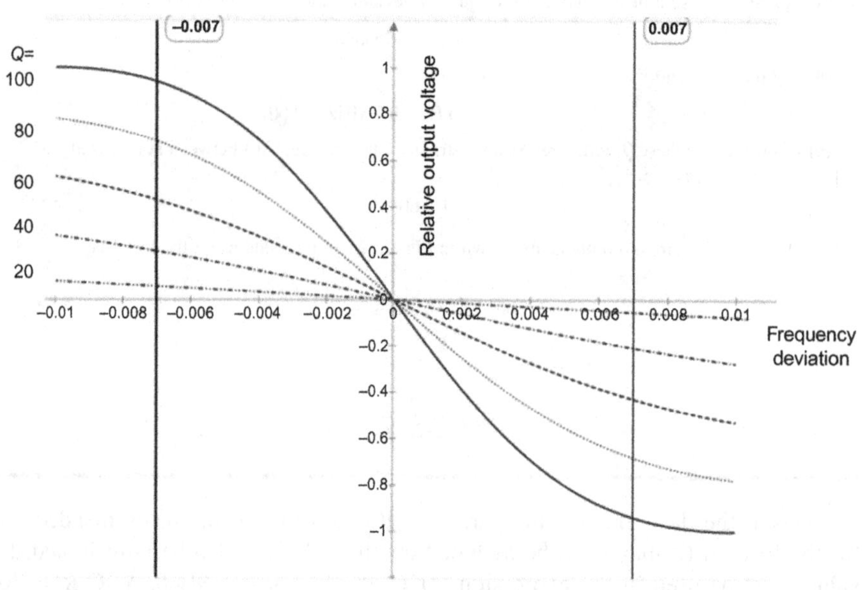


FIG. 7.13

FSK quadrature detector frequency response.

Also note that choosing a $Q$ too low will result in a lower baseband signal level out of the IC and a lower $S/N$ into the data comparator. Too high a $Q$ can cause unsymmetrical data due to frequency drift across the $S$ curve of Fig. 7.13, resulting in an increased error rate out of the data comparator.

Many receiver chips, such as Analog Devices ADF7020-1, have a digital FSK demodulator and external components are not necessary. Parameters are adjusted through software programming.

#### 7.8.2.1 Digital methods for frequency shift keying

A defining parameter for FSK is the modulation index. For binary modulation it is defined as

$$m = \frac{f_2 - f_1}{R} \tag{7.24}$$

where $f_2$ and $f_1$ are the high (usually mark) and low (usually space) frequencies on both sides of the carrier frequency, and $R$ is the maximum data rate. When $m = 1$ or a higher integer, the data can be detected optimally using two pairs of $I$ and $Q$ channels for energy detection (see Section 4.4.6 for a description of $I/Q$ demodulation) [6]. Abrupt changes in data levels cause spectrum widening well beyond $f_2 - f_1$. The resulting potential adjacent channel interference can be reduced by inserting a low pass filter in the data line before the FSK modulator in the transmitter. A Gaussian filter is especially suitable as it has a sharp cutoff beyond the passband and no overshoot. It is characterized by the product of its bandwidth B at 3 dB down times the bit period $T$, that is, $BT$. While the filter significantly reduces interfering sidebands, it also has the detrimental effect of causing inter-symbol distortion. FSK with a Gaussian filter is called GFSK, Gaussian frequency shift keying. The modulation index is not necessarily unity. For legacy Bluetooth, for example (Chapter 12), $m$ is nominally 0.3 and $BT = 0.5$ to keep the radiated spectrum within a span of 1 MHz with a data rate of 1 Mbps.

Improved spectral efficiency (data rate/occupied bandwidth) is provided by MSK (minimum shift keying). When coherent detection is employed, which implies phase synchronization between transmitter and receiver oscillators, the modulation index for optimum detection (orthogonal mark and space signals) is reduced to 0.5, that is, the difference between mark and space frequencies is one-half the data rate. Here, too, a Gaussian filter is used in the transmitter to limit the transmitted sidebands, creating GMSK (Gaussian medium shift keying). In addition to good spectral efficiency, the GSMK waveform envelope has constant amplitude which permits the use of efficient nonlinear amplification. GMSK is the modulation format of GSM cellular.

### 7.8.3 PSK

PSK is the most efficient modulation form from the point of view of $S/N$ or $Eb/N$ vs. BER, but it is rarely used, if at all, for the short-range security and control systems we have been dealing with. However, it is very often employed for modulating the pseudo-random spreading signal in direct-sequence spread-spectrum applications and in other high data rate systems.

Demodulation of PSK is more complicated than ASK and FSK because it is a coherent process: that is, it needs knowledge of the phase of the transmitted carrier signal, and this information is not directly available. If we input a coherent CW signal representing the RF carrier to one input of a phase comparator and the phase-modulated signal to the other input, the output of the phase comparator will be a baseband signal that reflects the changing phase of the data.

A method for demodulating binary PSK is presented in Fig. 7.14. In the squaring loop demodulator shown, the 180° phase changes of the modulated carrier are removed when doubling the IF signal input (doubling the frequency doubles the phase, and $2 \times 180° = 0°$).

This is shown at the top of the figure. A phase-locked loop at the doubled frequency provides a coherent carrier which is divided by two and applied to a phase comparator at the IF. The reconstructed data appears after a low-pass filter.

Another popular BPSK receiving circuit is the Costas demodulator. Its performance is similar to the squaring loop demodulator but frequency doubling is not required. Instead, the input is split and applied to two multiplier elements, whose local oscillator signals are in quadrature. The multiplier outputs, after filtering, are applied to a third multiplier, whose filtered output, free of phase changes, controls the VFO and phase locks its output to the input carrier. Fig. 7.15 shows the Costas demodulator.

In both types of demodulators, the absolute identity of a data bit as MARK or SPACE is not available—only the change from one state to the other. If a pulse width

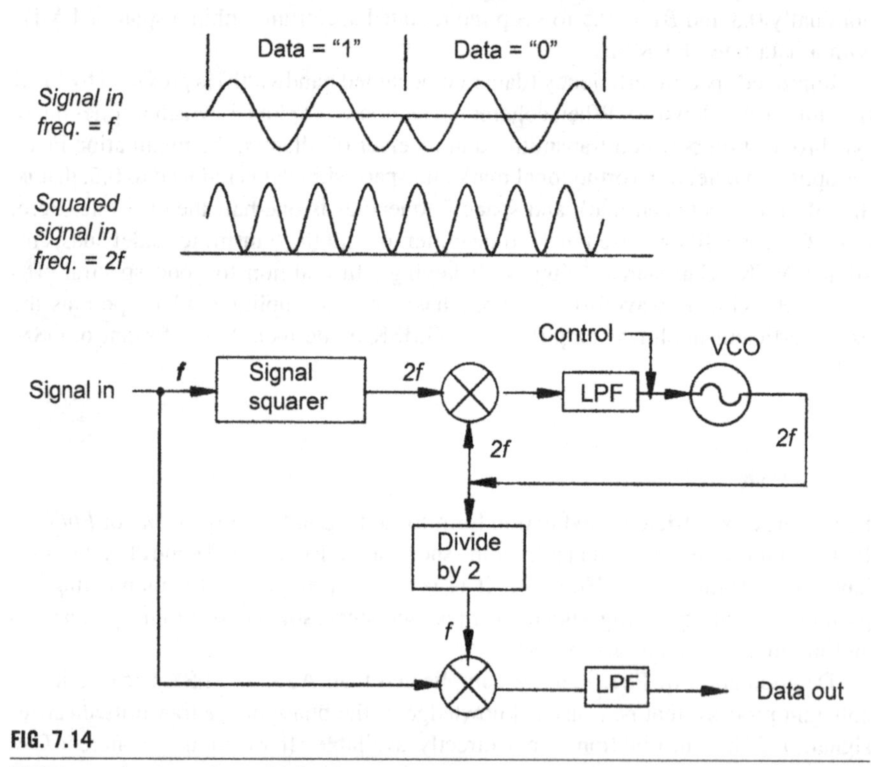


**FIG. 7.14**

BPSK squaring loop demodulator.

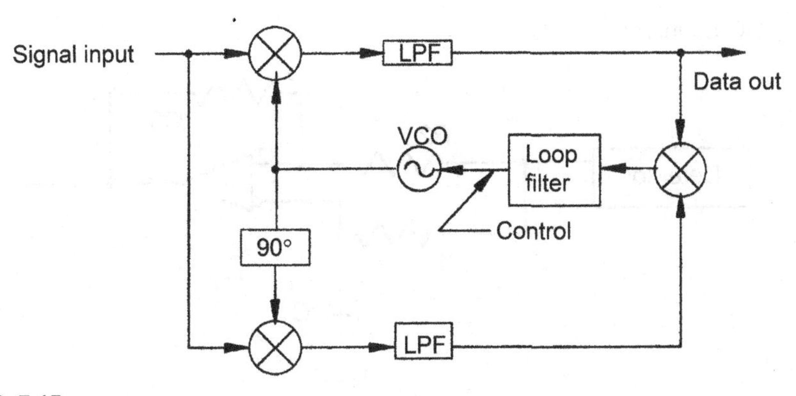


**FIG. 7.15**

BPSK Costas loop demodulator.

protocol is used, or code inversion modulation such as bi-phase mark or bi-phase space, the transmitted code is readily reproduced. Another way to retain bit identity is to use differential modulation, where the change or lack of change from one data bit to the next determines whether a transmitted bit is MARK or SPACE.

### 7.8.4 Data comparator

The output from the ASK, FSK, and PSK detectors described above is essentially an analog signal contaminated by noise, not useful for connection to digital logic circuits. An additional process, sometimes called data slicing, is required to output the data in the form of digital logic levels. This is done in a comparator, which compares the signal with a threshold level and outputs a logical one or zero depending on which side of the threshold the signal falls.

The problem is determining the threshold. When the detector output is unipolar—that is, referenced to ground as in ASK—the optimum threshold varies according to the strength of the signal and the level of the noise.

In the DC-coupled circuit of Fig. 7.16A, the comparator threshold is determined as the average voltage difference between zeros and ones. The time constant $R1 \times C$ should be several times the longest bit length. $R2$ and $R3$ provide hysteresis, reducing data chatter during bit changes.

Fig. 7.16B shows AC coupling via a capacitor. The threshold is derived from a fixed voltage point, in this case a voltage divider. Note that in Fig. 7.16B, as opposed to Fig. 7.16A, the polarity of the output pulses is reversed. AC coupling can be used for baseband data formats that have no DC component, like Manchester coding and bi-phase mark and space.

The capacitors in the data comparators of Fig. 7.16 take time to charge up to a new equilibrium after the signal has been absent for some time, or when the receiver is just turned on. Therefore the threshold level isn't optimum during the first few bits

DC coupling

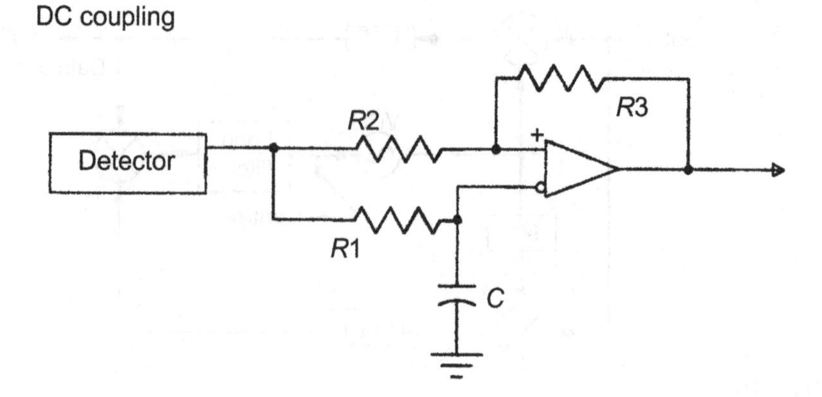

(A) AC coupling

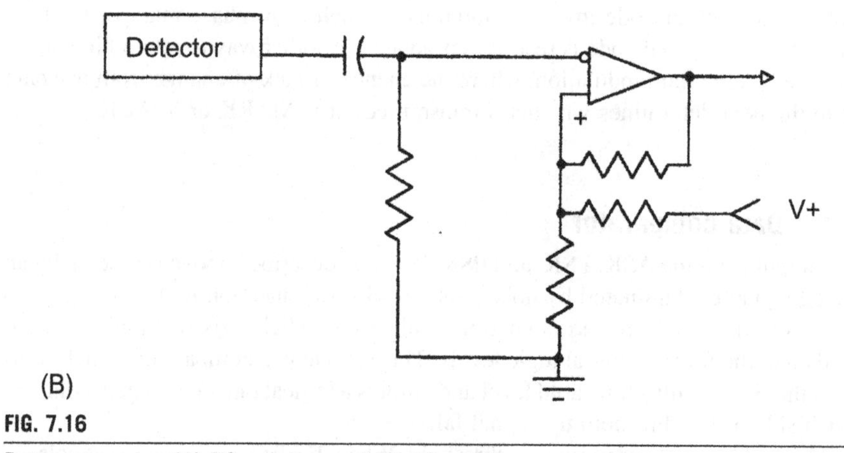


(B)

**FIG. 7.16**

Data comparator. (A) DC coupling and (B) AC coupling.

received and the error rate may be high. The data protocol should counter this effect by sending a simple preamble of alternating zeros and ones before the actual data, to give time for the threshold to reach its proper value.

## 7.9 Transmitter design

There are fewer design topologies in the short-range transmitter than in the receiver, and their influence on overall system performance is less, given a specific allowed power output. However, much more attention must be given in the transmitter to the restrictions placed on unlicensed equipment and failure to recognize and come to grips with potential problems early in the design may result in delays in getting the product to the market. Worse, it may result in a system that doesn't give the performance that even low-power short-range equipment can deliver.

### 7.9.1 Spurious radiation

Short-range unlicensed radio regulations, both in North America and in Europe, put more emphasis on reducing radiation outside of the communication channel than on frequency stability and in-channel interference. It's obviously desirable to use the maximum power allowed for best range and link reliability, but this aim is often thwarted because of an excessive level of radiated harmonics whose reduction is accompanied by a reduction of the desired signal power. It's not the purpose of this section to give detailed instructions on reducing spurious radiation, but rather to call attention to the importance of planning for it early in the design stage. Measures to consider in this regard are

- reduction of printed conductor lengths in the RF stage
- use of small SMT (surface mounted) inductors for low unwanted radiation
- isolating low-frequency circuits and power and battery leads from RF by effective capacitive bypassing and grounding
- providing a high $Q$ output coupling circuit to the antenna
- assuring a good impedance match to the antenna
- choosing and biasing the output transistor for low distortion
- incorporating a bandpass filter between the RF output transistor and the antenna
- shielding the RF stage, or the whole transmitter.

The last two measures usually go together, since they may be the only way to get sufficient isolation between the input of the bandpass filter and the output. Shielding and filters will be necessary to achieve maximum performance for transmitters certified under EN300 220 in Europe, where in most countries 10-mW output power is allowed, but only 250-nW for harmonics, a difference of 46 dB. In the United States and Canada, significantly less power is allowed (except for spread spectrum) under the unlicensed rules, and prudent circuit layout should be adequate to get maximum fundamental radiation while not exceeding spurious radiation limits.

## 7.10 Bandwidth

As mentioned above, restrictions on bandwidth are not severe for most types of unlicensed operation, a fact that allows the use of convenient and inexpensive SAW-controlled oscillators. However, there are exceptions, and there may be advantages to using narrow bandwidth unlicensed channels where the highest performance and reliability are essential. Examples of narrow bandwidth channels are 40.66 to 40.70 in the United States and some places in Europe, 173 to 173.35 MHz in the United Kingdom, and several channels in the European 868-870 MHz band.

The first requirement for meeting narrow bandwidth requirements is a stable frequency source. This can be provided only by a crystal oscillator. Bandwidth occupation depends on the required modulated bandwidth of the transmitted signal plus the stability of the frequency. Narrow-band bandwidth is typically 25 kHz. An FSK signal with data rate of 4 kb/s and deviation of 4.5 kHz will have a minimum bandwidth of approximately 13 kHz. That leaves 12 kHz for accuracy and drift. Thus, the crystal oscillator for 868 MHz must remain within 12 kHz/868 MHz = 14 ppm of the design center frequency.

## 7.11 Antenna directivity

Most short-range applications require nondirective antennas. The transmitter or receiver may be portable, and there is usually no way to align transmitting and receiving antennas. Most often the antennas are an integral part of the device and there is no transmission line or connector. During type testing, the maximum radiation is what counts as far as meeting the maximum power requirements, whereas it is the direction of minimum radiation that determines the ultimate performance for random installation orientations. Therefore, an antenna with significant directivity is wasteful of the potential of even the low power allowed for unlicensed devices.

Information on the radiation characteristics of the antennas used on unlicensed devices is given in Chapter 3. Usually there is no defined ground plane on the device, and the antenna is located in close proximity to other components which affect its directivity. Connection wires also have an effect and should be decoupled as well as possible from RF currents. While it is very difficult to predict the radiation pattern of a portable low power device, at least it should be tested and antenna orientation should be varied by trial and error to get a relatively omnidirectional pattern.

## 7.12 The power source

Most short-range devices, and usually transmitters, operate from batteries. The inconvenience of changing batteries is often a deterrent to the use of wireless devices. Security system sensors and portable control devices with very low current consumption rarely use rechargeable batteries. However, wireless telephones and other devices with medium to high current consumption use rechargeable batteries in preference to primary cell batteries, whose periodic replacement would be inconvenient and costly. Security system and other receivers require more energy than intermittently operated transmitters, and they normally have access to a mains power source, but they are almost universally backed up by secondary rechargeable batteries.

The various battery types available certainly affect wireless system design. Power supplies for portable equipment must be designed for specific kinds of batteries. First of all, batteries come in many different voltages. Unlike mains supply, battery terminal voltage and internal resistance change during battery life or discharge time. Requirements for lowest possible current drain also dictate the type of voltage regulation to be used, or often makes the designer do with no regulation at all. Thus, a handheld remote control transmitter may have good range just after the battery is replaced, but performance will often noticeably deteriorate due to reduced transmitter output power as the battery wears out. Many portable wireless devices have low-battery detection systems that indicate either visually or by signaling a control panel that the battery needs replacement. It is important for the wireless system designer to determine the appropriate battery type early in the design stage. We give here a brief description of several battery types used in short-range radio applications.

Batteries may be classified as primary and secondary. Primary batteries are not rechargeable, are inexpensive in respect to secondary rechargeable batteries and generally have superior performance or features in regard to lifetime and capacity relative to mass and volume. For example, Fig. 7.17 shows the superior energy to weight ratio of two primary battery types compared to secondary batteries. However, there are other characteristics the should be considered in choosing a battery, including power capability and endurance [7]. Nine-volt alkaline batteries have been very popular for portable control units and security sensors, as well as 12-V alkaline batteries for wireless keys, but lithium batteries are probably the most popular for new products. Most of the standard types of active electronic components and integrated circuits operate at 3 V and below, so the 3-V lithium cells can be used directly. Incorporation of lithium batteries in portable wireless equipment has extended the replacement period to 5 years and more. Lithium batteries are not homogeneous, and there are several chemical systems in use, each with its own performance and safety characteristics. These are some lithium primary battery types:

- Consumer grade *lithium iron disulfate* ($LiFeS_2$) cells can deliver high pulses, for example to power a camera flash. They are inexpensive but have a relatively high self discharge rate [8].
- *Manganese dioxide lithium cells* )$LiMNO_2$) are available in standard sizes. These cells have lower internal impedance than the other lithium battery types so they are well suited to applications needing relatively high continuous current or pulse requirements. Small button form cells have largely replaced 9 and 12 V alkaline batteries in keyless entry devices [9].
- *Lithium thionyl chloride* ($LiSOCl_2$) cells have long battery life and are particularly suitable for use in extreme environments. They are a preferred choice for remote wireless applications, including, for example, windshield mounted electronic toll tags and wireless sensors. Thionyl chloride lithium cells have the highest energy density of all lithium types. The cells are best suited to very low continuous current and moderate pulse demand applications and their long service life and low self-discharge rate make them appropriate in products having limited physical access, such as remote sensing systems. Thionyl chloride lithium cells have an open circuit voltage of 3.6 V [8].

The staple rechargeable (secondary) battery type in portable equipment has for many years been nickel cadmium. It has high energy density and tolerance for abuse, as well as being economical. Load voltage of a fully charged NiCd battery is between 1.2 and 1.3 V/cell. Its terminal voltage is fairly constant until discharge, when it reaches 1 V/cell. NiCd batteries have very low internal impedance so they

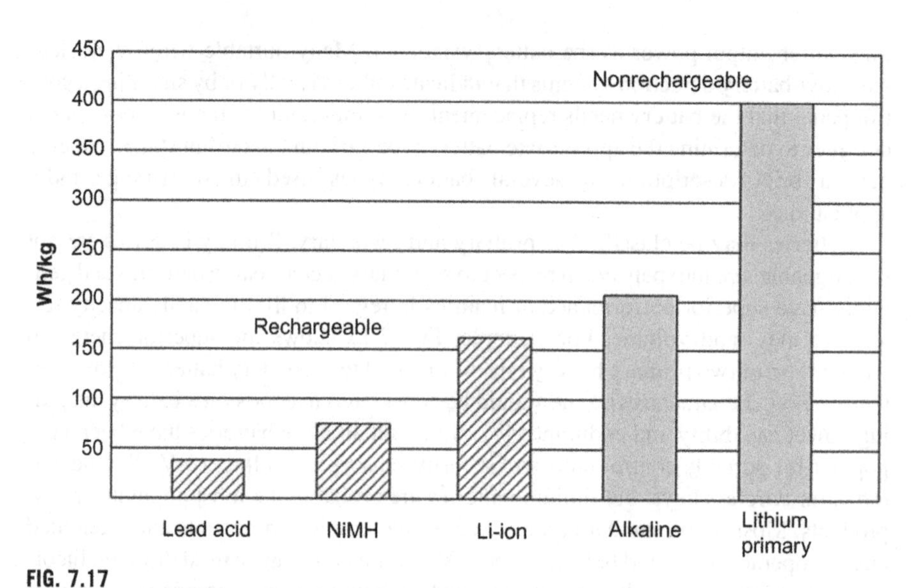


**FIG. 7.17**

Energy vs. weight comparison of secondary and primary batteries.

*Copied with permission from Cadex Electronics, Inc.*

can supply high current on demand. A known deficiency of NiCd batteries is the so-called memory effect. Batteries that are recharged often and fully after shallow discharge appear to "remember" their short discharge time and lose capacity. This effect is reversible, and may be corrected by several cycles of complete discharge and recharge.

Nickel metal hydride batteries have taken over many applications of nickel cadmium types. Having similar characteristics, including cell voltage and charging requirements, they don't have the capacity-reducing memory effect. Nickel metal hydride batteries are more expensive than similar capacity NiCd types. Nickel metal hydride batteries are reported to have higher internal leakage than NiCd and can withstand a lower number of charge/discharge cycles.

With a different battery chemistry, lithium-ion, is replacing both nickel cadmium and nickel metal hydride types for high-performance applications. Batteries based on lithium are lighter than competing types, and they have high cell voltage, over 3 V, like their primary cell counterparts. There are two lithium-ion technologies. One uses a liquid electrolyte, the other uses a solid polymer electrolyte. The solid polymer electrolyte has advantages as it can be made thin and flexible and can be configured in virtually any size. While more expensive than comparable NiCd batteries, the advantages of the solid polymer lithium-ion batteries make it the choice for many applications. Lithium-ion batteries require a special charging regimen, both to obtain maximum rated capacity and to prevent damage and insure safety with use. Fig. 7.18 shows a charge cycle [10]. It includes two periods of constant current followed by constant voltage.

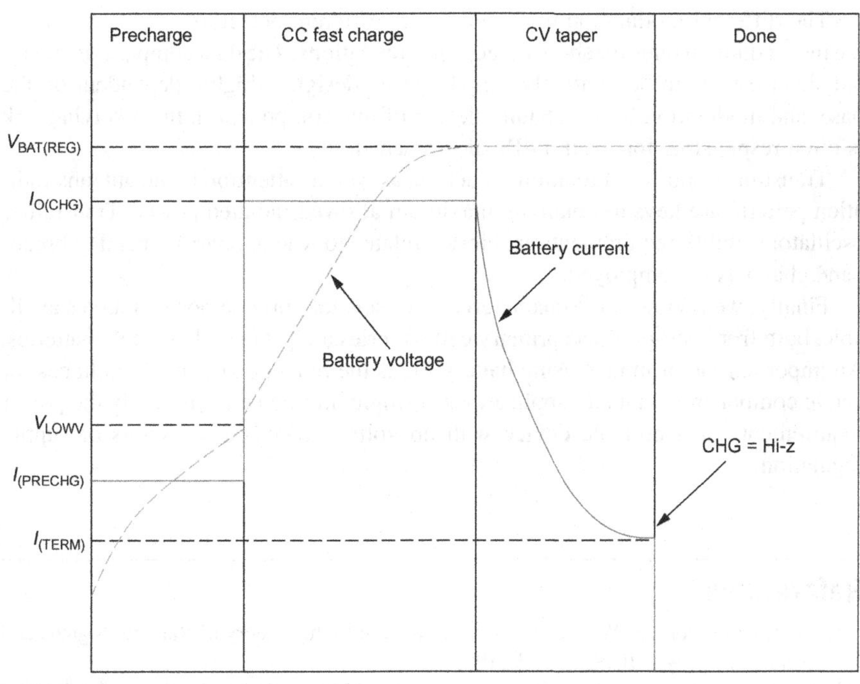


**FIG. 7.18**

Typical charge cycle for lithium-ion battery.

*Courtesy of Texas Instruments.*

## 7.13 Summary

Performance characteristics of a radio communication system are reviewed in detail in this chapter. We mentioned the conditions that affect communication range, and then dealt with the various factors that determine receiver sensitivity. Noise was presented as the ultimate limitation on communication range. We saw that external noise sources, principally man-made noise, must be taken into account to determine true sensitivity, which is a function of the relative intensity of noise received by the antenna to the noise generated in the receiver.

Preamplification is necessary in a receiver to reduce the effects of internal noise, but only up to a point. Excessive preamplification lowers the threshold of overloading, measured by compression and IMD, and reduces the dynamic range. The superheterodyne receiver topology, by far the most common in use, is susceptible to interference by signals on frequencies far removed from the desired frequency to which the receiver is tuned. We saw how to calculate these frequencies and discussed remedies for avoiding them, by filtering, shielding, and circuit layout. Since virtually every modern receiving device has digital logic circuits and microcomputers, the designer must be aware that radiation from those circuits to the receiver front end can reduce sensitivity.

The different modulation methods require different detectors and we discussed the most common ones for short-range communications. The data comparator, or signal slicer, seems to be relatively simple but its design is highly dependent on the baseband modulation format. Faulty design of this component in the receiving link is often responsible for several dBs of lost sensitivity.

Transmitter spurious radiation reduction, as well as attention to the antenna radiation pattern, are keys to obtaining maximum allowed radiated power. Transmitter oscillator stability requirements are closely related to whether narrow-band or broad-band channels are employed.

Finally, we reviewed the characteristics of several portable power sources available, both from one-time-use primary cell batteries and from rechargeable batteries. An important factor in increasing battery life is the reduction of supply voltages for active components. In many applications, a single lithium cell can supply the power requirements of a portable device with no voltage conversion circuits or supply regulation.

## References

[1] W.B. Davenport Jr., W.L. Root, An Introduction to the Theory of Random Signals and Noise, McGraw-Hill, New York, 1958.

[2] C. Razzell, System Noise-Figure Analysis for Modern Radio Receivers, Application Note 5594, Maxim Integrated Products, Inc, 2013.

[3] J.W.M. Rogers, C. Plett, I. Marsland, Radio Frequency System Architecture and Design, Artech House, 2013.

[4] Radio Noise, Recommendation ITU-R P.372-11 (09/2013).

[5] F.L. Dacus, J. Van Niekerk, S. Bible, Tracking Phase Noise in Short-Range Radios, Microwaves & RF, 2002, p.57.

[6] B. Sklar, Digital Communications Fundamentals and Applications, second ed., Prentice Hall, 2001.

[7] Battery University BU-106: Advantages of Primary Batteries, http:/batteryuniversity.com/learn/article/primary_batteries. Accessed 24 October 2018.

[8] S. Jacobs, Choosing the Right Battery to Power a Remote Wireless Device, Machine Design Magazine, 2015.

[9] Primary Lithium Cells, Sales Program and Technical Handbook, Varta Microbattery GmbH.

[10] bq2407x 1.5-A USB-Friendly Li-Ion Battery Charger and Power-Path Management IC, Texas Instruments SLUS810K, 2008, Revised March 2015.

CHAPTER

# System implementation

# 8

## 8.1 Introduction

The designer of a short-range radio system can take any of several approaches to implement their system. This chapter has examples of three levels of component integration for which devices are available from several manufacturers. These levels are:

- *Complete wireless modules*. They comprise a complete system; transmitter, receiver, or transceiver—and are packaged as a small assembled printed circuit board which can be inserted as a plug-in or solder-in component on the user's motherboard, or as a board sub-assembly to be wired to the user's baseband system. The system designer must add the antenna and the baseband application processing. Baseband inputs and outputs are standard digital levels, although some modules accept analog audio signals. There are modules available that have a standard or proprietary data interface, and do the lower protocol processing on board.
- *System on a chip (SOC)*. These are monolithic or hybrid wireless systems packaged as integrated circuits. They are almost as complete as the modules, but usually need a few more external components, such as surface acoustic wave or crystal frequency determining oscillator components, bandpass filters, and capacitors and resistors for parameter setting and DC filtering. The receivers amplify and demodulate the desired RF signal frequency and output the data in digital form. Transmitters accept digital or analog baseband inputs and output the RF signal at the required level, although external spurious radiation filtering is usually needed between the RF output and the antenna. Some devices have digital logic blocks or microcontrollers on the chip that perform baseband processing, including digital filtering and receive data clock regeneration.
- *Chips for wireless connectivity standards*. These are SOC that contain a complete RF transceiver section coupled with a microprocessor unit and peripheral components to carry out physical layer, or medium access control layer, functions of an industrial standard.

There are two other approaches to the design of short-range systems that we do not dwell on here. One is the use of discrete components and small scale ICs. This approach is rarely taken for receivers, except perhaps for lowest-cost, low-performance designs, including those based on superregenerative receivers. Low-cost transmitters for security systems and keyless entry can easily be designed from discrete



Short-range Wireless Communication. https://doi.org/10.1016/B978-0-12-815405-2.00008-7
© 2019 Elsevier Inc. All rights reserved.

components and usually are. When data encryption and other security measures must be included in a product, a mass produced low cost integrated circuit which may include RF signal handling will probably be the most practical solution.

At the opposite end of the application spectrum—sophisticated high data rate systems, often using spread-spectrum modulation—some designers will design their own proprietary application specific integrated circuit (ASIC) based device. This approach is viable for very large production quantities.

Consider the following points before deciding on which approach to take:

- overall cost, including development outlays and unit production cost,
- time to market,
- degree of satisfying particular system requirements with each approach,
- availability in-house of RF engineering proficiency and development and production test equipment, and
- possible supply problems due to choosing a single source for the main module or IC.

No matter which approach you take, you need basic RF know-how to evaluate and choose the devices to base the design on, develop or adopt appropriate coding protocols, design the antenna, submit the system to testing according to standards and regulations, present the product to the market and back it up for the customers.

Examples of devices from each of the three levels are given below, with comments on their specifications and unique features. An important parameter for choice—price—is not given but a prospective customer should be able to obtain it from the manufacturer or distributor. Remember that important performance specifications, such as sensitivity and range, should not be taken at face value but should be verified independently before using them to base a decision. While not necessarily the rule, some manufacturers engage in "specmanship," and a claimed high sensitivity may be either obviously impossible or may be stated without giving the error rate or signal-to-noise ratio on which it is based. Range specifications are most often abused, and should never be used for comparison between performance of products of different companies. In considering transmitter power output, remember that the regulations limit radiated power output, and that depends strongly on the antenna design. Declarations regarding strength of harmonics and other spurious radiation from transmitters are generally based on direct measurements at the antenna terminal, whereas their levels can be quite different in radiation measurements on the finished product.

Most of the companies that make modules and system ICs also offer evaluation boards and systems. Trying out a product before committing oneself to it is a good way to reduce the possibility of an expensive mistake.

The inclusion of a product below in no way indicates a recommendation or an endorsement of its characteristics. Many products not listed here may be more suitable for your purposes. All information is based on manufacturers' promotional materials. We tried to be accurate, but specs change, and the manufacturers should be consulted for authoritative information.

## 8.2 Wireless modules and chips

### 8.2.1 Circuit design, Japan

The company has products for unlicensed country and regional frequency bands from 335 through 868 MHz, and 2.4 GHz. Specific devices are designed for data links, audio and telecommand. The modules are designed to give maximum performance at low data rates and use stable synthesized crystal oscillators and narrow-band receivers. The form factor is metal cases with pins for soldering to a circuit board [1].

A representative transmitter-receiver pair and features:

*CDP-TX-02E-R transmitter*

- Frequency: 433-MHz band, 32 dip-switch selectable 25 kHz channels
- Modulation: FSK
- Output power: 10 or 1 mW selectable
- Maximum baud rate: 4800 bits/s
- Frequency control: Crystal-controlled synthesizer
- Stability: ±4 ppm over −10°C to +60°C.

*CDP-RX-02A receiver*

- Frequency, stability and demodulation: Compatible with transmitter
- Type: Double conversion superheterodyne
- Sensitivity: −120 dBm for 12 dB SINAD

### 8.2.2 Radiometrix, UK

A range of transmitter, receiver, and transceiver modules, operating on the popular UHF unlicensed bands and VHF frequencies under 1 GHz is offered by this company. Products differ in case style, performance, and baseband functions. An example is the NiM2B-434-650-10 transceiver module, which has high sensitivity for extended range in applications with data rates under 10 kbps. It is available for nominal supply voltages of 3.1 to 15 V. The module is shielded and can be soldered to or plugged into a motherboard. The module is shown in Fig. 8.1A and its block diagram in Fig. 8.1B [2]. These are the principle characteristics:

- Standard frequency 434.650-MHz, fractional N synthesizer controlled with temperature stabilized crystal oscillator
- FSK modulation, maximum data rate 10 Kbps
- Transmitter power output +10 dBm typical
- Receiver sensitivity −112 dBm at BER = $1 \times 10^{-6}$
- RSSI output
- Logic level and analog receiver outputs
- Built in Tx/Rx antenna switch
- Module dimensions 33 mm × 23 mm × 11 mm
- Complies with European standards EN 300 220-2, EN 301 489-3 (EMC)

Radiometrix also offers application boards containing modules similar to that described above, along with hardware and firmware for carrying out defined wireless communications functions for security and alarm systems, monitoring systems, remote control and industrial controls.

### 8.2.3 Linx, USA

Linx offer several ranges of RF data modules in different cost and performance categories. The company also supplies whip and board mounting helical antennas.

The HumPRC™ series devices are small circuit board transceiver modules with SMD packaging designed for automated assembly on final product circuit boards. See Fig. 8.2. This series is designed to compose a complete wireless remote control system [3].

Basic characteristics are:

- Operation on 900 and 868 MHz frequency bands
- Crystal controlled frequency synthesizer
- Frequency hopping spread spectrum on 900 MHz band for North America and frequency agility and listen before talk (LBT) features on 868 MHz in conformance with European regulations.
- 10 dBm output power

FIG. 8.1

Continued

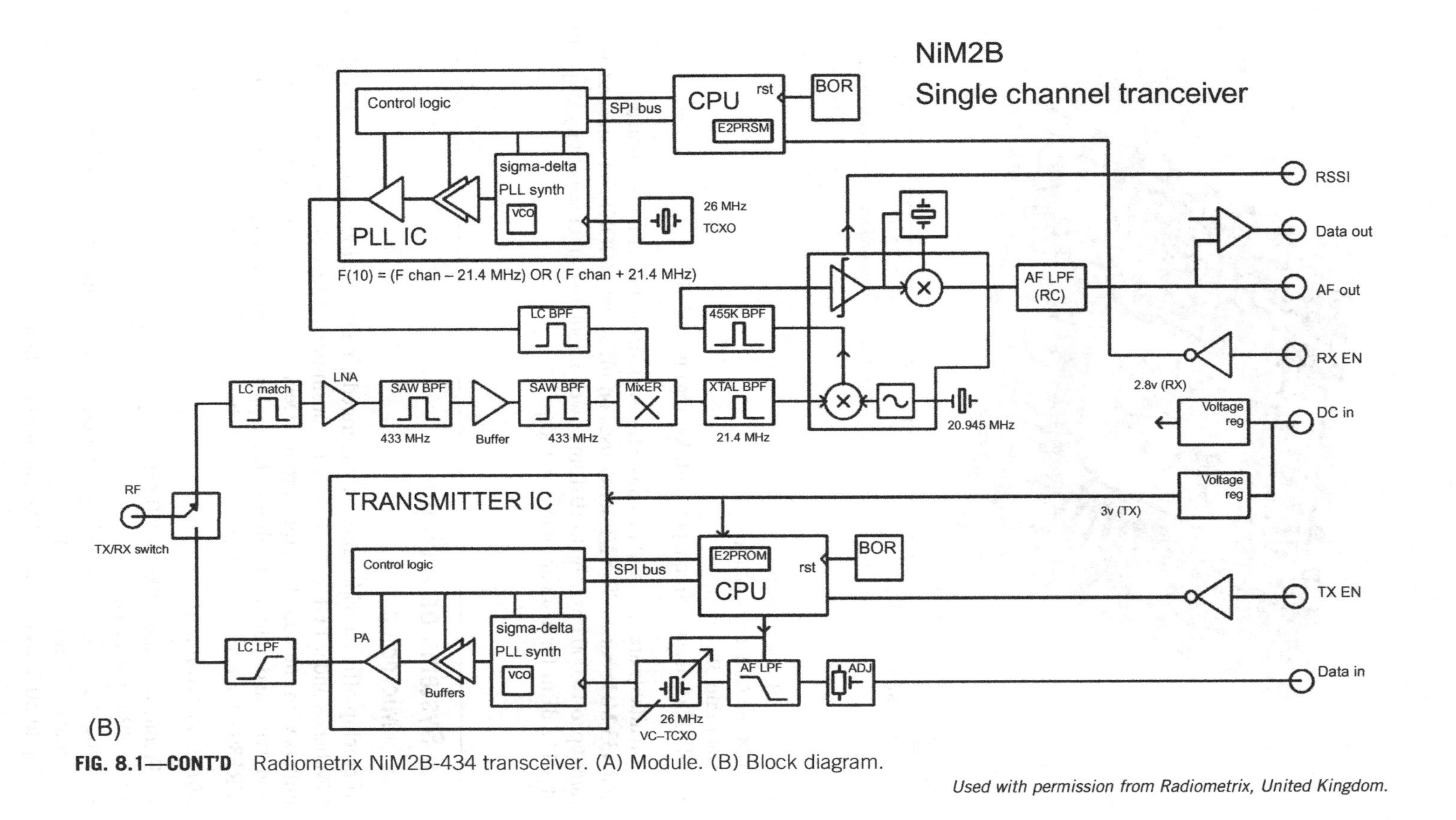


**FIG. 8.1—CONT'D** Radiometrix NiM2B-434 transceiver. (A) Module. (B) Block diagram.

*Used with permission from Radiometrix, United Kingdom.*

FIG. 8.2

Linx transceiver module.

*Used with permission from Linx Technologies, U.S.A.*

- −100 dBm sensitivity
- 128 bit AES encryption
- Data rate 38.4 kbps
- GFSK (Gaussian frequency shift keying) modulation.

Also available are encapsulated SMD transceiver modules for operation on 315, 418, and 433 MHz bands. The LT series uses on-off keying (OOK) and can send at data rates up to 10,000 bps at up to 10 dBm power output. Receiver sensitivity is specified at −112 dBm for a square wave at 1 kbps.

## 8.3 Systems on a chip

### 8.3.1 Microchip

Microchip offers a wide range of integrated circuit RF products for VHF and UHF ISM bands under 1 GHz. They include transmitters, receivers, transceivers and transmitters with embedded microcontroller. Devices differ by frequency range, modulation type and polling facilities. Examples are the MICRF112 transmitter and MRF39RA receiver [4]. These are their basic characteristics:

*MICRF112 transmitter*

- Frequency range 300 to 450 MHz, crystal control
- 10 dBm power out
- ASK/OOK (amplitude shift keying/on-off keyed) modulation, narrow band FSK (frequency shift keying)
- Up to 50 kbps data rate, Manchester encoding.

- Operating voltage 3.6 to 1.8 V
- Surface mounting packages 2 × 2 mm by 0.4 mm thick or 4.9 mm × 3 mm by 1.02 mm are specified.

*MRF39RA receiver*

- Covers popular sub 1 GHz bands 433, 868, and 915 MHz
- Sensitivity −120 dBm at 1.2 kbps FSK
- Demodulation FSK, GFSK, MSK, GMSK, OOK
- Bit rates 1.2 to 300 kbps for FSK
- Fractional-N PLL crystal controlled synthesizer with built in VCO
- Digital receiver; Analog to digital conversion after mixer.
- Zero IF non-image frequency architecture for frequency keying demodulation; low IF for ASK, OOK.
- Packet engine for packet message processing, performs AES decryption and checks CRC and other packet orientated tasks.
- Package 5 × 5 × 0.9 mm

### 8.3.2 Murata

Murata offers several RF transceiver chips for ISM bands. They are applicable for single and multichannel operation, have frequency hopping capability, a wide range of data rates, frequency and amplitude demodulation and wide operating supply voltage [5]. As an example, features of TRC103 are given below.

*TRC103 transceiver*

- Digital transmitter and receiver architecture.
- 863-870, 902-928, and 950-960 MHz.
- Modulation FSK or OOK with frequency hopping.
- Data rate up to 200 kbps.
- Low receiver current 3.3 mA.
- Receiver sensitivity −112 dBm FSK 2 kbps, BER $10^{-3}$
- 11 dBm transmit power.
- Operating supply voltage 2.1 to 3.6 V.

### 8.3.3 Nordic

Nordic offers SOC in the bands of 433, 868, and 915 MHz. The company's nRF9EF chip has a RF transceiver core, 8 bit CPU and RAM data and code memory plus on-chip analog and digital peripherals including a 10 bit analog to digital converter. User code is loaded from an external EEPROM. Fig. 8.3 shows the degree of system integration [6]. Suggested applications are alarm systems, keyless entry, home and industrial automation and RFID. Other characteristics are:

- Typical data rate 50 kbps, Manchester encoded
- GFSK modulation
- Maximum output power 10 dBm
- Sensitivity −100 dBm at 0.1% BER
- Supply voltage 1.9 to 3.6 V

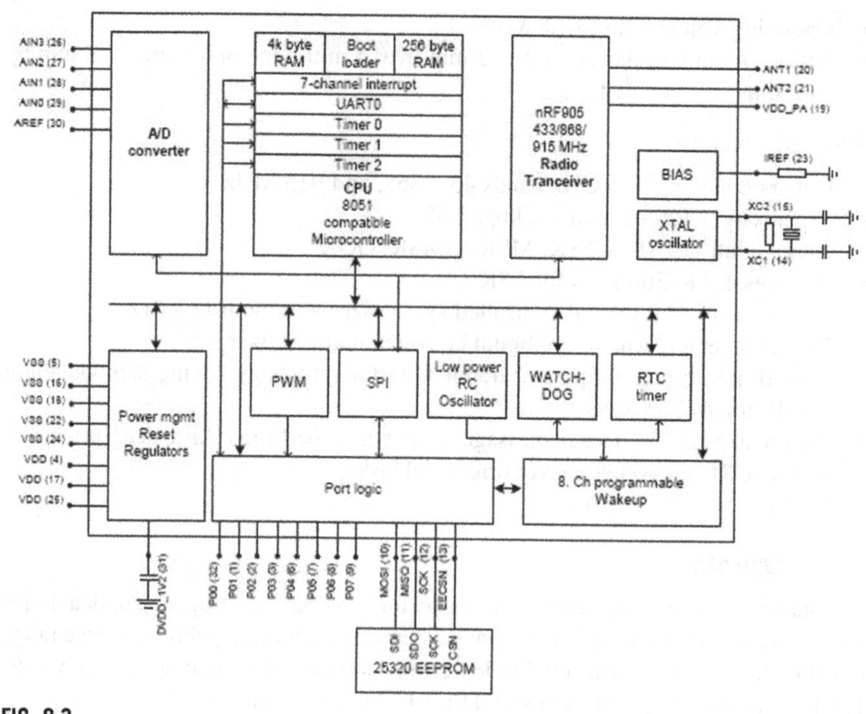


**FIG. 8.3**

Nordic nRF9E5 system on chip.

*Used with permission from Nordic Semiconductor.*

## 8.3.4 Melexis

Melexis RF products include receivers for tire pressure monitoring systems, transceivers aimed at keyless entry and low power tracking systems markets, and transmitters for various ISM applications. The company's multi-channel TH71221 FSK/FM/ASK transceiver can operate from 27 to 930 MHz, and therefore covers the 315, 433, 86, and 915 MHz ISM bands. It operates stand-alone on a fixed frequency with few external components, or in a programmable multi-channel mode with three wire serial control interface. In addition to digital FSK or ASK modulation, it can demodulate analog frequency modulated signals. The receiver has maximum rated sensitivity of $-107$ dBm for BER down to $3 \times 10^{-3}$ with 180 kHz IF bandwidth and $-105$ dBm for FSK with the same IF bandwidth [7].

## 8.3.5 Texas Instruments

TI has a wide range product portfolio which includes transceivers with and without built in microprocessor, and transmitters. Its low supply power CC1101 transceiver operates on ISM bands from 315 to 915 MHz, supports various modulation formats and a data rate up to 600 kbps. The chip supports packet oriented systems. Receiver sensitivity is rated at −116 dBm at 0.6 kbps, 433 MHz with 1% packet error rate. The transmitter has programmable output power up to 12 dBm.

TI's CC1125 is described as an ultra-high performance sub 1 GHz for narrow-band systems. It claims sensitivity of −129 dBm at 300 bps and −110 dBm at 50 kbps, and configurable data rates to 200 kbps. Programmable power output is up to 16 dBm. Similarly to CC1101, it supports packet oriented systems. Other built-in functions are antenna diversity support, support for retransmissions and auto-acknowledge of received packets. It also has built-in coding gain support [8].

### 8.3.6 Maxim

Maxim's MAX7037 transceiver operates on ISM bands from 315 to 930 MHz and has an integrated 8051 microcontroller, flash memory and sensor interface. Its fractional-N local oscillator frequency control allows multichannel operation and channel resolution of 244 and 488 Hz. Gross data rates are up to 125 kbps. The chip's mixed signal sensor interface has analog input/output and an 8 bit digital to analog converter. Targeted uses of the MAX7037 are automotive keyless entry, tire pressure monitor, building automation, smart meter and ultra low power sensor networks [9].

## 8.4 Chips for wireless connectivity standards

Several manufacturers supply chips and modules designed for specific connectivity standards. These standards are Bluetooth, Wi-Fi, and various derivatives of IEEE 802.15.4 including Zigbee and other industry variants. Also available are chips and modules that are the basis of dual systems, usually Bluetooth and Wi-Fi. Bluetooth and Wi-Fi products are aimed for installation in smartphones, while devices in all three categories are promoted for connectivity of Internet of things (IoT) networks. Some examples are given below.

### 8.4.1 Texas Instruments

The TI CC2564 is a complete Bluetooth BR (basic rate), EDR (extended data rate) and low energy chip with self contained physical layer functions. Multiple profiles contained in firmware in a coupled microcontroller are supported for a complete Bluetooth product solution. Chip hardware is designed to meet the RF specifications of Bluetooth version 4.2. TI also offers a CC2564MOD module that includes the microcontroller. Two variants are available, with integrated chip antenna and with interface for external antenna. Module footprint is 7 mm × 7 mm × 1.4 mm (typical).

A SOC product for 2.4 GHz IEEE 802.15.4 is the TI CC2538. It combines a powerful microcontroller with an IEEE 802.15.4 radio and can handle complex network stacks with security and over-the-air download. The chip supports Zigbee application profiles, and has hardware security accelerators to enable efficient authentication and encryption. It comes in an 8 × 8 mm SMD package.

## 8.4.2 Microchip

Microchip offers a wide range of Bluetooth chips and modules with varied capabilities to suit a range of applications including sensor devices, smart appliances health and fitness trackers, retail beacons and asset tracking. One example is IS1870 SOC for Bluetooth low energy (BLE) applications. The chip is designed to work with an external host microcontroller. However, the flexibility of the chip also enables the user to create a hostless implementation where a full application can be embed into the IS1870. Package size is 4 × 4 × 1 mm. Fig. 8.4 is a block diagram of the IS1870.

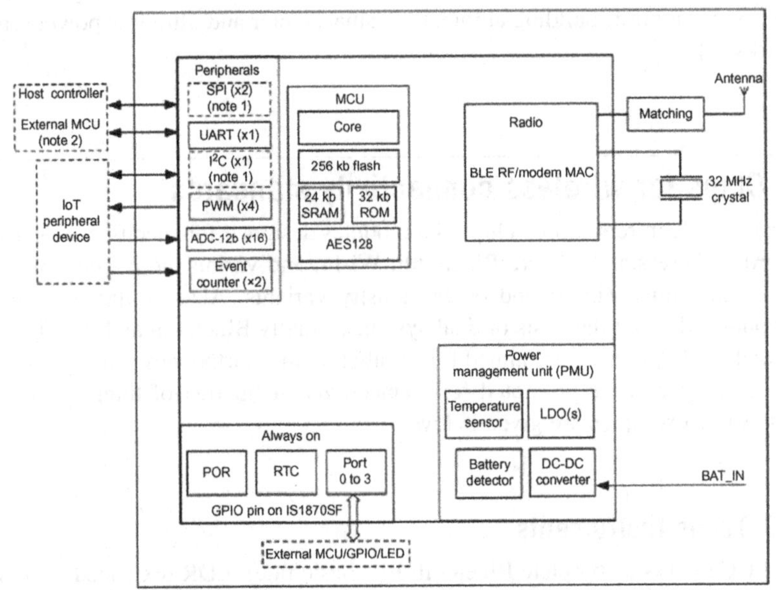


**Note 1:** Users can enable other peripheral (SPI and I²C) functions of the IS1870/71 IC by changing the default factory firmware. For more details, contact local Microchip representatives.
**2:** An external host MCU is required when using the default factory firmware.

**FIG. 8.4**

Microchip IS1870 Bluetooth integrated circuit.

*This book contains copyrighted material of Microchip Technology Incorporated replicated with permission. All rights reserved. No further replications may be made without Microchip Technology Inc.'s prior written consent.*

Microchip also offers compact modules for Zigbee on the 800/900 MHz bands as well as 2.4 MHz. Its ATZBX02564 module has Zigbee RF below 1 GHz and baseband functions, and an on board chip antenna. It is compatible with an IEEE 802.15.4/Zigbee stack that supports a self-healing, self-organizing mesh network while minimizing power consumption. Among applications stated for the product are building automation and monitoring, inventory management, environmental monitoring, water metering, industrial monitoring, including machinery condition and performance monitoring. and automated meter reading (AMR). Its under 1 GHZ frequency range gives it long range—over 6 km outdoor line-of-sight is claimed by the manufacturer. The board measures 18.8 × 13.5 × 0.2 mm.

### 8.4.3 Murata

Murata Bluetooth modules incorporate chips from other vendors. For example, Type MBN52832 module, shown in Fig. 8.5 is based on Nordic chipset nRF52832. This BLE module has an on-board antenna with provision for an external antenna. It is controlled by an ARM Cortex-M$ microprocessor and its size is 7.4 × 7.0 × 0.9 mm. This module supports Bluetooth version 5.0 Low Energy.

The company's Type 1EU module is designed for industrial automation wireless standard ISA100.11a at 2.4 GHz which is based on IEEE 802.15.4. The module, shown in Fig. 8.6, includes a radio transceiver, microcontroller and networking software.

**FIG. 8.5**

Murata MBN52832 Bluetooth module.

*Used with permission from Murata.*

**FIG. 8.6**

Murata type 1EU module for wireless industrial automation.

*Used with permission from Murata.*

## 8.5 Summary

This chapter has presented a sampling of devices that are available to the system designer for developing a short-range radio system. It is far from exhaustive and all of the companies mentioned have other devices which may be more suitable to the designer's need. Most of the companies also have application notes and demonstration boards which can help you decide what device to use and also assist in the design itself. While the availability of a wide range of standard hardware can greatly simplify the design and shorten time-to-market, the system designer must still prepare appropriate interface software and communication protocols, and he is responsible for getting the type approvals in the markets where the product will be sold.

## References

[1] Circuit Design, Japan, www.circuitdesign.co.jp.
[2] Radiometrix, UK, www.radiometrix.co.uk.
[3] Linx, U.S.A., www.linxtechnologies.com.
[4] Microchip, www.microchip.com.
[5] Murata, www.murata.com.
[6] Nordic, www.nordicsemi.com.
[7] Melexis, www.melexis.com.
[8] Texas Instruments, www.ti.com.
[9] Maxim, www.maximintegrated.com.

CHAPTER

# Introduction to information theory and coding 9

Up to now, all the chapters in this book have related directly to radio communication. This chapter is different. Information theory involves communication in general—on wires, fibers, or over the air—and it's applied to widely varied applications such as information storage on optical disks, radar target identification, and the search for extraterrestrial intelligence. In general, the goal of a communication system is to pass "information" from one place to another through a medium contaminated by noise, at a particular rate, and at a minimum specified level of fidelity to the source. Information theory gives us the means for quantitatively defining our objectives and for achieving them in the most efficient manner. The use of radio as a form of communication presents obstacles and challenges that are more varied and complex than a wired medium. A knowledge of information theory lets us take full advantage of the characteristics of the wireless interface.

In order to understand what information theory is about, you need at least a basic knowledge of probability. We start this chapter with a brief exposition of this subject. We've already encountered uses of probability theory in this book—in comparing different transmission protocols and in determining path loss with random reflections.

The use of coding algorithms is very common today for highly reliable digital transmission even with low signal-to-noise ratios. Error correction is one of the most useful applications of information theory.

Finally, information theory teaches us the ultimate limitations in communication—the highest rate that can be transmitted in a given bandwidth and a given signal-to-noise ratio.

## 9.1 Basics of probability

A common use of probability theory in communication is assessing the reliability of a received message transmitted over a noisy channel. Let's say we send a digital message frame containing 32 data bits. What is the probability that the message will be received in error—that is, that one or more bits will be corrupted? If we are using or are considering using an error-correcting code that can correct one bit, then we will want to know the probability that two or more bits will be in error. Another interesting question: what is the probability of error of a frame of 64 bits, compared to that of 32 bits, when the probability of error of a single bit is the same in both cases?



Short-range Wireless Communication. https://doi.org/10.1016/B978-0-12-815405-2.00009-9
© 2019 Elsevier Inc. All rights reserved.

In all cases we must define what is called an *experiment* in probability theory. This involves defining *outcomes* and *events* and assigning probability measures to them that follow certain rules.

We state here the three axioms of probability, which are the conditions for defining probabilities in an experiment with a finite number of outcomes. We also must describe the concept of *field* in probability theory. Armed with the three axioms and the conditions of a field, we can assign probabilities to the events in our experiments. But, first, let's look at some definitions [1].

An *outcome* is a basic result of an experiment. For example, throwing a die has six outcomes, each of which is a different number of dots on the upper face of the die when it comes to rest. An *event* is a set of one or more outcomes that has been defined, again according to rules, as a useful observation for a particular experiment. For an example, one event may be getting an odd number from a throw of a die and another event may be getting an even number. The outcomes are assigned probabilities and the events receive probabilities from the outcomes that they are made up of, in accordance with the three axioms. The term *space* refers to the set of all of the outcomes of the experiment. We'll define other terms and concepts as we go along, after we list the three axioms [2].

### 9.1.1 Axioms of probability

**I.** $P(A) \geq 0$ The probability of an event $A$ is zero or positive.

**II.** $P(S) = 1$. The probability of space is unity.

**III.** If $A \cdot B = 0$ then $P(A + B) = P(A) + P(B)$. If two events $A$ and $B$ are mutually exclusive, then the probability of their sum equals the sum of their individual probabilities.

The product of two events, shown in Axiom III as $A \cdot B$ and called *intersection,* is an event which contains the outcomes that are common to the two events. The sum of two events $A + B$, also called *union*, is an event which contains all of the outcomes in both component events.

*Mutually exclusive*, referred to in Axiom III, means that two events have no outcomes in common. This means that if in an experiment one of the events occurs, the other one doesn't. Returning to the experiment of throwing a die, the odd event and the even event are mutually exclusive.

Now we give the definition of a field, which tells us what events we must include in an experiment. We will use the term *complement* of an event, which is all of the outcomes in the space not included in the event. The complement of $A$ is $A'$. $A \cdot A' = 0$.

### 9.1.2 Definition of a field *F* [2]

**1.** If $A \in F$ then $A' \in F$

If event $A$ is contained in the field $F$, then its complement is also contained in $F$.

**2.** If $A \in F$ and $B \in F$ then $A + B \in F$

If the events $A$ and $B$ are each contained in $F$, then the event that is the sum of $A$ and $B$ is also contained in $F$.

Now we need one more definition before we get back down to earth and deal with the questions raised and the uses mentioned at the beginning of this section.

### 9.1.3 Independent events

Two events are called *independent* if the probability of their product (intersection) equals the product of their individual probabilities:

$$P(A \cdot B) = P(A) \times P(B)$$

This definition can be extended to three or more events. For three independent events $A$, $B$, $C$:

$$P(A \cdot B \cdot C) = P(A) \times P(B) \times P(C)$$

and

$$P(A \cdot B) = P(A) \times P(B); P(A \cdot C) = P(A) \times P(C); P(B \cdot C) = P(B) \times P(C)$$

Similarly, for more than three events the probability of the product of all events equals the product of the probabilities of each of the events, and the probability of the product of any lesser number of events equals the product of the probabilities of those events. If there are $n$ independent events, then the total number of equations like those shown above for three events that are needed to establish their independence is $2^n - (n + 1)$.

We now can look at an example of how to use probability theory.

**Example 9.1 Probability of correctly receiving a sequence**

Problem: What is the probability of correctly receiving a sequence of 12 bits if the probability of error of a bit is one out of one hundred, or $p_e = 0.01$? All bits are independent.

Solution: We look at the problem as an experiment in which we must define space, events according to the conditions of a field, and the probabilities of the events. We'll call the probability of a correct sequence $P_c$ and the probability of an error in the sequence $P_e$.

The space in our experiment contains all conceivable outcomes. Since we have a sequence of 12 bits, we can receive $2^{12} = 4096$ different sequences of bits, or words. The events in our experiment, conforming to the conditions for a field, are

**(1)** The reception of the correct word—that is, no bits are in error.
**(2)** The reception of an erroneous word—a sequence that has 1 or more bits in error.
**(3)** The reception of any word.
**(4)** The reception of no word.

The inclusion of event (3) is necessary because of condition 2 in the definition of a field, which says that the event that is the sum of events must also be in the field. The sum of events (1) and (2) is all of the outcomes, which is the space, and this is event (3). Event (4) is needed because of condition 1 of a field—the complement of any event must be included—and event (4) is the complement of event (3). The complements of events (1) and (2) are each other, so the requirements of a field are complied with.

Now we assign probabilities to the events. In the statement of the problem we designated the probability of a bit error as $p_e$. In the field of an individual bit, we have two events: bit error or no bit error. The sum of these two events is the bit space, whose probability is unity. It follows that the probability of no bit error + probability of bit error ($p_e$) = probability of bit space = 1. Thus, the probability of no bit error = $1 - p_e$.

Now, the bits in the received sequence are independent, so the probability of a particular sequence equals the product of the probabilities of each of its bits. In the case of the errorless sequence, the probability that each bit has no bit error is $1 - p_e$, so this sequence's probability is $(1 - p_e)^{12}$. Using the given bit error probability of 0.01, we find that:

Answer: The probability of correctly receiving the sequence is $0.99^{12}$ or approximately 88.6%.

How can we interpret this answer? If the sequence is sent only once for all time, it will either be received correctly or it won't. In this case, the establishment of a probability won't have much significance, except for the purpose of placing bets. However, if sequences are sent repeatedly, we will find that as the number of sequences increases, the percentage of those correctly received approaches 88.6.

## Example 9.2 Probability of a sequence error

Just as we found the probability of no bit error = $1 - p_e$, we find the probability of a sequence error is $1 - 0.886 = 0.114$. This is from axioms II and III and the fact that the sum of the two mutually exclusive events—incorrect and correct sequences—is space, whose probability is 1.

Fig. 9.1 gives a visual representation of this problem, showing space, the events, and the outcomes. Each outcome, representing one of the 4096 sequences, is assigned a probability $P_i$:

$$P_i = p_1 p_2 p_3 p_4 p_5 p_6 p_7 p_8 p_9 p_{10} p_{11} p_{12}$$

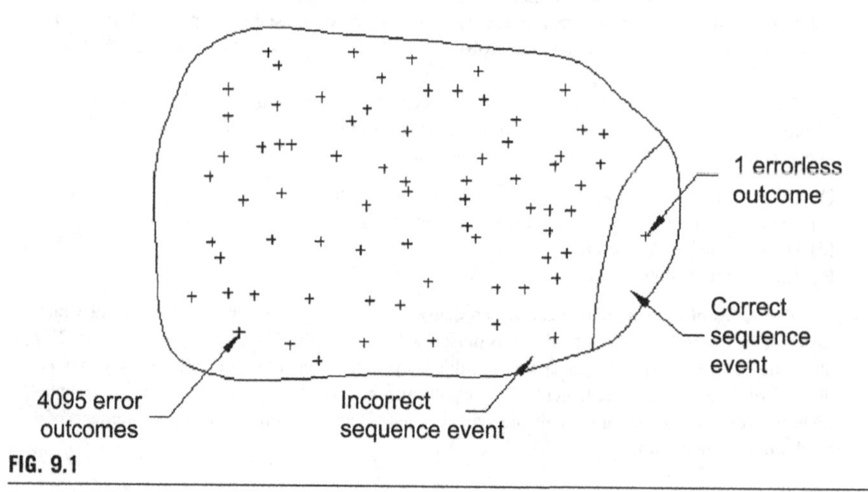


**FIG. 9.1**

A probability space.

where $p_1, p_2$, and so on equals either $p_e$ or $1-p_e$, depending on whether that specific bit in the sequence is in error or not. Sequences having the same number of error bits have identical probabilities.

**Example 9.3 Probability of one error in a sequence**

For example, an outcome that is a sequence having one bit in error has a probability of $p_e(1-p_e)^{11}$. There are 12 of these mutually exclusive sequences, so the probability of receiving a sequence having one and only one error bit is (from axiom III)

$$P_1 = 12p_e(1-p_e)^{11} = 0.107$$

### 9.1.4 Conditional probability

An important concept in probability theory is *conditional probability*. It is defined as follows:

Given an event $B$ with nonzero probability $P(B) > 0$, then the conditional probability of event $A$ assuming event $B$ is known is:

$$P(A|B) = P(A \cdot B)/P(B) \tag{9.1}$$

A consequence of this definition is that the probability of an event is affected by the knowledge of probability of another event. We see from the expression above that if $A$ and $B$ are mutually exclusive, the conditional probability is zero because $P(A{\cdot}B) = 0$. If $A$ and $B$ are independent, then the occurrence of $B$ has no effect on the probability of $A$: $P(A|B) = P(A)$. Conditional probabilities abide by the three axioms and the definition of a field.

So far we have discussed only sets of finite outcomes, but the theory holds when events are defined in terms of a continuous quantity as well. A space can be the set of all real numbers and subsets or events to which we can assign probabilities are intervals in this space. For example, we can talk about the probability of a train arriving between 1 and 2 p.m., or of the probability of a received signal giving a detector output above 1 V. The axioms and definition of a field still hold but we have to allow for the existence of infinite sums and products of events.

### 9.1.5 Density and cumulative distribution functions

In the problem above we found the probability, in a sequence of 12 bits, of getting no errors, of getting an error (one or more errors), and of getting an error in one bit only. We may be interested in knowing the probability of receiving exactly two bits in error, or any other number of error bits. We can find it using a formula called the binomial distribution [3]:

$$P_b(n) = \binom{m}{n} p^n q^{m-n} \tag{9.2}$$

where $P_b(n)$ is the probability of receiving exactly $n$ bits in error, $m$ is the total number of bits in the sequence, $p$ is the probability of one individual bit being in error and

$q = 1 - p$, the probability that an individual bit is correct. $\binom{m}{n}$ represents the number of different combinations of $n$ objects taken from a set of $m$ elements:

$$\binom{m}{n} = \frac{m!}{n!(m-n)!}$$

The notation ! is factorial—for example, $m! = m(m-1)(m-2)\ldots(2)(1)$.

Each time we send an individual sequence, the received sequence will have a particular number of bits in error—from 0 to 12 in our example. When a large number of sequences are transmitted, the frequency of having exactly $n$ errors will approach the probability given in Eq. (9.2). In probability theory, the entity that expresses the observed result of a random process is called a *random variable*. In our example, we'll call this random variable $N$. Thus, we could rewrite Eq. (9.2) as:

$$P_b(N = n) = \binom{m}{n} p^n q^{m-n} \tag{9.2a}$$

We represent uppercase letters as random variables and lowercase letters as real numbers. We may write $P(x)$ which means the probability that random variable $X$ equals real number $x$.

The random value can be any number that expresses an outcome of an experiment, in this case an integer 0 through 12. The random values are mutually exclusive events.

The probability function given in Eq. (9.2a), which gives the probability that the random variable equals a discrete quantity, is sometimes called the *frequency function*. Another important function is the *cumulative probability distribution function*, or *distribution function*, defined as

$$F(x) = \text{Prob}(X \leq x) \tag{9.3}$$

defined for any number $x$ from $-\infty$ to $+\infty$ and $X$ is the random value.

In our example of a sequence of bits, the distribution function gives the probability that the sequence will have $k$ or fewer bits in error, and its formula is

$$F(k) = \sum_{n=0}^{k} \binom{m}{n} p^n q^{m-n} \tag{9.4}$$

where $\Sigma$ is the symbol for summation.

The example which we used up to now involves a discrete random variable, but probability functions also relate to continuous random variables, such as time or voltage. One of these is the Gaussian probability function, which describes, for example, thermal noise in a radio receiver. The analogous type of function to the frequency probability function defined for the discrete variable is called a density function. The Gaussian probability density function is:

$$P(x) = \frac{1}{\sqrt{2\pi\sigma^2}} e^{\frac{-(x-a)^2}{2\sigma^2}} \tag{9.5}$$

where $\sigma^2$ is the variance and $a$ is the average (defined below).

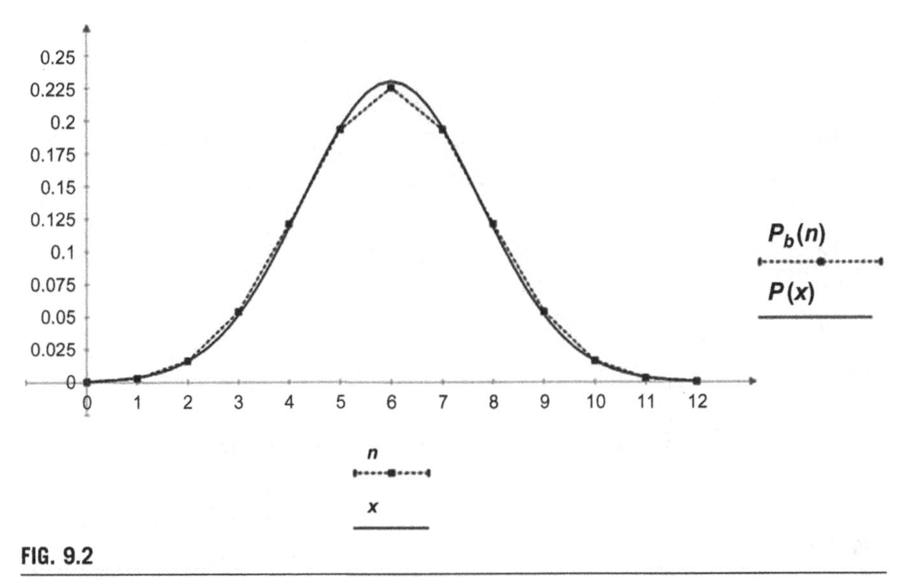


**FIG. 9.2**

Frequency and density functions. $P_b(n)$ shows discrete probability at the small squares for 0–12 errors in a sequence of 12 bits. $P(x)$ is a Gaussian density function. For both functions, average = 6, and standard deviation = 1.732.

Plots of the frequency function $P_b(n)$ ($p$ and $q = 1/2$) and the density function $P(x)$, with the same average and variance, are shown in Fig. 9.2. While these functions are analogous, as stated above, there are also fundamental differences between them. For one, $P(x)$ is defined on the whole real abscissa, from $-\infty$ (infinity) to $+\infty$ whereas $P_b(n)$ is defined only for the discrete values of $n = 0$ to 12. Second, points on $P(x)$ are not probabilities. A continuous random variable can have a probability greater than zero only over a finite interval. Thus, we cannot talk about the probability of an instantaneous noise voltage of 2 V, but we can find the probability of it being between, say, 1.8 and 2.2 V. Probabilities on the curve $P(x)$ are the area under the curve over the interval we are interested in. We find these areas by integrating the density curve between the endpoints of the interval, which may be plus or minus infinity.

The more useful probability function for finding probabilities directly for continuous random variables is the distribution function. Fig. 9.3 shows the Gaussian distribution function, which is the integral of the density function between $-\infty$ and $x$. All distribution functions have the characteristics of a positive slope and values of 0 and 1 at the extremities. To find the probability of a random variable over an interval, we subtract the value of the distribution function evaluated at the lower boundary from its value at the upper boundary. For example, the probability of the interval of 4 to 6 in Fig. 9.3 is

$$F_G(6) - F_G(4) = 0.5 - 0.124 = 0.376$$

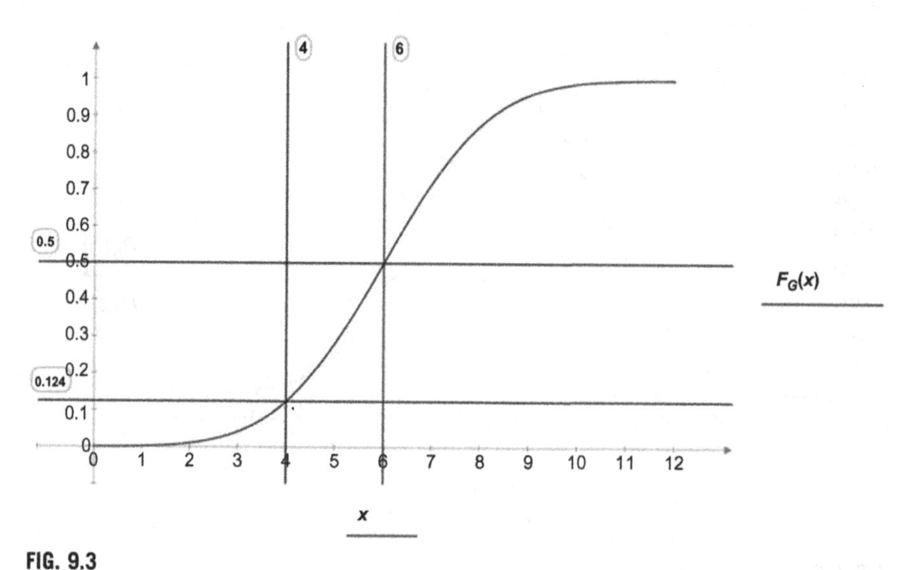


**FIG. 9.3**

Gaussian distribution function.

## 9.1.6 Average, variance, and standard deviation

While the distribution function is easier to work with when we want to find probabilities directly, the density function or the frequency function is more convenient to use to compute the statistical properties of a random variable. The two most important of these properties are the *average* and the *variance*.

The statistical average for a discrete variable is defined as

$$\overline{X}=\sum_i x_i P(x_i) \tag{9.6}$$

Writing this using the frequency function for the $m$-bit sequence Eq. (9.2) we have:

$$\overline{X}=\sum_n nP_m(n)$$

where $n$ ranges from 0 to $m$.

Calculating the average using $p=0.1$, for example, we get $X=1.2$ bits. In other words, the average number of bits in error in a sequence with a bit error probability of 0.1 is just over 1 bit.

The definition of the average of a continuous random variable is

$$\overline{X}=\int_{-\infty}^{\infty} xP(x)dx \tag{9.7}$$

If we apply this to the expression for the Gaussian density function in Eq. (9.5) we get $X=m$ as expected.

We can similarly find the average of a function of a random variable, in both the discrete and the continuous cases. When this function is expressed as $f(x)=x^n$, its average is called the *nth* moment of $X$. The first moment of $X$, for $n = 1$, is its average, shown above. The second moment of the continuous random variable $x$ is

$$\overline{X^2} = \int_{-\infty}^{\infty} x^2 P(x)dx \tag{9.8}$$

In the case where the random variable has a nonzero average, $a$, a more useful form of the second moment is the second-order moment about a point $a$, also called the second-order central moment, defined as

$$Var(X) = \int_{-\infty}^{\infty} (x-a)^2 P(x)dx \tag{9.9}$$

The second-order central moment of $X$ is called the variance of $X$, and its square root is the standard deviation. The standard deviation, commonly represented by the Greek letter sigma ($\sigma$), gives a measure of the form factor of a probability density function. Fig. 9.4 shows two Gaussian density functions, one with $\sigma = 1$ and the other with $\sigma = 2$. Both have the same average, $a = 6$.

The first and second moments are used all the time by electrical engineers when they are talking about voltages and currents, either steady-state or random. The first moment represents the DC level and the second moment is proportional to the power. The variance of a voltage across a unit resistance is its AC power, and the standard deviation is the RMS value of a current or voltage about its DC level.

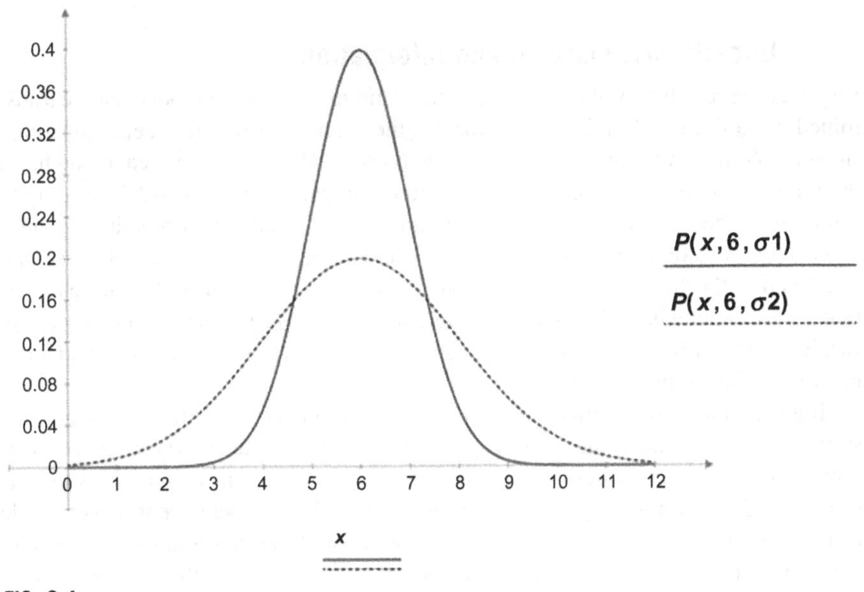


**FIG. 9.4**

Gaussian density function with different standard deviations: $\sigma_1 = 1$, $\sigma_2 = 2$. Average = 6.

Probability functions and curves are defined and displayed in **Worksheet "Probability.mcdx.**

## 9.2 Information theory

In 1948 C. E. Shannon published his Mathematical Theory of Communication, which was a tremendous breakthrough in the understanding of the possibilities of reliable communication in the presence of noise. A whole new field of study opened up, that of *information theory*, which deals with three basic concepts: the measure of information, the capacity of a communication channel to transfer information, and the use of coding as the means of achieving near error-free communication at rates that approach this capacity. Much of the significance of Shannon's work can be realized by considering this statement, which sums up what is called the fundamental theorem of information theory:

*It is possible to transmit information through a noisy communication channel at any rate up to the channel capacity with an arbitrarily small probability of error.*

The converse of this statement has also been proven:

*It is not possible to achieve reliable communication through a channel at a rate higher than the channel capacity.*

The key to reliable communication in the presence of noise is *coding*. We will look at an example of error correction coding after a brief review of the basics of information theory.

### 9.2.1 Uncertainty, entropy, and information

Engineers generally talk about *bit* rate as the number of binary symbols that are transmitted per unit time. In information theory, the bit has a different, deeper meaning. Imagine the transmission of an endless stream of digital ones where each one has a duration of one millisecond. Is the rate of communication 1000 bits/s? According to information theory, the rate is zero. A stream of ones, or any other repetitive pattern of symbols reveals nothing to the receiver, and even in the presence of noise and interference, the "message" is always detected with 100% certainty. The more uncertain we are about what is being transmitted, the more information we get by correctly receiving it. In information theory, the term "bit" is a unit used to measure a quantity of information or uncertainty.

Information theory defines mathematically the uncertainty of a message or a symbol. Let's say we want to send a message using a sequence (symbol) of three binary digits. We know that we can send up to eight different messages using this sequence ($2^3$). Each message has a particular probability of being sent. An example of this situation is a room with 8 patients in a hospital that has a nurse call system. When one of the patients presses a button next to his or her bed, a three-digit message is sent to the nurse's desk where it rings a bell and causes a display to show a number representing that patient. Depending on their condition, some patients may use the call system more than others. We present the probability of a patient pressing the nurse call button in the following table:

| Patient's name | Patient's number | Probability |
|---|---|---|
| John | 1 | 0.1 |
| Mary | 2 | 0.5 |
| Jane | 3 | 0.2 |
| Mike | 4 | 0.05 |
| Pete | 5 | 0.05 |
| Sue | 6 | 0.03 |
| Tom | 7 | 0.01 |
| Elaine | 8 | 0.06 |

The total probability of one of them having pressed the call button when the bell rings is, of course, one.

When the nurse hears the bell, she won't be surprised to see the number 2 on the display, since Mary requires assistance more than any of the other patients. The display of "7" though will be quite unexpected, since Tom rarely resorts to calling the nurse. The message indicating that Tom pressed his assistance button gives more information than the one triggered by Mary.

If we label the probability of an event, such as a patient pressing the assistance button, by $p_i$, we quantify the self-information of the event $i$ by

$$I_i = \log_2(1/p_i) \tag{9.10}$$

With $i$ referring to the number of the patient, we find that the self-information of Mary's signal is $I(2) = \log_2(2) = 1$ bit, and Tom's signal self-information is $I(7) = \log_2(100) = 6.64$ bits. The unit of self-information is the *bit* when the log is taken to the base 2. The self-information of the other patients' messages can be found similarly.

More important than the individual self-information of each of the messages is the average self-information of all the messages. This quantity is commonly labeled $H$ and is called the *uncertainty* of the message, or its *entropy*. The latter term was taken from a parameter in thermodynamics with similar properties. Taking the statistical average of the self information over all messages, we get

$$H = \sum_i p_i I(i) \tag{9.11}$$

With $i$ ranging from 1 to 8 and using the probabilities in the table above, we get for our present example

$$H = 0.1(3.322) + 0.5(1) + 0.2(2.322) + 0.05(2)(4.322) + 0.03(5.059) + 0.01(6.644) + 0.06(4.059)$$

$$H = 2.191 \text{ bits/message}$$

Now let's assume that all of the patients are equally likely to call the nurse, that is, the probability of each message is 1/8. The self-information of each message $I$(i) in this case is $\log_2(8) = 3$ bits. The entropy is $H_m = 8 \times 1/8 \times 3 = 3$ bits/message.

It turns out that this is the maximum possible entropy of the message—the case where the probability of each of the signals is equal.

Let's stretch our example of the nurse call system to an analogy with a continuous stream of binary digits. We'll assume the patients press their call buttons one after another at a constant rate.

In the case where the message probabilities are distributed as in the table, the entropy per digit is $H/3 = 0.73$ (since there are 3 binary digits per message). If each patient pressed his button with equal probability, the entropy per digit would be the maximum of $H_m/3 = 1$. So with the different probabilities as listed in the table, the system communicates only 73% of the information that is possible to send over the communication channel per unit time. Using coding, discussed below, we can match a source to a channel to approach the channel's capability, or capacity, to any extent that we want it to, and in so doing, to increase the data rate.

Up to now we have been talking about the entropy, or uncertainty, of a source, and have assumed that what is sent is what is received. At least of equal importance is to measure the entropy, and the information, involved with messages sent and received over a noisy channel. Because of the noise, the probabilities of the received messages might not be the same as those of the source messages, because the receiver will make some wrong decisions as to the identity of the source. We saw above that entropy is a function of probabilities, and in the case of communication over a noisy channel, several sets of probability functions can be defined.

On a discrete, memoryless channel (memoryless because the noise affecting one digit is independent of the noise affecting any other digit) we can present the effect of the noise as a matrix of conditional probabilities. Assume a source transmits symbols having one of four letters $x1$, $x2$, $x3$, and $x4$. $X$ is a random variable expressing the transmitted symbol and $Y$ is a random variable for the received symbol. If a symbol $x1$ happens to be sent, the receiver may interpret it as $y1$, $y2$, $y3$, or $y4$, depending on the effect of the random noise at the moment the symbol is sent. There is a probability of receiving $y1$ when $x1$ is sent, another probability of receiving $y2$ when $x1$ is sent, and so on for a matrix of 16 probabilities as shown below:

$$P(Y/X) = \begin{bmatrix} p(y1|x1) & p(y2|x1) & p(y3|x1) & p(y4|x1) \\ p(y1|x2) & p(y2|x2) & p(y3|x2) & p(y4|x2) \\ p(y1|x3) & p(y2|x3) & p(y3|x3) & p(y4|x3) \\ p(y1|x4) & p(y2|x4) & p(y3|x4) & p(y4|x4) \end{bmatrix} \tag{9.12}$$

This example shows a square matrix, which means that the receiver will interpret a signal as being one of those letters that it knows can be transmitted, which is the most common situation. However, in the general case, the receiver may assign a larger or smaller number of letters to the signal so the matrix doesn't have to be square. The conditional entropy of the output $Y$ when the input $X$ is known is:

$$H(Y|X) = -\sum_i p(x_i) \sum_j p(y_j|x_i) \log_2 p(y_j|x_i) \tag{9.13}$$

where the sums are over the number of source letters $x_i$ and received letters $y_j$, and to get units of bits the log is to the base 2. The expression has a minus sign to make the entropy positive, canceling the sign of the log of a fraction, which is negative.

If we look only at the $Y$s that are received, we get a set of probabilities $p(y1)$ through $p(y4)$, and a corresponding uncertainty $H(Y)$.

Knowing the various probabilities that describe a communication system, expressed through the entropies of the source, the received letters, and the conditional entropy of the channel, we can find the important value of the *mutual information* or *transinformation* of the channel. In terms of the entropies we defined above:

$$I(X;Y) = H(Y) - H(Y|X) \tag{9.14a}$$

The information associated with the channel is expressed as the reduction in uncertainty of the received letters given by a knowledge of the statistics of the source and the channel. The mutual information can also be expressed as

$$I(X;Y) = H(X) - H(X|Y) \tag{9.14b}$$

which shows the reduction of the uncertainty of the source by the entropy of the channel from the point of view of the receiver. The two expressions of mutual information are equal.

### 9.2.2 Capacity

In the fundamental theorem of information theory, referred to above, the concept of channel capacity is a key attribute. It is connected strongly to the mutual information of the channel. In fact, the capacity is the maximum mutual information that is possible for a channel having a given probability matrix:

$$C = \max I(X;Y) \tag{9.15}$$

where the maximization is taken over all sets of source probabilities. For a channel that has a symmetric noise characteristic, so that the channel probability matrix is symmetric, the capacity can be shown to be

$$C = \log_2(m) - h \tag{9.16}$$

where $m$ is the number of different sequences for each source symbol and $h = H(Y|X)$, a constant independent of the input distribution when the channel noise is symmetric. This is the case for expression (9.12), when each row of the matrix has the same probabilities except in different orders.

For the noiseless channel, as in the example of the nurse call system, the maximum information that could be transferred was $H_m = 3$ bits/message, achieved when the source messages all have the same probability. Taking that example into the frame of the definition of capacity, Eq. (9.16), the number of source sequences is the number of different messages, so for $m = 8$ (sequences)

$$C = \log_2 m = \log_2 8 = 3$$

This is a logical extension of the more general case presented previously. Here the constant $h = 0$ for the situation when there is no noise.

Another situation of interest for finding capacity on a discrete memoryless channel is when the noise is such that the output $Y$ is independent of the input $X$, and the conditional entropy $H(Y|X) = H(Y)$. The mutual information of such a channel $I(X;Y) = H(Y) - H(Y|X) = 0$ and the capacity is also zero.

## 9.3 Shannon-Hartley theorem

The notion of channel capacity and the fundamental theorem also hold for continuous, "analog" channels, where signal-to-noise ratio (S/N) and bandwidth ($B$) are the characterizing parameters. The capacity in bits per second in this case is given by the Hartley-Shannon law:

$$C = B \cdot \log_2\left(1 + \frac{S}{N}\right) \tag{9.17}$$

The extension from a discrete system to a continuous one is easy to conceive of when we consider that a continuous signal can be converted to a discrete one by sampling, where the sampling rate is a function of the signal bandwidth (at least twice the highest frequency component in the signal).

From a glance at Eq. (9.17) we see that bandwidth can be traded off for signal-to-noise ratio, or vice-versa, while keeping a constant capacity. Actually, it's not quite that simple since the signal-to-noise ratio itself depends on the bandwidth. We can show this relationship by rewriting Eq. (9.17) using $N_0$ = the noise density. $S$ = signal power. Substituting $N = N_0 \cdot B$:

$$C = B \cdot \log_2\left(1 + \frac{S}{N_0 B}\right) \tag{9.17a}$$

In this expression the tradeoff between bandwidth and signal-to-noise ratio (or transmitter power) is tempered somewhat but it still exists. The limit of $C$ as $B$ increases is

$$\lim_{B \to \infty} (C) = \frac{1}{\ln(2)} \left(\frac{P}{N_0}\right) \tag{9.18}$$

For example, given room temperature noise density of $-174$ dBm/Hz, received power of $-90$ dBm and bandwidth of 20 MHz, the limiting errorless information rate is 75 Mbps. The ultimate capacity attainable by increasing the bandwidth is 362 Mbps.

Increasing the bandwidth doesn't automatically allow sending at higher data rates while keeping a low probability of errors. Coding is necessary to keep the error rate

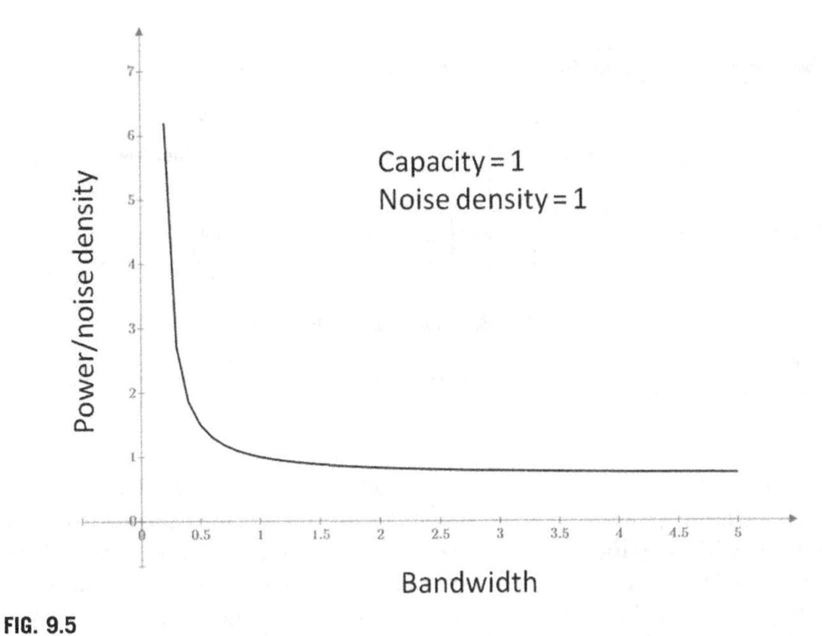


**FIG. 9.5**

Signal power vs. bandwidth at constant capacity.

down, and the higher bandwidth facilitates the addition of error-correcting bits in accordance with the coding algorithm.

In radio communication, a common aim is the reduction of signal bandwidth. Shannon-Hartley indicates that we can reduce bandwidth if we increase signal power to keep the capacity constant. This is fine, but Fig. 9.5 shows that as the bandwidth is reduced more and more below the capacity, large increases in signal power are needed to maintain that capacity.

## 9.4 Coding

A communication system can be represented conveniently by the block diagram in Fig. 9.6 [4]. Assume the source outputs binary data. If the data is analog, we could make it binary by passing it through an analog-to-digital converter. The modulator and demodulator act as interfaces between the discrete signal parts of the system and the waveform that passes information over the physical channel—the modulated carrier frequency in a wireless system, for example.

We can take the modulator and demodulator blocks together with the channel and its noise input and look at the ensemble as a binary discrete channel. Consider the

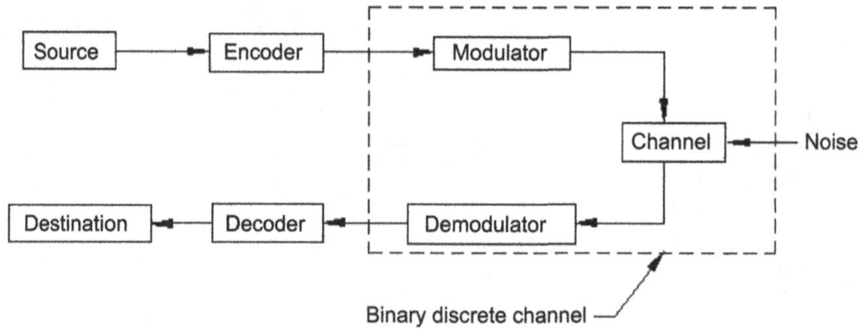


**FIG. 9.6**

Communication system.

encoder and decoder blocks as matching networks which process the binary data so as to get the most "power" out of the system, in analogy to a matching network that converts the RF amplifier output impedance to the conjugate impedance of the antenna in order to get maximum power transfer. "Power," in the case of the communication link, may be taken to mean the data rate and the equivalent probability of error. A perfect match of the encoder and decoder to the channel gives a rate of information transfer equaling the capacity of the equivalent binary channel with an error rate approaching zero.

### 9.4.1 Noiseless coding

When there's no noise in the channel (or relatively little) there is no problem of error rate but coding is still needed to get the maximum rate of information transfer. We saw above that the highest source entropy is obtained when each message has the probability. If source messages are not equi-probable, then the encoder can determine the message lengths that enter the binary channel so that highly probable messages will be short, and less probable messages will be long. On the average, the channel rate will be obtained.

For example, let's say we have four messages to send over a binary channel and that their rates of occurrence (probabilities) are 1/2, 1/4, 1/8, and 1/8. The following table shows two coding schemes that may be chosen for the messages $m1$, $m2$, $m3$, and $m4$.

| Message | Probability | Code A | Code B |
|---|---|---|---|
| $m1$ | 0.5 | 00 | 0 |
| $m2$ | 0.25 | 01 | 10 |
| $m3$ | 0.125 | 10 | 110 |
| $m4$ | 0.125 | 11 | 111 |

One sequence of messages that represents the probabilities is $m1 \cdot m1 \cdot m1 \cdot m1 \cdot m2 \cdot m2 \cdot m3 \cdot m4$. The bit streams that would be produced by each of the two codes for this sequence is:

$$\text{CODE A}: 0000000001011011$$

$$\text{CODE B}: 00001010110111$$

Code A needs 16 binary digits, to send the message stream while Code B needs only 14 digits. Thus, using Code B, we can make better use of the binary channel than if we use Code A, since, on the average, it lets us send 14% more messages using the same digit rate.

We can determine the best possible utilization of the channel by calculating the entropy of the messages which is, from Eq. (9.11):

$$H = 0.5 \times \log_2(1/0.5) + 0.25 \times \log_2(1/0.25) + 0.25 \times \log_2(1/0.125) + 0.25 \times \log_2(1/0.125)$$

$$H = 1.75\ \text{bits/message}$$

In the example, we sent 8 messages, which can be achieved using a minimum of $8 \times H = 14$ bits of information. This we get using Code B.

The capacity of the binary symmetric channel, which sends ones and zeros with equal probability is from Eq. (9.16)

$C = \log 2(2) - 0 = 1$ information bit/digit.

Each binary digit on the channel is a bit, so Code B fully utilizes the channel capacity. If each message had equal probability of 0.25, $H$ would be $\log_2(4) = 2$ bits per message (Eqs. 9.10, 9.11). Again consider sending a stream of 8 messages, with equal probability of each message: $m1$ $m1$ $m2$ $m2$ $m3$ $m3$ $m4$ $m4$. The bit streams produced for each code are:

$$\text{Code A}: 0000010110101111$$

$$\text{Code B}: 001010110110111111$$

This time the best code to use gives $m \times H = 8 \times 2 = 16$ bits, which matches code A. The point is that the chosen code matches the entropy of the source to get the maximum rate of communication.

There are several schemes for determining an optimum code when the message probabilities are known. One of them is Huffman's minimum-redundancy code. This code, like Code B in the example above, is easy to decode because each word in the code is distinguishable without a deliminator and decoding is done "on the fly." When input probabilities are not known, such codes are not applicable. An example of a code that works on a bit stream without dependency on source probabilities or knowledge of word lengths is the Lempel-Ziv algorithm, used for file compression. Its basic idea is to look for repeated strings of characters and to specify where a previous version of the string started and how many characters it has.

The various codes used to transfer a given amount of information in the shortest time for a given bit rate are often called compression schemes, since they also can be viewed as reducing the average length of symbols or number of digits needed to represent this information. One should remember that when the symbols in the input message stream are randomly distributed, coding will not make any difference from the point of view of compression or increasing the message transmission rate.

### 9.4.2 Error detection and correction

A most important use of coding is on noisy channels where the aim is to reduce the error rate while maintaining a given message transmission rate. We've already learned from Shannon that it's theoretically possible to reduce the error rate to as close to zero as we want, as long as the signal rate is below channel capacity. There's more than one way to look at the advantage of coding for error reduction. For example, if we demand a given maximum error rate in a communication system, we can achieve that goal without coding by increasing transmitter power to raise the signal-to-noise ratio and thereby reduce errors. We can also reduce errors by reducing the transmission rate, allowing us to reduce bandwidth, which will also improve the signal-to-noise ratio (since noise power is proportional to bandwidth). So using coding for error correction lets us either use lower power for a given error rate, or increase transmission speed for the same error rate. The reduction in signal-to-noise ratio that can be obtained through the use of coding to achieve a given error rate compared to the signal-to-noise ratio required without coding for the same error rate is defined as the *coding gain*.

A simple and well-known way to increase transmission reliability is to add a parity bit at the end of a block, or sequence, of message bits. The parity bit is chosen to make the number of bits in the message block odd or even. If we send a block of seven message bits, say 0110110, and add an odd parity bit (to make the sum of all bits odd), and the receiver gets the message 01001101, it knows that the message has been corrupted by noise, although it doesn't know which bit is in error. This method of error detection is limited to errors in an odd number of bits, since if, say, two bits were corrupted, the total number of "1" bits is still odd and the errors are not noticed. When the probability of a bit error is relatively low, by far most errors will be in one bit only, so the use of one parity bit is justified.

Another simple way to provide error detection is by logically adding together the contents of a number of message blocks and appending the result as an additional block in the message sequence. The receiver performs the same logic summing operation as the transmitter and compares its computation result to the block appended by the transmitter. If the blocks don't match, the receiver knows that one or more bits in the message blocks received are not correct. This method gives a higher probability of detection of errors than does the use of a single parity bit, but if an error is detected, more message blocks must be rejected as having suspected errors.

When highly reliable communication is desired, it's not enough just to know that there is an error in one or more blocks of bits, since the aim of the communication is to receive the whole message reliably. A very common way to achieve this is by an automatic repeat query (ARQ) protocol. After sending a message block, or group of blocks, depending on the method of error detection used, the transmitter stops sending and listens to the channel to get confirmation from the receiver. After the receiver notes that the parity bit or the error detection block indicates no errors, it transmits a short confirmation message to the transmitter. The transmitter waits long enough to receive the confirmation. If the message is confirmed, it transmits the next message. If not, it repeats the previous message.

ARQ can greatly improve communication reliability on a noisy communication link. However, there is a price to pay. First, the receiver must have a transmitter, and the transmitter a receiver. This is OK on a two-way link but may be prohibitive on a one-way link such as exists in most security systems. Second, the transmission rate will be reduced because of the necessity to wait for a response after each short transmission period. If the link is particularly noisy and many retransmissions are required, the repetitions themselves will significantly slow down the communication rate. In spite of its limitations, ARQ is widely used for reliable communications and is particularly effective when combined with a forward error correction (FEC) scheme as discussed in the next section.

#### 9.4.2.1 Forward error correction (FEC)

Just as adding one odd or even parity bit allows determining if there has been an error in reception, adding additional parity bits can tell the location of the error. For example, if the transmitted message contains 15 bits, including the parity bits, the receiver will need enough information to produce a four-bit word to indicate that there are no errors, (error word 0000), or which bit is in error (one out of 15 error words 0001 through 1111). The receiver might be able to produce this four-bit error detection and correction word from four parity bits (in any case, no less than four parity bits), and then the transmitter could send 11 message information bits plus four parity bits and get a capability of detecting and correcting an error in any one bit.

R.W. Hamming devised a relatively simple method of determining parity bits for correcting single-bit errors. If $n$ is the total number of message bits in a sequence to be transmitted, and $k$ is the number of parity bits, the relationship between these numbers permitting correction of one digit is [3, p. 171]

$$2^k \geq n+1 \tag{9.19}$$

If we represent $m$ as the number of information bits ($n = m + k$) we can find $n$ and $k$ for several values of $m$ in the following table.

| $m$ | 4 | 8 | 11 | 26 | 57 |
|---|---|---|---|---|---|
| $k$ | 3 | 4 | 4 | 5 | 6 |
| $n$ | 7 | 12 | 15 | 31 | 63 |

As an example, one possible set of rules for finding the four parity bits for a total message length of 12 bits (8 information bits), derived from Hamming's method, is as follows.

We let $x1$, $x2$, $x3$, and so on, represent the position of the bits in the 12-bit word. Bits $x1$, $x2$, $x4$, and $x8$ are chosen in the transmitter to give even parity when the bits are summed up using binary arithmetic as shown in the four equations below (in binary arithmetic, $0 + 0 = 0$, $0 + 1 = 1$, $1 + 0 = 1$, $1 + 1 = 0$). Let $s1$ through $s4$ represent the results of the equations.

$$x1 + x3 + x5 + x7 + x9 = 0 \qquad \text{s1}$$

$$x2 + x3 + x6 + x7 + x10 + x11 = 0 \qquad \text{s2}$$

$$x4 + x5 + x6 + x7 + x12 = 0 \qquad \text{s3}$$

$$x8 + x9 + x10 + x11 + x12 = 0 \qquad \text{s4}$$

In this scheme, $x1$, $x2$, $x4$, and $x8$ designate the location of the parity bits. If we number the information bits appearing, say, in a byte-wide register of a microcomputer, as $m1$ through $m8$, and we label the parity bits $p1$ through $p4$, then the transmitted 12-bit code word would look like this:

$$p1 \cdot p2 \cdot m1 \cdot p3 \cdot m2 \cdot m3 \cdot m4 \cdot p4 \cdot m5 \cdot m6 \cdot m7 \cdot m8$$

Those parity bits are calculated by the transmitter for even parity as shown in the four equations.

Assume, as an example, that transmitted message bits $m1$ through $m8$ are 1 0 0 1 1 1 0 1. Table 9.1 shows the resulting parity bits $p1$ through $p4$ (underlined) which make $s1$ through $s4$ equal to zero. Now the transmitted code word, $x1$ through $x12$, is 1 1 1 0 0 0 1 1 1 1 0 1.

When the receiver receives a code word, it computes the four equations and produces what is called a *syndrome*—a four-bit word composed of $s4 \cdot s3 \cdot s2 \cdot s1$. $s1$ is 0 if the first equation = 0 and 1 otherwise, and so on for $s2$, $s3$, and $s4$. If the syndrome word is 0000, there are no single-bit errors. If there is a single-bit error, the value of the syndrome points to the location of the error in the received code word, and

**Table 9.1** Parity bits for correcting single-bit errors

| | x1 | x2 | x3 | x4 | x5 | x6 | x7 | x8 | x9 | x10 | x11 | x12 |
|---|---|---|---|---|---|---|---|---|---|---|---|---|
| | p1 | p2 | m1 | p3 | m2 | m3 | m4 | p4 | m5 | m6 | m7 | m8 |
| | 1 | 1 | 1 | 0 | 0 | 0 | 1 | 1 | 1 | 1 | 0 | 1 |
| s1 | 1 | | 1 | | 0 | | 1 | | 1 | | | |
| s2 | | 1 | 1 | | | 0 | 1 | | | 1 | 0 | |
| s3 | | | | 0 | 0 | 0 | 1 | | | | | 1 |
| s4 | | | | | | | | 1 | 1 | 1 | 0 | 1 |

complementing that bit performs the correction. For example, if bit 5 is received in error, the resulting syndrome is 0101, or 5 in decimal notation. This is evident from Table 9.1. An error in any bit column switches parity bits so that the decimal equivalent of *s*1 *s*2 *s*3 *s*4 points to the position of the error.

Note that different systems could be used to label the code bits. For example, the four parity bits could be appended after the message bits. In this case a lookup table might be necessary in order to show the correspondence between the syndrome word and the position of the error in the code word.

The one-bit correcting code just described gives an order of magnitude improvement of message error rate compared to transmission of information bytes without parity digits, when the probability of a bit error on the channel is around $10^{-2}$ (depends on code word length). However, one consequence of using an error-correcting code is that, if message throughput is to be maintained, a faster bit rate on the channel is required, which entails reduced signal-to-noise ratio because of the wider bandwidth needed for transmission. This rate, for the above example, is 12/8 times the uncoded rate, an increase of 50%. Even so, there is still an advantage to coding, particularly when coding is applied to larger word blocks.

Error probabilities are usually calculated on the basis of independence of the noise from bit to bit, but on a real channel, this is not likely to be the case. Noise and interference tend to occur in bursts, so several adjacent bits may be corrupted. One way to counter the noise bursts is by *interleaving* blocks of code words. For simplicity of explanation, let's say we are using four-bit code words. Interleaving the code words means that after forming each word with its parity bits in the encoder, the transmitter sends the first bit of the first word, then the first bit of the second word, and so on. The order of transmitted bits is best shown as a matrix, as in the following table, where the *a*'s are the bits of the first word, and *b*, *c*, and *d* represent the bits of the second, third, and fourth words.

| | | | |
|---|---|---|---|
| a1 | a2 | a3 | a4 |
| b1 | b2 | b3 | b4 |
| c1 | c2 | c3 | c4 |
| d1 | d2 | d3 | d4 |

The order of transmission of bits is $a1 \cdot b1 \cdot c1 \cdot d1 \cdot a2 \cdot b2 \ldots d3 \cdot a4 \cdot b4 \cdot c4 \cdot d4$. In the receiver, the interleaving is decomposed, putting the bits back in their original order, after which the receiver decoder can proceed to perform error detection and correction.

The result of the interleaving is that up to four consecutive bit errors can occur on reception while the decoder can still correct them using a one-bit error correcting scheme. A disadvantage is that there is an additional delay of three word durations until decoded words start appearing at the destination.

FEC schemes which deal with more than one error per block are much more complicated, and more effective, than the Hamming code described here.

### 9.4.3 Convolutional coding

Convolutional coding is a widely used coding method which is not based on blocks of bits but rather the output code bits are determined by logic operations on the present bit in a stream and a small number of previous bits. In the encoder, data bits are input to a shift register of length $K$, called the constraint length. As each bit enters at the left of the register, the previous bits are shifted to the right while the oldest bit in the register is removed. Two or more binary summing operations, let's say $r$, create code bits which are output during one data flow period. Therefore, the code bit rate is $1/r$ times the data rate and the encoder is called a rate $1/r$ convolutional encoder of constraint length $K$. Also needed to completely define the encoder are the connections from stages in the shift register to the $r$ summing blocks. These are generator vectors each of which may be simply expressed as a row of K binary digits. The $r$ binary adders create even parity bits at their outputs; that is, connections to an odd number of logic "ones" result in an output of "one," otherwise the output is "zero."

Fig. 9.7 shows an example with $K=3$, $r=2$, and the generator vectors are chosen as [1 1 1] and [1 0 1]. Discrete sampling times are labeled $n$. The data stream enters on the left and the present bit at time $n$, the most recent bit $n-1$ and the next earliest bit at $n-2$ occupy the shift register. Two parity bits are switched out in the interval between $n$ and $n-1$ from the upper adder and then the lower one. When the next data bit arrives, the shift register moves its contents to the right. The $K-1$ earlier bits, in this case two, determine the state of the encoder. They are shown in gray in Fig. 9.7. There are $2^{K-1}$ states. For each encoder state there are two possibilities of

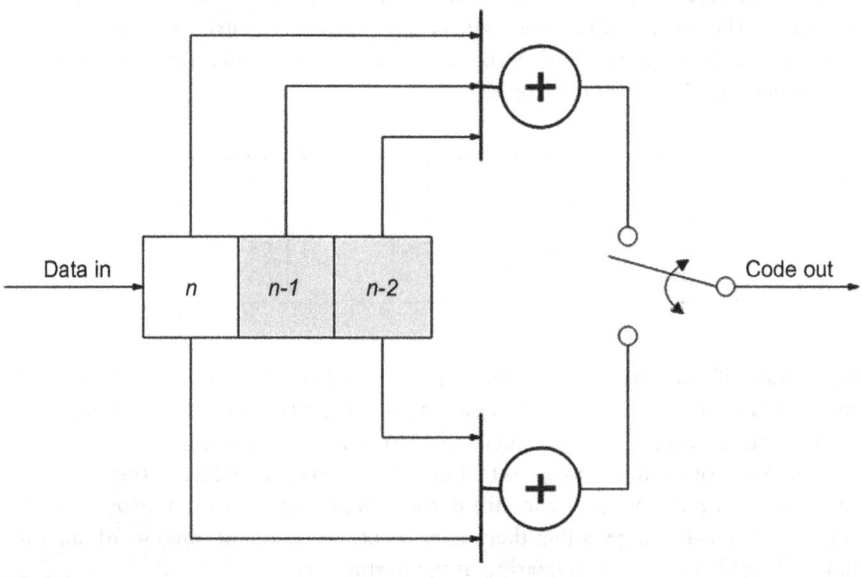


**FIG. 9.7**

Convolutional encoder.

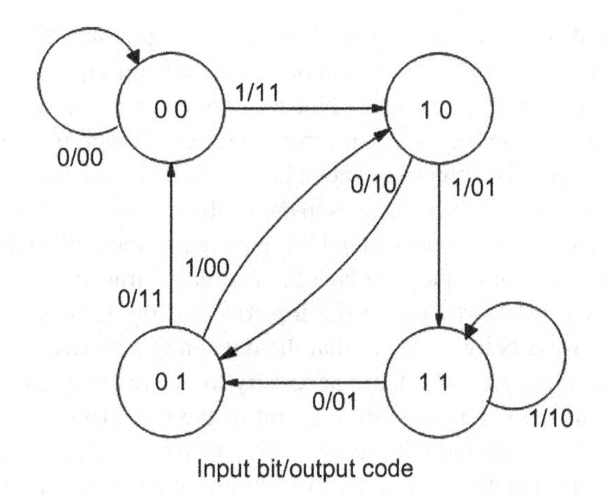


**FIG. 9.8**

Convolutional encoder state diagram.

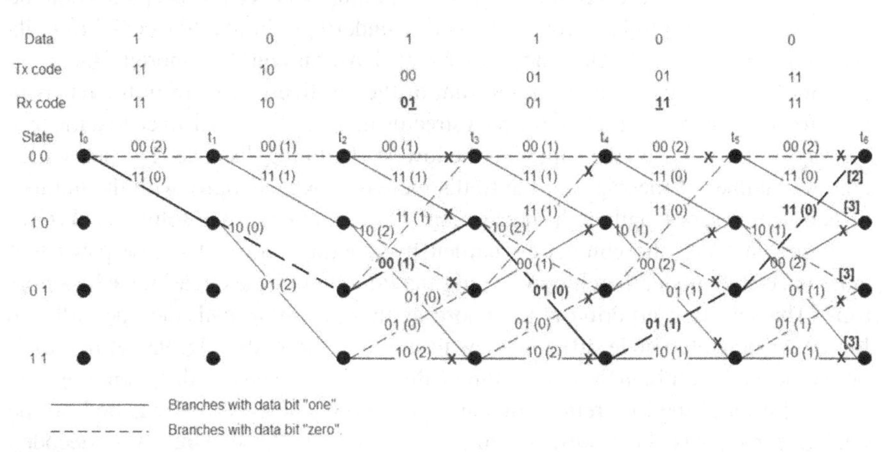


**FIG. 9.9**

Convolutional decoder trellis diagram.

output code bits, depending on whether the input bit is "zero" or "one." The progression of states in time, then, are a function of the data stream. Fig. 9.8 is a *state diagram* of our example. Each state is shown inside a circle and the change from one state to another is shown by an arrow, identified by the input bit, slash, output code bits. You can see that encoding can be done by relatively simple hardware.

The decoder estimates the data stream on the basis of the received code bit sequence and knowledge of the encoder state diagram, exemplified by Fig. 9.8. The progression of states in time for all message streams can be shown by a *trellis diagram*, like Fig. 9.9 which is a continuation of our previous example. We use this diagram to describe the decoding process.

Convolutional coding is based on the fact that every possible coded message must traverse through a definitive progression of states, and consequently, of $r$-tuple code words, in our case with $r = 2$, bit pairs. Noise and interference on the communication channel may cause some bits to be in error. The trellis diagram shows *all possible transmitted messages*. The number of code bits in the received data stream (message) that differ from any one of the messages in the trellis is called the *Hamming distance*. The task of the decoder is to find a coded bit sequence in the trellis that has the lowest Hamming distance. That sequence gives the estimated transmitted message.

Fig. 9.9 shows a data stream on the top row and the transmitted and received codes in the two rows below it. Note that the data ends with two consecutive zero's ($K - 1$ in the general case) which are necessary to insure complete decoding and to have a flushed shift register when the first bit of the next data stream arrives. Each branch in the trellis is labeled with its code bit pair followed in parenthesis with the number of code bits that differ from the corresponding bits in the received code. Solid branch lines are ones and dashed lines are zeros. The transmitted message path which matches the data stream in the first row on the top is shown with bold lines. If there were no errors in the received message, the Hamming distance for this path would be zero. However, due to bit errors at $t_2$ and $t_4$, underlined in the "Rx code" row, its Hamming distance is 2. Does any other path have an equal or shorter Hamming distance? The receiver can check the sum of the bit divergences from the received code for all branches of each of the data streams in the trellis and deduce that the true message corresponds to the path with the lowest Hamming distance. The problem is that the number of messages in the trellis increases exponentially with the number of sample times, or length of the data stream. One commonly used solution is *Viterbi decoding*. It eases the computing burden by deleting one of the two paths that enter each state node, thereby preventing the doubling of messages at each sample time. The selection criterion is to choose as the remaining path the one with the lowest accumulated code differences with the received code. Fig. 9.9 shows an **X** on the deleted path branches. The sum of the code divergences, the Hamming distance, of each of the four remaining paths is shown in brackets at the $t_6$ nodes. The bold line path has the lowest Hamming distance, so the message was decoded successfully.

The maximum number of errors that can be corrected is a function of the constraint length and the selected generator vectors. A path length of at least several constraint lengths is necessary to realize that number. For our example with $K = 3$, two errors over a length of at around 20 code bits should be correctable, but that also depends on how the errors are distributed [5]. In the example, the coding worked with a shorter message.

The above explanation of convolution decoding dealt with hard decisions, that is, code differences were between trellis branch codes and discrete logic signal levels. However, the receiver detector may output multiple level digital words for each symbol that reflect a degree of confidence whether the bit is a "zero" or "one." In this case the Hamming distant can't be used and a *Euclidean distance* between a noisy point (the received symbol) and the trellis point has to be calculated. This soft-decision

Viterbi decoding is otherwise the same as the hard-decision method we described and should give better results since more information about the signal is available.

As we have seen, the basic code rate of a convolutional encoder with $r = 2$ is ½ and it gives good error correcting performance. In many situations, particularly when there is a high signal-to-noise ratio, a high data rate is desirable even at the expense of error correction. To increase the code rate, *puncturing* is used which is a procedure for omitting some of the code bits to get a higher data rate while maintaining proportionally less error correction capability. In the decoder, dummy bits are inserted in place of the omitted bits and the Viterbi algorithm is carried out as described. For example, in OFDM IEEE 802.11 a $K = 7$ convolutional encoder can produce code rates of ½ (basic), 2/3 or ¾, plus 5/6 for the high-throughput and very high throughput physical layers (see Chapter 11).

## 9.5 Summary

In this chapter we reviewed the basics of probability and coding. Axioms of probability were presented and commonly used terms associated with random variables were defined, including average, variance and standard deviation. We explained some basic concepts of information theory, among them information, entropy and capacity which are meaningful for noise-free and noisy channels. Through the use of coding, which involves inserting redundancy in the transmitted data, it's possible to approach error-free communication on noisy channels up to the channel capacity. A description and an example of FEC was presented and the use of interleaving to facilitate error correction in the presence of bursty interference was explained. We also explained convolution coding, where code bits are calculated with the passing of each data bit in contrast to block coding where parity bits are formed on a given length of data. In general, coding is a very important aspect of radio communication and its use and continued development are a prime reason why wireless systems can continue to replace wires while making more effective use of the available spectrum and limited transmitter power.

Although effective and efficient coding algorithms are quite complicated, the great advances in integrated circuit miniaturization and logic circuit speed in recent years have allowed incorporating error-correction coding in a wide range of wireless communications applications. Many of these applications are in products that are mass produced and relatively low cost—among them are cellular telephones and wireless local area networks. This trend will certainly continue and will give even the most simple short-range wireless products "wired" characteristics from the point of view of communication reliability.

This chapter has discussed matters which are not necessarily limited to radio communication, and where they do apply to radio communication, not necessarily to short-range radio. However, anyone concerned with getting the most out of a short-range radio communication link should have some knowledge of information theory and coding. The continuing advances in the integration of complex logic circuits and digital signal processing are bringing small, battery-powered wireless devices closer to the theoretical bounds of error-free communication predicted by Shannon more than 70 years ago.

## References

[1] W.B. Davenport Jr., W.L. Root, An Introduction to the Theory of Random Signals and Noise, McGraw-Hill, 1958.
[2] A. Papoulis, Probability, Random Variables, and Stochastic Processes, McGraw-Hill, 1965.
[3] F.M. Reza, An Introduction to Information Theory, McGraw-Hill, 1961.
[4] R.G. Gallager, Information Theory and Reliable Communication, Wiley, 1968.
[5] B. Sklar, Digital Communications—Fundamentals and Applications, second ed., Prentice Hall, 2001.

CHAPTER

# Regulations and standards 10

## 10.1 Introduction

Most short-range radio equipment falls under the category of unlicensed equipment. That means that you don't need a license to use the equipment. However, in most cases, the manufacturer does have to get approval from the relevant authorities to market the devices. The nature of this approval differs from country to country, and equipment which meets approval requirements in one country may not meet them in another. Differences in requirements may be a question of frequency bands, transmitter powers, spurious radiation limits and measurement techniques.

From the point of view of radio regulations, the world is divided into three regions, shown in Fig. 10.1. International agreements classify frequency allocations according to these regions, but within each of them, every country is sovereign and makes its own rules controlling radio emissions and the procedures to follow for getting approval. Generally speaking, similar equipment can be sold in the United States, Canada, and the other countries of the Americas, whereas this equipment has to be modified for marketing in Europe. We focus in this chapter on the radio regulations in the United States and in Europe.

In addition to mandatory requirements that concern the communication qualities of radio equipment, there are also regulations which aim to limit interference from any equipment that radiates electromagnetic waves, intentionally or not. Of major importance is the EMC Directive in the European Community. Radio equipment marketed in the EC must prove the non-interference of its radiation as well as its resistance to mis-operation caused by radiation of other devices.

There are also non-mandatory standards in individual countries that are not a prerequisite for putting a product on sale, but will affect to one degree or another the ability to attract customers. These standards are promulgated by organizations that have the consumer's interest in mind. Examples are the underwriters laboratories (UL) in the United States and the National Security Inspectorate (NSI) in the United Kingdom.

In the following sections, we review the regulations and standards affecting short-range radio equipment in the United States and Europe. The information given here is necessarily incomplete, and the referenced sources should be consulted to get a more complete understanding of possibilities and restrictions. Anyone contemplating developing any radio transmitting device should take into account very early in



Short-range Wireless Communication. https://doi.org/10.1016/B978-0-12-815405-2.00010-5
© 2019 Elsevier Inc. All rights reserved.

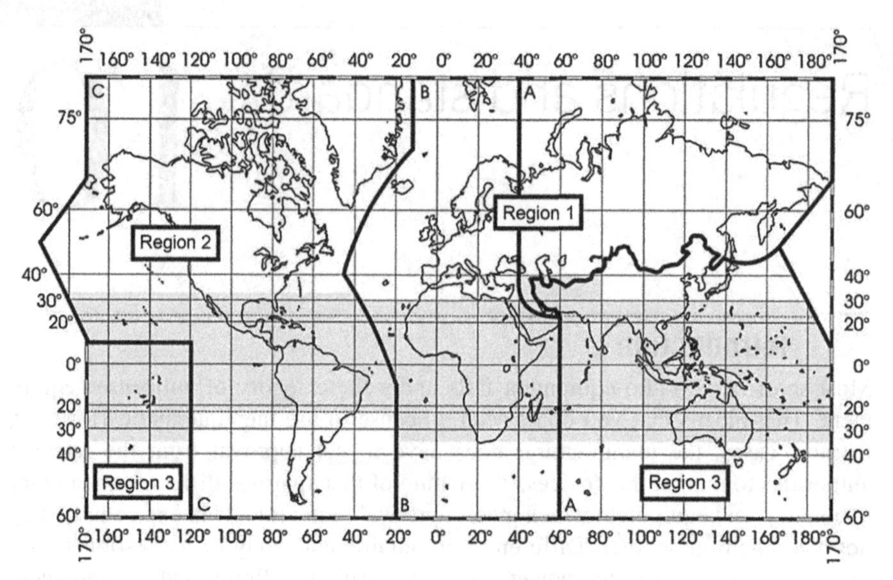


**FIG. 10.1**

Division of world into three regions for radio spectrum allocation.

the design the fact that the final product has to meet specific requirements depending on the countries where it will be sold. Usually, the equipment will have to be tested at an approved testing laboratory. Note that regulations undergo changes from time to time, so the relevant official documents should be consulted for authoritative, up-to-date information.

## 10.2 FCC regulations

Title 47 of the United States Electronic Code of Federal Regulations, Chapter I, sets out the Federal Communications Commission Rules and Regulations relating to radio and telecommunication equipment [1]. These requirements are both technical and administrative. The regulations are divided into parts, each part dealing with a particular class of equipment. Most of the short-range devices that we are concerned with in this book fall under Part 15, which among other things details the requirements for low-power, unlicensed devices. We will also take a look at other Parts which relate to short-range radio equipment, some of which must be licensed. To start with, we overview Part 2, which has information relating to all of the other parts. Sentences and paragraphs within quotation marks are copied directly from the specification.

### 10.2.1 Part 2—Frequency allocations and radio treaty matters: General rules and regulations

Section 2.1 gives terms and definitions that are referred to throughout the FCC rules. Some of the important ones for interpreting the requirements for short-range radio are listed in Appendix 10.A of this chapter.

Subpart B—allocation, assignment, and use of radio frequencies—forms the bulk of Part 2. Presented in table form are the assignment to various user classes of frequencies in the range of 9 kHz to 400 GHz. The table includes frequency uses for the three world regions, and reference to the relevant part number for each user class.

Subpart C sets out the nomenclature for defining emission, modulation, and transmission characteristics. It also defines bandwidths.

Subpart J defines equipment authorization procedures. These may be verification, declaration of conformity, and certification. Verification and declaration of conformity are procedures where the manufacturer, or defined responsible party, makes measurements or takes steps to insure that the equipment complies with the appropriate technical standards. Submittal of samples or test data to the FCC is not required unless specifically requested by the FCC. Receivers generally fall under these categories. In the case of transmitters, a certification procedure must be followed, and the applicant must submit representations and test data to get equipment authorization by the FCC. In practice, what this means is that the equipment has to be tested for compliance with the technical standards by a test facility approved by the FCC. The application for certification is then submitted, along with the test report, by the testing laboratory as a service, or by the applicant himself.

### 10.2.2 Part 15—Radio frequency devices

Part 15 of the FCC regulations "…sets out the regulations under which an intentional, unintentional, or incidental radiator may be operated without an individual license. It also contains the technical specifications, administrative requirements and other conditions relating to the marketing of part 15 devices." [2, paragraph 15.1(a)] We review here the paragraphs of Part 15 that are relevant to short-range radio devices as discussed in this book. Some parts of the regulations are not mentioned, particularly those that refer to non-communication applications. The source Part 15 specifications should be consulted for relevance to special applications not widely used [2].

Radio receivers are considered unintentional radiators, inasmuch as they produce radio-frequency signals as part of their operation but without the intention to radiate them. The local oscillators (one or more) of superheterodyne receivers produce a frequency in the vicinity of the frequency received (removed from it by the value of the IF) and could interfere with other radio receivers. Superregenerative receivers produce a broadband noisy carrier approximately centered on the received frequency. Tuned radio frequency receivers, including the ASH receiver, do not generate a radio frequency for their operation, but if they contain a microprocessor they are classified as a digital device whose radiation is regulated.

Receiver radiation must meet the following field strength limits, according to Paragraph 15.109:

| Frequency of emission (MHz) | Field strength (μV/m) |
| --- | --- |
| 30-88 | 100 |
| 88-216 | 150 |
| 216-960 | 200 |
| Above 960 | 500 |

These field strengths are measured at a distance of 3 m, with the receiver antenna connected.

If the receiver antenna is connected through terminals, the power at the terminals without the antenna may be measured, instead of measuring the radiation from the antenna. The maximum power allowed at the antenna terminals is 2 nW. If this option for checking RF at the antenna is chosen, the limits in paragraph 15.109 must still be checked, with the antenna terminals terminated in the nominal impedance of the antenna.

We now look into the various technical requirements for certifying "intentional radiators" according to Part 15. A common requirement for all Part 15 devices is this:

> *An intentional radiator shall be designed to ensure that no antenna other than that furnished by the responsible party shall be used with the device. The use of a permanently attached antenna or of an antenna that uses a unique coupling to the intentional radiator shall be considered sufficient to comply with the provisions of this section. The manufacturer may design the unit so that a broken antenna can be replaced by the user, but the use of a standard antenna jack or electrical connector is prohibited... (Section 15.203).*

Devices that are connected to an AC power line must also comply with limits on conducted radiation up to 30 MHz.

The least restrictive requirements for unlicensed devices regarding their use, the nature of their signals, and frequencies are specified in Section 15.209 reproduced in part below. Subparagraphs (e) and (f) are not copied here.

#### 10.2.2.1 Paragraph 15.209 radiated emission limits: General requirements

**(a)** Except as provided elsewhere in this subpart, the emissions from an intentional radiator shall not exceed the field strength levels specified in the following table [Part 15]:

| Frequency (MHz) | Field strength (μV/m) | Measurement distance (m) |
| --- | --- | --- |
| 0.009-0.490 | 2400/F (kHz) | 300 |
| 0.490-1.705 | 24000/F (kHz) | 30 |
| 1.705-30.0 | 30 | 30 |

Continued

| Frequency (MHz) | Field strength (μV/m) | Measurement distance (m) |
|---|---|---|
| 30-88 | 100[a] | 3 |
| 88-216 | 150[a] | 3 |
| 216-960 | 200[a] | 3 |
| Above 960 | 500 | 3 |

[a] *Except as provided in paragraph (g), fundamental emissions from intentional radiators operating under this section shall not be located in the frequency bands 54-72, 76-88, 174-216, or 470-806 MHz. However, operation within these frequency bands is permitted under other sections of this part, e.g., Sections 15.231 and 15.241.*

**(b)** In the emission table above, the tighter limit applies at the band edges.

**(c)** The level of any unwanted emissions from an intentional radiator operating under these general provisions shall not exceed the level of the fundamental emission. For intentional radiators which operate under the provisions of other sections within this part and which are required to reduce their unwanted emissions to the limits specified in this table, the limits in this table are based on the frequency of the unwanted emission and not the fundamental frequency. However, the level of any unwanted emissions shall not exceed the level of the fundamental frequency.

**(d)** The emission limits shown in the above table are based on measurements employing a CISPR quasi-peak detector except for the frequency bands 9-90 kHz, 110-490 kHz, and above 1000 MHz. Radiated emission limits in these three bands are based on measurements employing an average detector.

.................................................................................................................

**(g)** Perimeter protection systems may operate in the 54-72 and 76-88 MHz bands under the provisions of this section. The use of such perimeter protection systems is limited to industrial, business and commercial applications.

The field strengths for frequencies between 216 and 960 MHz, which are our primary interest, are quite low and suffice for reliable communication only over a few meters, which may well be sufficient for some classes of short-range devices—a wireless mouse or keyboard, for example. The 200 μV/m limit at 3 m translates to radiated power of only 12 nW.

Subsection (d) refers to a CISPR quasi-peak detector. This instrument has specified attack times, decay times, and 6 dB bandwidths as well as overload factors which are shown as a function of frequency range in Table 10.1 [3].

On specified frequency bands, higher power is allowed, although with restrictions as to use. Below are highlights of the regulations, with comments, for short-range communicating devices, presented according to paragraph or section numbers of Part 15.

#### *10.2.2.2 Paragraph 15.229 operation within the band 40.66-40.70*

Operation on this band is possible with a field strength limitation of 1000 μV/m at 3 m, as compared to 100 μV/m at 3 m according to Section 15.209. Oscillator stability must be within 100 ppm. The equivalent radiated power is 300 nW. Potential

**Table 10.1** CISPR 16 quasi-peak detector specifications

| | 9-150 kHz | 0.15-30 MHz | 30-1000 MHz |
|---|---|---|---|
| 6 dB bandwidth (kHz) | 0.2 | 9 | 120 |
| Charge time constant (ms) (attack time) | 45 | 1 | 1 |
| Discharge time constant (ms) (decay time) | 500 | 160 | 550 |
| Predetector overload factor (required excess linearity) (dB) | 24 | 30 | 43.5 |

users should note that the background noise at this low VHF frequency is comparatively high, compared to the more popular UHF bands. Note that there are no restrictions to the type of operation according to this section, so, for example, continuous voice transmissions could be used, which are prohibited under Section 15.231.

#### *10.2.2.3 Paragraph 15.231 periodic operation in the band 40.66-40.70 and above 70 MHz*

Most of the short-range wireless links for security and medical systems and keyless entry are certified to the requirements in this section. Considerably more field strength is allowed than in 15.209, and there is a wide choice of frequencies. However, the transmission duty cycle is restricted and voice and video transmissions are not allowed. This is not a serious impediment for security systems, except in regard to sending periodic supervision or status signals, which are limited to 2 seconds per hour. The provisions of this section are reproduced below.

*Section 15.231 Periodic operation in the band 40.66-40.70 MHz and above 70 MHz*

**(a)** The provisions of this section are restricted to periodic operation within the band 40.66-40.70 MHz and above 70 MHz. Except as shown in paragraph (e) of this section, the intentional radiator is restricted to the transmission of a control signal such as those used with alarm systems, door openers, remote switches, etc. Continuous transmissions, voice, video and the radio control of toys are not permitted. Data are permitted to be sent with a control signal. The following conditions shall be met to comply with the provisions for this periodic operation:

- **(1)** A manually operated transmitter shall employ a switch that will automatically deactivate the transmitter within not more than 5 s of being released.
- **(2)** A transmitter activated automatically shall cease transmission within 5 s after activation.
- **(3)** Periodic transmissions at regular predetermined intervals are not permitted. However, polling or supervision transmissions, including data, to determine system integrity of transmitters used in security or safety applications are allowed if the total duration of transmissions does not exceed more than 2 seconds per hour for each transmitter. There is no limit on the number of individual transmissions, provided the total transmission time does not exceed 2 seconds per hour.

**(4)** Intentional radiators which are employed for radio control purposes during emergencies involving fire, security, and safety of life, when activated to signal an alarm, may operate during the pendency of the alarm condition.

**(5)** Transmission of set-up information for security systems may exceed the transmission duration limits in Paragraphs (a)(1) and (a)(2) of this section, provided such transmissions are under the control of a professional installer and do not exceed 10 s after a manually operated switch is released or a transmitter is activated automatically. Such set-up information may include data.

**(b)** In addition to the provisions of Section 15.205, the field strength of emissions from intentional radiators operated under this section shall not exceed the following:

| Fundamental frequency (MHz) | Field strength of fundamental (μV/m) | Field strength of spurious emissions (μV/m) |
|---|---|---|
| 40.66-40.70 | 2,250 | 225 |
| 70-130 | 1,250 | 125 |
| 130-174 | [a]1,250 to 3,750 | [a]125 to 375 |
| 174-260 | 3,750 | 375 |
| 260-470 | [a]3,750 to 12,500 | [a]375 to 1,250 |
| Above 470 | 12,500 | 1,250 |

[a] *Linear interpolations.*

**(1)** The above field strength limits are specified at a distance of 3 m. The tighter limits apply at the band edges.

**(2)** Intentional radiators operating under the provisions of this section shall demonstrate compliance with the limits on the field strength of emissions, as shown in the above table, based on the average value of the measured emissions. As an alternative, compliance with the limits in the above table may be based on the use of measurement instrumentation with a CISPR quasi-peak detector. The specific method of measurement employed shall be specified in the application for equipment authorization. If average emission measurements are employed, the provisions in Section 15.35 for averaging pulsed emissions and for limiting peak emissions apply. Further, compliance with the provisions of Section 15.205 shall be demonstrated using the measurement instrumentation specified in that section.

**(3)** The limits on the field strength of the spurious emissions in the above table are based on the fundamental frequency of the intentional radiator. Spurious emissions shall be attenuated to the average (or, alternatively, CISPR quasi-peak) limits shown in this table or to the general limits shown in Section 15.209, whichever limit permits a higher field strength.

**(c)** The bandwidth of the emission shall be no wider than 0.25% of the center frequency for devices operating above 70 MHz and below 900 MHz. For devices operating above 900 MHz, the emission shall be no wider than 0.5% of the center frequency. Bandwidth is determined at the points 20 dB down from the modulated carrier.

**(d)** For devices operating within the frequency band 40.66-40.70 MHz, the bandwidth of the emission shall be confined within the band edges and the frequency tolerance of the carrier shall be ±0.01%. This frequency tolerance shall be maintained for a temperature variation of −20 degrees to +50°C at normal supply voltage, and for a variation in the primary supply voltage from 85% to 115% of the rated supply voltage at a temperature of 20°C. For battery operated equipment, the equipment tests shall be performed using a new battery.

**(e)** Intentional radiators may operate at a periodic rate exceeding that specified in paragraph (a) of this section and may be employed for any type of operation, including operation prohibited in paragraph (a) of this section, provided the intentional radiator complies with the provisions of paragraphs (b) through (d) of this section, except the field strength table in paragraph (b) of this section is replaced by the following:

| Fundamental frequency (MHz) | Field strength of fundamental (μV/m) | Field strength of spurious emission (μV/m) |
|---|---|---|
| 40.66-40.70 | 1,000 | 100 |
| 70-130 | 500 | 50 |
| 130-174 | 500 to 1,500[a] | 50 to 150[a] |
| 174-260 | 1,500 | 150 |
| 260-470 | 1,500 to 5,000[a] | 150 to 500[a] |
| Above 470 | 5,000 | 500 |

[a] *Linear interpolations.*

In addition, devices operated under the provisions of this paragraph shall be provided with a means for automatically limiting operation so that the duration of each transmission shall not be greater than 1 s and the silent period between transmissions shall be at least 30 times the duration of the transmission but in no case less than 10 s.

*Comments on Section 15.231*

**(a)** Section 15.205 referred to in the present section contains a table of restricted frequency bands where fundamental transmissions are not allowed and where spurious limits may be lower than those presented above. This table is reproduced in Appendix 10.B.

**(b)** The field strength limits in this section are based on average measurements. Thus, when ASK is used, peak field strength may exceed average field strength by a factor that is the inverse of the duty cycle of the modulation. However, Section 15.35 limits the peak-to-average ratio to 20 dB. The method of determining the average field strength is stated in Section 15.35 as follows:

*... when the radiated emission limits are expressed in terms of the average value of the emission, and pulsed operation is employed, the measurement field strength shall be determined by averaging over one complete pulse train, including blanking intervals, as long as the pulse train does not exceed 0.1 seconds. As an alternative (provided the transmitter operates for longer than 0.1 seconds) or in cases where the pulse train exceeds 0.1 seconds, the measured field strength shall be determined from the average absolute voltage during a 0.1 second interval during which the field strength is at its maximum value. The exact method of calculating the average field strength shall be submitted with any application for certification or shall be retained in the measurement data file for equipment subject to Supplier's Declaration of Conformity.*

#### 10.2.2.4 Paragraph 15.235 operation within the band 49.82-49.90 MHz

*Section 15.235 operation within the band 49.82-49.90 MHz*

**(a)** The field strength of any emission within this band shall not exceed 10,000 μV/m at 3 m. The emission limit in this paragraph is based on measurement instrumentation employing an average detector. The provisions in Section 15.35 for limiting peak emissions apply.

**(b)** The field strength of any emissions appearing between the band edges and up to 10 kHz above and below the band edges shall be attenuated at least 26 dB below the level of the unmodulated carrier or to the general limits in Section 15.209, whichever permits the higher emission levels. The field strength of any emissions removed by more than 10 kHz from the band edges shall not exceed the general radiated emission limits in Section 15.209. All signals exceeding 20 μV/m at 3 m shall be reported in the application for certification.

This low VHF band has an advantage over the operation permitted under Section 15.231 because it can be used for continuous transmissions for either analog or digital communication. Permitted average field strength at 3 m is 10,000 μV/m, which is capable of attaining a range of several hundred meters if an efficient receiving antenna is used. The allowed bandwidth is only 80 kHz so transmitters and receivers must have crystal-controlled oscillators.

#### 10.2.2.5 Paragraph 15.237 operation in the bands 72.0-73, 74.6-74.8, and 75.2-76

This section is restricted to auditory assistance devices, defined as

An intentional radiator used to provide auditory assistance communications (including but not limited to applications such as assistive listening, auricular training, audio description for the blind, and simultaneous language translation) for:

**(1)** Persons with disabilities: In the context of part 15 rules (47 CFR part 15), the term "disability," with respect to the individual, has the meaning given to it by section 3(2)(A) of the Americans with Disabilities Act of 1990 (42 U.S.C. 12102 (2)(A)), i.e., a physical or mental impairment that substantially limits one or more of the major life activities of such individuals;

**(2)** Persons who require language translation; or

**(3)** Persons who may otherwise benefit from auditory assistance communications in places of public gatherings, such as a church, theater, auditorium, or educational institution (Section 15.3 (a)).

Maximum allowed bandwidth is 200 kHz and the limit on average field strength is 80 mV/m at 3 m.

#### *10.2.2.6 Paragraph 15.241 operation in the band 174-216 MHz*

Operation under this Section is restricted to biomedical telemetry devices, defined as "An intentional radiator used to transmit measurements of either human or animal biomedical phenomena to a receiver" (Section 15.3(b)).

Maximum bandwidth is 200 kHz and the average field strength limit is 1500 μV/m at 3 m.

#### *10.2.2.7 Paragraph 15.242 operation in the bands 174-216 and 470-668 MHz*

This section also is dedicated to biomedical telemetry devices but they must be employed solely on the premises of health care facilities, which includes hospitals and clinics. Signals of bandwidths of 200 kHz and up to 200 mV/m at 3 m are allowed. This section defines in detail the requirements for avoiding interference with television stations and radio astronomy observatories.

#### *10.2.2.8 Paragraph 15.247 operation within the bands 902-928, 2400-2483.5, and 5725-5850 MHz*

This section defines the requirements for devices using spread spectrum and other wideband modulation in the ISM bands. Devices designed to comply with the requirements of this section have the potential for highest performance, in respect to range and interference immunity, of all unlicensed equipment. We present here a summary of the major points of this section. For details, refer to the source.

Both frequency hopping and what is called digital modulation arc allowed, as well as a combination of the two methods. Digital modulation is defined as (Section 15.403(f)): "The process by which the characteristics of a carrier wave are varied among a set of predetermined discrete values in accordance with a digital modulating function…"

Previous to the revision dated August 2002, this section referred to direct sequence spread-spectrum modulation but that term was dropped so as not to restrict the introduction of new types of wideband modulation methods. In particular, the modulation systems introduced to increase data rates in WLANs according to IEEE 802.11 revisions a, *b*, *g*, and later do not fit the classical definition of direct sequence spread-spectrum but were deemed to be suitable for inclusion in this FCC section (see Chapter 11). The technical requirements of this section differ according to the type of spectrum-spreading (frequency-hopping or digital) and the particular frequency band.

**(1)** Frequency-hopping spread-spectrum. The general requirement is "The system shall hop to channel frequencies that are selected at the system hopping rate from a pseudo randomly ordered list of hopping frequencies. Each frequency must be used equally on the average by each transmitter. The system receivers shall have input bandwidths that match the hopping channel bandwidths of their corresponding transmitters and shall shift frequencies in synchronization with the transmitted signals." A system may choose and adapt hopsets to avoid interference with other users within the spectrum band.

| | |
|---|---|
| Minimum hopping channel separation | Minimum 25 kHz, but not less than the channel bandwidth at 20 dB down. Alternatively, frequency hopping systems operating in the 2400-2483.5 MHz band may have hopping channel carrier frequencies that are separated by 25 kHz or two-thirds of the 20 dB bandwidth of the hopping channel, whichever is greater, provided the systems operate with an output power no greater than 125 mW |
| Minimum number of pseudorandomly ordered hopping frequencies | (a) 900 MHz band—50 for hopping channel BW less than 250 kHz and 25 for BW greater than 250 kHz. Max. BW 500 kHz at 20 dB down<br>(b) 2.4 GHz band—15 non-overlapping channels. Frequency hopping systems may avoid or suppress transmissions on a particular hopping frequency provided that a minimum of 15 channels are used<br>(c) 5.7 GHz band—75 hopping frequencies. Max BW 1 MHz at 20 dB down |
| Occupancy time on channel frequency | (a) 900 MHz band—Max. 0.4 s within 20 s period for BW less than 250 kHz, and 0.4 s within 10 s for BW greater than 250 kHz<br>(b) 2.4 GHz band—Max. 0.4 s within 0.4 s times number of hopping channels used<br>(c) 5.7 GHz band—Max. 0.4 s within 30 s period |
| Maximum peak output power | (a) 900 MHz band—1 watt when using at least 50 hopping channels; 0.25 watt with less than 50 but at least 25 hopping channels<br>(b) 2.4 GHz band—1 watt when using at least 75 hopping channels, otherwise 0.125 watt<br>(c) 5 GHz band—1 watt<br>NOTE: For all bands, except for 2.4 and 5 GHz band fixed point-to-point applications, when using directional antennas the maximum peak power limit is reduced by the amount that the antenna gain exceeds 6 dBi |

**(2)** Digital modulation.

| | |
|---|---|
| Minimum bandwidth | 500 kHz at 6 dB down |
| Maximum peak output power | 1 W |
| Peak power spectral density to the antenna | 8 dBm in any 3-kHz band |

#### 10.2.2.9 Paragraph 15.249 operation within the bands 902-928, 2400-2483.5, 5725-5875, and 24.0-24.25 GHz

This subsection regulates operation in upper UHF and microwave bands without specifying the type of use or modulation constraints. Relevant provisions principally relating to the three lower bands are reproduced below.

*Section 15.249 Operation within the bands 902-928 MHz, 2400-2483.5 MHz, 5725-5875 MHz, and 24.0-24.25 GHz*

**(a)** Except as provided in paragraph (b) of this section, the field strength of emissions from intentional radiators operated within these frequency bands shall comply with the following:

| Fundamental frequency | Field strength of fundamental (mV/m) | Field strength of harmonics (μV/m) |
|---|---|---|
| 902-928 MHz | 50 | 500 |
| 2400-2483.5 MHz | 50 | 500 |
| 5725-5875 MHz | 50 | 500 |
| 24.0-24.25 GHz | 250 | 2500 |

**(b)** Relates to fixed point-to-point operation on the 24.05 to 24.25 GHz band.
**(c)** Field strength limits are specified at a distance of 3 m.
**(d)** Emissions radiated outside of the specified frequency bands, except for harmonics, shall be attenuated by at least 50 dB below the level of the fundamental or to the general radiated emission limits in Section 15.209, whichever is the lesser attenuation.
**(e)** As shown in Section 15.35(b), for frequencies above 1000 MHz, the field strength limits in paragraphs (a) and (b) of this section are based on average limits. However, the peak field strength of any emission shall not exceed the maximum permitted average limits specified above by more than 20 dB under any condition of modulation.

If contemplating the use of this section, note the following:

**(1)** Harmonic radiation must be reduced 40 dB compared to the maximum allowed fundamental, and not 20 dB as on the lower UHF bands.

**(2)** The field strength on the 900-MHz band is determined by a quasipeak or peak detector, and is not an average value as in Section 15.231 referenced above. This reduces the possible advantage of using ASK with low duty cycle (LDC).

**(3)** The detector bandwidth for 902-928 MHz band measurements is 120 kHz. Since field strength is measured over this bandwidth, signals with wider bandwidths can have higher total field strength, or equivalent power, depending on the degree that their bandwidth is greater than 120 kHz and on the shape of the signal's spectrum.

#### *10.2.2.10 Subpart E—Unlicensed National Information Infrastructure devices (U-NII)*

A special category was created in FCC Part 15 for "Intentional radiators operating in the frequency bands 5.15-5.35 and 5.47-5.85 GHz that use wideband digital modulation techniques and provide a wide array of high data rate mobile and fixed communications for individuals, businesses, and institutions." Our interest in U-NII primarily relates to their use in WLANs, specifically in connection with specification IEEE 802.11a and later amendments (see Chapter 11). While the wide growth of WLAN has occurred with equipment operating in the 2.4 to 2.4835 GHz ISM band, problems of interference with other types of wireless services and restricted bandwidth are making the 5 GHz U-NII an attractive alternative for continued development. With a more restricted definition of allowed applications and a total bandwidth of 580 MHz as compared to the 83.5 MHz available on 2.4 GHz, U-NII offers an outlet for high performance wireless network connectivity not possible on lower frequency bands.

The principle characteristics of U-NII devices as put forth in Section 15.407 of FCC Part 15 are summarized in Table 10.2.

#### *10.2.2.11 Subpart F—Ultra-wideband (UWB) operation*

As of the FCC Part 15 revision of August 2002, a new category of short-range communication was included in the standard—ultra-wideband (UWB). This category differs from all the other intentional radiation sections by being defined in a much wider spectrum allocation in relation to the median frequency of the band. Chapter 11 of this book gives a description of UWB transmissions.

The bandwidth of UWB is defined by the FCC as the difference between the two frequencies on both sides of the frequency of maximum radiation at which the radiated emission is 10 dB down. The bandwidth is based on the whole transmission system, including the antenna. If the upper and lower 10 dB down frequencies are $f_H$ and $f_L$ respectively, then the fractional bandwidth equals $2(f_H - f_L)/(f_H + f_L)$. The center frequency is defined as $f_C = (f_H + f_L)/2$. A transmitter that falls in the category of UWB has a fractional bandwidth equal to or greater than 0.2 or a UWB bandwidth equal to or greater than 500 MHz, regardless of fractional bandwidth.

FCC Part 15 Subpart F specifies several categories of UWB systems, some of which must be licensed and are intended for specific users. For example, UWB imaging systems, used for through the wall imaging, ground penetration radar, medical

**Table 10.2** Maximum power limits for U-NII band

| | 5.15-5.25 GHz | 5.25-5.35 GHz and 5.47–5.725 GHz | 5.725-5.85 GHz |
|---|---|---|---|
| Maximum power to antenna | 1 W for access points, 250 mW for client devices. Outdoor: maximum e.i.r.p.[a] 125 mW at elevation angle greater than 30° from horizon | Lesser of250 mW or 11 dBm + 10 log (emission bandwidth in MHz at 26 dB down) | 1 W |
| Maximum power spectral density | Access point: 17 dBm in any 1 MHz band Client: 11 dBm in any 1 MHz band | 11 dBm in any 1 MHz band | 30 dBm in any 500 kHz band |
| Antenna gain restriction | Maximum allowed power is reduced by the amount of antenna gain over 6 dBi. For fixed point-to-point, up to 23 dBi gain is allowed without power restriction | Maximum allowed power is reduced by the amount of antenna gain over 6 dBi | Maximum allowed power is reduced by the amount of antenna gain over 6 dBi. For fixed point-to-point, power reduction is not required with high gain antennas |

[a] *NOTE: e.i.r.p. = equivalent isotropic radiated power.*

imaging (detects location or movement of objects within a living body), and surveillance, may be marketed only to law enforcement, rescue, medical, or certain industrial or commercial entities who are solely eligible to operate the equipment. Our concern is with two more general categories—indoor and handheld UWB systems.

Both indoor and handheld devices must have UWB bandwidths contained in the frequency range of 3,100 to 10,600 MHz. Radiation at or below 960 MHz must not exceed the levels in Section 15.209 (see above). Above 960 MHz, emission limits for the two categories are according to the following table:

| | Indoor | Handheld |
|---|---|---|
| Frequency in MHz | Average e.i.r.p. in dBm | Average e.i.r.p. in dBm |
| 960-1610 | –75.3 | -75.3 |
| 1610-1990 | –53.3 | -63.6 |
| 1990-3100 | –51.3 | -61.3 |
| 3100-10600 | –41.3 | -41.3 |
| Above 10600 | –51.3 | -61.3 |

The resolution bandwidth of the measuring device is 1 MHz.

In order to specifically protect GPS (global positioning system) receivers from UWB interference, radiation around GPS channels is checked with narrow-resolution bandwidth as follows:

| Frequency in MHz | e.i.r.p. in dBm |
|---|---|
| 1164-1240 | −85.3 |
| 1559-1610 | −85.3 |

This table applies to both indoor and handheld devices. Resolution bandwidth is no less than 1 kHz.

Peak emissions within a 50 MHz bandwidth around the frequency of highest radiation are limited to 0 dBm e.i.r.p.

### 10.2.3 Part 74 experimental radio, auxiliary, special broadcast and other program distributional services

Subpart H regulates low-power auxiliary stations, defined as follows:

> Low-power auxiliary station. *An auxiliary station authorized and operated pursuant to the provisions set forth in this subpart. Devices authorized as low power auxiliary stations are intended to transmit over distances of approximately 100 m for uses such as wireless microphones, cue and control communications, and synchronization of TV camera signals.*

Frequency assignments listed in Section 74.802 cover defined frequency bands from 26.1 up to 7125 MHz. The regulations specify requirements to avoid interference to TV channels.

Users of devices covered in this part must obtain a license from the FCC. Eligible for licenses are radio and television station licensees, cable television station operators that produce live programs, television broadcasting and motion picture producers, licensees of stations in the Broadband Radio Service, large venue owners or operators (who run major events or productions using 50 or more auxiliary station devices), and professional sound companies (that routinely provide audio services using 50 or more low power auxiliary station devices). Authorization for temporary use of low power auxiliary station operation is also provided for in the regulations.

Maximum allowed power output is 1 W, except for the 1435-1525 MHz band where it is 250 mW. Bandwidth and emission masks are specified for analog and digital modulation and depend on the frequency band used.

### 10.2.4 Part 95 personal radio service

Personal radio services comprise citizens band radio, radio control, and other licensed and unlicensed services described in Part 95. Here is a brief description of services that are relevant to short-range wireless communication as covered in this book.

#### 10.2.4.1 Subpart B—Family radio service (FRS)

The family radio service is "A short-distance two-way voice communication service, with limited data applications, between low power hand-held radios, for facilitating individual, family, group, recreational and business activities." It is allotted 22 channels, each having a channel bandwidth of 12.5 kHz. Channels 1 through 7 have center frequencies between 462.5625 and 462.7125 MHz and are allowed a maximum effective radiated power (e.r.p.) of 2 W. Channels 8 through 14 have center frequencies between 467.5625 and 467.7125 MHz and are allowed a maximum power of 0.5 W. 2 W power is allowed on channels 15 through 22 with center frequencies between 462.5500 and 462.7250 MHz. Phase or frequency modulation of voice or limited data (location and brief text) is authorized. Occupied bandwidth must not exceed 12.5 kHz. Maximum peak frequency deviation is 2.5 kHz, and modulating audio frequencies must be below 3.125 kHz. The antenna must be a non-removable, integral part of the device.

#### 10.2.4.2 Subpart C—Radio control radio service

The radio control radio service (RCRS) is "A non-commercial short-distance radio service for wirelessly controlling the operation of devices, including, but not limited to, model vehicles such as aircraft and surface craft." It is only to be used for one-way communications. Frequency bands used depend on the type of device being controlled, as follows:

**(a)** *Control of model aircraft and surface crafts and devices.*
  **(1)** RCRS channels in the 72 MHz frequency band may be used only to control and operate model aircraft.
  **(2)** RCRS channels in the 75 MHz frequency band may be used only to control and operate model surface craft.
  **(3)** RCRS channels in the 26-28 MHz frequency band may be used to control or operate any kind of device.

**(b)** *Telecommand.* Any RCRS channel may be used by the operator to turn on and/or off a device at a remote location.

**(c)** *Telemetry.* Any RCRS channel in the 26-28 MHz frequency band may be used to transmit a signal from a sensor at a remote location that turns on and/or off an indicating device for the operator.

Channels are allocated around center frequencies in the 26-28 MHz band, from 26.995 to 27.255 MHz, in the 72 MHz band from 72.01 to 72.99 MHz, and in the 75 MHz band from 75.41 to 75.99 MHz. Occupied bandwidth may not exceed 8 kHz. Any type of modulation can be used, but voice transmissions are not allowed. Mean transmitter output power in the 72 and 75 MHz bands must not exceed 0.75 W. Maximum power in the 26-28 MHz band is 4 W, except for 27.255 MHz where up to 25 watts is allowed. The antenna must be an integral part of the transmitter.

#### 10.2.4.3 Subpart G—Low power radio service

The low power radio service (LPRS) is "A short-distance voice and data communication service for providing auditory assistance to persons with disabilities (and others), health care related communications, law enforcement tracking, and for certain other purposes." It is used for assistive listening devices, audio description for the blind, simultaneous language translation, for health care related communications for the ill, and for law enforcement tracking signals. It is not intended for two-way voice communications. 260 channels between 216 and 217 MHz are allotted for this service, classified in three groups: standard band channels of 25 kHz bandwidth, extra band channels of 50 kHz bandwidth, and narrow band channels with 5 kHz bandwidth. Maximum transmitting power for the uses mentioned above is 100 mW.

#### 10.2.4.4 Subpart H—Medical services

Part 95 regulates low power transmissions for medical services, in two categories. Subpart H defines the wireless medical telemetry service (WMTS). Subpart I relates to Medical Device Radio Communications Service.

The WMTS is a short-distance data communication service for the transmission of physiological parameters and other patient medical information via radiated electromagnetic signals. Operation is in the 608-614, 1395-1400, and 1427-1432 MHz frequency bands. Only authorized health care providers are eligible to operate transmitters in this service without an FCC issued license. Field strength limits are as follows:

**(a)** For WMTS transmitter types operating in the 608-614 MHz band, the field strength of the transmitted signal must not exceed 200 mV/m, measured at a distance of 3 m, using instrumentation with a CISPR quasi-peak detector.
**(b)** For WMTS transmitter types operating in the 1395-1400 and 1427-1432 MHz bands, the field strength of the transmitted signal must not exceed 740 mV/m, measured at 3 m, using instrumentation with an averaging detector and a 1 MHz reference bandwidth.

The medical device radio (MedRadio) Service is "An ultra-low power radio service for the transmission of non-voice data for the purpose of facilitating diagnostic and/or therapeutic functions involving implanted and body-worn medical devices." Transmitters in the MedRadio service may be operated only by duly authorized health care professionals, and individuals using them under the direction of a health care professional. This includes medical devices that have been implanted in or placed on the body of the individual by, or under the direction of, a duly authorized health care professional. Other than for testing and demonstration, "MedRadio programmer/control transmitters may transmit only non-voice data containing operational, diagnostic and therapeutic information associated with a medical implant device or medical body-worn device that has been implanted or placed on the person by or under the direction of a duly authorized health care professional." MedRadio transmitters operate in the 401-406, 413-419, 426-432, 438-444, 451-457, and 2360-2400 MHz bands. Allowed transmitter radiated power output varies according to frequency band segment and is specified in specification paragraph 95.2567.

### 10.2.5 Test method for Part 15

Paragraph 15.31 references measurement procedures for determining compliance with the technical requirements. The principle of field strength measurements is shown in Fig. 10.2. Measurements are preferably made in an FCC-approved open field test site but they can also be carried in an anechoic chamber. The equipment under test (EUT) is placed on a wooden turntable, 0.8 m high, located 3 m from the test antenna. Measurements are taken on a field strength measuring receiver or spectrum analyzer. Test instrument bandwidth must be at least 100 kHz for signals below 1 GHz and 1 MHz for signals above 1 GHz. When peak readings are taken, the average value for digital signals can be calculated from the baseband pulse diagram. Reading is taken at the fundamental frequency and all spurious radiation frequencies.

Field strength readings are taken at all practical equipment orientations by rotating the turntable. The height of the test antenna is varied from between 1 and 4 m and readings are taken for both vertical and horizontal antenna polarization. At each emission frequency, the maximum reading is recorded for all possibilities of equipment orientation, test antenna height and polarization.

Note the importance, in developing a short-range radio transmitter, of obtaining as far as possible omnidirectional radiation characteristics of the built-in antenna. Since the power of the device is limited at its position of maximum radiation, a transmitter with unintentional directional characteristics will give less than maximum range when the orientation of the transmitter relative to the receiver is uncontrollable, as in most applications.

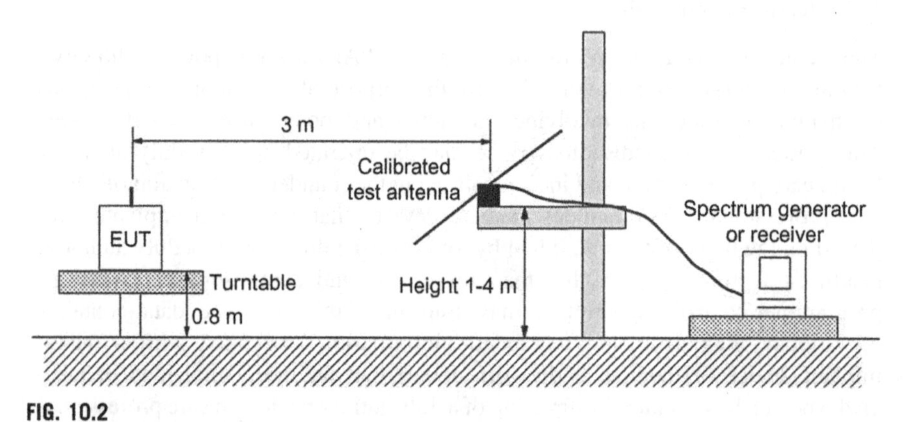


**FIG. 10.2**

Equipment arrangement for FCC Part 15 tests.

## 10.3 European radiocommunication regulations

The European Conference of Postal and Telecommunications Administrations (CEPT) is the regional regulatory telecommunication organization for Europe. It was established by 19 countries in 1959, and has 48 member countries as of 2017. The European Radiocommunications Committee (ERC) is one of three committees that form CEPT. One of its principal aims is the harmonization of frequency use and regulation regimes in Europe and the fostering of world-wide harmonization of frequencies. Associated with CEPT is the European Telecommunications Standards Institute (ETSI), which prepares the technical standards on which European telecommunication regulations are based.

### 10.3.1 CEPT/ERC recommendation 70-03

Table 10.3 shows frequency bands, maximum power, and maximum occupied bandwidth that are designated by ERC for common applications of short-range devices. The information comes from CEPT/ERC Recommendation 70-03 [4]. Applications covered are telemetry, telecommand, alarms, and data transfer in general, including local area networks (LAN) where noted. Frequency ranges between 26 MHz and 9 GHz are shown in the table, although lower and higher frequency bands are available. User licenses are not required. It is important to realize that not all frequencies and technical parameters are available for use in all CEPT countries, and some countries authorize operation on frequencies not listed or with different technical parameters. The individual regulating bodies in each country where marketing is contemplated should be consulted to get the latest information for that country.

ERC Recommendation (70-03) defines regulatory parameters of short-range devices categorized in 13 annexes as follows:

Annex 1. Non-specific short range devices (SRD). Includes telemetry, telecommand, alarms and data in general.
Annex 2. Tracking, tracing and data acquisition. Among applications and devices covered are meter reading, data acquisition and sensors and actuators in data networks, medical body area network systems (MBANS), wireless industrial applications.
Annex 3. Wideband data transmission systems. Covers data networks in the 863-868 MHz band and radio local area networks (RLAN) in the 2400-2483.5 MHz band.
Annex 4. Railway applications.
Annex 5. Transport and traffic telematics.
Annex 6. Radio determination applications. Includes equipment for detecting movement and alert.
Annex 7. Alarms. Includes social alarms and alarms for security and safety.
Annex 8. Model control. Model control equipment is used for the purpose of controlling the movement of a model in the air, on land or over or under water.

Table 10.3 Frequency bands and uses

| Annex | Frequency band (MHz) | Maximum Power (mW e.r.p., e.i.r.p. above 1 GHz) | Channel bandwidth (kHz) | Maximum duty cycle (%)[a] | Comments |
|---|---|---|---|---|---|
| 1 | 26.957-27283 | 10 | – | – | |
| 1 | 40.66-40.7 | 10 | – | – | |
| 1 | 138.2-138.45 | 10 | – | 1.0 | |
| 1,2 | 169.4-169.475 | 500 | 50 | 1, 10 (Annex 2) | Meter reading (Annex 2) |
| 1 | 169.4-169.4875 | 10 | – | 0.1 | |
| 1 | 433.050-434.790 | 10 | – | 10 | |
| 1 | 862-870 | 25 | Depends on subband | 0.1 | FHSS, DSSS, narrow band and wideband in parts of band range |
| 3 | 863-868 | 25 | 1 | 10 (access points) or 2.8 | Data networks |
| 2 | 865-868 | 500 | 200 | 10 | Data networks. Four 200 kHz segments allocated |
| 2 | 870-875.6 | | 200 | 2.5 | Data networks |
| 1 | 868-868.6 | 25 | – | 1 | |
| 7 | 868.6-868.7 | 10 | 25 | 1 | |
| 1 | 868.7-869.2 | 25 | | 0.1 | |
| 7 | 869.2-869.25 | 10 | 25 | 0.1 | [b]Social alarms |
| 7 | 869.25-869.3 | 10 | 25 | 0.1 | |
| 7 | 869.3-869.4 | 10 | 25 | 1 | |
| 1 | 869.4-869.65 | 500 | | 10 | |
| 7 | 869.65-869.7 | 25 | 25 | 10 | |
| 1 | 869.7-870 | 5; 25 | | 1 for 25 mW power | |
| 1 | 915-921 | 25 | 200 | 0.1 | |

| | | | | | |
|---|---|---|---|---|---|
| | 915.2-920.8 | 25 | 600 or 400 (4 freq.) | 1 | |
| 1 | 2400-2483.5 | 10 | – | – | |
| 3 | 2400-2483.5 | 100 | | | Local area networks |
| 2 | 2483.5-2500 | 1 | 3 | 10 + LBT/AFA | Medical body area network in healthcare facilities |
| 2 | 2483.5-2500 | 10 | 3 | 2 + LBT/AFA | Medical body area network in home |
| 1 | 5725-5825 | 25 | – | – | |
| 2 | 5725-5875 | 400 | Between 1 and 20 MHz | | Wireless industrial applications. Adaptive power control (APC) required |
| 1 | 6000-8500 | −41.3 dBm/MHz, 0 dBm peak in 50 MHz | – | – | Source: ECC decision (06)04 |
| 1 | 3100-4800, 6000-9000 | – | – | – | UWB. Details in ECC decisions (06) 04, (12)03 (aircraft) |

[a] *For Annex 1, when listen before talk (LBT) with adaptive frequency agility (AFA) are used, the duty cycle limit does not apply.*
[b] *Social alarms: "devices that allow reliable communication including portable equipment which allows a person in distress in a limited area to initiate a call for assistance by a simple manipulation" [5].*

Annex 9. Inductive applications. Includes car immobilisers, some RFID, animal identification, automatic road tolling and other very short range applications.
Annex 10. Radio microphone applications including assistive listening devices, wireless audio and multimedia streaming systems.
Annex 11. Radio frequency identification (RFID) applications. Includes automatic article identification, asset tracking, access control, anti-theft systems, location systems.
Annex 12. Active medical implants and their associated peripherals.

Table 10.3 is a partial list only and does not include all possibilities and limitations of use. The source [70-03] should be reviewed to get the complete picture of frequencies and operating characteristics and limitations.

### 10.3.2 ETSI specifications

We review below the ETSI specifications that are relevant to short-range radio communications. Each country in the European Community may or may not base their regulations wholly on these specifications, so it's necessary to check with the regulating authorities in each country where the product will be sold.

#### *10.3.2.1 EN 300 220-1 technical characteristics and test methods for radio equipment to be used in the 25-MHz to 1000-MHz frequency range with power levels ranging up to 500 mW [5]*

The document details test methods for a range of short-range devices and applications within the frequency range and power levels of its title.

Measured parameters, and brief comments, are:

| Effective radiated power (e.r.p.) | Includes antenna gain over dipole |
|---|---|
| Maximum effective radiated power spectral density | Highest power density in 100 kHz bandwidth |
| Duty cycle | Transmitting time per observation time (typically 1 h) |
| Occupied bandwidth | Bandwidth containing 99% of total mean power |
| Frequency error | Deviation of nominal carrier frequency over extreme temperature and voltage |
| Tx out of band emissions | Emissions outside operating channel and within operational frequency band |
| Spurious emissions | Emissions outside operational frequency band |
| Transient power | Out of operating channel emissions resulting from switching transmitter on and off |
| Adjacent channel power | Power falling in neighboring channels |
| TX behavior under low voltage conditions | Checks frequency deviation and emissions due to lower than low voltage level |

*Continued*

| Effective radiated power (e.r.p.) | Includes antenna gain over dipole |
|---|---|
| Adaptive power control | Tests minimum power level using adaptive power control (APC) |
| Rx sensitivity level | Measures best receiver sensitivity |
| Adjacent channel selectivity | Checks receiver sensitivity with interfering signal in adjacent channel |
| Receiver saturation at adjacent channel | Checks receiver capability with strong signal on adjacent channel |
| Spurious response rejection | Measures receiver operation degradation due to signals on spurious response frequencies |
| Blocking | Measures receiver operation degradation due to strong signals not on spurious response frequencies |
| Behavior at high wanted signal level | Measures receiver operation ability at high level of wanted signal |
| Polite spectrum access | Tests receiver operation using clear channel assessment feature and adaptive frequency agility, if available |

The method used to measure transmitter emissions depends on whether the device has an antenna connector or an integral antenna that cannot be disconnected. In the former case, radiation characteristics can be measured directly across a dummy load on the RF output connector.

We describe here the principle of the method used to determine power output, both fundamental and spurious, by measuring e.r.p. (refer to the standard for authoritative details).

The test site is arranged as shown in Fig. 10.3. It may be an open area, a semi anechoic room or a fully anechoic room. In the first two cases a conducting surface that simulates as far as practical a ground plane of infinite extent is part of the test site. The recommended distance between the EUT and the test antenna is 3 or 10 m.

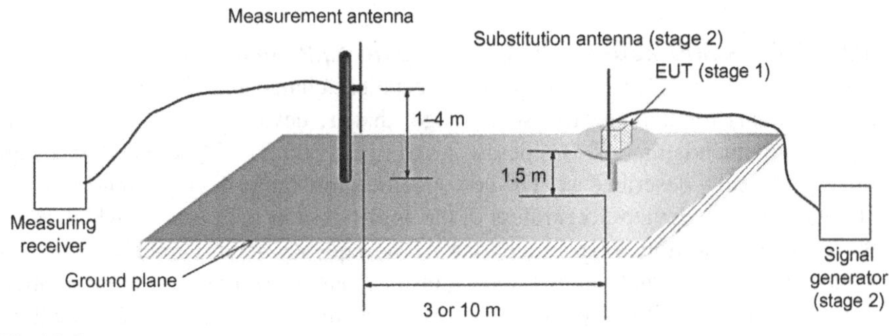


**FIG. 10.3**

EN300 220 test site.

Table 10.4 Spurious output limits

| State | 47 MHz to 74 MHz<br>87.5 to 118 MHz<br>174 to 230 MHz<br>470 to 790 MHz | Other frequencies below 1000 MHz | Frequencies above 1000 MHz |
|---|---|---|---|
| Transmitter mode | −54 dBm | −36 dBm | −30 dBm |
| Receiver and all other modes | −57 dBm | −57 dBm | −47 dBm |

It is not critical. Reflecting objects around the site must be far enough away so that they don't influence the measurements. The e.r.p. at the fundamental frequency and harmonics and spurious frequencies is determined in two stages as follows.

The EUT is powered and the receiving instrument, receiver or spectrum analyzer, indicates the level received by the measurement antenna. A CW carrier from the transmitter is preferred but if data modulation is used, the bandwidth of the measuring receiver is set to 120 kHz for frequencies under 1 GHz and 1 MHz for frequencies over 1 GHz (spurious radiation). A peak reading (CW) or quasi-peak (modulated) is used. The equipment is rotated and the test antenna height is varied between 1 and 4 m to find the maximum level, which is noted. The signal strength survey is conducted for both vertical and horizontal polarization.

In the second stage, the EUT is replaced by a calibrated antenna connected to a signal generator. The height of the test antenna is adjusted for maximum output. Then the generator output is adjusted so that the receiver indicates the same maximum level as obtained in the first stage. The power level of the generator is recorded. The test is carried out for vertical and horizontal polarization. The higher of the recorded generator power outputs, plus the antenna gain relative to a dipole, is the e.r.p. of the EUT.

For fundamental frequency power measurements, the e.r.p. shall not exceed the maximum value as shown in Table 10.3 or otherwise permitted by the relevant authority. For spurious outputs the limits are as shown in Table 10.4 [5, Paragraph 5.9.2].

#### *10.3.2.2 EN 300 220-2 non-specific radio equipment* [6]

EN 300 220 has a Part 2 that specifies technical characteristics and methods of measurements for non-specific short range, that is, devices listed in Annex 1 of ERC Recommendation (70-03) below 1 GHz [EN 300 220-2]. These devices are more specifically described as "The non-specific short-range device category covers all kinds of radio devices, regardless of the application or the purpose, which fulfill the technical conditions as specified for a given frequency band. Typical uses include telemetry, telecommand, alarms, data transmissions in general and other applications" [6]. The specifications in EN 300 220-2 are those harmonized throughout the European Union (EU). The document also includes conformance parameters in force in only some of the countries of the EU.

#### *10.3.2.3 EN 300 440 short range devices (SRD); radio equipment to be used in the 1 GHz to 40 GHz frequency range [7]*

This is a generic standard for short-range radio equipment transmitting on frequencies above 1 GHz. It covers non-specific Short Range Devices, including alarms, telecommand, telemetry, data transmission in general, RFID devices, and radiodetermination devices, which include detection, movement and alert applications. These devices operate in the bands 2.4 to 2.4835 GHZ, 2.446 to 2.454 GHZ (RFID), and 5.725 to 5.875 GHz, plus higher frequencies that are mostly used for radio determination devices. UWB devices are not covered in this document.

EN 300 440 describes measurement requirements and methods for the following parameters.

Transmitter requirements:

- Equivalent isotropically radiated power (e.i.r.p.). Two measurement methods are given, depending on bandwidth and whether modulation is spread spectrum.
- Range of operating frequencies. This is particularly applicable to frequency hopping spread spectrum (FHSS).
- Unwanted spurious emissions.
- Duty cycle.
- Additional requirements for FHSS equipment. These state that FHSS has to use a least 20 channels hopping over 90% of the assigned frequency band and that the dwell time on any channel must not exceed 1 s.

Receiver requirements:

- Sensitivity, meeting the criteria of data bit error rate of $10^{-2}$, message acceptance ratio of 80%, and an appropriate false alarm rate.
- Adjacent channel selectivity for receivers classified as highly reliable.
- Blocking resistance.
- Spurious radiation.

The specification describes spectrum access techniques based on listen before talk (LBT) and adaptive frequency agility (AFA) (not mandatory). The maximum time for a single transmission, the on-time, is 2 s. Minimum off-time is 25 ms. Acknowledge transmissions and maximum transmitter on-time are also defined. There is no requirement for a listen time before an acknowledge transmission.

RFID systems at around 2.45 GHz have power limits of 500 mW e.i.r.p. (default) and 4 W e.i.r.p. The higher limit applies to fixed indoor installations and the specification describes methods for ensuring that an indoor RFID reader reverts to the default power if the installation is changed or the system is moved.

Specification EN 300 440 gives details concerning the presentation of equipment for test purposes, ways to test depending on whether the device has an external 50 Ω RF connector or an integral antenna, and prescribes normal and extreme environmental test conditions including temperature and supply voltage. It also describes test signals and modulation.

#### *10.3.2.4 ETS 300 328 wideband transmission systems operating in the 2.4 GHz ISM band* [8]

This specification regulates wide-band data transmission equipment in accordance with ERC Recommendation 70-03, Annex 3, that operates in the band 2.4 to 2.4835 GHz [EN 300 328]. Examples are IEEE 802.11 (Wi-Fi), Bluetooth, and Zigbee. Two types of wide band data transmission are categorized in the document: FHSS, and other types of wideband modulation which include DSSS and OFDM. Equipment is also classified as adaptive and non-adaptive. An adaptive mode is defined as a mode in which equipment can adapt its medium access to its radio environment by identifying other transmissions present in the band. When interference is detected on a channel on which a device intends to transmit, the device can wait for the interference to subside, or move to another channel. Interference is detected using clear channel access (CCA) as part of a listen before transmit (LBT) procedure. The mechanism which mitigates potential interference by avoiding use of frequencies upon detection of other transmissions on those frequencies is called detect and avoid (DAA). FHSS equipment with adaptive mode can modify its hopping sequence to avoid transmitting on occupied frequencies. The specification also categorizes receivers into three categories for the purpose of defining blocking specifications, according to the power of the associated transmitter, and a medium utilization (MU) factor, which for non-adaptive devices is a measure of the amount of power and time resources used by the equipment.

Transmitter and receiver performance limits are specified in respect to the characterization of a device according to the categories described above.

An overview of some performance parameters are given here. For definitive details the latest version of EN 300 328 should be consulted.

**A.** Frequency hopping equipment.

- RF output power: 20 dBm e.i.r.p. maximum.
- Hopping frequency separation (separation between two adjacent hopping frequencies: 100 kHz minimum).
- Number of hops: (1) Non-adaptive equipment: either 5, or the result of 15 MHz divided by the minimum hopping frequency separation in MHz, whichever is greater. (2) Adaptive equipment: either 15, or the result of 15 MHz divided by the minimum hopping frequency separation in MHz, whichever is greater.
- MU factor. MU is defined by the formula MU = ($P$/100 mW) × DC, where $P$ is the RF output power, applicable for output power above 10 dBm, and DC is the duty cycle. The limit is maximum MU of 10%.
- Occupied channel bandwidth (contains 99% of signal power during one frequency hop): 5 MHz maximum.
- Other limits are transmitter out-of-band emissions due to modulation, transmitter spurious emissions, receiver spurious emissions, and receiver blocking (limits depend on receiver category).

**B.** Other types of wide band modulation.

- RF output power: 20 dBm e.i.r.p. maximum.
- Power spectral density: 10 dBm/MHz.
- Duty cycle. Performed only for non-adaptive equipment. Transmission time shall be less than 10 ms within a 1 s observation period.
- Adaptivity. Defined for adaptive equipment with RF power greater than 10 dBm e.i.r.p. Requirements depend on the DAA mechanism which is adopted to identify other transmissions within its occupied channel bandwidth. Limits apply to the time a channel must remain unavailable due to detected signal energy above a threshold, and to channel occupancy time and idle period.
- Occupied channel bandwidth: Less than 20 MHz for non-adaptive equipment and e.i.r.p. greater than 10 dBm. Otherwise, 2.4 to 2.4835 GHz.
- Transmitter and receiver unwanted emissions limits are defined by their frequency separation from the allocated band (out of band domain) and by stated frequency ranges between 30 MHz and 12.75 GHz (spurious domain).
- Receiver blocking. Limits depend on receiver category. They are applicable only for adaptive equipment with output power over 10 dBm. Their purpose is to assure that channel occupancy can be detected in the presence of nearby potentially blocking transmissions on the edges of the communication frequency band.

#### *10.3.2.5 EN 301 893 5 GHz RLAN (radio local area networks) [9]*

Technical parameters for use of IEEE 802.11 type WLAN stations specified for 5 GHz operation and 20 MHz channel bandwidth is governed by this document. Operating frequencies and power output limits are shown in Table 10.5. The table refers to TPC (transmit power control) which is a technique in which the transmitter output power is controlled resulting in reduced interference to other systems.

**Table 10.5** 5 GHz RLAN frequency and power output requirements

| Operating bands (GHz) | 5.150 to 5.350 | 5.470 to 5.725 |
|---|---|---|
| Channels | 10 × 20 MHz | 12 × 20 MHz |
| RF output power (dBm) | | |
| with TPC | 23 | 30 |
| without TPC[a] | 20 | 27 |
| Output power density (dBm/MHz) | | |
| with TPC | 10 | 17 |
| without TPC[a] | 7 | 14 |

[a] *In the band 5.15 to 5.25 GHz the power output and power density limits are the same as for stations without TPC.*

Standard EN 301 893 mandates a dynamic frequency selection (DFS) function to

- detect interference from radar systems (radar detection) and to avoid co-channel operation with these systems;
- provide on aggregate a near-uniform loading of the spectrum (*uniform spreading*).

It also gives requirements for a channel access mechanism, called adaptivity. Two types of channel access are described: frame based equipment (FBE) and load based equipment (LBE). Both require a LBT procedure to check channel occupancy before transmitting. FBE prescribes a fixed frame time interval occurring periodically during which a device may transmit if a *clear channel assessment* (CCA) check shows that the channel energy level is below a given limit. LBE is a random access technique that includes a backoff period during which a CCA indicates that the channel is unoccupied before an initiating station can transmit. The Standard defines four priority classes to differentiate access to and use of the channel.

Channels may be bonded, or grouped such that total transmission bandwidths of 20, 40, 80, or 160 MHz can be accommodated for a wide range of data rates and throughputs.

#### 10.3.2.6 EN 302 065-1 ultra-wideband [10]

Specification EN 302 065-1 applies to transceivers, transmitters and receivers utilizing UWB technologies and used for short-range applications (UWB). It relates to fixed (indoor only), mobile or portable applications with a dedicated external antenna or integral antenna. Operation is permitted over the range of 30 MHz to 10.6 GHz but there are two intended, and preferred, ranges of operation: 3.1 to 4.8 GHz and 6.0 to 9 GHz over which higher power may be used for practical communication. Here are some of the major characteristics of UWB in the specification.

- Operating bandwidth: minimum 50 MHz at 10 dB down
- Maximum mean power spectral density: −41.3 dBm/MHz in the in the intended ranges and significantly less outside of these bands, provided "mitigation techniques" are used, referring to DAA measures and LDC. In the band 6 to 8.5 GHz the limit is −41.3 dBm/MHz without mitigation techniques.
- Maximum peak power: 0 dBm over 50 MHz in the in the intended ranges and significantly less outside of these bands, provided "mitigation techniques" are used, as described above. In the band 6 to 8.5 GHz the limit is –0 dBm over 50 MHz without mitigation techniques.
- Other emissions, outside of operating ranges: no greater than −36 dBm/100 kHz from 30 MHz to 1 GHz and −30 dBm/1 MHz from 1 to 40 GHz. However, in the frequency ranges 47 to 74, 87.5 to 118, 174 to 230, and 470 to 862 MHz the limit value is −54 dBm/100 kHz.
- Spectrum access requirements: DAA, which involves sensing other radio systems that are transmitting in the operating band and ceasing transmission until the channel is clear. Another mitigation technique is LDC which allows less than 18 s transmitter on time per hour and more than 950 ms off time per second.

#### *10.3.2.7 ETSI EN 302 208 RFID equipment [11]*

The subject of this standard is RFID equipment in the bands 865 to 868 and 915 to 921 MHz. The band 915 to 921 MHz has limited implementation in the European Union. Types of equipment covered by the standard are

- fixed interrogators (readers);
- portable interrogators;
- batteryless (passive) tags;
- battery assisted tags;
- battery powered (active) tags.

Four 200 kHz channels are specified for interrogator transmission between 865.6 to 867.6 MHz at a maximum power of 2 W e.r.p. Interrogator receive and tag transmit and receive is specified between 865.2 and 868 MHz. Between 916.1 and 920.1 MHz on four 400 kHz channels the interrogator power limit is 4 W e.r.p. The interrogator receive band is 915.3 to 925.0 MHz whereas tag transmit and receive is between 915.3 and 920.9 MHz.

### 10.3.3 **Directive 2014/53/EU** [12]

The title of this document includes "...the harmonization of the laws of the Member States relating to the making available on the market of radio equipment and repealing Directive 1999/5/EC" [12]. The Directive's stated aim is the establishment of a regulatory framework for making available on the market and putting into service radio equipment in the European Union. Among the essential requirements are the protection of health and safety of persons and domestic animals and the protection of property and the effective use and support of radio spectrum in order to avoid harmful interference. Other essential requirements are that the radio equipment:

- works with accessories, in particular common chargers;
- can be connected to work via networks with other radio equipment;
- does not harm the network or its functioning nor misuses network resources;
- incorporates safeguards to ensure that the personal data and privacy of the user and the subscriber are protected;
- supports certain features ensuring protection from fraud;
- supports certain features ensuring access to emergency services;
- supports certain features in order to facilitate its use by users with a disability;
- radio equipment supports certain features in order to ensure that software can only be loaded into the radio equipment where the compliance of the combination of the radio equipment and software has been demonstrated.

The directive stipulates that manufacturers of radio equipment provide Member States and the European Commission with information on compliance with the essential requirements. This information results from a conformity assessment and is given in the form of a statement of compliance. The document details the procedures for making the conformity assessment and the content of the declaration of conformity that asserts the fulfillment of the essential requirements. It also gives the rules for affixing a CE marking on the equipment to indicate its conformity, the benefit of consumers and public authorities.

## 10.4 Summary

This chapter presents standards and regulations applying to short-range wireless communication in the United States and in the European Community. U.S. FCC regulations pertaining to ISM (industrial, scientific, medical) use from Part 15, and relevant provisions from Part 74 dealing with low power auxiliary stations and Part 95, the personal radio service, are quoted and explained. The method of measuring field strength to determine transmitter power limits is explained.

Features of relevant European standards are described. Covered are characteristics and test methods for low power equipment in the 25 to 1000 MHz frequency range and separately, equipment operating on frequencies above 1000 MHz. The special requirements for widely used devices in the 2.4 GHz band, including Wi-Fi, Bluetooth, and Zigbee, are presented, along with radiation testing methods. UWB specifications are reviewed. Following are RFID equipment frequency ranges and power limits. In conclusion, essential requirements of Directive 2014/53/EU, which deals with harmonization and marketing, are listed.

Two appendices to this chapter include a list of definitions of common terms and a list of restricted frequencies from the FCC regulations.

## 10.A Terms and definitions (FCC Part 2 subpart A Section 2.1)

The following terms and definitions were selected from the source as relevant to the context of this book or of special interest, and particularly those whose meanings are important to understand equipment specifications.

*Assigned frequency band.* The frequency band within which the emission of a station is authorized; the width of the band equals the necessary bandwidth plus twice the absolute value of the frequency tolerance. Where space stations are concerned, the assigned frequency band includes twice the maximum Doppler shift that may occur in relation to any point of the Earth's surface.

*Carrier Power (of a radio transmitter).* The average power supplied to the antenna transmission line by a transmitter during one radio frequency cycle taken under the condition of no modulation (RR).

*Class of emission.* The set of characteristics of an emission, designated by standard symbols, e.g., type of modulation, modulating signal, type of information to be transmitted, and also if appropriate, any additional signal characteristics.

*Duplex operation.* Operating method in which transmission is possible simultaneously in both directions of a telecommunication channel.

*Effective radiated power (e.r.p.) (in a given direction).* The product of the power supplied to the antenna and its gain relative to a half-wave dipole in a given direction.

*Equivalent isotropically radiated power (e.i.r.p.).* The product of the power supplied to the antenna and the antenna gain in a given direction relative to an isotropic antenna (absolute or isotropic gain).

*Frequency hopping systems.* A spread spectrum system in which the carrier is modulated with the coded information in a conventional manner causing a conventional spreading of the RF energy about the frequency carrier. The frequency of the carrier is not fixed but changes at fixed intervals under the direction of a coded sequence. The wide RF bandwidth needed by such a system is not required by spreading of the RF energy about the carrier but rather to accommodate the range of frequencies to which the carrier frequency can hop. The test of a frequency hopping system is that the near term distribution of hops appears random, the long term distribution appears evenly distributed over the hop set, and sequential hops are randomly distributed in both direction and magnitude of change in the hop set.

*Gain of an antenna.* The ratio, usually expressed in decibels, of the power required at the input of a loss free reference antenna to the power supplied to the input of the given antenna to produce, in a given direction, the same field strength or the same power flux-density at the same distance. When not specified otherwise, the gain refers to the direction of maximum radiation. The gain may be considered for a specified polarization.

Note: Depending on the choice of the reference antenna a distinction is made between:

**(1)** Absolute or isotropic gain ($G_i$), when the reference antenna is an isotropic antenna isolated in space.

**(2)** Gain relative to a half-wave dipole ($G_d$), when the reference antenna is a half-wave dipole isolated in space whose equatorial plane contains the given direction.

**(3)** Gain relative to a short vertical antenna ($G_v$), when the reference antenna is a linear conductor, much shorter than one quarter of the wavelength, normal to the surface of a perfectly conducting plane which contains the given direction.

*Hybrid spread spectrum systems.* Hybrid spread spectrum systems are those which use combinations of two or more types of direct sequence, frequency hopping, time hopping and pulsed FM modulation in order to achieve their wide occupied bandwidths.

*Industrial, scientific and medical (ISM) (of radio frequency energy) applications.* Operation of equipment or appliances designed to generate and use locally radio-frequency energy for industrial, scientific, medical, domestic or similar purposes, excluding applications in the field of telecommunications.

*Interference.* The effect of unwanted energy due to one or a combination of emissions, radiations, or inductions upon reception in a radiocommunication system, manifested by any performance degradation, misinterpretation, or loss of information which could be extracted in the absence of such unwanted energy.

*International telecommunication union (ITU).* An international organization within the United Nations System where governments and the private sector coordinate global telecom networks and services. The ITU is headquartered in Geneva, Switzerland and its internet address is *www.itu.int*.

*Left-hand (or anti-clockwise) polarized wave.* An elliptically or circularly-polarized wave, in fixed plane, normal to the direction of propagation, whilst looking in the direction of propagation, rotates with time in a left hand or anti-clockwise direction.

*Mean power (of a radio transmitter).* The average power supplied to the antenna transmission line by a transmitter during an interval of time sufficiently long compared with the lowest frequency encountered in the modulation taken under normal operating conditions.

*Necessary bandwidth.* For a given class of emission, the width of the frequency band which is just sufficient to ensure the transmission of information at the rate and with the quality required under specified conditions.

*Occupied bandwidth.* The width of a frequency band such that, below the lower and above the upper frequency limits, the mean powers emitted are each equal to a specified percentage $\beta/2$ of the total mean power of a given emission.

Note: Unless otherwise specified in an ITU-R Recommendation for the appropriate class of emission, the value of $\beta/2$ should be taken as 0.5% (RR).

*Out-of-band emission.* Emission on a frequency or frequencies immediately outside the necessary bandwidth which results from the modulation process, but excluding spurious emissions.

*Peak envelope power (of a radio transmitter).* The average power supplied to the antenna transmission line by a transmitter during one radio frequency cycle at the crest of the modulation envelope taken under normal operating conditions.

*Power.* Whenever the power of a radio transmitter, etc. is referred to it shall be expressed in one of the following forms, according to the class of emission, using the arbitrary symbols indicated:

**(1)** Peak envelope power (*PX* or *pX*)
**(2)** Mean power (*PY* or *pY*)
**(3)** Carrier power (*PZ* or *pZ*)

Note 1: For different classes of emission, the relationships between peak envelope power, mean power and carrier power, under the conditions of normal operation and of no modulation, are contained in ITU-R Recommendations which may be used as a guide.

Note 2: For use in formulae, the symbol p denotes power expressed in watts and the symbol P denotes power expressed in decibels relative to a reference level (RR).

*Radiodetermination.* The determination of the position, velocity and/or other characteristics of an object, or the obtaining of information relating to these parameters, by means of the propagation properties of radio waves.

*Right-hand (or clockwise) polarized wave.* An elliptically or circularly-polarized wave, in which the electric field vector, observed in any fixed plane, normal to the direction of propagation, whilst looking in the direction of propagation, rotates with time in a right-hand or clockwise direction.

*Software defined radio.* A radio that includes a transmitter in which the operating parameters of frequency range, modulation type or maximum output power (either radiated or conducted), or the circumstances under which the transmitter operates in accordance with commission rules, can be altered by making a change in software without making any changes to hardware components that affect the radio frequency emissions. In accordance with Section 2.944 of this part, only radios in which the software is designed or expected to be modified by a party other than the manufacturer and would affect the above-listed operating parameters or circumstances under which the radio transmits must be certified as software defined radios.

*Spread spectrum systems.* A spread spectrum system is an information bearing communications system in which: (1) Information is conveyed by modulation of a carrier by some conventional means, (2) the bandwidth is deliberately widened by means of a spreading function over that which would be needed to transmit the information alone. (In some spread spectrum systems, a portion of the information being conveyed by the system may be contained in the spreading function.)

*Spurious emission.* Emission on a frequency or frequencies which are outside the necessary bandwidth and the level of which may be reduced without affecting the corresponding transmission of information. Spurious emissions include harmonic emissions, parasitic emissions, intermodulation products and frequency conversion products, but exclude out-of-band emissions.

*Time hopping systems.* A time hopping system is a spread spectrum system in which the period and duty cycle of a pulsed RF carrier are varied in a pseudorandom manner under the control of a coded sequence. Time hopping is often used effectively with frequency hopping to form a hybrid time-division, multiple-access (TDMA) spread spectrum system.

*Unwanted emissions.* Consist of spurious emissions and out-of-band emissions.

## 10.B Restricted bands of operation for emissions from intentional radiators (FCC Part 15 Section 15.205)

With exceptions, only spurious emissions are permitted in any of the frequency bands listed below. See source for exceptions and field strength limits and measuring conditions.

| MHz | MHz | MHz | GHz |
|---|---|---|---|
| 0.090-0.110 | 16.42-16.423 | 399.9-410 | 4.5-5.15 |
| 0.495-0.505 | 16.69475-16.69525 | 608-614 | 5.35-5.46 |
| 2.1735-2.1905 | 16.80425-16.80475 | 960-1240 | 7.25-7.75 |
| 4.125-4.128 | 25.5-25.67 | 1300-1427 | 8.025-8.5 |

*Continued*

| MHz | MHz | MHz | GHz |
|---|---|---|---|
| 4.17725-4.17775 | 37.5-38.25 | 1435-1626.5 | 9.0-9.2 |
| 4.20725-4.20775 | 73-74.6 | 1645.5-1646.5 | 9.3-9.5 |
| 6.215-6.218 | 74.8-75.2 | 1660-1710 | 10.6-12.7 |
| 6.26775-6.26825 | 108-121.94 | 1718.8-1722.2 | 13.25-13.4 |
| 6.31175-6.31225 | 123-138 | 2200-2300 | 14.47-14.5 |
| 8.291-8.294 | 149.9-150.05 | 2310-2390 | 15.35-16.2 |
| 8.362-8.366 | 156.52475-156.52525 | 2483.5-2500 | 17.7-21.4 |
| 8.37625-8.38675 | 156.7-156.9 | 2690-2900 | 22.01-23.12 |
| 8.41425-8.41475 | 162.0125-167.17 | 3260-3267 | 23.6-24.0 |
| 12.29-12.293 | 167.72-173.2 | 3332-3339 | 31.2-31.8 |
| 12.51975-12.52025 | 240-285 | 3345.8-3358 | 36.43-36.5 |
| 12.57675-12.57725 | 322-335.4 | 3600-4400 | Above 38.6 |
| 13.36-13.41 | | | |

# References

[1] United States Electronic Code of Federal Regulations, Title 47, Chapter I.

[2] FCC Regulations, Part 15, Radio Frequency Devices.

[3] IEEE Std 802.15.4-2006, Part 15.4: Wireless Medium Access Control (MAC) and Physical Layer (PHY) Specifications for Low-Rate Wireless Personal Area Networks (WPANs), Table F.2 (CISPR).

[4] ERC Recommendation (70-03), 13 October 2017.

[5] ETSI EN 300 220-1 V3.1.1 (2017-02), Short Range Devices (SRD) Operating in the Frequency Range 25 MHz to 1 000 MHz; Part 1: Technical Characteristics and Methods of Measurement.

[6] ETSI EN 300 220-2 V3.1.1 (2017-02), Short Range Devices (SRD) Operating in the Frequency Range 25 MHz to 1 000 MHz; Part 2: Harmonized Standard Covering the Essential Requirements of Article 3.2 of Directive 2014/53/EU for Nonspecific Radio Equipment.

[7] ETSI EN 300 440 V2.1.1 (2017-03) Short Range Devices (SRD); Radio equipment to be Used in the 1 GHz to 40 GHz Frequency Range; Harmonised Standard Covering the Essential Requirements of Article 3.2 of Directive 2014/53/EU.

[8] Draft ETSI EN 300 328 V2.2.0 (2017-11) Wideband Transmission Systems; Data Transmission Equipment Operating in the 2.4 GHz ISM Band and Using Wide Band Modulation Techniques; Harmonised Standard for Access to Radio Spectrum.

[9] ETSI EN 301 893 V2.1.1 (2017-05) 5 GHz RLAN; Harmonised Standard Covering the Essential Requirements of Article 3.2 of Directive 2014/53/EU.

[10] ETSI EN 302 065-1 V2.1.1 (2016-11) Short Range Devices (SRD) Using Ultra Wide Band technology (UWB); Harmonized Standard Covering the Essential Requirements of Article 3.2 of the Directive 2014/53/EU; Part 1: Requirements for Generic UWB Applications.

[11] Draft ETSI EN 302 208 V3.2.0 (2018-02) Radio Frequency Identification Equipment Operating in the Band 865 MHz to 868 MHz With Power Levels Up To 2 W and in the Band 915 MHz to 921 MHz With Power Levels Up To 4 W; Harmonized Standard for Access to Radio Spectrum.
[12] Directive 2014/53/EU of the European Parliament and of the Council of 16 April 2014, From Official Journal of the European Union.

CHAPTER

# Wireless local area networks

# 11

An important factor in the widespread penetration of short-range devices into the office and the home is the basing of the most popular applications on industry standards. In this chapter and the next, we take a look at some of these standards and the applications that have emerged from them. Those covered pertain to Wi-Fi, Bluetooth, Zigbee and others based on IEEE 802.15.4. In order to be successful, a standard has to be built so that it can keep abreast of rapid technological advancements by accommodating modifications that don't obsolete earlier devices that were developed to the original version. A case in point is the competition between the WLAN (wireless local area network) standard that was developed by the HomeRF Working Group based on the SWAP (shared wireless access protocol) specification, and IEEE specification 802.11, commonly known as Wi-Fi. The former used frequency-hopping spread-spectrum exclusively, and although some increase of data rate was provided for beyond the original 1 and 2 Mbps, it couldn't keep up with Wi-Fi, which incorporated new bandwidth efficient modulation methods to increase data rates 50-fold while maintaining compatibility with first generation DSSS terminals.

Most of the pervasive WLANs are designed for operation on the 2.4 GHz ISM band, available for license-free operation in North America and Europe, as well as virtually all other regions in the world. However, The 5 GHz band has become a favorite for advanced systems because of its higher bandwidth and reduced interference from other unlicensed users. Most systems have provisions for handling errors due to interference, but when the density of deployment of one or more systems is high, throughput, voice intelligibility, or quality of service (QoS) in general is bound to suffer. We will look at some aspects of this problem and methods for solving it. This chapter will also describe multi-antenna systems—MIMO, communication security, and location awareness, as they apply to IEEE 802.11. Wi-Fi is the term adopted by the Wi-Fi Alliance association in year 2000 to identify its technical work and we use it herein synonymously with the generic IEEE 802.11 WLAN. From its self-identification, the Wi-Fi Alliance "defines innovative, standards-based Wi-Fi technologies and programs, certifies products that meet quality, performance, security, and capability standards, provides industry thought leadership, and advocates globally for fair spectrum rules"[1] By promoting compatibility of products from different manufacturers, the organization has contributed, together with IEEE, to making Wi-Fi/IEEE 802.11 the most widely deployed wireless communication system in the world.



**Short-range Wireless Communication. https://doi.org/10.1016/B978-0-12-815405-2.00011-7**
© 2019 Elsevier Inc. All rights reserved.

## 11.1 Introduction

One of the most prevalent deployments of short-range radio communication is WLANs. While the advantage of a wireless versus wired LAN is obvious, the early versions of WLAN had considerably inferior data rates making conversion to wireless not necessarily worthwhile, particularly when portability is not an issue. However, advanced modulation techniques have allowed wireless throughputs to approach and even exceed those of wired networks, and the popularity of highly portable laptop and handheld computers, along with the decrease in device prices, have made computer networking a common occurrence in multi-computer offices and homes.

There are still three prime disadvantages of wireless networks as compared to wired: range limitation, susceptibility to electromagnetic interference, and security. Direct links may be expected to perform at a top range of 50 to 100 m depending on frequency band and surroundings. Longer distances and obstacles will reduce data throughput. Greater distances between network participants are achieved by installing additional access points (APs) to bridge remote network nodes. Reception of radio signals may be interfered with by other services operating on the same frequency band and in the same vicinity.

Wireless transmissions are subject to eavesdropping, and a standardized security implementation in Wi-Fi called WEP (wired equivalent privacy) was found to be compromised with relative ease by persistent and knowledgeable hackers. More sophisticated encryption techniques have been instituted that greatly increase, but do not eliminate absolutely, the security risk of wireless communication although they may be accompanied by reduction of convenience in setting up connections and possibly in performance.

Various systems of implementation are used in wireless networks. They may be based on an industrial standard which facilitates compatibility between devices by different manufacturers, or a proprietary design. The latter would primarily be used in a special purpose network, such as in an industrial application where all devices are made by the same manufacturer and where performance may be improved without the limitations and compromises inherent in a widespread standard. In this chapter we deal with Wi-Fi, based on IEEE 802.11 with amendments and revisions.

As mentioned above, the Wi-Fi Alliance promotes Wi-Fi, and certifies devices to ensure their interoperability. The original link specification is continually updated by IEEE working groups to incorporate technical improvements and feature enhancements that are agreed upon by a wide representation of potential users and industry representatives. 802.11 is the predominant industrial standard for WLAN and products adhering to it are acceptable for marketing all over the world, albeit with features and characteristics conforming to the various regulatory organizations.

IEEE 802.11 covers the data link layer of lower-level software, the physical layer hardware definitions, and the data flow and management interfaces between them. The connection between application software and the wireless hardware is the MAC (medium access control). It is built so that the upper application software doesn't have to know what wireless technique is being used—the MAC interface firmware takes care of that. In fact, application software doesn't have to know that a wireless connection is being used at all and mixed wired and wireless links can coexist in the same network. The original basic specification defined three types of wireless communication techniques: DSSS (direct sequence spread spectrum), FHSS (frequency-hopping spread spectrum) and IR (infra-red). FHSS and IR early fell by the wayside, and DSSS has given way to OFDM, although specification amendments and revisions have maintained compatibility with earlier DSSS modes.

Wireless communication according to 802.11 is conducted on the 2.400 to 2.4835 GHz frequency band and on frequencies between 5 and 5.8 GHz that are authorized for unlicensed equipment operation in the United States and Canada and most European and other countries. A few countries allow unlicensed use in only in portions of these bands. Amendments to the original document add increased data rates and other features while retaining compatibility with equipment using the DSSS physical layer of the basic specification. Amendment 802.11a specified considerably higher rate operation in bands of frequencies between 5.2 and 5.8 GHz. The higher data rates were made available on the 2.4 GHz band by 802.11g that has backward compatibility with 802.11b.

Changes to 802.11 which reflect technological improvements and expansion of application areas are issued from time to time in the form of amendments to the current specification. Every four years or so amendments are consolidated into the current specification as a revision. The revision which serves as a reference for this edition of the book is IEEE 802.11-2016, which includes the basic performance amendments through 802.11ac [2].

## 11.2 Network architecture

Wi-Fi architecture is very flexible, allowing considerable mobility of stations and transparent integration with wired IEEE networks. The transparency comes about because upper application software layers (see below) are not dependent on the actual physical nature of the communication links between stations. Also, all IEEE LAN stations, wired or wireless, use the same 48-bit addressing scheme so an application only has to reference source and destination addresses and the underlying lower level protocols will do the rest.

Three Wi-Fi network configurations are shown in Figs. 11.1–11.3. Fig. 11.1 shows two unattached basic service sets (BSS), each with two stations (STA). The BSS is the basic building block of an 802.11 WLAN. A station can make ad hoc connections with other stations within its wireless communication range but not with those in another BSS that is outside of this range. In order to interconnect terminals that are not in direct range one with the other, the distributed system shown in Fig. 11.2 is needed. Here, terminals that are in range of a station designated as an AP can communicate with other terminals not in direct range but who are associated with the same or another AP. Two or more such APs communicate between

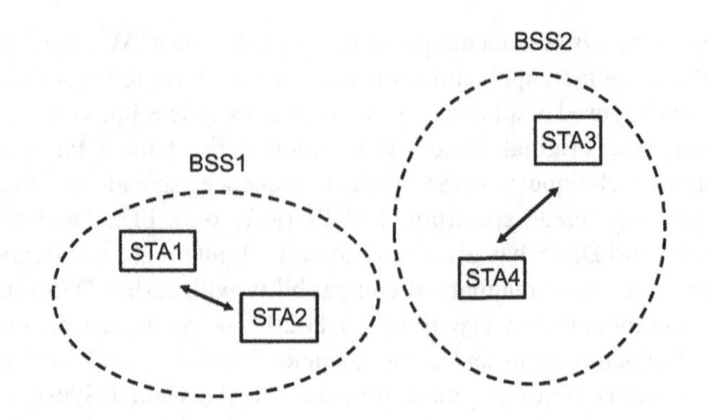


FIG. 11.1
Basic service set.

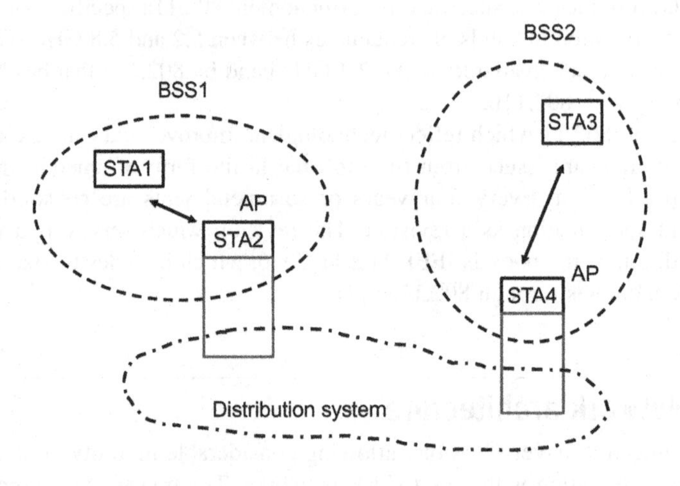


FIG. 11.2
Distribution system and access points.

themselves either by a wireless or wired medium, and therefore data exchange between all terminals in the network is supported. The important thing here is that the media connecting the STAs with the APs, and connecting the APs among themselves are totally independent. Note that the STA is an addressable destination, not necessarily at a fixed location. STA's may have varied characteristics and functions. A STA may be, for example, an AP terminal, a mobile terminal, or it may have another specified function.

A network of arbitrary size and complexity can be maintained through the architecture of the extended service set (ESS), shown in Fig. 11.3. Here, STAs have full mobility and may move from one BSS to another while remaining in the network. Fig. 11.3 shows another element type—a portal. The portal is a gateway between

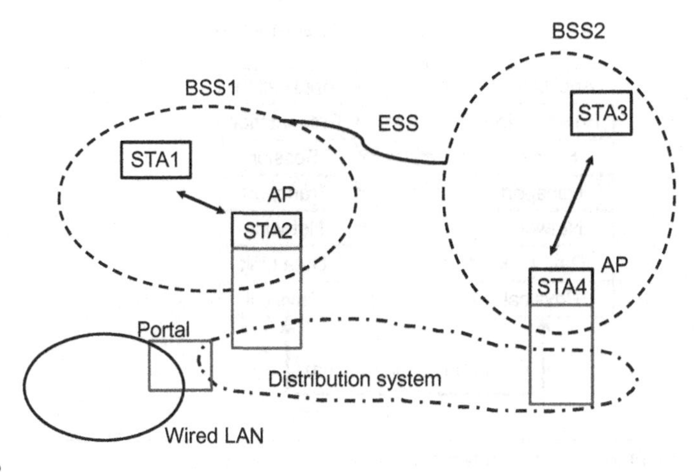


FIG. 11.3

Extended service set.

the WLAN and a wired LAN. It connects the medium over which the APs communicate to the medium of the wired LAN—coaxial cable or twisted pair lines, for example.

In addition to the functions Wi-Fi provides for distributing data throughout the network, two other important services, although optionally used, are provided. They are authentication and encryption. Authentication is the procedure used to establish the identity of a station as a member of the set of stations authorized to associate with another station. Encryption applies coding to data to prevent an eavesdropper from intercepting it. 802.11 details the implementation of these services in the MAC. Further protection of confidentiality may be provided by higher software layers in the network that are not part of 802.11.

The operational specifics of WLAN are described in IEEE 802.11 in terms of defined protocols between lower-level software layers. In general, networks may be described by the communication of data and control between adjacent layers of the Open System Interconnection Reference Model (OSI/RM), shown in Fig. 11.4, or the peer-to-peer communication between like layers of two or more terminals in the network. The bottom layer, physical, represents the hardware connection with the transmission medium that connects the terminals of the network—cable modem, radio transceiver and antenna, infrared transceiver, or power line transceiver, for example. The software of the upper layers is wholly independent of the transmission medium and in principle may be used unchanged no matter what the nature of the medium and the physical connection to it. IEEE 802.11 is concerned only with the two lowest layers, physical and data link.

IEEE 802.11 prescribes the protocols between the MAC sublayer of the data link layer and the physical layer, as well as the electrical specifications of the physical layer. Fig. 11.5 illustrates the relationship between the physical and MAC layers

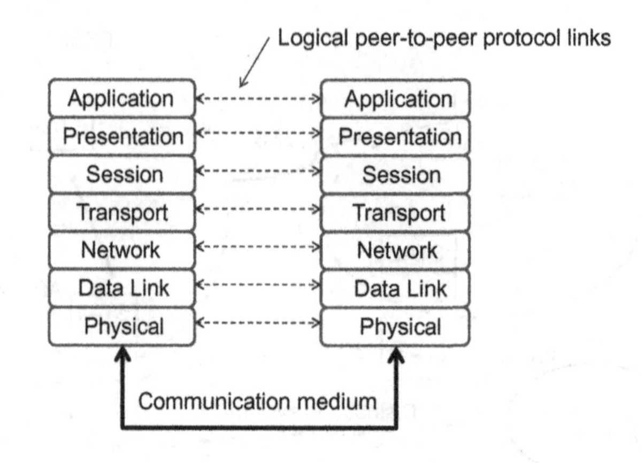


**FIG. 11.4**

Open system interconnection reference model.

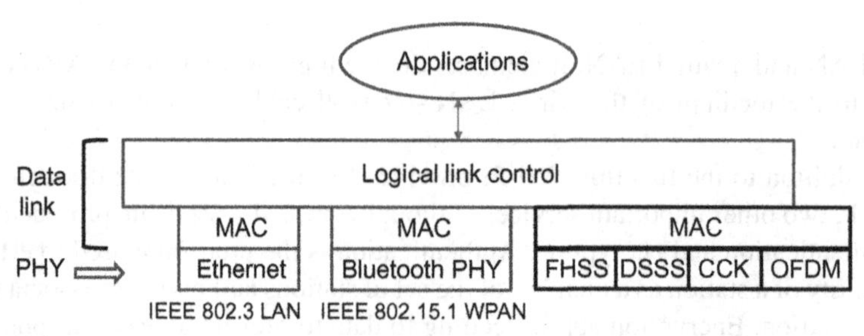


**FIG. 11.5**

Data link and physical layers (PHY).

of several types of networks with upper-layer application software interfaced through a commonly defined logical link control (LLC) layer. The LLC is common to all IEEE local area networks and is independent of the transmission medium or medium access method. Thus, its protocol is the same for wired local area networks and the various types of wireless networks.

The MAC service is the essence of the WLAN. Its implementation may be by high-level digital logic circuits or a combination of logic and a microcontroller or a digital signal processor. While the PHY of IEEE 802.11 describes wireless signal characteristics such as data rates and modulation techniques (DSSS, CCK, OFDM), the MAC station service consists of the following functions:

- Authentication
- Deauthentication
- Data confidentiality

- Delivery of data packets to and from higher protocol levels, which includes control of access to the physical medium
- Dynamic frequency selection (DFS)
- Transmit power control (TPC)
- QoS support
- Radio measurement

## 11.3 Medium access

An important attribute of any communications network is the method of access to the medium. 802.11-2016 prescribes three possibilities: distributed coordination function (DCF), hybrid coordination function (HCF), and mesh coordination function (MCF). A point coordination function (PCF) which is an optional polling method prescribed for 802.11, is now obsolete.

The fundamental access method in IEEE 802.11 is the DCF, which is known as CSMA/CA (carrier sense multiple access with collision avoidance). It is based on a procedure during which a station wanting to transmit may do so only after listening to the channel and determining that it is not busy. If the channel is busy, the station must wait until the channel is idle. In order to minimize the possibility of collisions when more than one station wants to transmit at the same time, each station waits a random time-period, called a back off interval, before transmitting, after the channel goes idle. Fig. 11.6 shows how this method works. Figure sections are shown as a numeral in a circle.

In the figure, a previous transmission is recognized by a station that is attempting to transmit. The station may start to transmit if the channel is idle, determined by a carrier sense mechanism, for a period of at least a duration of an interframe space (IFS) since the end of any other transmission (section 1 of the figure). Several different duration IFS's are defined, in order to give access priority to different types of frame exchanges. Data frame and management frame exchanges require the use of the DIFS (distributed coordination function interframe space), shown in the figure. If the channel is busy, as shown in section 2 of the figure, it must defer access and enter

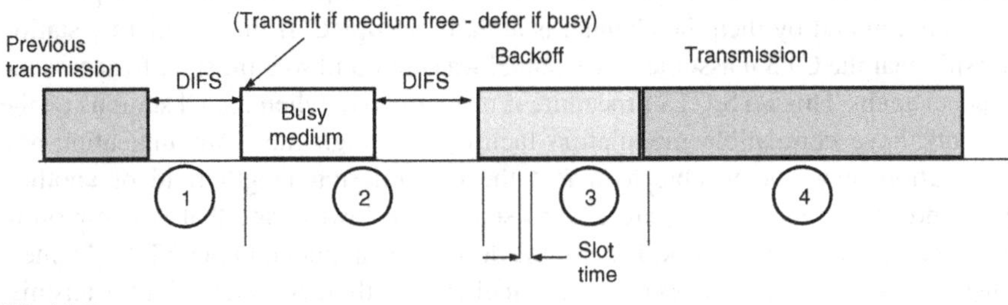


**FIG. 11.6**

CSMA/CA access method.

a back off procedure. The station waits until the channel is idle, and then waits an additional period of DIFS. Now it computes a time-period called a back off window that equals a pseudo-random number multiplied by a constant called the "slot time." As long as the channel is idle, as it is in section 3 of the figure, the station may transmit its frame at the end of the back off window, section 4. During every slot time of the back off window the station senses the channel, and if it is busy, the counter that holds the remaining time of the back off window is frozen until the channel becomes idle and the back off counter resumes counting down.

At the conclusion of a received frame and ascertaining through a frame check sequence field (FCS) that it is not in error, the receiving station sends a short acknowledgement (ACK) to the sender. The waiting period for sending the ACK is a short interframe space (SIFS) of less duration than the DIFS, thereby giving priority to the acknowledgement. If the sender does not hear an ACK during a specified waiting period after concluding his transmission, he assumes the frame was not received and must send it again.

In waiting for a channel to become idle, a transmission contender doesn't have to listen continuously. When one station hears another station access the channel, it notes the frame length field that is transmitted on every frame and updates a memory location called a network allocation vector (NAV) which gives the total occupation time for the station presently using the channel. After taking into account the time of the acknowledgement transmission that replies to a data transmission, the time until the channel will become idle is known even without physically sensing it. This is called a virtual carrier sense mechanism.

The procedure shown in Fig. 11.6 may not work well under some circumstances. For example, if several stations are trying to transmit to a single AP, two or more of them may be positioned such that they all are in range of the AP but not of each other. In this case, a station sensing the activity of the channel may not hear another station that is transmitting on the same network. To get around this "hidden node" problem, a refinement of the described CSMA/SA procedure can be used. A station thinking the channel is clear sends a short RTS (request to send) control frame to the AP. It then waits to receive a CTS (clear to send) reply from the AP, which is in range of all contenders for transmission, before sending its data transmission. RTS and CTS transmissions access the channel after an SIFS, so other stations waiting to transmit cannot interfere because they have to wait the longer DIFS time after the previous transmission and by then the channel is already occupied. If the originating station doesn't hear the CTS it assumes the channel was busy and so it must try to access the channel again. This RTS/CTS procedure is also effective when not all stations on the network have compatible modulation facilities for high rate communication and one station may not be able to detect the transmission length field of another. RTS and CTS transmissions are always sent at the lowest rate that is common to all participants in the network. Data frame duration is included in the RTS/CTS messages so stations waiting to access a channel can set their NAV and defer transmissions even when they do not hear both sides of the communication.

Use of RTS/CTS is optional and its use depends on frame length. For short data or management frames it would not be used because the overhead of the RTS/CTS transmissions reduces the benefit of priority access to the medium. Above a certain frame length threshold, the RTS/CTS option can increase throughput.

### 11.3.1 Hybrid coordination function

The HCF is a medium access procedure associated with QoS requirements that gives differentiated priorities to transmitted traffic through four different access classes. It works essentially by defining backoff times within an IFS that is shorter than the DIFS and can give controlled access to applications that require regular time slots, for example voice and multimedia streaming while maintaining protection against transmission collisions. HCF controlled channel access (HCCA) is an aspect of HCF that gives higher priority channel access to non-AP stations. It works essentially as a polling mechanism. Stations may transmit multiple frames during a polling period.

### 11.3.2 Mesh coordination function (MCF)

The MCP is a MAC mechanism for mesh networks. It has a contention based channel access procedure and controlled channel access where management frames are used to make reservations for future transmissions. Neighboring stations hear these reservations and do not transmit during the reserved periods, thereby reducing potential congestion.

## 11.4 Physical layer

The discussion so far on the services and the organization of the WLAN did not depend explicitly on the details of the wireless connection between the members of the network but those details affect the nature and quality of the services that the network can provide. 802.11 and its amendments specify various bit rates, modulation methods, and operating frequency channels on three frequency bands (plus TV white space), which we discuss in this section. The distinctions of the types of physical layers covered by IEEE 802.11-2016 are shown in Table 11.1.

Details are given below.

### 11.4.1 IEEE 802.11 basic

The original version of the 802.11 specification prescribes three different air interfaces, each having two data rates. One is infrared and the others are based on FHSS and direct-sequence spread-spectrum, each supporting raw data rates of 1 and 2 Mbps. Below is a short description of the IR and FHSS links, for historical interest since these modes are now obsolete. A more detailed review of DSSS follows.

**Table 11.1** IEEE 802.11-2016 physical layers

| Name | Amendment | Frequencies (GHz) | Bandwidth (MHz) | Modulation | MIMO streams | Data rate (Mb/s) |
|---|---|---|---|---|---|---|
| DSSS | — | 2.4 | 20 | DSSS | — | 1, 2 |
| HR/DSSS (high rate) | 802.11b | 2.4 | 20 | CCK complementary code keying | — | 5.5, 11 |
| OFDM | 802.11a | 5 | 20 (also 10, 5) | OFDM | — | Up to 54 |
| ERP (extended rate) | 802.11g | 2.4 | 20 | OFDM, CCK, DSSS | — | Up to 54 |
| HT (high throughput) | 802.11n | 5, 2.4 | 20, 40 | OFDM + backward compatible | 4 | 600 |
| DMG (directional multi-gigabit) | 802.11ad | 60 | 2160 | SC Single carrier | Beam forming | 8,085 max |
| VHT (very high throughput) | 802.11ac | 5 | 20, 40, 80, 160 | OFDM | 8, MU-MIMO 4 × 4 (8 max) | 6,933.3 max |
| TVHT (television VHT) | 802.11af | 0.540-0.790 TV white space | 6, 7, 8; × 1, 2, 4 | OFDM | 4 | 568.9 max |

#### 11.4.1.1 Infrared PHY

Infrared communication links have some advantages over radio wave transmissions. They are completely confined within walled enclosures and therefore eavesdropping concerns are greatly relieved, as are problems from external interference. Also, they are not subject to intentional radiation regulations. The IEEE 802.11 IR physical layer is based on diffused infrared links, and the receiving sensor detects radiation reflected off ceilings and walls, making the system independent of line-of-site. The range limit is on the order of 10 m. Baseband pulse position modulation is used, with a nominal pulse width of 250 nsec. The IR wavelength is between 850 and 950 nM. The 1 Mbps bit rate is achieved by sending symbols representing 4 bits, each consisting of a pulse in one of 16 consecutive 250 ns slots. This modulation method is called 16-PPM. Optional 4-PPM modulation, with four slots per two-bit symbol, gives a bit rate of 2 Mbps.

Although part of the original IEEE 802.11 specification and having what seems to be useful characteristics for some applications, products based on the infrared physical layer for WLAN were not generally commercially available. However, point-to-point, very short-range infrared links using the IrDA (infrared data association) standard were widespread. These links worked reliably line-of-site at 1 m and may be still found, for example, in desktop and notebook computers, handheld PC's, printers, cameras and toys. Data rates range from 2400 bps to 16 Mbps. Bluetooth devices have taken over many of the applications but for some cases IrDA imbedding still has an advantage because of its much higher data rate capability. We may expect that 802.11 DMG (see table) will fill the role that IR has had for very short range high speed connections to computer peripherals but with data rates several orders of magnitude greater.

#### 11.4.1.2 FHSS PHY

While overshadowed by the DSSS PHY, acquaintance with the FHSS option in 802.11 may still be of interest. In FHSS WLAN, transmissions occur on carrier frequencies that hop periodically in pseudo-random order over almost the complete span of the 2.4 GHz ISM band. This span in North America and most European countries is 2.400 to 2.4835 GHz, and in these regions there are 79 hopping carrier frequencies from 2.402 to 2.480 GHz. The dwell on each frequency is a system-determined parameter, but the recommended dwell time is 20 ms, giving a hop rate of 50 hops per second. In order for FHSS network stations to be synchronized, they must all use the same pseudo-random sequence of frequencies, and their synthesizers must be in step, that is, they must all be tuned to the same frequency channel at the same time. Synchronization is achieved in 802.11 by sending the essential parameters—dwell time, frequency sequence number, and present channel number—in a frequency parameter set field that is part of a beacon transmission sent periodically on the channel. A station wishing to join the network can listen to the beacon and synchronize its hop pattern as part of the network association procedure.

The FHSS physical layer uses GFSK (Gaussian frequency shift keying) modulation, and must restrict transmitted bandwidth to 1 MHz at 20 dB down (from peak carrier).

This bandwidth holds for both 1 Mbps and 2 Mbps data rates. For 1 Mbps data rate, nominal frequency deviation is ±160 kHz. The data entering the modulator is filtered by a Gaussian (constant phase delay) filter with 3 dB bandwidth of 500 kHz. Receiver sensitivity must be better than −80 dBm for a 3% frame error rate. In order to keep the same transmitted bandwidth with a data rate of 2 Mbps, four-level frequency shift-keying is employed. Data bits are grouped into symbols of two bits, so each symbol can have one of four levels. Nominal deviations of the four levels are ±72 and ±216 kHz. A 500 kHz Gaussian filter smoothes the four-level 1 megasymbols per second at the input to the FSK modulator. Minimum required receiver sensitivity is −75 dBm.

Although development of Wi-Fi for significantly increased data rates has based on DSSS, FHSS does have some advantageous features. Many more independent networks can be collocated with virtually no mutual interference using FHSS than with DSSS. As we will see later, only three independent DSSS networks can be collocated. However, 26 different hopping sequences (North America and Europe) in any of three defined sets can be used in the same area with low probability of collision. Also, the degree of throughput reduction by other 2.4 GHz band users, as well as interference caused to the other users is lower with FHSS. FHSS implementation may at one time also have been less expensive. However, the updated versions of 802.11—specifically 802.11a, 802.11b, and 802.11g—have all based their methods of increasing data rates on the broadband channel characteristics of DSSS in 802.11, while being downward compatible with the 1 and 2 Mbps DSSS modes (except for 802.11a which operates on a different frequency band). Bluetooth has some of the characteristics of 802.11 FHSS but has advanced well beyond the capabilities of the earlier standard.

#### *11.4.1.3 DSSS PHY*

The channel characteristics of the DSSS physical layer in 802.11 are retained in the high data rate updates of the specification. This is natural, since systems based on the newer versions of the specification must retain compatibility with the basic 1 and 2 Mbps physical layer. The channel spectral mask is shown in Fig. 11.7, superimposed on the simulated spectrum of a filtered 1 Mbps transmission. It is 22 MHz wide at the −30 dB points. Fourteen channels are allocated in the 2.4 GHz ISM band, whose center frequencies are 5 MHz apart, from 2.412 to 2.484 GHz. The highest channel, number fourteen, is designated for Japan where the allowed band edges are 2.471 and 2.497 GHz. In the United States and Canada, the first eleven channels are used. Fig. 11.8 shows how channels one, six and eleven may be used by three adjacent independent networks without co-interference. When there are no more than two networks in the same area, they may choose their operating channels to avoid a narrow-band transmission or other interference on the band.

In 802.11 DSSS, a pseudo-random bit sequence phase modulates the carrier frequency. In this spreading sequence, bits are called chips. The chip rate is 11 megachips per second (Mcps). Data is applied by phase modulating the spread carrier. There are eleven chips per data symbol. The chosen pseudo-random sequence is a

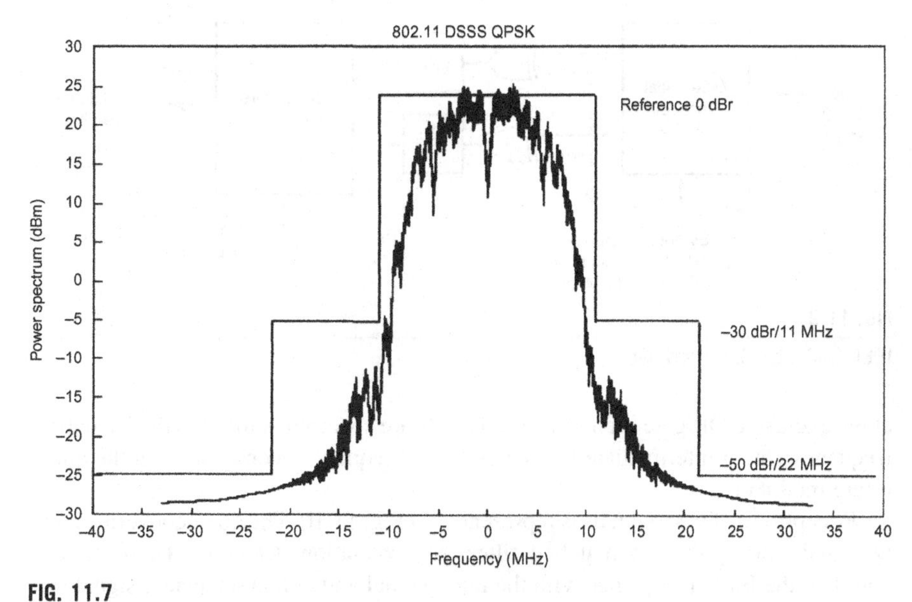


**FIG. 11.7**

802.11 DSSS spectral mask and example spectrum.

**FIG. 11.8**

DSSS non-interfering channels.

Barker sequence, represented as 1,−1,1,1,−1,1,1,1,−1,−1,−1. Its redeeming property is that it is optimally detected in a receiver by a matched filter or correlation detector. Fig. 11.9 is one possible implementation of the modulator. The DSSS PHY specifies two possible data rates—1 and 2 Mbps. The differential encoder takes the data stream and produces two output streams at 1 Mbps that represent changes in data polarity from one symbol to the next. For a data rate of 1 Mbps, differential binary phase shift keying is used. The input data rate of 1 Mbps results in two identical output data streams that represent the changes between consecutive input bits. Differential quadrature phase shift keying handles 2 Mbps of data. Each sequence of two input bits creates four permutations on two outputs. The differential encoder outputs the differences from symbol to symbol on the lines that go to the inputs of the exclusive OR gates shown in Fig. 11.9. The outputs on the $I$ and $Q$ lines are the Barker sequence of 11 Mcps inverted or sent straight through, at a rate of 1 Msps (mega-symbols per second), according to the differentially encoded data

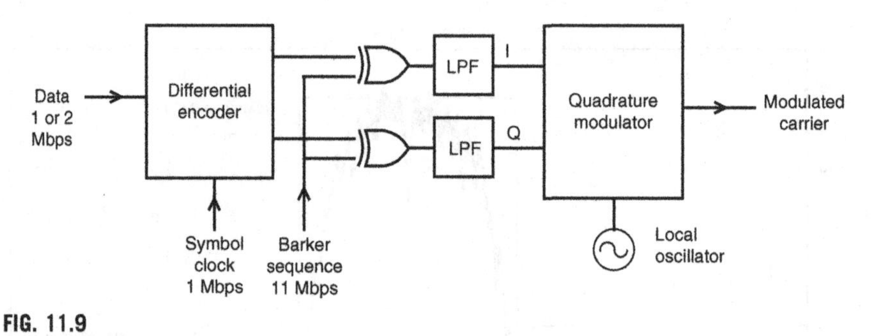


**FIG. 11.9**

IEEE 802.11 DSSS modulation.

at the exclusive OR gate inputs. These outputs are spectrum shifted to the RF carrier frequency (or an intermediate frequency for subsequent upconversion) in the quadrature modulator.

Reception of DSSS signals is represented in Fig. 11.10. The downconverted $I$ and $Q$ signals are applied to matched filters or correlation detectors. These circuits correlate the Barker sequence with the input signal and output an analog signal that represents the degree of correlation. The following differential decoder performs the opposite operation of the differential encoder described above and outputs the 1 or 2 Mbps data.

The process of despreading the input signal by correlating it with the stored spreading sequence requires synchronization of the receiver with transmitter timing and frequency. To facilitate this, the transmitted frame starts with a synchronization field (SYNC), shown at the beginning of the physical layer protocol data unit in Fig. 11.11. Then a start frame delimiter (SFD) marks out the commencement of

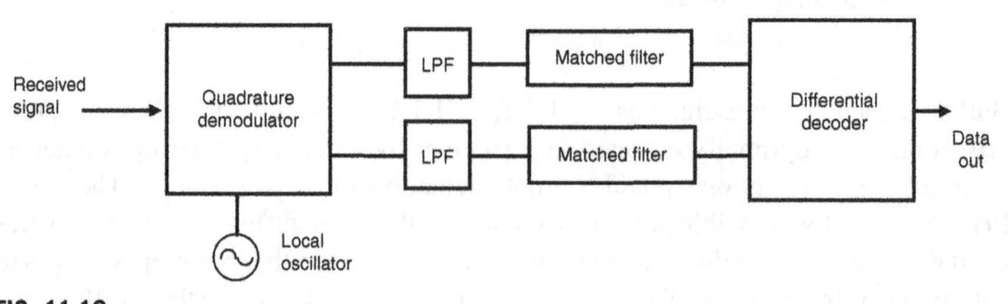


**FIG. 11.10**

IEEE 802.11 reception.

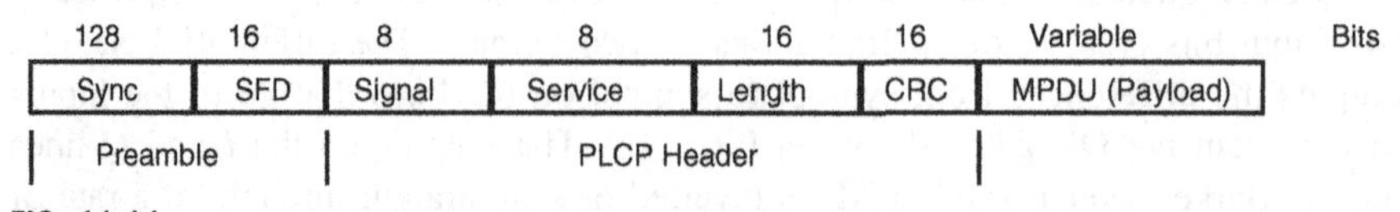


**FIG. 11.11**

IEEE 802.11 frame format.

the following information bearing fields. All bits in the indicated preamble are transmitted at a rate of 1 Mbps, no matter what the subsequent data rate will be. The signal field specifies the data rate of the following fields in the frame so that the receiver can adjust itself accordingly. The next field, SERVICE, contains all zeros for devices that are only compliant with the basic version of 802.11, but some of its bits are used in devices conforming with updated versions. The value of the length field is the length, in microseconds, required to transmit the data-carrying field labeled MPDU (MAC protocol data unit). An error check field, labeled CRC, protects the integrity of the SIGNAL, SERVICE, and LENGTH fields. The last field MPDU (MAC protocol data unit) is the data passed down from the MAC to be sent by the physical layer, or to be passed up to the MAC after reception. All bits in the transmitted frame are pseudo-randomly scrambled to ensure even power distribution over the spectrum. Data are returned to its original form by descrambling in the receiver.

### 11.4.2 High rate DSSS

The "b" amendment to the original 802.11 specification supports a higher rate physical layer for the 2.4 GHz band. It is this 802.11b version that provided the impetus for Wi-Fi proliferation. With it, data rates of 5.5 and 11 Mbps were enabled, while retaining downward compatibility with the original 1 and 2 Mbps rates. The slower rates may be used not only for compatibility with devices that aren't capable of the extended rates, but also for fall back when interference or range don't provide the required signal-to-noise ratio for communication using the higher rates.

As previously stated, the increased data rates specified in 802.11b do not entail a larger channel bandwidth. Also, the narrow-band interference rejection, or jammer resisting qualities of direct sequence spread-spectrum are retained. The classical definition of processing gain for DSSS as being the chip rate divided by the data bandwidth doesn't apply here. In fact, the processing gain requirement that for years was part of the FCC Rules paragraph 15.247 definition of DSSS was deleted in an update from August 2002, and at the same time reference to DSSS was replaced by "digital modulation."

The mandatory high-rate modulation method of 802.11b is called complementary code keying (CCK). An optional mode called packet binary convolutional coding (PBCC) was also described in the specification but it is no longer applicable. Although there are similarities in concept, the two modes differ in implementation and performance. First the general principle of high-rate DSSS is presented below, applying to both CCK and PBCC, then the details of CCK are given.

As in the original 802.11, a pseudo-random noise sequence at the rate of 11 Mcps is the basis of high-rate transmission in 802.11b. It is this 11 Mcps modulation that gives the 22 MHz null-to-null bandwidth. However, in contrast to the original specification, the symbol rate when sending data at 5.5 or 11 Mbps is 1.375 Msps. Eight chips per symbol are transmitted instead of eleven chips per symbol for data rates of 1 or 2 Mbps. In "standard" DSSS as used in 802.11, the modulation, BPSK or QPSK, is applied to the group of eleven chips constituting a symbol. The series of eleven chips in the symbol is always the same (the Barker sequence previously defined) while their phase as a whole is modified in accordance with the data. In contrast, high-rate DSSS uses a different 8-chip sequence in each symbol, depending on the sequence of data bits that each symbol represents. Quadrature modulation is used, and each chip has an $I$ value and a $Q$ value which represent a complex number having a normalized amplitude of one and some angle, $\alpha$, where $\alpha =$ arctangent ($Q/I$). $\alpha$ can assume one of four values divided equally around 360 degrees. Since each complex bit has four possible values, there are a total of $4^8 = 65536$ possible 8-bit complex words. For the 11 Mbps data rate, 256 out of these possibilities are actually used—which one being determined by the sequence of 8 data bits applied to a particular symbol. Only 16-chip sequences are needed for the 5.5 Mbps rate, determined by four data bits per symbol. The high-rate algorithm describes the manner in which the 256 code words, or 16 code words, are chosen from the 65536 possibilities. The chosen 256 or 16 complex words have the very desirable property that when correlation detectors are used on the $I$ and $Q$ lines of the received signal, down converted to baseband, the original 8-bit (11 Mbps rate) or 4-bit (5.5 Mbps rate) sequence can be decoded correctly with high probability even when reception is accompanied by noise and other types of channel distortion.

The concept of CCK modulation and demodulation is shown in Figs. 11.12 and 11.13. It's explained below in reference to a data rate of 11 Mbps. The multiplexer of Fig. 11.12 takes a block of eight serial data bits, entering at 11 Mbps, and outputs them in parallel, with updates at the symbol rate of 1.375 MHz. The six latest data bits determine 1 out of 64 ($2^6$) complex code words. Each code word is a sequence of eight complex chips, having phase angles $\alpha_1$ through $\alpha_8$ and a magnitude of unity. The first two data bits, $d_0$ and $d_1$, determine an angle, $\alpha_8'$, which, in the code rotator (see Fig. 11.12), rotates the whole code word relative to $\alpha_8$ of the previous code word. This angle of rotation becomes the absolute angle $\alpha_8$ of the present code word. The normalized $I$ and $Q$ outputs of the code rotator, which after filtering are input to a quadrature modulator for up conversion to the carrier (or intermediate) frequency, are:

$$I_i = \cos(\alpha_i), Q_i = \sin(\alpha_i)\ i = 1 \ldots 8$$

Fig. 11.13 is a summary of the development of code words a for 11 Mbps rate CCK modulation. High rate modulation is applied only to the payload—MPDU in Fig. 11.11. The code word described in Fig. 11.13 is used as shown for the first symbol and then every other symbol of the payload. However, it is modified by adding 180° to each element of the code word of the second symbol, fourth symbol, and so on.

The development of the symbol code word or chip sequence may be clarified by an example worked out per Fig. 11.13. Let's say the 8-bit data sequence for a symbol is $\boldsymbol{d} = d_0 \ldots d_7 = 1\,0\,1\,0\,1\,1\,0\,1$. From the phase table of Fig. 11.13 we find the angles $\varphi: \varphi_1 = 180°$, $\varphi_2 = 180°$, $\varphi_3 = -90°$, $\varphi_4 = 90°$ . Now summing up these values to

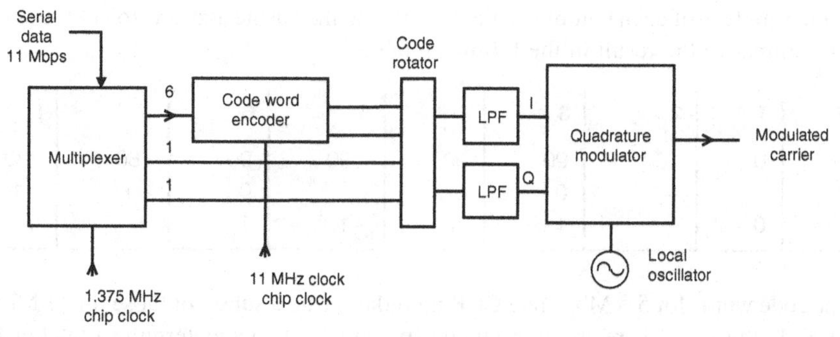


**FIG. 11.12**

High-rate modulator—11 Mbps.

Data symbol: $d_0\ d_1\ d_2\ d_3\ d_4\ d_5\ d_6\ d_7$

| Phase table | |
|---|---|
| $d_i\ d_{i+1}$ | $\varphi$ |
| 0 0 | $0^0$ |
| 1 0 | $180^0$ |
| 0 1 | $90^0$ |
| 1 1 | $-90^0$ |

Phase $(d_0, d_1) = \varphi_1$
Phase $(d_2, d_3) = \varphi_2$
Phase $(d_4, d_5) = \varphi_3$
Phase $(d_6, d_7) = \varphi_4$

$\alpha_1 = \varphi_1 + \varphi_2 + \varphi_3 + \varphi_4$
$\alpha_2 = \varphi_1 + \varphi_3 + \varphi_4$
$\alpha_3 = \varphi_1 + \varphi_2 + \varphi_4$
$\alpha_4 = \varphi_1 + \varphi_4 + 180^0$
$\alpha_5 = \varphi_1 + \varphi_2 + \varphi_3$
$\alpha_6 = \varphi_1 + \varphi_3$
$\alpha_7 = \varphi_1 + \varphi_2 + 180^0$
$\alpha_8 = \varphi_1$

$I_i = \cos(\alpha_i)$
$Q_i = \sin(\alpha_i)$
$i = 1 \ldots 8$

**FIG. 11.13**

Derivation of code word.

get the angle $\alpha_i$ of each complex chip, then taking the cosine and sine to get $I_i$ and $Q_i$, we summarize the result in the following table:

| $i$ | 1 | 2 | 3 | 4 | 5 | 6 | 7 | 8 |
|---|---|---|---|---|---|---|---|---|
| $\alpha$ | 0 | 180 | 90 | 90 | −90 | 90 | 180 | 180 |
| $I$ | 1 | −1 | 0 | 0 | 0 | 0 | −1 | −1 |
| $Q$ | 0 | 0 | 1 | 1 | −1 | 1 | 0 | 0 |

The code words for 5.5 Mbps rate CCK modulation are a subset of those for 11 Mbps CCK. In this case, there are four data bits per symbol which determine a total of 16 complex chip sequences. Four 8-element code words (complex chip sequences) are determined using the last two data bits of the symbol, $d_2$ and $d_3$. The arguments (angles) of these code words are shown in Table 11.2. Bits $d_0$ and $d_1$ are used to rotate the code words relative to the preceding code word as in 11 Mbps modulation and shown in the phase table of Fig. 11.13. Code words are modified by 180° every other symbol, as in 11 Mbps modulation.

The concept of CCK decoding for receiving high rate data is shown in Fig. 11.14. For the 11 Mbps data rate, a correlation bank decides which of the 64 possible codes best fits each received 8-bit symbol. It also finds the rotation angle of the whole code relative to the previous symbol (one of four values). There are a total of 256 (64 × 4) possibilities and the chosen one is output as serial data. At the 5.5 Mbps rate there are four code words to choose from and after code rotation a total of 16 choices from which to decide on the output data.

To maintain compatibility with earlier non-high-rate systems, the DSSS frame format shown in Fig. 11.11 is retained in 802.11b. The 128-bit preamble and the

**Table 11.2** 5.5 Mbps CCK decoding

| $d_3, d_2$ | $\alpha_1$ | $\alpha_2$ | $\alpha_3$ | $\alpha_4$ | $\alpha_5$ | $\alpha_6$ | $\alpha_7$ | $\alpha_8$ |
|---|---|---|---|---|---|---|---|---|
| 00 | 90 | 0 | 90 | 180 | 90 | 0 | −90 | 0 |
| 10 | −90 | 180 | −90 | 0 | 90 | 0 | −90 | 0 |
| 01 | −90 | 0 | −90 | 180 | −90 | 0 | 90 | 0 |
| 11 | 90 | 180 | 90 | 0 | −90 | 0 | 90 | 0 |

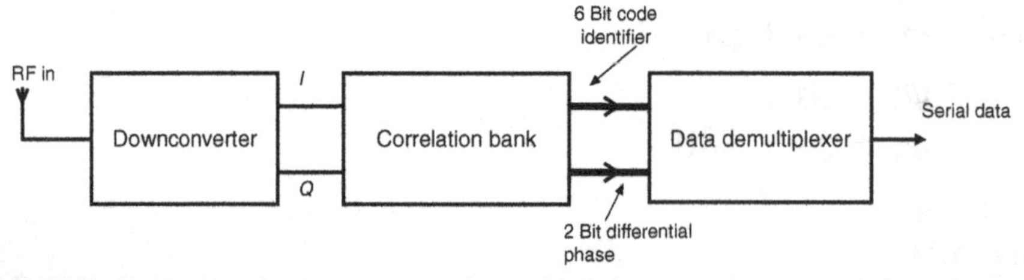


**FIG. 11.14**

CCK decoding.

header are transmitted at 1 Mbps while the payload MPDU can be sent at a high rate of 5.5 or 11 Mbps. The long and slow preamble reduces the throughput and cancels some of the advantage of the high data rates. 802.11b defines an optional short preamble and header which differ from the standard frame by sending a preamble with only 72 bits and transmitting the header at 2 Mbps, for a total overhead of 96 μs instead of 192 μs for the long preamble and header. Devices using this option can only communicate with other stations having the same capability.

Use of higher data rates entails some loss of sensitivity and hence range. The minimum specified sensitivity at the 11 Mbps rate is −76 dBm for a frame-error rate of 8% when sending a payload of 1024 bytes, as compared to a sensitivity of −80 dBm for the same frame-error rate and payload length at a data rate of 2 Mbps.

### 11.4.3 802.11a and OFDM

In the search for ways to communicate at even higher data rates than those applied in 802.11b, a completely different modulation scheme, OFDM (orthogonal frequency division multiplexing) was adopted for 802.11a. It is not DSSS yet it has a channel bandwidth similar to the DSSS systems already discussed. The 802.11a amendment is defined for channel frequencies between 5.2 and 5.85 GHz, obviously not compatible with 802.11b signals in the 2.4 GHz band. However, since the channel occupancy characteristics of its modulation are similar to that of DSSS Wi-Fi, the same system was adopted in IEEE 802.11g for enabling the high data rates of 802.11a on the 2.4 GHz band, while allowing downward compatibility with transmissions conforming to 802.11b.

802.11a specifies data rates of 6, 9, 12, 18, 24, 36, 48, and 54 Mbit/s. As transmitted data rates go higher and higher, the problem of multipath interference becomes more severe. Reflections in an indoor environment can result in multipath delays on the order of 100 ns but may be as long as 250 ns, and a signal with a bit rate of 10 Mbps (period of 100 ns) can be completely overlapped by its reflection. When there are several reflections, arriving at the receiver at different times, the signal may be mutilated beyond recognition. The OFDM transmission system goes a long way to solving the problem. It does this by sending the data partitioned into symbols whose period is several times the expected reflected path length time differences. The individual data bits in a symbol are all sent in parallel on separate subcarrier frequencies within the transmission channel. Thus, by sending many bits during the same time, each on a different frequency, the individual transmitted bit can be lengthened so that it won't be affected by the multipath phenomenon. The higher bit rates are accommodated by representing a group of data bits as the phase and amplitude of a symbol sent on a particular subcarrier. A subcarrier modulated using quadrature phase shift keying (QPSK) can represent two data bits per symbol and 64-QAM (quadrature amplitude modulation) can present six data bits as a single symbol on a subcarrier.

Naturally, transmitting many subcarriers on a channel of given width brings up the problem of interference between those subcarriers. There will be no interference between them if all the subcarriers are orthogonal—that is, if the integral of any two different subcarriers over the symbol period is zero. It is easy to show that this condition exists if the frequency difference between adjacent subcarriers is the inverse of the symbol period.

In OFDM, the orthogonal subcarriers are generated mathematically using the inverse Fourier transform (IFT), or rather its discrete equivalent, the inverse discrete Fourier transform (IDFT). The IDFT may be expressed as:

$$x(n) = \frac{1}{N}\sum_{m=0}^{N-1} X(m)[\cos(2\pi mn/N) + j\sin(2\pi mn/N)]$$

$x(n)$ are complex sample values in the time domain, $n = 0 \ldots N-1$, and $X(m)$ are the given complex values, representing magnitude and phase, for each subcarrier frequency in the frequency domain. $N$ is the number of subcarriers. The IDFT expression indicates that each sample of the time domain signal is the sum of $N$ harmonically related sine and cosine waves each of whose magnitude and phase is given by $X(m)$. We can relate the right side of the expression to absolute frequency by multiplying the arguments $2\pi mn/N$ by $f_1/f_s$ to get

$$x(n) = \frac{1}{N}\sum_{m=0}^{N-1} X(m)[\cos(2\pi m f_1 n t_s) + j\sin(2\pi m f_1 n t_s)] \quad (11.1)$$

where $f_1$ is the fundamental subcarrier and the difference between adjacent subcarriers, and $t_s$ is the sample time $1/f_s$. In 802.11a OFDM, the sampling frequency $f_s$ is 20 MHz and $N = 64$, so $f_1 = 312.5$ kHz. Symbol time is $Nt_s = 64/f_s = 3.2$ μs.

In order to prevent intersymbol interference, 802.11a inserts a guard time of 0.8 μs in front of each symbol, after the IDFT conversion. During this time, the last 0.8 μs of the symbol is copied in front of its beginning, so the guard time is also called a circular prefix. Thus, the extended symbol period that is transmitted is 3.2 + 0.8 = 4 μs. The guard time segment is deleted after reception and before reconstruction of the transmitted data.

Although Eq. (11.1), where $N = 64$, indicates 64 possible subcarriers, only 48 are used to carry data, and four more for pilot signals to help the receiver phase lock to the transmitted carriers. The remaining carriers that are those at the outside of the occupied bandwidth, and the DC term ($m = 0$ in Eq. 11.1), are null. It follows that there are 26 ((48 + 4)/2) carriers on each side of the nulled center frequency. Each channel width is 312.5 kHz, so the occupied channels have a total width of 53 × 312.5 kHz = 16.5625 MHz.

For accommodating a wide range of data rates, four modulation schemes are used—BPSK, QPSK, 16-QAM and 64-QAM, requiring 1, 2, 4, and 6 data bits per symbol, respectively. Forward error correction (FEC) coding is employed with OFDM, which entails adding code bits in each symbol. Three coding rates: 1/2, 2/3, and 3/4, indicate the ratio of data bits to the total number of bits per symbol for different degrees of coding performance. FEC permits reconstruction of the correct message in the receiver, even when one or more of the 48 data channels have selective interference that would otherwise result in a lost symbol. Symbol bits are

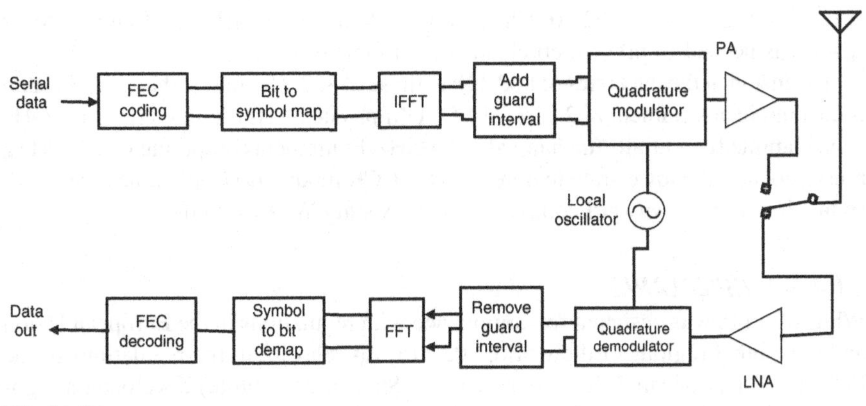


**FIG. 11.15**

OFDM system block diagram.

interleaved so that even if adjacent subcarrier bits are demodulated with errors, the error correction procedure will still reproduce the correct symbol. A block diagram of the OFDM transmitter and receiver is shown in Fig. 11.15. Blocks FFT and IFFT indicate the fast Fourier transform and its inverse instead of the mathematically equivalent (in terms of results) discrete Fourier transform and IFDT that we used above because it is much faster to implement. Table 11.3 lists the modulation type and coding rate used for each data rate, and the total number of bits per OFDM symbol, which includes data bits and code bits. The data rate in the first column is the result of multiplying the data bits per OFDM symbol (last column) by the transmitted symbol rate which is the inverse of the extended symbol period of 4 μs.

The frequency ranges in the 5 GHz band in accordance with FCC paragraphs 15.401–15.407 for unlicensed national information infrastructure (U-NII) devices

**Table 11.3** OFDM characteristics according to data rate

| Data rate (Mbps) | Modulation | Coding rate | Coded bits per subcarrier | Coded bits per OFDM symbol | Data bits per OFDM symbol |
|---|---|---|---|---|---|
| 6 | BPSK | 1/2 | 1 | 48 | 24 |
| 9 | BPSK | 3/4 | 1 | 48 | 36 |
| 12 | QPSK | 1/2 | 2 | 96 | 48 |
| 18 | QPSK | 3/4 | 2 | 96 | 72 |
| 24 | 16-QAM | 1/2 | 4 | 192 | 96 |
| 36 | 16-QAM | 3/4 | 4 | 192 | 144 |
| 48 | 64-QAM | 2/3 | 6 | 288 | 192 |
| 54 | 64-QAM | 3/4 | 6 | 288 | 216 |

are given in Section 10.2.2.10. Channel allocations are 5 MHz apart and 20 MHz spacing is needed to prevent co-channel interference.

Extension of the data rates of 802.11b to those of 802.11a, but on the 2.4 GHz band is covered in amendment 802.11g. The OFDM physical layer defined for the 5 GHz band is applied essentially unchanged to 2.4 GHz. Equipment complying with 802.11g must also have the lower-rate features and the CCK modulation technique of 802.11b so that it will be downward compatible with existing Wi-Fi systems.

#### *11.4.3.1 HIPERLAN/2*

While 802.11b was designed for compliance with regulations in the European Union and most other regions of the world, 802.11a was relevant to the regulations of the FCC. ETSI (European Telecommunications Standards Institute) developed a high-speed wireless LAN specification, called HIPERLAN/2 (high performance local area network), which met the European regulations and in many ways went beyond the capabilities of 802.11a. HIPERLAN/2 defined a physical layer essentially identical to that of 802.11a, using coded OFDM and the same data rates up to 54 Mbps. However, its second layer software level is very different from the 802.11 MAC and the two systems are not compatible. Features of HIPERLAN/2 that were necessary to meet European regulations, specifically DFS and TPC, as well as QoS features and high data security, have since been incorporated into IEEE 802.11 which has become the leading technology for WLAN the world over.

### 11.4.4 IEEE 802.11 throughput improvements

Two parameters of the 802.11 communication link that were modified for performance improvements are bandwidth and number of data streams. Both are responsible for the significantly increased data rate of the high throughput (HT) physical layer, the 802.11n amendment, over the basic OFDM specifications, 802.11a and 802.11g. HT PHY doubled bandwidth to 40 MHz and introduced MIMO, with a maximum of 4 signal streams. A third contributor to increased data rate is the modulation level. The VHT PHY, 802.11ac, uses up to 256-QAM, for 8 data bits per symbol, compared to 64-QAM, 6 bits per symbol, in the HT PHY. An increase in modulation level also increases susceptibility to noise and interference. The later amendments specified improved FEC, which also facilitated a higher code rate, which is the ratio of data bits to total bits (including code bits) in a code block. Still another factor in increased throughput is reduced overhead. Maintaining backward compatibility requires longer preambles and transmitting the frame header at the basic low data rate, needed for collision avoidance when a network is open to older devices conforming to older versions of the specification. HT PHY and VHT PHY have provision for "greenfield" operational mode, where a more streamlined frame header is used, assuming all devices on the network comply with the latest specification revision. Another way to reduce overhead is by using a shorter cyclic prefix on the OFDM symbol. HT and VHT PHY can optionally have a cyclic prefix of 400 ns instead of the legacy length of 800 ns.

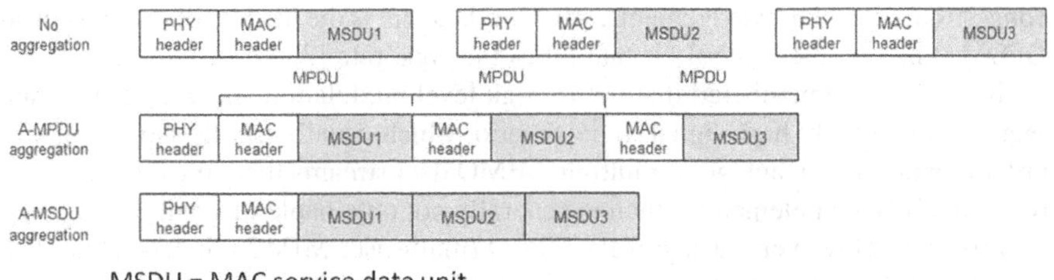


**FIG. 11.16**

Frame aggregation.

Still another throughput improvement technique in HT and VHT is packet aggregation, where data packets are combined in a single frame to substantially increase the data to header size ratio as shown in Fig. 11.16 [3]. The data, handed down to the data link layer from applications is designated MSDU (MAC Service Data Unit). MPDU (MAC Protocol Data Unit) is a subframe that includes the MSDU and a MAC header. The MAC sub-layer adds the MAC header, which includes addressing information and other functional fields relating to the data. Every transmitted frame must have a PHY header, which consists of synchronization and MIMO streaming fields. Not shown in Fig. 11.16 are the acknowledgement transmissions which follow each frame, so the impact on throughput is greater than is apparent in the figure. Although A-MSDU aggregation appears to yield the highest throughput, its advantage is reduced when frame errors occur. The whole A-MSDU frame must be retransmitted when there is an error, whereas in A-MPDU aggregation, only the MSDU and its header (the MPDU) that contained the error has to be retransmitted, with a lower reduction in throughput.

Table 11.4 shows a selection of parameters for the OFDM, HT, and VHT specifications and how they affect data rate. The difference between throughput and data rate should be noted. Throughput depends on frame length, header length, network

**Table 11.4** Comparison of some OFDM, HT, and VHT parameters

| PHY | Modulation | Code rate | Bandwidth (MHz) | MIMO streams | Data rate (Mb/s) |
|---|---|---|---|---|---|
| OFDM | 16-QAM | 1/2 | 20 | — | 24 |
| OFDM | 64-QAM | 3/4 | 20 | — | 54 |
| HT | 16-QAM | 1/2 | 20 | 1 | 26 |
| HT | 64-QAM | 5/6 | 20 | 1 | 65 |
| HT | 64-QAM | 5/6 | 20 | 2 | 130 |
| HT | 64-QAM | 5/6 | 40 | 2 | 270 |
| HT | 64-QAM | 5/6 | 40 | 4 | 540 |
| VHT | 256-QAM | 5/6 | 40 | 4 | 720 |
| VHT | 256-QAM | 5/6 | 80 | 4 | 1560 |

*Note: cyclic prefix is 800 ns for all entries.*

congestion, and acknowledgements, whereas data rate is the net rate of data bits in an OFDM symbol—the symbol bit rate times the code rate [3, 4].

It should be remembered that use of high level modulation and a high code rate requires a relatively high signal to noise ratio, which usually means short range. In order to gain full advantage of multiple MIMO data streams there must be low correlation of antenna elements which is generally not obtainable in small devices like smartphones. However, through MU-MIMO (multi-user MIMO), independent data streams can be transmitted simultaneously to several separated devices.

## 11.4.5 MIMO in 802.11

### 11.4.5.1 Estimating channel state at the receiver

The key to reaping the advantages of multiple antenna elements on the transmitter side, receiver side, or both, is knowledge of the propagation medium, represented by matrix **H** in Chapter 2 (Eqs. 2.20, 2.21). A receiver can construct this matrix by comparing received symbols at each of its receiving antenna elements with known symbols transmitted in turn from each of the transmitting antenna elements. For example, consider three transmitting antennas and two receiving antennas. Fig. 11.17 shows the propagation path constants when only antenna TX1 is transmitting a symbol

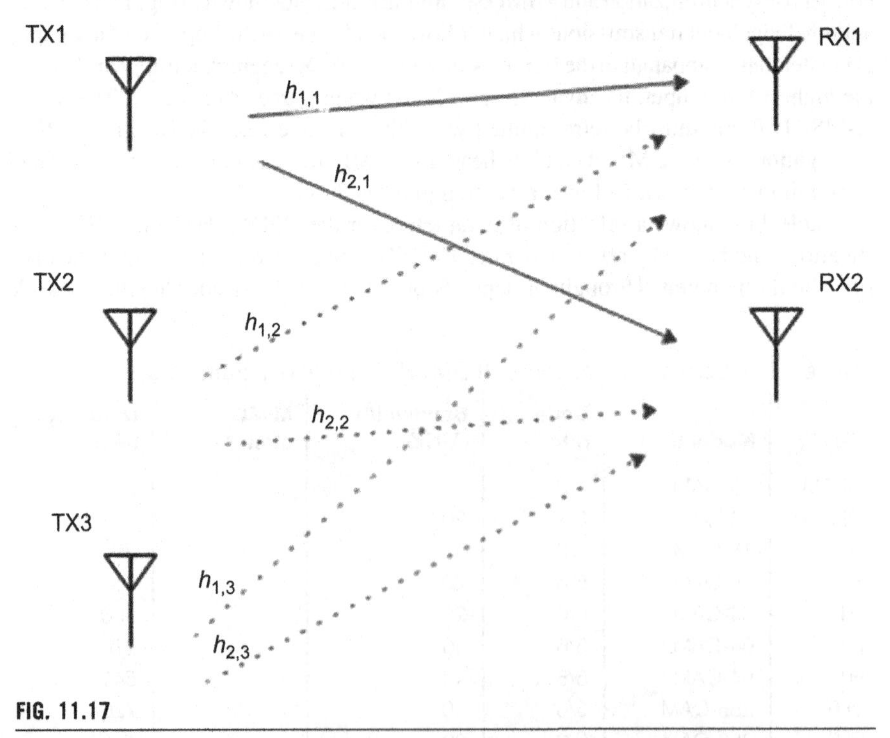


**FIG. 11.17**

Channel matrix. Solid line paths show channel coefficients when only TX1 is transmitting.

stream that is known to the receiver. In general, for each antenna element transmitting individually, the received signal $r$ at each receiving antenna element is:

$$r_i = h_{i,j}s + n_i \tag{11.2}$$

where subscripts $i$ and $j$ refer to the specific receiver and transmitter antenna element respectively, $s$ is the transmitted symbol and $n$ is noise. This, and the following expressions, refer to the channel of one OFDM subcarrier, since $h$ is a function of frequency and is different in each subchannel.

From sequential transmissions over each antenna element of a training symbol known to the receiver, the receiver can estimate each element of **H** without interference from the signals from the other antenna elements as

$$\hat{h}_{i,j} = r_i s^{-1} = h_{i,j}s \cdot s^{-1} + n_i s^{-1} = h_{i,j} + n_i s^{-1} \tag{11.3}$$

However, in IEEE 802.11 HT and VHT physical layers which use MIMO, the channel matrix **H** is found by sending a sequence of training symbols over all antenna elements at the same time yet avoiding interference between the spatial streams [2, 5, 6]. Fig. 11.18 shows a simplified PPDU (PHY protocol data unit) frame, taken from HT but applying in principle to VHT as well. The first two fields, labeled STF (short training field) and LTF (long training field) are used for synchronizing the frame to allow OFDM demodulation of the following SIG (signal) field, which gives the parameters, including the modulation and coding scheme, bandwidth, and frame length, that are needed to interpret the whole frame. Following SIG, one, two, or four long training field symbols for HT (up to eight for VHT), corresponding to the number of space-time streams (with the exception of four symbols for three spatial streams), are sent over all MIMO antenna elements. They are labeled $s_1$ through $s_{NLTF}$. Each of these LTF symbols is a modification of a long training symbol, which we label as a vector **S**, which has a value of "1" or "−1" for each of 56 sub-carrier data elements of the OFDM symbol. Fig. 11.19 shows the flow of the training

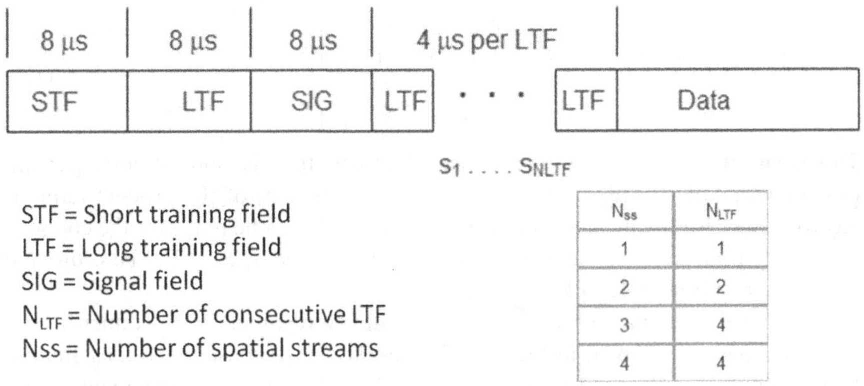

| $N_{ss}$ | $N_{LTF}$ |
|---|---|
| 1 | 1 |
| 2 | 2 |
| 3 | 4 |
| 4 | 4 |

**FIG. 11.18**

IEEE 802.11 HT PPDU frame showing training sequences.

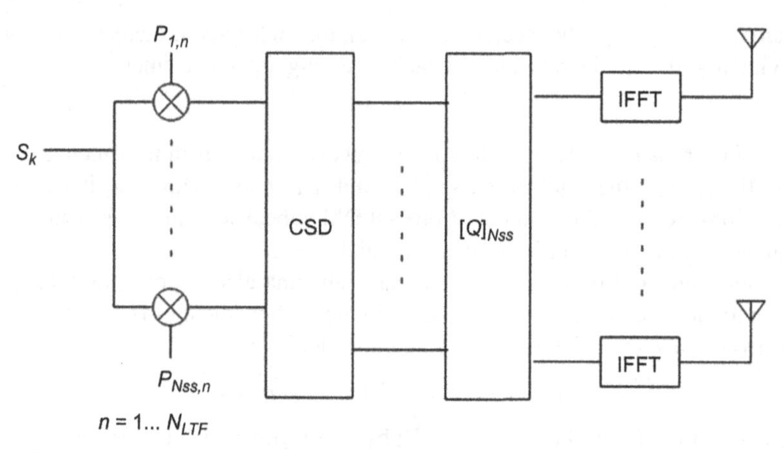


**FIG. 11.19**

Flow of training signals to antenna elements.

symbols for a particular subcarrier, subscript $k$. The parallel paths are the spatial streams in the frequency domain of the particular $k$ symbol. The vector composed of all 64 subcarrier elements (56 data element and 8 null elements) for each spatial stream are inverse fast Fourier transformed (IFFT) to the time domain, then upconverted to the carrier frequency and transmitted on an antenna element. The block labeled CSD (cyclic shift diversity) shifts the phase of each symbol in each spatial stream in order to prevent unintentional beamforming. Block $[Q]$ produces intentional beamforming, which is optionally used, by adjusting phases and magnitudes of the spatial streams. Otherwise, that block passes each data stream to the IFFT block unchanged. The multiplying factors, elements of a matrix **P**, are what allows the receiver to obtain an estimation of each element in the channel matrix for each sub-carrier. The matrix **P** for HT 802.11 is:

$$\mathbf{P} = \begin{bmatrix} 1 & -1 & 1 & 1 \\ 1 & 1 & -1 & 1 \\ 1 & 1 & 1 & -1 \\ -1 & 1 & 1 & 1 \end{bmatrix} \tag{11.4}$$

This is an orthogonal matrix with the characteristic that the sum of corresponding product terms of any two rows is zero, whereas the sum of the product terms of any row times itself equals the number of columns. The same holds for the columns. A possibly truncated version of **P**, henceforth designated as $\mathbf{P}_{\mathrm{Nss}}$, has Nss rows and $N_{\mathrm{LTF}}$ columns (see table in Fig. 11.18).

The signal flow diagram in Fig. 11.19 should be regarded in two dimensions—time and space. As shown in Fig. 11.18, there are $N_{\mathrm{LTF}}$ consecutive long training field symbols of duration 4 μs each. In Fig. 11.19 the $n$th training element in time $s_n$ of the $k$th subcarrier, with a value of $S_k =$ "1" or "−1" depending on $k$, enters on the left and is used to create Nss spatial streams after multiplication by an element of matrix $\mathbf{P}_{\text{Nss}}$. A column of $\mathbf{P}_{\text{Nss}}$ holds multipliers for each spatial stream. For example, for the first LTF, $n = 1$, $s_1$ is multiplied by "1" for the first three spatial streams and "$-1$" for the last (the first column of $\mathbf{P}_{\text{Nss}}$).

The frequency domain sub-carrier symbol detected at each receiver antenna element for a given LTF is

$$r_{i,n} = [h_{i,1}, \ldots, h_{i,\text{NTX}}] \cdot S_k \cdot [\mathbf{P}_{\text{Nss}}]_{(:,\,n)} + n_i \tag{11.5}$$

where $i$ is the index of the receiver antenna element, NTX is the number of transmitting antennas (here the number of spatial streams), $n$ is the index of the training frame and $n_i$ is the noise at the $i$th receiving antenna. $[\mathbf{P}_{\text{Nss}}]_{(:,n)}$ means the $n$th column of $\mathbf{P}_{\text{Nss}}$. Note that the path coefficient $h$ is constant from one symbol period (4 μs) to the next and is expected to be constant for the whole frame.

The total training information for a particular subcarrier symbol at all receiver antenna elements and over all training symbol time sequences can be expressed in matrix form as

$$\mathbf{r}_k = \mathbf{H}_k \cdot S_k \cdot \mathbf{P}_{\text{Nss}} + \mathbf{N}_k \tag{11.6}$$

where $\mathbf{r}_k$ and $\mathbf{N}_k$ are matrices whose rows correspond to the number of receiver antenna elements and columns correspond to the number of symbols in a training sequence.

In order to estimate $\mathbf{H}_k$, $\mathbf{r}_k$ is multiplied by the inverse (or pseudo inverse if the matrix is not square) of the matrix of the training signals $S_k\mathbf{P}_{\text{Nss}}$, Because of the orthogonal nature of $\mathbf{P}_{\text{Nss}}$ and the fact that its elements are real, its inverse can be expressed as $\mathbf{P}_{\text{Nss}}^{\text{T}}$ times a scaling factor of $1/N_{\text{LTF}}$. Superscript T stands for *transpose*. The estimate of $\mathbf{H}$ is then, comparable to Eq. (11.3)

$$\hat{\mathbf{H}}_k = \mathbf{r}_k \cdot \frac{\mathbf{P}_{\text{Nss}}^{\text{T}}}{S_k \cdot N_{\text{LTF}}} = \mathbf{H}_k + \mathbf{N}_k \cdot \frac{\mathbf{P}_{\text{Nss}}^{\text{T}}}{S_k \cdot N_{\text{LTF}}} \tag{11.7}$$

As an example, consider three transmitter antenna elements and two receiver antenna elements as depicted in Fig. 11.17. For the three spatial streams, $N_{\text{LTF}} = 4$. Let $S_k = 1$. From Eqs. (11.4), (11.6) (after truncation of $\mathbf{P}$):

$$\mathbf{r}_k = \begin{bmatrix} h_{1,1} & h_{1,2} & h_{1,3} \\ h_{2,1} & h_{2,2} & h_{2,3} \end{bmatrix} \cdot 1 \cdot \begin{bmatrix} 1 & -1 & 1 & 1 \\ 1 & 1 & -1 & 1 \\ 1 & 1 & 1 & -1 \end{bmatrix} + \begin{bmatrix} n_{1,1} & n_{1,2} & n_{1,3} & n_{1,4} \\ n_{2,1} & n_{2,2} & n_{2,3} & n_{4,4} \end{bmatrix}$$

Inserting $\boldsymbol{r}_k$ in Eq. (11.7):

$$\hat{\mathbf{H}}_k = \begin{bmatrix} h_{1,1} & h_{1,2} & h_{1,3} \\ h_{2,1} & h_{2,2} & h_{2,3} \end{bmatrix} \cdot \begin{bmatrix} 1 & -1 & 1 & 1 \\ 1 & 1 & -1 & 1 \\ 1 & 1 & 1 & -1 \end{bmatrix} \cdot \begin{bmatrix} 1 & 1 & 1 \\ -1 & 1 & 1 \\ 1 & -1 & 1 \\ 1 & 1 & -1 \end{bmatrix} \cdot \frac{1}{4}$$

$$+ \begin{bmatrix} n_{1,1} & n_{1,2} & n_{1,3} & n_{1,4} \\ n_{2,1} & n_{2,2} & n_{2,3} & n_{2,4} \end{bmatrix} \cdot \begin{bmatrix} 1 & 1 & 1 \\ -1 & 1 & 1 \\ 1 & -1 & 1 \\ 1 & 1 & -1 \end{bmatrix} \cdot \frac{1}{4}$$

$$\hat{\mathbf{H}}_k = \begin{bmatrix} h_{1,1} & h_{1,2} & h_{1,3} \\ h_{2,1} & h_{2,2} & h_{2,3} \end{bmatrix} \cdot \begin{bmatrix} 1 & & \\ & 1 & \\ & & 1 \end{bmatrix}$$

$$+\frac{1}{4} \cdot \begin{bmatrix} n_{1,1}-n_{1,2}+n_{1,3}+n_{1,4} & n_{1,1}+n_{1,2}-n_{1,3}+n_{1,4} & n_{1,1}+n_{1,2}+n_{1,3}-n_{1,4} \\ n_{2,1}-n_{2,2}+n_{2,3}+n_{2,4} & n_{2,1}+n_{2,2}-n_{2,3}+n_{2,4} & n_{2,1}+n_{2,2}+n_{2,3}-n_{2,4} \end{bmatrix}$$

$$\mathbf{H}_k = \begin{bmatrix} h_{1,1} & h_{1,2} & h_{1,3} \\ h_{2,1} & h_{2,2} & h_{2,3} \end{bmatrix} + \text{noise}$$

This result shows that the channel state has been estimated without interchannel interference. The noise statistics are not changed by this procedure [7, p. 214].

#### *11.4.5.2 Transmitter channel state information (T-CSI)*

For some MIMO modes including beamforming and spatial division multiplexing (SDM) discussed in Chapter 2, channel state information (CSI) represented by the channel matrix H is needed by the transmitter to predistort the emitted signals in order to obtain the MIMO advantage [2, 8]. We have just seen how the channel propagation matrix H is estimated in the receiver. There are two ways for the transmitter to get this information: implicit feedback and explicit feedback. Both ways are used for HT; VHT uses only explicit feedback. For implicit feedback, Fig. 11.20, the transmitter requests sounding frames, which contain a sequence of training symbols but no data, from the receiver. Through these signals, the transmitter can determine the channel state represented by the propagation matrix. For explicit feedback, Fig. 11.21, the receiver estimates the channel state as described in the previous section and sends this information to the transmitter. The problem with implicit feedback is that the resulting raw CSI is not the same as what would be measured by the

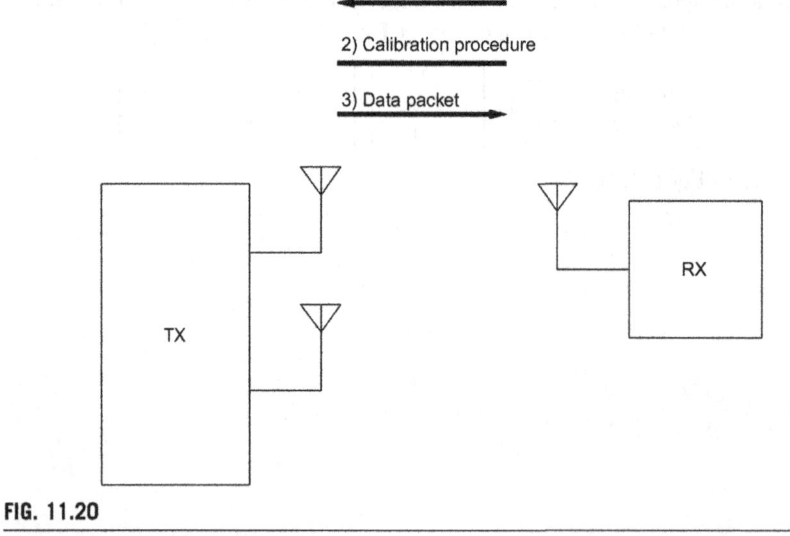


**FIG. 11.20**

Implicit feedback. *(After [9]).*

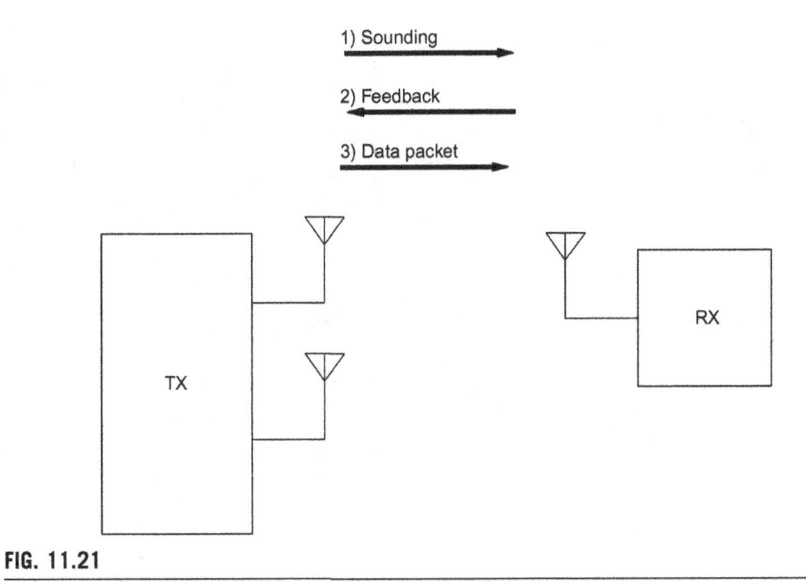


**FIG. 11.21**

Explicit feedback. *(After [9]).*

receiver. Although the channel between transmitting and receiving antennas is symmetrical, differences in the transmitting and receiving hardware of the terminals distort the amplitude and phases measured. In order to remove this distortion a calibration procedure is needed, which is described in [2] for HT. Since the receiver determines the CSI in explicit feedback, it is just what the transmitter needs for beamforming, but the defining information must be sent back to the transmitter. In order to reduce the amount of data to be transferred, the CSI is compressed into what is specified as the V matrix. This requires more back and forth transmissions, but the CSI is apt to be more accurate than when implicit feedback is used.

#### *11.4.5.3 Multi-user MIMO (MU-MIMO)*

While VHT has quantitative advantages over HT in bandwidth, modulation modes, number of spatial streams and consequently throughput, downlink multiuser MIMO (DL-MU-MIMO) is specified only for VHT. Fig. 11.22 shows an illustrative example. Two client receivers at 802.11 stations each receive on the downlink two independent data streams from the transmitting AP, all on the same carrier frequency and during the same time. Note that all receivers receive signals from all transmitting antennas. The data streams are separated using spatial multiplexing as explained in Chapter 2. The challenge here is to receive the required CSI from each of the client receivers. Explicit feedback is used, so the protocol has to coordinate the sending of sounding frames to each of the clients and reception of the CSI. The basis of this protocol is as follows [10]. The AP sends a null data packet announcement frame that notifies the relevant receiving stations that a null data sounding packet follows. The recipient stations will each process the sounding packet to estimate the channel

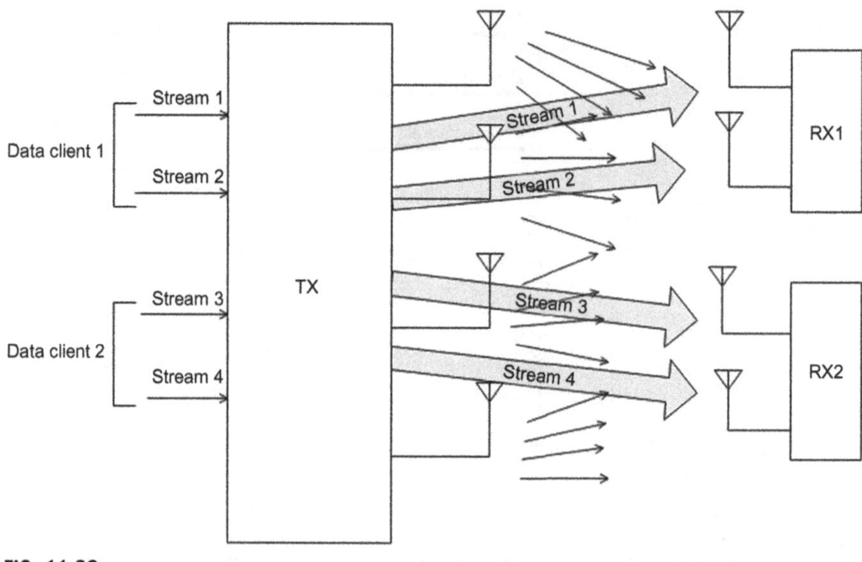


**FIG. 11.22**

Multi-user MIMO.

matrix and will send back to the AP, using polled access, their CSI report using the prescribed compressed format. The API uses the information it receives for the *Q* block of Fig. 11.19 to create the spatial multiplexing beams for the multiple users. Note that MU-MIMO works only in the downlink direction, from AP to client station, in the VHT specification.

## 11.4.6 IEEE 802.11ax

The successor to VHT in 802.11, draft amendment IEEE 802.11ax, has not yet been published as this book is being written. It is called high efficiency (HE) wireless. Its major features compared to VHT are shown in Table 11.5.

**Table 11.5** 802.11ax compared to VHT

| | VHT | 802.11ax |
|---|---|---|
| Bands | 5 GHz | 2.4 and 5 GHz |
| Bandwidth | 20, 40, 80, 80 + 80, 160 MHz | 20, 40, 80, 80 + 80, 160 MHz |
| FFT size | 64, 128, 256, 512 | 256, 512, 1024, 2048 |
| Highest modulation mode | 256-QAM | 1024-QAM |
| MIMO | DL MU-MIMO | DL and UL MU-MIMO |
| Access method | CSMA/CA | OFDMA |
| Data rates | 433 MB/s, 80 MHz, 1 spatial stream | 600.4 Mb/s (80 MHz, 1 spatial stream) |

The comparison in the table doesn't give the whole picture. An impetus for 802.11ax is to provide high throughput in a dense Wi-Fi environment where there may be many APs in a relatively small area. To this end, new technologies were introduced to optimize average per-user throughput, instead of increasing the physical layer transmission rate and peak throughput under a single-user scenario [11]. Also, features are included for reduced power consumption. Technologies and features of HE 802.11 include:

- Orthogonal frequency division multiple access (OFDMA) PHY
- Uplink as well as downlink MIMO
- Spatial reuse
- Trigger frame
- Power saving with target wake time (TWT)
- Station-to-station (S2S) operation

OFDMA is the multiple access means used in LTE (long term evolution) cellular communication. In 802.11ax it substitutes frequency division multiplex for the collision avoidance with random stand-off used in earlier versions of 802.11. In addition, OFDMA allocates transmission bandwidth according to the data rate requirements of each station instead of using a common bandwidth for all, to a greater degree than is achieved with the multiple of 20 MHz bandwidths of HT and VHT 802.11 (amendments *n* and *ac*). It does this by allocating blocks of OFDM subchannels, called resource units (RU), to different stations in a BSS according to their needs. In order to provide a larger number of subchannels to allocate among stations, subcarrier spacing is decreased by four, from 312.5 to 78.125 kHz, increasing by four the number of subchannels per overall channel width. Correspondingly, the OFDM symbol length is increased by four to maintain orthogonality, leaving more room for inter-symbol interference compensation that is necessary particularly for longer reflections in outdoor environments.

IEEE 802.11ax supports uplink as well as downlink multi-user MIMO (MU-MIMO). Each of up to eight users can get up to four time-space streams, but the total number of streams is limited to eight. Combining OFDMA and MU-MIMO enables two-dimensional scheduling, frequency and spatial.

For multi-user simultaneous uplink transmissions frequency and time synchronization has to be exact. To this end, a new control frame, called a trigger frame, is introduced. The AP notifies in the trigger frame which stations are to respond and allows those stations to synchronize and to insure orthogonality between them. After receiving equal length PPDUs (PHY protocol data units) from the stations, the AP can send a common acknowledgement frame to all of them, thereby also reducing the acknowledgement overhead.

In dense WLAN deployments, spatial reuse is limited by overlapping independent BSSs. Stations associated with one AP, and the AP itself, can be heard by adjacent APs and their associated stations. This causes frequent transmission deferrals, as in the description above of the CSMA/CA access method, which of course reduces throughput. 802.11ax proposes to adjust the threshold power level for detecting interference on the channel, as well as controlling transmitter power output for minimum interference.

IEEE 802.11ax has new power saving schemes. A TWT (target wake time) mechanism allows stations to stay in power saving mode for a long time without listening for a beacon. Stations are scheduled to wake up at different times to minimize contention between them. Another power saving measure is for a station to identify a received packet as not in the same BSS, and to enter the "doze" state until the end of the frame. Operation mode indication (OMI) is the name of a mechanism for reducing power by changing PHY parameters like bandwidth and number of spatial streams.

S2S operations, including Wi-Fi Aware and Wi-Fi Direct, can increase contention in the vicinity of a BSS due to the lack of coordination between the transmissions. A quiet time period (QTP) is proposed during which only the S2S stations transmit frames and other non-participating stations remain quiet.

IEEE 802.11ax proposal includes operation on the 2.4 and 5 GHz band, as well as newly assigned channels at 6 GHz, and so can have complete backward compatibility when necessary.

### 11.4.7 IEEE 802.11ah

IEEE 802.11ah amendment specifies a new physical layer for sub 1 GHZ license-exempt bands. Designated Wi-Fi HaLow by the Wi-Fi Alliance and S1G by IEEE, it offers the possibility of extended range for WLAN although throughput is limited compared to the 2.4 and 5 GHz bands due to reduced bandwidth. Among the potential use cases are wireless sensor networks and utility meter monitoring. Modulation is based on OFDM, but the PHY is not compatible with 2.4 and 5 GHz systems [12].

The physical layer (PHY) of 802.11ah is based on VHT of IEEE 802.11-2016 but with a sampling frequency reduced by 1/10 to give bandwidths of 2, 4, 8, and 16 MHz on ISM frequencies below 1 GHz [13]. In addition, 1 MHz bandwidth operation is defined for increased sensitivity and consequently range, but at reduced data rates. Sub-carrier separation is 31.25 kHz, 1/10 that of VHT, HT and 802.11a. Cyclic prefix length may be 4 or 8 μs. Modulation schemes are BPSK, QPSK, 16-QAM, 64-QAM, and 256-QAM, with coding rates of ½ to 5/6, as for VHT. Up to four data streams for SU-MIMO (single-user MIMO) and MU-MIMO (multi-user MIMO) are specified with a maximum of three streams per user for MU-MIMO. Maximum data rates are between 150 kbps and 234 Mbps, according to bandwidth, modulation level, coding rate, cyclic prefix length and number of spatial streams. At 2 MHz bandwidth, one spatial stream, 256-QAM modulation and coding rate of 3/4, raw data rate is 7.8 Mbps [14]. Operating frequency bands are 902 to 928 MHz in the United States and 863 to 868 MHz in Europe.

The MAC layer of 802.11ah has special features to meet the particular requirements of sub 1 GHz applications. With its increased range and use in large multinode networks, a larger address space is accommodated. A 2.4 and 5 GHz 802.11 network supports up to 2007 network addresses (association identifiers—AID) whereas 802.11ah can assign up to 8191 addresses. With a larger number of nodes, medium access collisions become more likely. A restricted access window (RAW) mechanism divides the network devices into groups and assigns specific access periods (RAW) for each group such that contention in a group is significantly less than if all devices had access to the medium at all times. The RAW mechanism is also used to reduce energy consumption. Since a node knows when it can access the medium, it can plan its sleep intervals accordingly. The node will wake up at a scheduled time when the AP sends a beacon indicating which nodes have messages waiting for them. A node expecting a message, or having data to send, will then wake up during its RAW to access the channel. Variations of the access and sleep mechanism are explained in [14] and in the 802.11ah amendment [15].

## 11.5 IEEE 802.11 Wi-Fi certified location

The Wi-Fi Alliance has announced that Wi-Fi includes advanced capabilities to bring location determination indoors [16]. The location service is based on the fine timing measurement (FTM) protocol in IEEE 802.11-2016, which can deliver meter-level accuracy for indoor device location data. FTM calculates an accurate distance from a mobile device to an AP, which is configured with its exact location, including geospatial coordinates (latitude, longitude, and altitude) and a civic address. This can give precise location determination, even in multilevel structures. More information on location is given in Chapter 14.

## 11.6 Wi-Fi security

Security in Wi-Fi networks, as specified in IEEE 802.11, has gone through major enhancements since the originally defined wireless equivalent privacy (WEP). Since 2006, all Wi-Fi devices certified by the Wi-Fi Alliance implement WPA2 [17]. In 2017 the Wi-Fi Alliance introduced Wi-Fi Certified WPA3 as a new generation of Wi-Fi security with new capabilities to enhance Wi-Fi protection in personal and enterprise networks. We discuss here the evolution of 802.11 security from WEP since understanding the basic architecture will help learning the more complicated and much more secure arrangement that has been in force for over 20 years.

First, let's examine some basic security concepts, shown in Fig. 11.23.

*Confidentiality* is keeping the data secret from the unintended listeners on the network. It is maintained by *encryption*, where a secret key known only to the two sides maps the message (plaintext) to a form unintelligible to a third party (encrypted text). In Fig. 11.23, Eve (the eavesdropper) is monitoring the communication medium between Alice (side A) and Bob (side B), but is unable to decipher the messages between them.

*Integrity* is insuring that the received data is the data that was actually sent, and was not modified or replaced in transit. It is implemented using a *one-way hash*

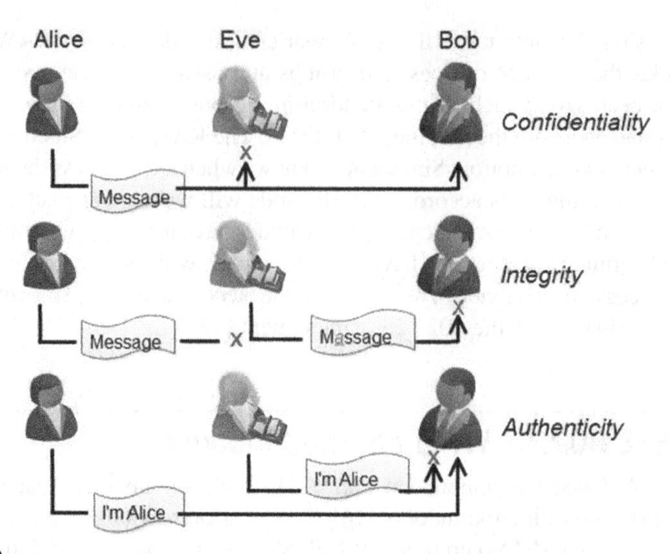


**FIG. 11.23**

Security concepts.

*function* (similar purpose to a CRC -cyclic redundancy check). A cryptographic hash function is an algorithm that takes an arbitrary block of data and returns a fixed-size bit string, the (**cryptographic**) **hash value**, such that any (accidental or intentional) change to the data will (with very high probability) change the hash value. The data to be encoded is often called the *message*. Eve has intercepted the message in the (second row in Fig. 11.23) and changed the first "e" to "a". Although "Massage" has a valid meaning, Bob knows that the message he received was modified, due to integrity protection.

*Authenticity* is ascertaining the identity of the end point to ensure that the end point is the intended entity to communicate with. Use of a digital signature with challenge-response schemes prove authenticity. These schemes involve public key cryptography (PKC) (asymmetric keys). Referring to row 3 of Fig. 11.23, when setting up a communication link or associating with a network Bob needs to be sure the intended opposite terminal, Alice, is who he thinks it is, and not an imposter (Eve). Often authentication is performed for one side of the link only, but for higher security each side will authenticate the other.

Two more security concepts are *non-repudiation* and *service reliability* [18]. *Non-repudiation* is the receiver's ability to prove that the sender did in fact send a given message (bank client cannot deny that she did not withdraw money from her account). PKC, digital signature, or time stamping, can prevent such fraud. *Service reliability* is the ability to protect the communication session against denial of service attacks. These are difficult to combat. Countermeasures include reservation of redundant resources, identification of the attack source and selective blocking.

**Table 11.6** Security threats and countermeasures

| Threat | Category | Countermeasure |
|---|---|---|
| Denial of service | Service reliability | Multiple resources, source tracing |
| Eavesdropping | Confidentiality | Encryption |
| Man-in-the-middle | Authentication, confidentiality | Authentication, encryption |
| Masquerading | Authentication | Authentication |
| Message modification | Integrity | Hash function |
| Message replay | Authentication | Time stamp, session numbering |
| Traffic analysis | Confidentiality | Steganography |

Table 11.6 lists some threats and their categories, and countermeasures.

The word steganography, in the last row under **Countermeasure**, means *the science of hiding information*. It involves hiding the fact that a message is sent—for example hiding a message in insignificant pixels of a JPEG image, or hiding one message inside another (i.e., in a WEB page) [18].

The principles of symmetric and asymmetric (public) key cryptography are shown in Fig. 11.24. For symmetric key cryptography (SKC), both terminals, Alice and Bob, share a common password, or secret key. In the simplified example in Fig. 11.24, that secret key is XOR'ed with the plaintext at Alice's terminal, creating the ciphertext. Bob performs exactly the same process, XOR'ing the secret key with ciphertext, which produces the original plaintext. There are other algorithms for using a common key to encrypt and decrypt a message. For all of them and for the asymmetric key cryptography as well, it is preferable that the security of the cryptosystem resides in the secrecy of the keys rather than with the supposed secrecy of the crypto algorithm. This means that it should be virtually impossible to decrypt a ciphertext to plaintext if the decryption key is unknown, even if the full details of the encryption and decryption algorithms are known.

The major problem with SKC is dissemination of the keys. Both sides use the same key, so how is the key sent from the key originating side to others that need it? If it is sent over a communications medium it is susceptible to interception. In a small place like an office or a home, it can be transferred personally but for large area distribution this is not an option. PKC gets around this problem.

PKC (Fig. 11.24B) is an asymmetric mechanism. Two keys are created by an appropriate key generation algorithm, for example RSA, (Rivest-Shamir-Adleman protocol) [18]. Either key can decrypt a ciphertext created with the other. The public encryption key can be openly distributed. The private decryption key is secret, and is not disseminated. The public/private keys are a pair. Asymmetric algorithms are slow—not practical for lots of data. They are used to distribute secret symmetrical keys, which are then used in a symmetric algorithm for encrypting/decrypting

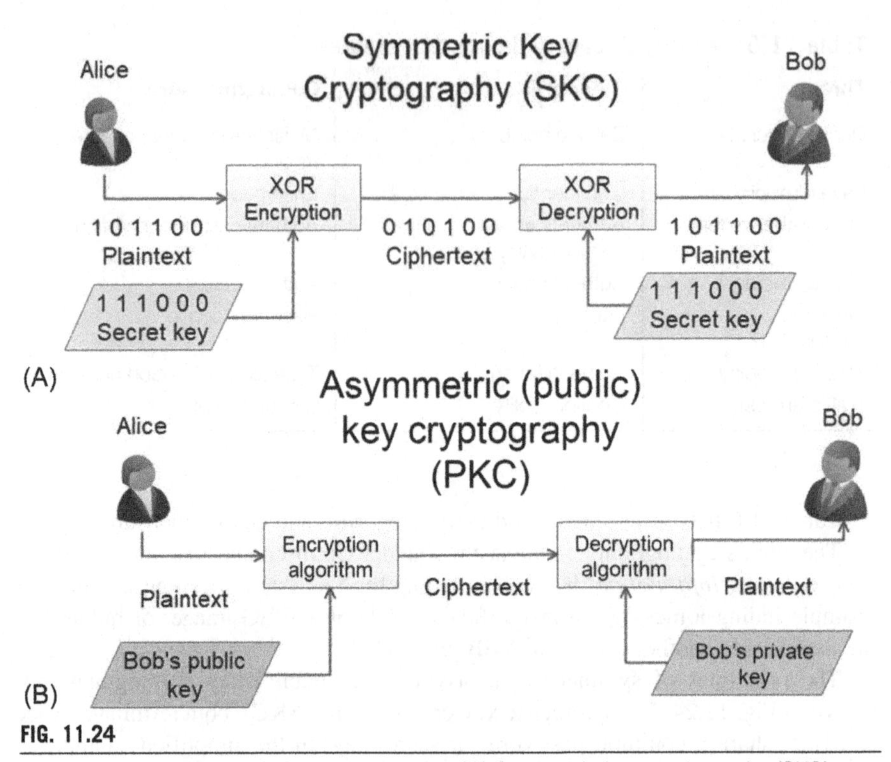


**FIG. 11.24**

Symmetric and asymmetric key cryptography. (A) Symmetric key cryptography (SKC). (B) Asymmetric (Public) key cryptography (PKC).

messages or streams of data. In Fig. 11.24B, Alice encodes a plaintext using Bob's public key. Only Bob can decode it, because only he has the private key. Eve, from Fig. 11.23 *Confidentiality*, can eavesdrop on the message, but she can't decode it, even if she has Bob's public key. As mentioned, Alice can send a symmetric key to Bob through PKC which can be then be used to efficiently communicate messages between them.

A random number sent back and forth is used in an authentication process. The authenticator (Bob) sends a random number in the clear to Alice. Alice encrypts the number with her private key which serves as a *digital signature*. Bob decodes the message with Alice's public key. If the message (random number) is correct, Bob knows it can only be from Alice. This message itself is not secure, as Eve can decode it also if she has Alice's public key. But it serves its purpose of authenticating that Alice is the origin of the message.

PKC is also used for non-repudiation, Bob sends a message encrypted with Alice's public key (which can contain for example access to a large sum of money) to Alice. When the message is acknowledged, by withdrawing money from a bank account for example, Alice can't say she didn't receive it (the message or the money) since only she could have decrypted the message using her private key.

Of course it is very important that the owner of a public key be verifiable. The secure distribution of public keys is done utilizing specific certificates. Secure public key management is accomplished with a public key infrastructure (PKI), which contains catalogs with public keys and their owners, as well as such information as the validity period of the keys. A certificate for public keys is a document that confirms the connection between the public key and the key-owner. Each certificate includes the name of the authority that issued it, the name of the entity to which the certificate was issued, the entity's public key, and time stamps that indicate the certificate's expiration date.

### 11.6.1 WEP

Now that the basics are understood, we move on to discuss confidentiality in 802.11. The original 802.11 security architecture and protocol is called WEP. It aimed to give 802.11 the same security as traditional wired (Ethernet) networks. Here's how it works, following the circled numbers in Fig. 11.25.

**(1)** Calculate ICV (Integrity Check Value) using CRC-32 algorithm.
**(2)** Select one of 4 pre-established secret keys.
**(3)** Create 24 bit IV (Initialization Vector) and concatenate it with the key—used to insure a different key sequence for each message.
**(4)** Use the *seed* (2+3) in an encryption algorithm (Pseudo Random Noise Generator—type RC4) to get keystream which is XOR'd with ICV plus plaintext, giving ciphertext.
**(5)** Concatenate the ciphertext to a plaintext header containing the three byte IV plus a one byte key ID indicating which of 4 keys used and the 802.11 header.

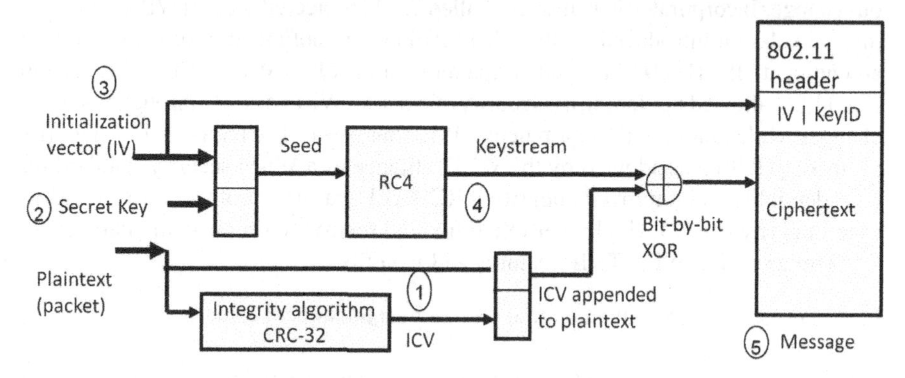


**FIG. 11.25**

Wireless equivalent privacy (WEP).

WEP was found to have multiple weaknesses that made it imperative to revamp the security architecture of 802.11.

**(1)** IV is transmitted in the clear so attacker knows 3 bytes (24 bits) of the 64 or 128 bit seed—40 (or 104) bits of the key usually don't change. This makes the seed a weak key. By accumulating enough packets, the secret key can be found.
**(2)** Additionally, a short IV and no rules on how it is created means that the seed, thus the keystream, will be repeated after a relatively short time (hours). Two encrypted packets with common keys can be unraveled if one plaintext is known or can be guessed. Reuse of the seed allows logical cancelling out of the keystream to expose the message.
**(3)** The integrity algorithm is not secure (CRC-32 uses no key) and allows substitution or modification of a data packet without being detected.
**(4)** The 802.11 header is not encrypted—packets can be maliciously diverted without detection by changing the destination address.
**(5)** There is no protection against replay attacks.

In order to distinguish between levels of security in Wi-Fi implementations, IEEE 802.11 defined two classes of security algorithms: algorithms for creating and using a *robust security network association* (RSNA) which includes TKIP and CCMP (within WPA2), and pre-RSNA of which WEP is a part. These terms are explained below.

### 11.6.2 WPA

The second phase of Wi-Fi security was based on existing WEP hardware with numerous changes incorporated in firmware. Called Wi-Fi protected access (WPA), it was an interim solution introduced by the Wi-Fi Alliance in anticipation of completion of amendment IEEE 802.11i and implementation of WPA2. WPA, shown in Fig. 11.26, allowed upgrading the security of existing Wi-Fi devices through firmware changes while retaining the equipment's WEP hardware. The temporal key integrity protocol (TKIP) was adopted by the Wi-Fi alliance as a Wi-Fi security standard for confidentiality and improved integrity. MICHAEL was new protocol for MIC (Message Integrity Check) with higher reliability and relatively simple computations.

Here are some WPA/TKIP features and benefits:

**(1)** May use 802.1X authentication and key-establishment (enterprise deployments).
**(2)** Creates PTK (pair-wise transient keys) per session (station connection to AP) based on a secret key plus access point (AP) and client station (STA) MAC addresses and nonces (one time random numbers) to insure one-time key sequences, and key mixing functions to reduce exposure of the master key (instead of the simple concatenation used in WEP). In Fig. 11.27, a shared secret, which may be a password or an authentication key obtained through an

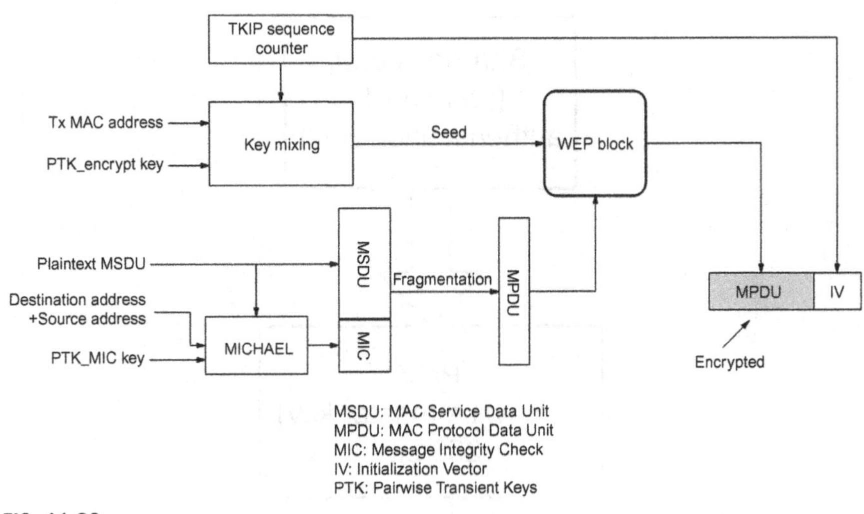


**FIG. 11.26**

WPA with temporal key integrity protocol (TKIP).

authentication server over the network, is transformed to a pair-wise master key (PMK) of 256 bits. The PTK is formed using a pseudo-random function acting on the PMK, session terminal addresses and nonces. It contains up to four keys: encryption and integrity keys for the STA/AP data flow and two keys for communication with an out-of-local-network server based on the EAPoL (Extensible Authentication Protocol over LAN) used by enterprise networks.

**(3)** Initialization vector (IV) increased to 56 bits (of which 48 used) to insure stronger keys.

**(4)** Specifies incrementation of IV for each packet—no repetition. Protects against replay.

**(5)** MICHAEL integrity check much safer than CRC-32 (which is still used within WEP hardware). MIC (message integrity check) computation includes packet destination and source addresses to protect against redirection attacks.

While, as described above, TKIP is an RSNA algorithm, its use, as is the use of WEP, is deprecated [2].

### 11.6.3 WPA2

The outstanding feature of WPA2 is the use of the Advanced Encryption Standard AES. Counter mode cipher block chaining message authentication code protocol (CCMP) provides the highest level of confidentiality, integrity and replay protection available in the 802.11 standard [19]. WPA2 is shown in Fig. 11.28. It uses essentially the same key-establishment process and key hierarchy architecture as WPA (Fig. 11.26). An exception is that the same key is used for confidentiality and

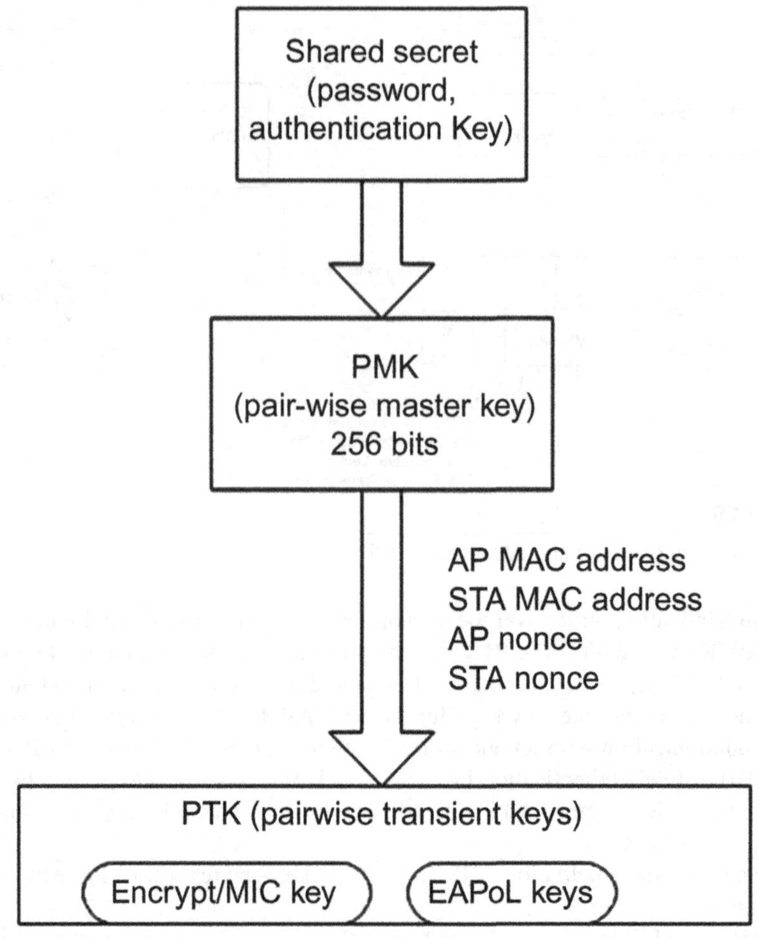


**FIG. 11.27**

Transient key formation.

integrity; there are no separate encryption and MIC keys. In contrast to RC4 of WEP which is a stream cipher, AES is a block cipher, although it does use a key stream to encrypt each block. CCMP achieves confidentiality in counter mode by taking each of consecutive 128 bit blocks of packet plaintext and XORing it with a keystream formed from encryption of a counter, incremented for each block, using the temporal key. To create the MIC used to insure integrity, cipher block chaining message mode is used. Each block is XOR'ed with the ciphertext of the previous block and encrypted with the temporal key. The process is started with an initializing vector (IV) created with a counter that is incremented for each packet. MIC is 64 bits of the last ciphertext block. Because of the chaining of blocks, a change in one or more bits of the message will cause a large difference between the MIC sent with the packet and the MIC constructed at the receiver.

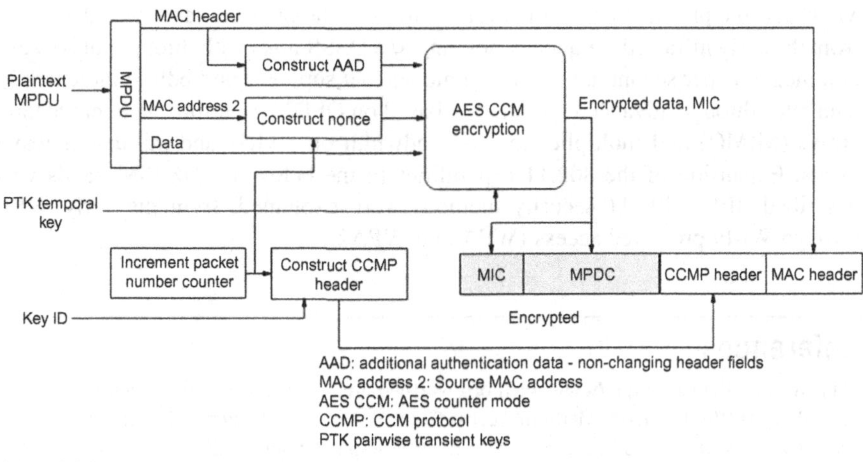


**FIG. 11.28**

WPA2.

Fig. 11.28 shows the main blocks of WPA2. Note that in addition to the message data, fields of the MPDU header, notably the address fields and QoS field, formed in AAD (additional authentication data), are encrypted. The packet number counter, which is incremented on each packet is used in composing a nonce, which never repeats itself in a session. The AES CCM (Counter mode with CBC-MAC) block includes both the counter and cipher block chaining message modes for confidentiality and integrity. The CCMP header and the MAC header which are transmitted in the clear, let the receiver compose the IV which it needs to check integrity and to make the counter used for decryption. The fact that addresses are encrypted from AAD assures that diversion of packets caused by hostile change of packet header addresses can be detected in the receiver.

WPA2 gives similar protection as WPA with TKIP. It gives superior security, however, by using the AES security standard, which is stronger than RC4 for encryption and gives better integrity protection than MICHAEL used in WPA.

The Wi-Fi Alliance announced that new capabilities under WPA3 are available from 2018. Among them are protection of users who choose weak passwords, simplification of the configuration process for devices with limited display interfaces such as sensors and IoT modules, improved privacy on open networks, and stronger security for government, defense and industrial networks through new protocols using a 192 bit security suite [20].

## 11.7 Summary

The technical capabilities of WLANs have been updated continuously since the emergence of IEEE standard 802.11 in the last decade of the 20th century. In this chapter we reviewed 802.11 from the point of view of network architecture,

MAC and the physical layer (PHY). Evolution of the physical layer was described from the early infra-red, frequency hopping and DSSS links with throughputs of several megabits per second up to the techniques that support hundredfold increases in data rates through advanced modulation based on OFDM, multiple element antenna arrays (MIMO) and multiplication of bandwidth on 5 GHz and millimeter wave bands. Expansion of the 802.11 capabilities to the below 1 GHz ISM bands was described. IEEE 802.11 security methods were explained, from the early WEP through Wi-Fi protected access (WPA and WPA2).

## References

[1] Wi-Fi Alliance, https:/www.wi-fi.org/who-we-are, 2018. Accessed 24 October 2018.
[2] Part 11: Wireless LAN Medium Access Control (MAC) and Physical Layer (PHY) Specifications, IEEE Std 802.11™-2016, IEEE Computer Society, 2016.
[3] 802.11ac: The Fifth Generation of Wi-Fi, Technical White Paper. Cisco, 2017.
[4] 802.11ac In-Depth, White Paper. Aruba Networks, 2014.
[5] T. Paul, T. Ogunfunmi, Wireless LAN comes of age: understanding the IEEE 802.11n amendment, IEEE Circ. Syst. Mag. 8 (1) (First Quarter) (2008).
[6] R.P.F. Hoefel, IEEE 802.11n: on performance of channel estimation schemes over OFDM MIMO spatially-correlated frequency selective fading TGn channels, in: XXX SIMPÓSIO BRASILEIRO DE TELECOMUNICAÇÕES—SBrT'12, 13-16 DE SETEMBRO DE 2012, BRASÍLIA, DF, 2012.
[7] A. Sibille, C. Oestges, A. Zanella, MIMO From Theory to Implementation, Elsevier, 2011.
[8] 802.11ac In Depth, Aruba Networks, www.arubanetworks.com, White Paper2014. Accessed 24 October 2018.
[9] R.U. Nabar, MIMO in WiFi systems, in: Smart Antenna Workshop, August 1, 2014. https://web.stanford.edu/~apaulraj/workshop70/pdf/MIMO_WiFi_Nabar.pdf. Accessed 24 October 2018.
[10] O. Bejarano, E.W. Knightly, IEEE 802.11ac: from channelization to multi-user MIMO, IEEE Commun. Mag. (2013).
[11] D.-j. Deng, Y.-p. Lin, X. Yang, J. Zhu, Y.-b. Li, J. Luo, K.-c. Chen, IEEE 802. 11ax: Highly Efficient WLANs for Intelligent Information Infrastructure, IEEE Commun. Mag. 55 (12) (2017) 52–59, https://doi.org/10.1109/MCOM.2017.1700285.
[12] W. Sun, M. Choi, S. Choi, IEEE 802.11ah: a long range 802.11 WLAN at sub 1 GHz, J. ICT Stand. 1 (2013) 83–108, https://doi.org/10.13052/jicts2245-800X.125_c. River Publishers.
[13] P.C. Jain, S. Taneeru, Performance evaluation of IEEE 802.11ah protocol in wireless area network, in: 2016 Int. Conf. Micro-Electronics Telecommun. Eng, 2016, pp. 578–583.
[14] H. Wang, A.O. Fapojuwo, A survey of enabling technologies of low power and long range machine-to-machine communications, IEEE Commun. Surv. Tutor. 19 (4) Fourth Quarter. (2017) 2621–2639.
[15] IEEE Std 802.11ah-2016, Part 11: Wireless LAN Medium Access Control (MAC) and Physical Layer (PHY) Specifications Amendment 2: Sub 1 GHz License Exempt Operation, IEEE Computer Society, 2016.

[16] Wi-Fi Certified Location. Indoor location over Wi-Fi, Wi-Fi Alliance, 2017.
[17] Wi-Fi Security, Wi-Fi-Alliance, https://www.wi-fi.org/discover-wi-fi/security, 2018. Accessed 24 October 2018.
[18] P. Chandra, Bulletproof Wireless Security, Elsevier, 2005.
[19] K. Benton, The Evolution of 802.11 Wireless Security, INF 795, UNLV Informatics-Spring, 2010.
[20] Wi-Fi Alliance® Introduces Security Enhancements, https://www.wi-fi.org/news-events/newsroom/wi-fi-alliance-introduces-security-enhancements (retrieved 17 July 2018), 2018.

CHAPTER

# Wireless personal area networks

# 12

This chapter discusses device and network standards for wireless personal area networks (WPAN), which typically serve a communication range of 10 m. Although most applications probably involve point-to-point connections between computers and printers or cameras, or cell phones and headphones, in recent years WPAN has marked its place as the communications backbone of wireless sensor networks (WSN) and the Internet of things (IoT). Our discussion starts with the ubiquitous Bluetooth, after which are presented the prominent features of Zigbee, 6LoWPAN, Thread, WirelessHART, Z-Wave DASH7, and ANT. Several of these use the IEEE 802.15.4 physical layer (PHY) and differ in network architecture and applications. Concluding the chapter, ultra-wideband (UWB) is described, although it relates less to networks compared to the previous devices referred to, and mostly serves as a high data rate technology within WPAN distances, as well as the basis for high resolution indoor distance measurement.

## 12.1 Bluetooth

There are two sources of the Bluetooth specification. One is the Bluetooth Special Interest Group (Bluetooth SIG). The current version at this writing is Version 5, which is the reference for most of the following discussion [1]. It is arranged in seven volumes covering topics from architecture through wireless coexistence. Volume 1, an overview of architecture and terminology, is a reasonably readable summary of Bluetooth technology. The other volumes detail two broad areas: (1) each of the three low level controllers, BR/EDR (basic rate/enhanced data rate), AMP (alternate MAC PHY, the adaptation of IEEE 802.11 for high data rates), and low energy (Bluetooth LE, or BLE) and (2) the host (high level) core system, host controller interface (HCI), and finally wireless coexistence. The other Bluetooth source specification is IEEE 802.15.1. It is basically a rewriting of the original SIG core specification, made to fit the format of IEEE communications specifications in general. The latest published version is IEEE 802.15.1-2005 which has not been updated, and indeed there is no need to do so since the Bluetooth SIG specification has a wider scope of coverage, from the physical layer through standard application protocols (profiles) than the MAC-PHY coverage normally adhered to in IEEE specifications.

Fig. 12.1 gives an orientation to the Bluetooth architecture, with main logical blocks and data traffic flow. Most of the material on Bluetooth in this chapter



Short-range Wireless Communication. https://doi.org/10.1016/B978-0-12-815405-2.00012-9
© 2019 Elsevier Inc. All rights reserved.

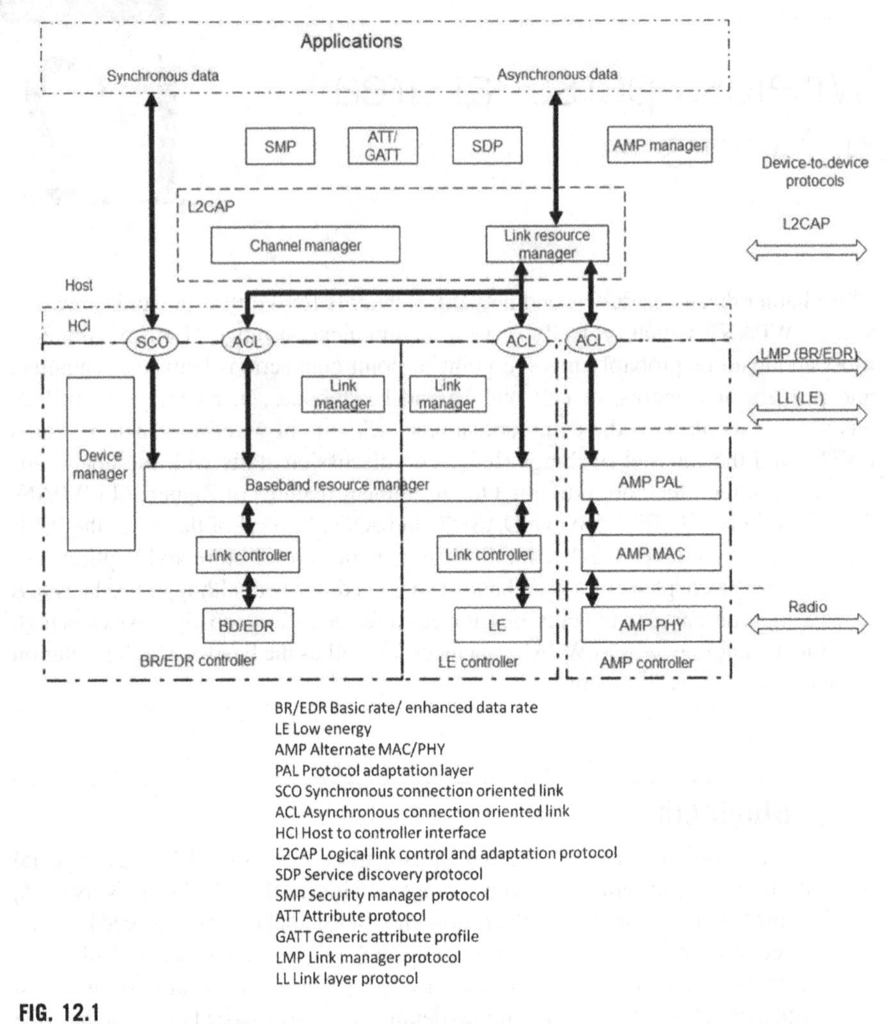


**FIG. 12.1**

Bluetooth core architecture.

concerns the bottom, physical layer. Section 12.1.6 describes Bluetooth profiles. While new acronyms are spelled out in Fig. 12.1 for convenience, terms that are relevant to our discussion are explained later in this chapter.

We start with a description of concepts and operation details of classic, or legacy Bluetooth. Following that, we cover the evolution of Bluetooth through enhanced data rate (EDR), alternate MAC PHY (AMP) for high speed data, Bluetooth low energy (BLE), and the latest (at this writing), Bluetooth 5, for increased speed or range.

### 12.1.1 Legacy Bluetooth

Bluetooth is an example of a WPAN, as opposed to the larger range wireless local area network (WLAN). It's based on the creation of ad hoc, or temporary, on-the-fly connections between digital devices associated with an individual person and located in the vicinity of around 10 m from him. Bluetooth devices in a network have the function of a master or a slave, and all communication is between a master and one or more slaves, never directly between slaves. The basic Bluetooth network is called a piconet. It has one master and from one to seven slaves. A scatternet is an interrelated network of piconets where any member of a piconet may also belong to an adjacent piconet. Thus, conceptually, a Bluetooth network is infinitely expandable. Fig. 12.2 shows a scatternet made up of three piconets. A node may be a slave in one piconet and a master in another. A device may be a master in one piconet only.

The basic RF communication characteristics of Bluetooth are shown in Table 12.1.

A block diagram of a Bluetooth transceiver is shown in Fig. 12.3. It's divided into three basic parts: RF, baseband, and application software. A Bluetooth chip set will usually include the RF and baseband parts, with the application software being contained in the system's computer or controller. The user data stream originates and terminates in the application software. The baseband section manipulates the data and forms frames or data bursts for transmission. It also controls the frequency synthesizer according to the Bluetooth frequency-hopping protocol. The blocks in Fig. 12.3 are general and various transmitter and receiver configurations are adopted by different manufacturers. The Gaussian low-pass filter block before the modulator, for example, may be implemented digitally as part of a complex signal *I/Q* modulation unit or it may be a discrete element filter whose output is applied to the frequency control line of a VCO. Similarly, the receiver may be one of several types, as discussed in Chapter 6. If a superheterodyne configuration is chosen, the filter at the output of the downconverter will be a bandpass type. A direct conversion receiver

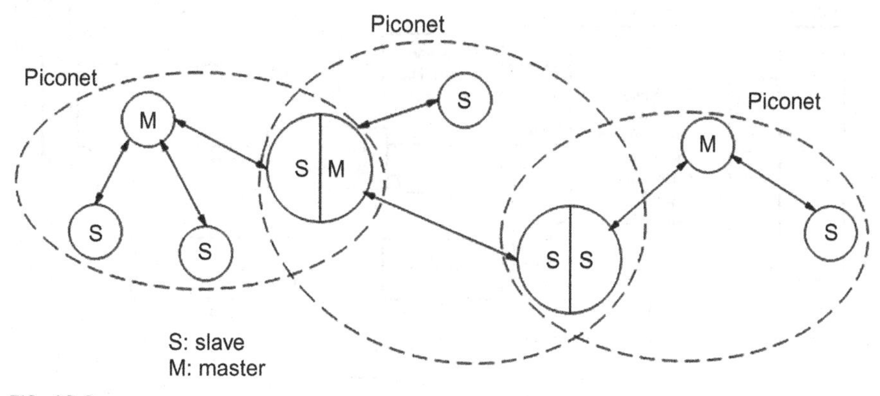


**FIG. 12.2**

Bluetooth scatternet.

**Table 12.1** Bluetooth communication characteristics

| Characteristic | Value | Comment |
|---|---|---|
| Frequency band | 2.4 to 2.483 GHz | May differ in some countries |
| Frequency hopping spread spectrum (FHSS) | [a]79 1-MHz channels from 2402 to 2480 MHz | May differ in some countries |
| Hop rate | 1600 hops per second | |
| Channel bandwidth | 1 MHz | 20 dB down at ±500 kHz edges |
| Modulation | Gaussian frequency shift keying (GFSK) | |
| | Filter BT = 0.5 | Gaussian filter bandwidth = 500 kHz |
| | Nominal modulation index = 0.32 | |
| Symbol rate | 1 Mbps | |
| Transmitter maximum power: | | |
| Class 1 | 100 mW | Power control required |
| Class 2 | 2.5 mW | Must be at least 0.25 mW |
| Class 3 | 1 mW | No minimum specified |
| Receiver sensitivity | −70 dBm for BER = 0.1% | |

[a] *The hopping pattern can be adapted—on a per-slave basis—to exclude a portion of the frequencies that are used by interfering devices. The adaptive hopping technique improves Bluetooth co-existence with static (non-hopping) ISM systems when they are co-located.*

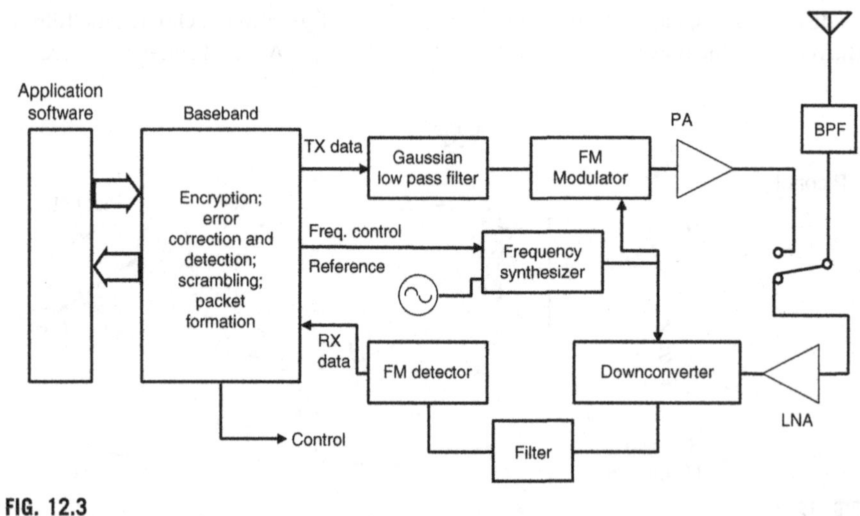


**FIG. 12.3**

Bluetooth transceiver.

will use low pass filters in complex $I$ and $Q$ outputs of the downconverter. While different manufacturers employ a variety of methods to implement the Bluetooth radio, all must comply with the same strictly defined Bluetooth specification, and therefore the actual configuration used in a particular chipset should be of little concern to the end user.

The Bluetooth protocol has a fixed-time slot of 625 μs, which is the inverse of the hop rate given in Table 12.1. A transmission burst may occur within a duration of one, three, or five consecutive slots on one hop channel. As mentioned, transmissions are always between the piconet master and a slave, or several slaves in the case of a broadcast, or point-to-multipoint transmission. All slaves in the piconet have an internal timer synchronized to the master device timer, and the state of this timer determines the transmission hop frequency of the master and that of the response of a designated slave. Fig. 12.4 shows a sequence of transmissions between a master and two slaves. Slots are numbered according to the state, or phase, of the master clock, which is copied to each slave when it joins the piconet. Note that master transmissions take place during even numbered clock phases and slave transmissions during odd numbered phases. Transmission frequency depends on the clock phase, and if a device makes a three or five slot transmission (slave two in the diagram), the intermediate frequencies that would have been used if only single slots were transmitted are omitted (*f*4 and *f*5 in this case). Note that transmissions, except on intermediate slots, do not take up a whole slot. Typically, a single-slot transmission burst lasts 366 μs, leaving 259 μs for changing the frequency of the synthesizer, phase locked loop settling time, and for switching the transceiver between transmit and receive modes.

There are two different types of wireless links associated with a Bluetooth connection. An asynchronous connectionless link (ACL) is used for packet data transfer while a synchronous connection oriented (SCO) link is primarily used for voice. There are two major differences between the two link types. When an SCO link is established between a master and a slave, transmissions take place on dedicated slots with a constant interval between them. Also, unlike an ACL link, transmitted frames are not repeated in the case of an error in reception. Both of these conditions are necessary because voice is a continuous real-time process whose data rate cannot be randomly varied without affecting intelligibility. On the other hand, packet data

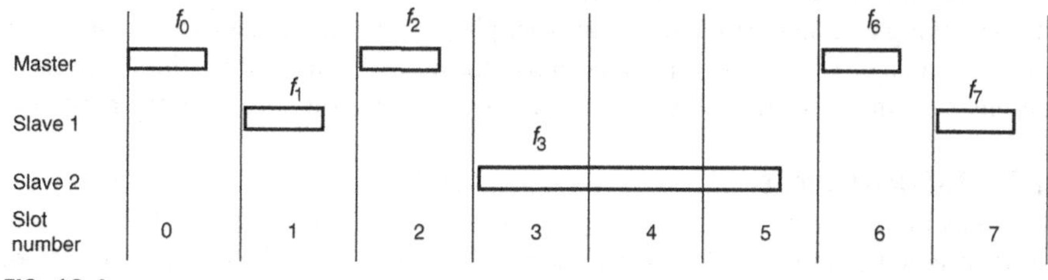


**FIG. 12.4**

Bluetooth timing.

transmission can use a handshaking protocol to regulate data accumulation and the instantaneous rate is not usually critical. Thus, for ACL links the master has considerable leeway in proportioning data transfer with the slaves in its network. An ARQ (automatic repeat request) protocol is always used, in addition to optional error correction, to ensure the highest reliability of the data transfer.

Bluetooth was conceived for employment in mobile and portable devices, which are more likely than not to be powered by batteries, so power consumption is an important issue. In addition to achieving low power consumption due to relatively low transmitting power levels, Bluetooth incorporates power saving features in its communication protocol. Low average power is achieved by reducing the transmission duty cycle, and putting the device in a low-power standby mode for as long a period as possible relative to transmit and receive times while still maintaining the minimum data flow requirements. Bluetooth has three modes for achieving different degrees of power consumption during operation: sniff, hold, and park. Even in the normal active mode, some power saving can be achieved, as described below.

#### *12.1.1.1 Active mode*

During normal operation, a slave can transmit in a particular time slot only if it is specifically addressed by the master in the proceeding slot. As soon as it sees that its address is not contained in the header of the master's message, it can "go to sleep," or enter a low-power state until it's time for the next master transmission. The master also indicates the length of its transmission (one, three, or five slots) in its message header, so the slave can extend its sleep time during a multiple slot interval.

#### *12.1.1.2 Sniff mode*

In this mode, sleep time is increased because the slave knows in advance the time interval between slots during which the master may address the slave. If it's not addressed during the agreed slot, it returns to its low-power state for the same period and then wakes up and listens again. When it is addressed, the slave continues listening during subsequent master transmission slots as long as it is addressed, or for an agreed time-out period.

#### *12.1.1.3 Hold mode*

The master can put a slave in the hold mode when data transfer between them is being suspended for a given period of time. The slave is then free to enter a low-power state, or do something else, like participate in another piconet. It still maintains its membership in the original piconet, however. At the end of the agreed time interval, the slave resynchronizes with the traffic on the piconet and waits for instructions from the master.

#### *12.1.1.4 Park mode*

Park has the greatest potential for power conservation, but as opposed to hold and sniff, it is not a directly addressable member of the piconet. While it is outside of direct calling, a slave in park mode can continue to be synchronized with the piconet and can rejoin it later, either on its own initiative or that of the master, in a manner that is faster than if it had to join the piconet from scratch. In addition to saving power, park mode can also be considered a way to virtually increase the network's capacity from eight devices to 255, or even more. When entering park mode, a slave gives up its active piconet address and receives an 8-bit parked member address. It goes into low-power mode but wakes up from time to time to listen to the traffic and maintain synchronization. The master sends beacon transmissions periodically to keep the network active. Broadcast transmissions to all parked devices can be used to invite any of them to rejoin the network. Parked units themselves can request re-association with the active network by way of messages sent during an access window that occurs a set time after what is called a "beacon instant." A polling technique is used to prevent collisions.

Park mode was removed in Version 5 of the Bluetooth specification.

#### *12.1.1.5 Packet format*

In addition to the data that originates in the high-level application software, Bluetooth packets contain fields of bits that are created in the baseband hardware or firmware for the purpose of acquisition, addressing, and flow control. Packet bits are also subjected to data whitening (randomization), error-correction coding, and encryption as defined for each particular data type. Fig. 12.5 shows the standard packet format.

The *access code* is used for synchronization, d-c level compensation, and identification. Each Bluetooth device has a unique address, and it is the address of the device acting as master that is used to identify transmitted packets as belonging to a specific piconet. A 64-bit synchronization word sandwiched between a four-bit header and four-bit trailer, which provide d-c compensation, is based on the master's address. This word has excellent correlation properties so when it is received by any of the piconet members it provides synchronization and positive identification that the packet of which it is a part belongs to their network. All message packets sent by members of the piconet use the same access code.

The *header* contains six fields with link control information. First, it has a three-bit active member address which indicates to which of the up to seven slaves a master's message is destined. An all zero address signifies a broadcast message to all slaves in the piconet. The next field has four bits that define the type of packet being sent. It specifies, for example, whether one, three, or five slots are occupied, and the level of error correction applied. The remaining fields involve flow control (handshaking), error detection and sequencing. Since the header has prime importance in the packet, it is endowed with forward-error correction having a redundancy of times three.

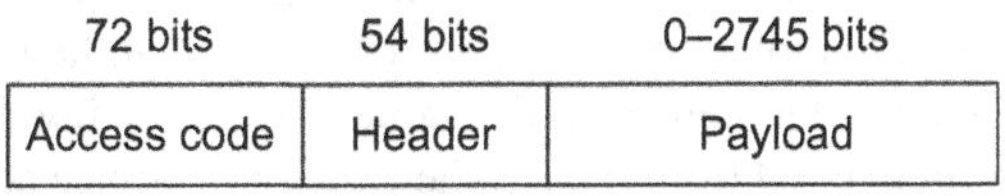

| 72 bits | 54 bits | 0–2745 bits |
|---|---|---|
| Access code | Header | Payload |

**FIG. 12.5**

Bluetooth packet.

Following the header in the packet is the *payload*, which contains the actual application or control data being transferred between Bluetooth devices. The contents of the payload field depend on whether the link is an ACL or SCO. The payload of ACL links has a payload header field that specifies the number of data bytes and also has a handshaking bit for data buffering control. A CRC (cyclic redundancy check) field is included for data integrity. As stated above, SCO links don't retransmit packets so they don't include a CRC. They don't need a header either because the SCO payload has a constant length.

The previous packet description covers packets used to transfer user data, but other types of packets exist. For example, the minimum length packet contains only the access code, without the four-bit trailer, for a total of 68 bits. It's used in the inquiry and paging procedures for initial frequency-hopping synchronization. There are also NULL and POLL packets that have an access code and header, but no payload. They're sent when slaves are being polled to maintain synchronization or confirm packet reception (in the case of NULL) in the piconet but there is no data to be transferred.

#### 12.1.1.6 Error correction and encryption

The use of forward error correction (FEC) improves throughput on noisy channels because it reduces the number of bad packets that have to be retransmitted. In the case of SCO links that don't use retransmission, FEC can improve voice quality. However, error correction involves bit redundancy so using it on relatively noiseless links will decrease throughput. Therefore, the application decides whether to use FEC or not.

As already mentioned, there are various types of packets, and the packet type defines whether or not FEC is used. The most redundant FEC method is always used in the packet header, and for the payload in one type of SCO packet. It simply repeats each bit three times, allowing the receiver to decide on the basis of majority rule what data bit to assign to each group of incoming bits.

The other FEC method, applied in certain type ACL and SCO packets, uses what's called a (15, 10) shortened Hamming code. For every ten data bits, five parity bits are generated. Since out of every 15 transmitted bits only ten are retrieved, the data rate is only two-thirds what it would be without coding. This code can correct all single errors and detect all double errors in each 15-bit code word.

Wireless communication is susceptible to eavesdropping so Bluetooth incorporates optional security measures for authentication and encryption. Authentication is a procedure for verifying that received messages are actually from the party we expect them to be and not from an outsider who is inserting false messages. Encryption prevents an eavesdropper from understanding intercepted communications, since only the intended recipient can decipher them. Both authentication and implementation routines are implemented in the same way. They involve the creation of secret keys that are generated from the unique Bluetooth device address, a PIN (personal identification number) code, and a random number derived from a random or pseudo-random process in the Bluetooth unit. Random numbers and keys are changed frequently. The length of a key is a measure of the difficulty of cracking a code.

Authentication in Bluetooth uses a 128-bit key, but the key size for encryption is variable and may range from 8 to 128 bits.

#### *12.1.1.7 Inquiry and paging*

A distinguishing feature of Bluetooth is its ad hoc protocol and connections are often required between devices that have no previous knowledge of their nature or address. Also, Bluetooth networks are highly volatile, in comparison to WLAN for example, and connections are made and dissolved with relative frequency. To make a new connection, the initiator—the master—must know the address of the new slave, and the slave has to synchronize its clock to the master's in order to align transmit and receive channel hop-timing and frequencies. The inquiry and paging procedures are used to create the connections between devices in the piconet.

By use of the inquiry procedure, a connection initiator creates a list of Bluetooth devices within range. Later, desired units can be summoned into the piconet of which the initiator is master by means of the paging routine. As mentioned previously, the access code contains a synchronization word based on the address of the master. During inquiry, the access code is a general inquiry access code (GIAC) formed from a reserved address for this purpose. Dedicated inquiry access codes (DIAC) can also be used when the initiator is looking only for certain types of devices. Now a potential slave can lock on to the master, provided it is receiving during the master's transmission time and on the transmission frequency. To facilitate this match-up, the inquiry procedure uses a special frequency hop routine and timing. Only 32 frequency channels are used and the initiator transmits two burst hops per standard time slot instead of one. On the slot following the transmission inquiry bursts, the initiator listens for a response from a potential slave on two consecutive receive channels whose frequencies are dependent on the previously transmitted frequencies.

When a device is making itself available for an inquiring master, it remains tuned to a single frequency for a period of 1.28 s and at a defined interval and duration scans the channel for a transmission. At the end of the 1.28-s period, it changes to another channel frequency. Since the master is sending bursts over the whole inquiry frequency range at a fast rate—two bursts per 1250 μs interval—there's a high probability the scanning device will catch at least one of the transmissions while it remains on a single frequency. If that channel happens to be blocked by interference, then the slave will receive a transmission after one of its subsequent frequency changes. When the slave does hear a signal, it responds during the next slot with a special packet called FHS (frequency hop synchronization) in which is contained the slave's Bluetooth address and state of its internal clock register. The master does not respond but notes the slave's particulars and continues inquiries until it has listed the available devices in its range. The protocol has provisions for avoiding collisions from more than one scanning device that may have detected a master on the same frequency and at the same time.

The master makes the actual connection with a new device appearing in its inquiry list using the page routine. The paging procedure is quite similar to that of the inquiry. However, now the master knows the paged device's address and can use it to form the synchronization word in its access code. The designated slave does its page scan while expecting the access code derived from its own address. The

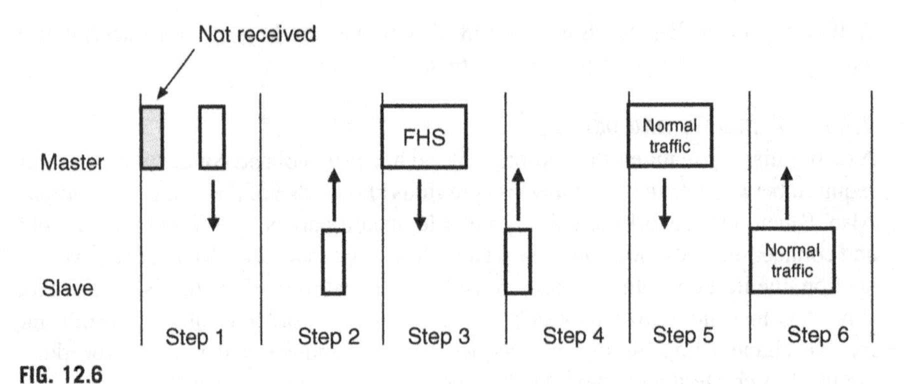


**FIG. 12.6**

Paging transmissions.

hopping sequence is different during paging than during inquiry, but the master's transmission bursts and the slave's scanning routine are very similar.

A diagram of the page state transmissions is given in Fig. 12.6. When the slave detects a transmission from the master (Step 1), it responds with a burst of access code based on its own Bluetooth address (Step 2). The master then transmits the FHS (Step 3), giving the slave the access code information (based on the master's address), timing and piconet active member address (between one and seven) needed to participate in the network. The slave acknowledges FHS receipt in Step 4. Steps 5 and 6 show the beginning of the network transmissions which use the normal 79 channel hopping-sequence based on the master's address and timing.

### 12.1.2 Enhanced data rate

Enhanced data rate operation, included in the Basic Rate/Enhanced Data Rate (BR/EDR) PHY, supports, in addition to the basic rate of 1 Mbps, a gross air bit rate of 2 or 3 Mbps. The packet format is different from that of the basic rate that was shown in Fig. 12.5. The EDR packet format is shown in Fig. 12.7. The access code and packet header transmission is at the basic rate of 1 Mbps with GFSK (Gaussian frequency shift keying). Following are a guard field of 5 μs duration to let the transmitter adjust to the new rate, then a synchronization sequence (Sync) consisting of a reference symbol followed by ten symbols with defined phases relative to the reference symbol. The enhanced data rates of 2 or 3 Mbps use differential phase shift keying (DPSK). The modulation format for 2 Mbps is pi/4 rotated differential encoded

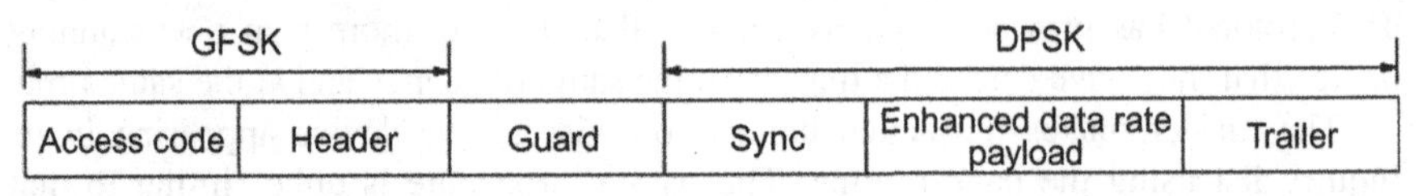


**FIG. 12.7**

Enhanced data rate packet format.

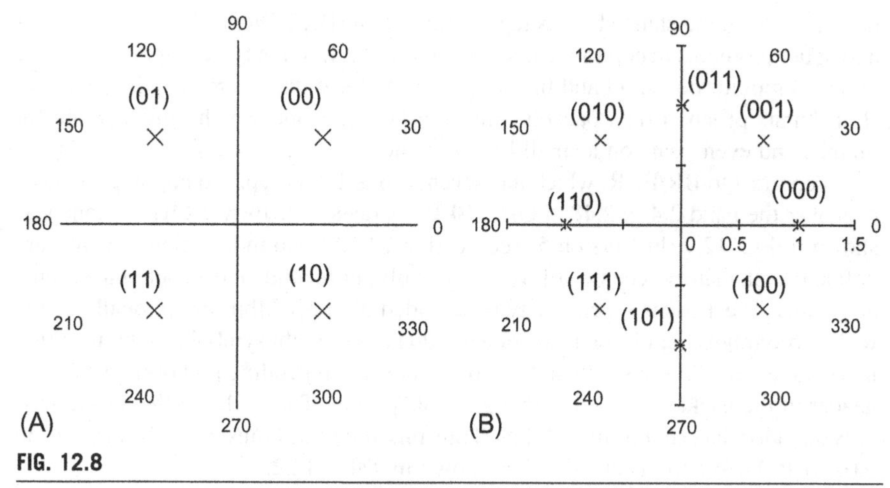


**FIG. 12.8**

EDR constellations. (A) pi/4-DQPSK 2 Mbps. (B) 8DPSK 3 Mbps.

quaternary phase shift keying (pi/4-DQPSK) and it is differential encoded 8-ary phase shift keying (8DPSK) for 3 Mbps. Fig. 12.8 shows constellation diagrams for the two enhanced data rates. Bit sequences for each symbol are shown in parentheses. The fact that both rates are differential encoded means that a symbol phase is the **difference** compared to the phase of the preceding symbol, not the absolute values shown in the figure. This simplifies the receiver which doesn't have to be phase locked to the transmitter in order to estimate the symbol phase. Both of the enhanced data rate modulation formats have a specified 2 MHz bandwidth, determined by a symbol rate of 1 Mbps. The spectrum mask shows 26 dB below carrier at ±1 MHz. The typical bandwidth, however, should be somewhat better, at around 26 dB down at carrier frequency ±700 kHz [1].

## 12.1.3 High speed operation

Using an Alternate MAC PHY (AMP) Bluetooth is capable of data rates up to 54 Mbps through an IEEE 802.11 physical connection. A protocol adaptation layer (PAL) provides the interface between Bluetooth and the 802.11 device. The Bluetooth radio, considered the primary radio, performs discovery, association, establishes the connection and maintains it. Once a high level connection through L2CAP (see Fig. 12.1) has been established between Bluetooth devices, the AMP managers can discover the alternate physical layer means for moving data traffic. Support for 802.11 Extended Rate PHY (ERP, 802.11b) is mandatory but other PHY types may be supported.

## 12.1.4 Bluetooth low energy

Bluetooth low energy (BLE) is intended for applications that have different requirements than the features provided by BR/EDR, using a physical layer with similarities to, but not compatible with, BR/EDR. However, BLE is easily integrated in the same device, or even the same chip, as legacy Bluetooth (BR/EDR) and BLE have lower and upper protocols in common, as shown in Fig. 12.1. Some typical applications are in health monitoring, sport and fitness, proximity location, and many others that use short bursts of small data packets, one way beacons, and which must operate for months and even years on a small lithium battery.

In contrast to BR/EDR, which has seventy nine 1 MHz spaced hopping frequencies over the band 2.4 to 2.4835 GHz, BLE operates over forty 2 MHz channels as shown in Fig. 12.9. In Version 5 (see Section 12.1.5) two modulation schemes are defined: a mandatory scheme at 1 Msps (symbols per second) and an optional scheme at 2 Msps. The 1 Msps scheme may be uncoded giving 1 Mbps or optionally coded with two symbols per bit for a data rate of 500 kbps or eight symbols per bit resulting in a data rate of 125 kbps. When the two symbols per bit coding is chosen, the access header in the packet is coded with 8 symbols per bit. The 2 Msps scheme supports only uncoded data supporting 2 Mbps. Transmission power must be in the range from 0.01 to 100 mW, and is classified as shown in Table 12.2.

Modulation type is GFSK with bandwidth- bit period product BT=0.5 and an modulation index (instantaneous frequency deviation from the carrier divided by one half the symbol rate) between 0.45 and 0.55. Bandwidth is specified as having spurious power of no more than -20 dBm at 2 MHz offset from the channel center frequency for the 1 Msym/s rate and—20 dBm at 4 MHz offset for the 2 Msym/s rate [1].

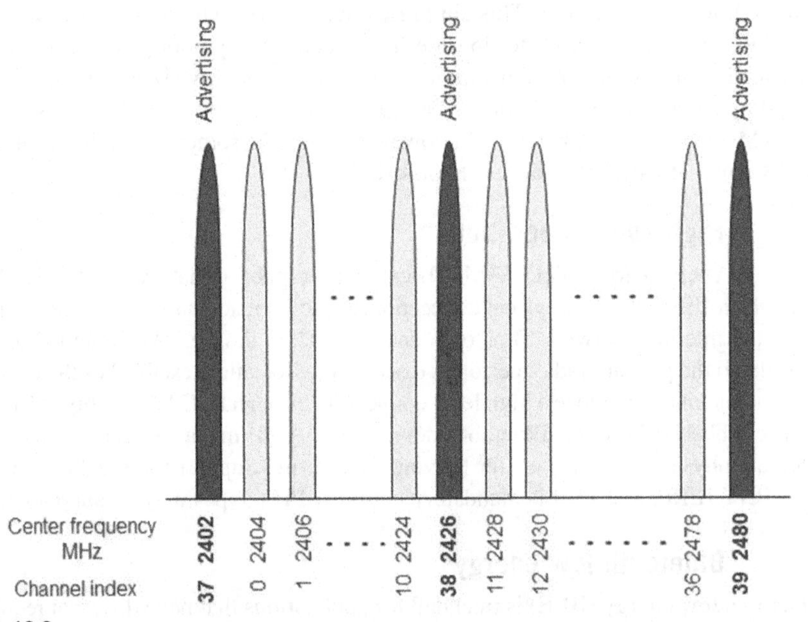


**FIG. 12.9**

Bluetooth low energy frequency channels.

**Table 12.2** BLE power classes

| Power class | Maximum output power | Minimum output power in class |
|---|---|---|
| 1 | 100 mW (+20 dBm) | 10 mW (+10 dBm) |
| 1.5 | 10 mW (+10 dBm) | 0.01 mW (−20 dBm) |
| 2 | 2.5 mW (+4 dBm) | 0.01 mW (−20 dBm) |
| 3 | 1 mW (0 dBm) | 0.01 mW (−20 dBm) |

BLE devices can have four different roles: broadcaster, observer, peripheral, and central. The role of a device dictates the requirements of its Controller (see LE controller in Fig. 12.1). A *Broadcaster* is designed for transmitter only applications, for example one-way beacons, whereas an *Observer* supports receiving only. *Peripheral* devices have low complexity and support a single connection. The most complex role is that of a *Central* device which can initiate and control multiple connections with peripherals. A smartphone is an example of a Central device, while a Peripheral device may be a low power sensor that connects to the central device, or a low power beacon.

The time domain physical channel is sub-divided into time units known as events. Data are transmitted between LE devices in packets that are positioned in these events. There are four types of events: advertising, extended advertising, periodic advertising, and connection events. Advertising and connection events are explained here.

Advertising packets are continuously transmitted from a peripheral device in order to be seen by other devices. Devices that transmit advertising packets on one or more of the three advertising PHY channels are referred to as advertisers. Advertisement provides a way for devices to broadcast their presence while making possible the setting up of connections. It can include a list of supported services and transmitter power levels. Parameters associated with the advertisement event are:

**(a)** advertisement interval—the interval between advertisement events which can be 20 ms to 10240 ms.
**(b)** which of three channels, index numbers 37, 38, and 39, or all of them, are used for the advertisement packets.
**(c)** discoverability—how the advertiser is visible to other devices.
**(d)** connectablility—whether or not the advertiser can be connected to other devices point-to-point.
**(e)** payload—number of bytes, 0 to 31 that can be included in an advertisement packet.

Devices that receive advertising packets on the advertising channels without the intention to connect to the advertising device are referred to as scanners. Scanning can be passive or active. In passive scanning the scanner listens for advertisements packets from one channel to another. Active scanning means that the scanner sends a scan request packet to the advertiser to get more information about its services.

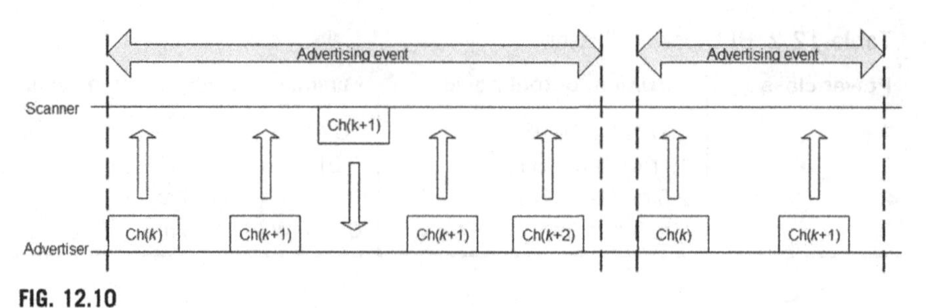


**FIG. 12.10**

BLE advertising events.

Transmissions on the advertising PHY channels occur in *advertising* events. See Fig. 12.10. At the start of each advertising event, the advertiser sends an advertising packet corresponding to the advertising event type. Depending on the type of advertising packet, the scanner may make a request to the advertiser on the same advertising PHY channel which may be followed by a response from the advertiser on the same advertising PHY channel. The advertising PHY channel changes on the next advertising packet sent by the advertiser in the same advertising event. The advertiser may end the advertising event at any time during the event. The first advertising PHY channel is used at the start of the next advertising event.

A *connection* is a continuing periodical data exchange of packets between two devices. Connections allow application data to be sent from one device to another. A connection event is a periodical exchange of data at certain specific points in time, shown in Fig. 12.11. First, a peripheral device sends connectable advertising packets periodically during advertising events. A central device, called *initiator*, scans the frequencies for connectable advertising packets, and when suitable, initiates a connection. Once the connection is established through acceptance by the advertiser, the initiator becomes the master device of a piconet and the advertiser becomes the slave device. The master manages the timing and initiates the beginning of each connection event. Adaptive frequency hopping (AFH) is used for robust transmission and reception. The connection event starts when the master sends a packet to the slave at

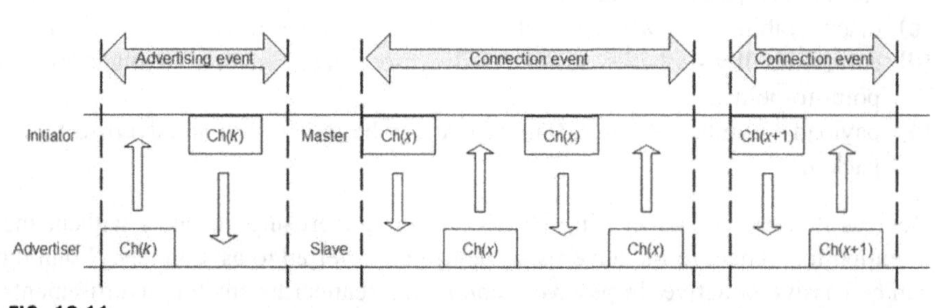


**FIG. 12.11**

BLE connection events.

a defined connection interval. The slave can respond 150 μs after it has received a packet from the master. If the slave has no data to send it can skip a certain number of connection events defined by a slave latency parameter. If no packets are received by the master or slave within the time defined by the supervision timeout, the connection is terminated. A key advantage of this arrangement is power saving - two devices can power up, exchange data, and then go to sleep until the next connection event.

Device master/slave pairs in a piconet each use a specific frequency hopping pattern, which is algorithmically determined by a field contained in the connection request sent by an initiating device. The hopping pattern used in LE is a pseudo-random ordering of the 37 non-advertising frequencies in the 2.5 GHz band (Fig. 12.9). The hopping pattern can be adapted to exclude a portion of the frequencies that are used by interfering devices. The adaptive hopping technique improves Bluetooth co-existence with static (non-hopping) ISM systems when these are co-located and there is access to information about the local radio environment, or interference is detected by other means.

Above the physical channel there are concepts of links, channels and associated control protocols (shown in Fig. 12.1). The hierarchy is physical channel, physical link, logical transport, logical link, and L2CAP channel. Within a physical channel, a physical link is formed between devices. As in legacy Bluetooth, the active physical link provides bidirectional packet transport between the master and slave devices. An LE physical channel can include multiple slave devices. There is a physical link between each slave and master. Slaves are permitted to have physical links to more than one master at a time and there is virtually no limit to the number of slaves associated with a master. A device is permitted to be master in one piconet and slave in another at the same time. Physical links are not formed directly between the slaves in a piconet. Role changes between a master and slave device are not presently supported. The advertising and periodic physical links provide a unidirectional packet transport from the advertiser to a potentially unlimited number of scanners or initiators.

The physical link is used as a transport for one or more logical links that support asynchronous traffic. Traffic on logical links is multiplexed onto the physical link assigned by a scheduling function in the Baseband Resource Manager (Fig. 12.1).

A control protocol for the link and physical layers is carried over logical links in addition to user data. This is the link layer protocol (LL). The Link Layer function uses the LL protocol to control the operation of devices in the piconet and provide services to manage the lower architectural layers (PHY and LL). Just as in BR/EDR, above the link layer the L2CAP layer provides a channel based abstraction to applications and services. It carries out fragmentation and de-fragmentation of application data and multiplexing and de-multiplexing of multiple channels over a shared logical link. L2CAP has a protocol control channel that is carried over the primary ACL logical transport.

In addition to L2CAP, LE provides two additional protocol layers that reside on top of L2CAP. The security manager protocol (SMP) uses a fixed L2CAP channel to implement the security functions between devices. The other is the attribute protocol

(ATT) that provides a method to communicate small amounts of data over a fixed L2CAP channel. The attribute protocol is also used by devices to determine the services and capabilities of other devices. The attribute protocol may also be used over BR/EDR.

### 12.1.5 Bluetooth 5

Three basic new performance features for BLE were made available in the Core Package Version 5.0 specification published 6 December 2016 [1].

1. *Increased data rate.* An optional capability of increasing the raw data rate to 2 Mbps was introduced. Not only can more data be transmitted in a given time, but transmission time for a given amount of data is reduced, thereby reducing the duty cycle and consequentially the power consumption. Range is reduced by approximately 20% [2].
2. *Increased range.* Two coding modes can be selected for use at the 1 Msps rate. Based on a FEC convolution encoder longer range is achieved through the ability to operate without error with reduced signal to noise ratio. In what is called an $S = 2$ coding scheme, two symbols are used for each bit, providing redundancy for error correction. The bit rate is reduced to 500 Mbps. In the $S = 8$ scheme, eight symbols make one bit, giving a bit rate of 125 kbps. An approximate range multiplication of two is gained by $S = 2$ and of 4 by $S = 8$ [2].
3. *Advertising extensions.* Major extensions to the advertising mode allow much larger amounts of data to be broadcast in connectionless scenarios, and also give significant improvements regarding contention and duty cycle. Bluetooth beacons can broadcast many types of information, beyond an ID and location parameter. Advertising packets can be up to 255 bytes long, with payloads sent on the 37 channels previously reserved for connection events only, and multiple packets can be chained and sent on different channels, for even greater total payload lengths. Small headers are used, reducing the duty cycle. For situations where rapid recognition and response to advertising packets is required, the minimum advertising interval was reduced to 20 ms.

In addition to the above, Bluetooth 5 includes means to identify potential interference from adjacent bands, such as those used by cellular networks (LTE). A system called slot availability masks was introduced, so that Bluetooth can indicate the availability of its time slots and synchronize in an optimum manner with the traffic in adjacent bands [2].

### 12.1.6 Bluetooth profiles

Bluetooth profiles assure interoperability of applications in Bluetooth. The profiles define the required functions and features of each layer from PHY to L2CAP and other protocols outside the core specification. A profile defines vertical interactions

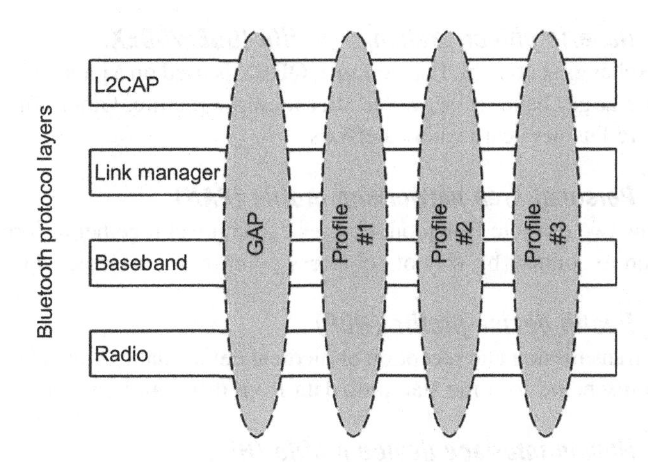


**FIG. 12.12**

Bluetooth profiles.

between layers as well as peer-to-peer interactions of specific layers between devices (Fig. 12.12) [1].

The generic access profile (GAP), the base profile which all Bluetooth devices implement, defines the basic requirements of a Bluetooth device. It describes methods for device discovery, connection establishment, security, authentication, association models and service discovery. In LE, GAP defines the four specific roles of broadcaster, observer, peripheral and central, which are described above. Another inherent profile is generic attribute profile (GATT). GATT defines the way that two Bluetooth Low Energy devices send and receive standard messages.

For two Bluetooth devices to be compatible, they must support the same profiles. There are a number of Bluetooth BR/EDR adopted profiles which describe commonly used application types. Bluetooth LE application developers can use a comprehensive set of adopted profiles, or they can use GATT to create new profiles. A sampling of the more common profiles are described briefly here [3].

#### 12.1.6.1 Serial port profile (SPP)

Emulates a RS-232 serial port. SPP provides an easy way to substitute Bluetooth wireless connectivity between devices which would normally be connected by RS-232 connectors and cables.

#### 12.1.6.2 Hands-free profile (HFP)

Used to place and receive calls for a hands-free device, for example, between a mobile phone and a car display. It provides remote control functions, including volume, and transfer of voice data using an audio codec. A profile with similar uses but less control functionability is the headset profile (HSP).

#### 12.1.6.3 Generic object-exchange profile (GOEP/OBEX)

Used for exchanging content. For example, OBEX is used on a phone camera to send a photo to a target handset or printer after taking a picture. Other objects for data exchange are business cards and calendars.

#### 12.1.6.4 Personal area networking profile (PAN)

Defines how two or more Bluetooth devices can connect together to form an ad hoc network and to connect by way of an access point to a remote network.

#### 12.1.6.5 Health device profile (HDP)

Facilitates transmission and reception of medical device data. It includes a time synchronization scheme for time stamping data from different body worn sensors.

#### 12.1.6.6 Human interface device profile (HID)

Enables mice and keyboards to use Bluetooth. It also provides support for simple buttons and indicators, and gaming devices. It is notable for low latency and reduced power consumption.

#### 12.1.6.7 Advanced audio distribution profile (A2DP)

Enables high quality mono or stereo music to be streamed to a headset. A2DP implementations must support a basic codec known as the sub-band codec (SBC). The profile include other, optional, codecs, including MPEG-1, MPEG-2 audio, MPEG-2 AAC and MPEG4 AAC, and Sony's ATRAC codecs. A related profile is the audio video remote-control profile (AVRCP) which allows track, volume and play commands to be controlled wirelessly from the headset.

## 12.2 Conflict and compatibility between Bluetooth and Wi-Fi

With the steep rise of Bluetooth product sales and the already large and growing use of WLAN, there is considerable concern about mutual interference between Bluetooth-enabled and Wi-Fi devices. Both occupy the 2.4 to 2.4835 GHz unlicensed band and use wideband spread-spectrum modulating techniques. They will most likely be operating concurrently in the same environments, particularly office/commercial but also in the home. Much of the material in this section is from [4]. It is also relevant to IEEE 802.15.4 devices in the 2.4 GHz band.

### 12.2.1 The interference problem

Interference can occur when a terminal of one network transmits on or near the receiving frequency of a terminal in another collocated network with enough power to cause an error in the data of the desired received signal. Although they operate on the same frequency band, the nature of Bluetooth and Wi-Fi signals are very different. Bluetooth has a narrowband transmission of 1 MHz bandwidth which hops

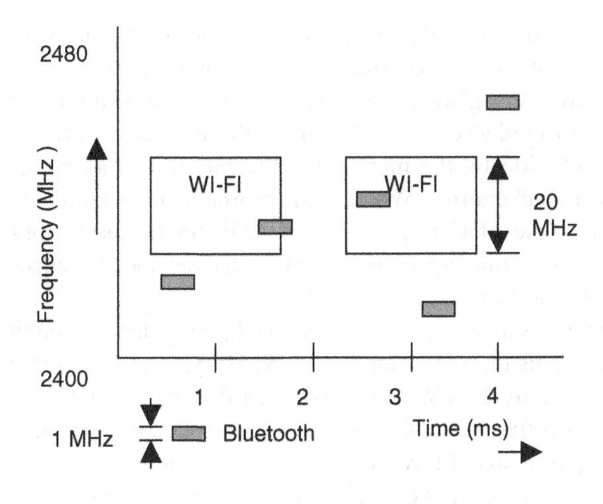


**FIG. 12.13**

Wi-Fi and Bluetooth spectrum occupation.

around pseudo-randomly over an 80 MHz band while Wi-Fi (using DSSS) has a broad, approximately 20 MHz, bandwidth that is constant in some region of the band. The interference phenomenon is apparent in Fig. 12.13. Whenever there is a frequency and time coincidence of the transmission of one system and reception of the other, it's possible for an error to occur. Whether it does or not depends on the relative signal strengths of the desired and undesired signals. These in turn depend on the radiated power outputs of the transmitters and the distance between them and the receiver. When two terminals are very close (on the order of centimeters), interference may occur even when the transmitting frequency is outside the bandwidth of the affected receiver.

Bluetooth and Wi-Fi systems are not synchronous and interference between them has to be quantified statistically [4]. We talk about the probability of a packet error of one system caused by the other system. The consequence of a packet error is that the packet will have to be retransmitted once or more until it is correctly received, which causes a delay in message delivery and data throughput. Voice transmissions generally don't allow packet retransmission because throughput cannot be delayed, so interference results in a decrease in message quality.

Following are parameters that affect interference between Bluetooth and Wi-Fi:

- *Frequency and time overlap.* A collision occurs when the interferer transmits at the same time as the desired transmitter and is strong enough to cause a bit or symbol error in the received packet.
- *Packet length.* The longer the packet length of the Wi-Fi system, relative to a constant packet length and hop rate of Bluetooth, the longer the victim may be exposed to interference from one or more collisions and the greater the probability of a packet error.

- *Bit rate.* Generally, the higher the bit rate, the lower the receiver sensitivity and therefore the more susceptible the victim will be to packet error for given desired and interfering signal strengths. On the other hand, higher bit rates usually result in reduced packet length, with the opposite effect.
- *Use factor.* Obviously, the more often the interferer transmits, the higher the probability of packet error. When both communicating terminals of the interferer are in the interfering vicinity of the victim the use factor is higher than if the terminals are further apart and one of them does not have adequate strength to interfere with the victim.
- *Relative distances and powers.* The received power depends on the power of the transmitter and its distance. Generally, Wi-Fi systems use more power than Bluetooth, typically 20 mW compared to 1 mW. Bluetooth Class 1 systems may transmit up to 100 mW, but their output is controlled to have only enough power to give a required signal level at the receiving terminal.
- Signal-to-interference ratio of the victim receiver, SIR, for a specified symbol or frame error ratio.
- Type of modulation, and whether error-correction coding is used.

A general configuration for the location of Wi-Fi and interfering Bluetooth terminals is given in Fig. 12.14. In this discussion, only transmissions from the access point to the mobile terminal are considered. We can get an idea of the vicinity around the Wi-Fi mobile terminal in which operating Bluetooth terminals will affect transmissions from the access point to the mobile terminal by examining the following parameters:

*C/cc, C/ac*—Ratio of signal carrier power to co-channel or adjacent channel interfering power for a given bit or packet error rate (probability).
$P_{WF}$, $P_{BT}$—Wi-Fi and Bluetooth radiated power outputs.

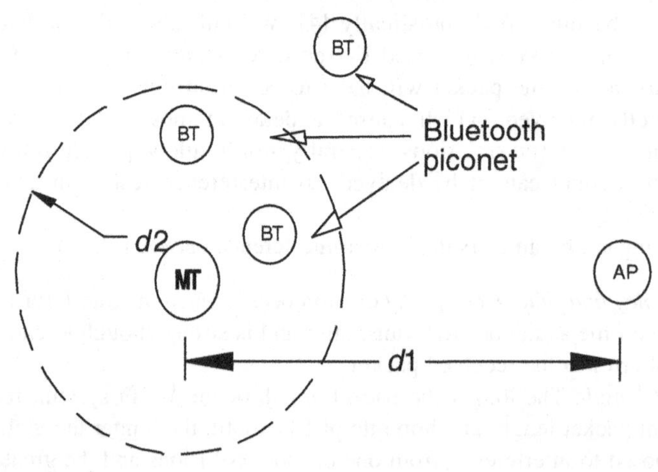


**FIG. 12.14**

Importance of relative terminal location.

$PL = Kd^r$—Path loss which is a function of distance $d$ between transmitting and receiving terminals, and the propagation exponent $r$. $K$ is a constant.
$d1$—Distance between Wi-Fi mobile terminal and access point.
$d2$—Radius of area around mobile terminal within which an interfering Bluetooth transmitter signal will increase the Wi-Fi bit error rate above a certain threshold.
$PR_{WF}$, $PR_{BT}$—Received powers from the access point and from the Bluetooth interfering transmitters.

$d2$ as a function of $d1$ is found as follows, using power in dBm:

**(1)** $PR_{WF} = P_{WF} - 10\log(Kd1^r)$; $PR_{BT} = P_{BT} - 10\log(Kd2^r)$
**(2)** $Clcc = PR_{WF} - PR_{BT} = P_{WF} - 10\log(Kd1^r) - P_{BT} + 10\log(Kd2^r)$
**(3)** $(Clcc - P_{WF} + P_{BT})/10r = \log(d2/d1)$

Then:

$$d2 = d1 \cdot 10^{(CI_{cc} - P_{WF} + P_{BT})/(10 \cdot r)} \quad (12.1)$$

As an example, the interfering area radius $d2$ will be calculated from Eq. (12.1) using the following system parameters:

$$Clcc = 10\,\text{dB}, P_{WF} = 13\,\text{dBm}, P_{BT} = 0\,\text{dBm}, r = 2\,(\text{free space})$$

Solving Eq. (12.1): $d2 = d1 \times 0.71$.

In this case, if a Wi-Fi terminal is located 15 m from an access point, for example, all active Bluetooth devices within a distance of 10.6 m from it have the potential of interfering. Only co-channel interference is considered. Adjacent channel interference, if significant, would increase packet error probability because many more Bluetooth hop channels would cause symbol errors. However, the adjacent channel $Clac$ is on the order of 45 dB lower than $Clcc$ and would be noticed only when Bluetooth is several centimeters away from the Wi-Fi terminal.

The effect of an environment where path loss is greater than in free space can be seen by using an exponent $r = 3$. For the same Wi-Fi range of 15 m, the radius of Bluetooth interference becomes 11.9 m.

While Eq. (12.1) does give a useful insight into the range where Bluetooth devices are liable to deteriorate Wi-Fi performance, its development did involve simplifications. It considered that the signal-to-interference ratio that causes the error probability to exceed a threshold is constant for all wanted signal levels, which isn't necessarily so. It also implies a step relationship between signal-to-interference ratio and performance degradation, whereas the effect of changing interference level is continuous. The propagation law used in the development is also an approximation.

### 12.2.2 Methods for improving Bluetooth and Wi-Fi coexistence

By dynamically modifying one or more system operating parameters according to detected interference levels, coexistence between Bluetooth and Wi-Fi can be improved. Some of these methods are discussed below.

#### 12.2.2.1 Power control

Limiting transmitter power to the maximum required for a satisfactory level of performance will reduce interference to collocated networks. Power control is mandatory for Class 1 Bluetooth systems, where maximum power is 100 mW. The effect of the power on the interference radius is evident in Eq. (12.1). For example, in a Bluetooth piconet established between devices located over a spread of distances from the master, the master will use only the power level needed to communicate with each of the slaves in the network. Lack of power control would mean that all devices would communicate at maximum power and the collocated Wi-Fi system would be exposed to a high rate of interfering Bluetooth packets.

#### 12.2.2.2 Adaptive frequency hopping

Wi-Fi and Bluetooth share approximately 25% of the total Bluetooth hop-span of 80 MHz. Probably the most effective way to avoid interference between the two systems is to restrict Bluetooth hopping to the frequency range not used by Wi-Fi. When there is no coordination or cooperation between collocated networks, the Bluetooth piconet master senses the presence of Wi-Fi transmissions and modifies the frequency-hopping scheme of the network accordingly. A minimum of 20 hop frequencies must be used.

There are situations where AFH may not be effective or may have a negative effect. When two or more adjacent Wi-Fi networks are operating concurrently, they will utilize different 22 MHz sections of the 2.4 GHz band—three non-overlapping Wi-Fi channels are possible. In this case, Bluetooth may not be able to avoid collisions while using a minimum of 20 hop frequencies. In addition, if there are several Bluetooth piconets in the same area, collisions among themselves will be much more frequent than when the full 79 channel hopping sequences are used.

#### 12.2.2.3 Packet fragmentation

The two interference-avoiding methods described above are applicable primarily for action by the Bluetooth network. One method that the Wi-Fi network can employ to improve throughput is packet fragmentation. By fragmenting data packets and sending more, but shorter transmission frames, each transmission will have a lower probability of collision with a Bluetooth packet. Although reducing frame size increases the percentage of overhead bits in the transmission, when interference is heavy the overall effect may be higher throughput than if fragmentation was not used. Increasing bit rate for a constant packet length will also result in a shorter transmitted frame and less exposure to interference.

The methods mentioned above for reducing interference presume no coordination between the two different types of collocated wireless networks. However, devices are now being produced, in laptop and notebook computers for example, that include both Wi-Fi and Bluetooth, sometimes even in the same chipset. In this case collaboration is possible in the device software to prevent inter-network collisions.

## 12.3 IEEE 802.15.4

IEEE specification 802.15.4 is usually associated with Zigbee, but it should be seen as a separate entity. It was drawn up by a working group of the IEEE as a low data rate WPAN which describes the physical and data link layers that typically address remote monitoring and control applications. An association of committed companies, the Zigbee Alliance, has defined the network, security, and application layers above the 802.15.4 physical and medium access control (MAC) layers, and deals with interoperability certification and testing. The reference specification for the descriptions below is IEEE 802.15.4-2015 [5].

The distinguishing features of 802.15.4 to which the IEEE standard addresses itself are

- Low data rates—raw rates between 20 and 250 kbps
- Low power consumption—several months up to two years on standard primary batteries
- Network topology appropriate for multisensor monitoring and control applications
- Low complexity for low cost and ease of use
- Very high reliability and security

These lend themselves to wide-scale use embedded in consumer electronics, home and building automation and security systems, industrial controls, PC peripherals, medical and industrial sensor applications, toys and games and similar applications. It's natural to compare Zigbee, which is based on 802.15.4, with the other WPAN standard, Bluetooth, and there will be some overlap in implementations. However, the two systems are quite different, as is evident from the comparison in Table 12.3.

### 12.3.1 Architecture

The basic protocol architecture of Zigbee and 802.15.4, shown in Fig. 12.15, is similar to that of other IEEE standards, Wi-Fi and Bluetooth for example. On the bottom are the physical layers, showing two alternative options for the RF transceiver functions of the specification. Both of these options are not expected to exist in a single device, and indeed their transmission characteristics—frequencies, data rates, modulation system—are quite different. However, the embedded firmware and software layers above them will be essentially the same no matter what physical layer is applied. Just above the physical layers (PHY) is the MAC layer which is responsible for management of the physical layer and delivering data to and from the applications through the layer above it. Among its functions are channel access, keeping track of slot times, and message delivery acknowledgement. The higher layers are not part of IEEE 802.15.4. The principle functions of the network layer are forming a network, managing association with the network and disassociation from it, and routing. Its functions and the application services above it are defined by the Zigbee Alliance.

**Table 12.3** Comparison of Zigbee and Bluetooth

| | Bluetooth | Zigbee |
|---|---|---|
| Transmission scheme | FHSS (frequency hopping spread spectrum) | DSSS (direct sequence spread spectrum) |
| Modulation | GFSK (Gaussian frequency shift keying), $\pi/4$ Quadrature and 8 angle phase shift keying | O-QPSK (offset quadrature phase shift keying) or BPSK (binary phase shift keying) |
| [a]Frequency band | 2.4 GHz | 2.4 GHz, 915 MHz, 868 MHz |
| Raw data bit rate | 1, 2, 3 Mbps | 250, 100, 40, or 20 kbps (depends on frequency band and modulation) |
| Power output | Maximum 100, 2.5, or 1 mW, depending on class | Minimum capability 0.5 mW; maximum as allowed by local regulations |
| Minimum sensitivity | −70 dBm for 0.1% BER | −85 dBm (2.4 GHz) or −92 dBm (915/868 MHz) for packet error rate <1% |
| Network topology | Master-slave 8 active nodes, no fixed limit for LE | Star or peer to peer up to 64k active nodes |

[a] *IEEE 802.15.4 also offers additional frequency bands, specifically for China and Japan and for UWB (ultra-wideband), and other modulation types.*

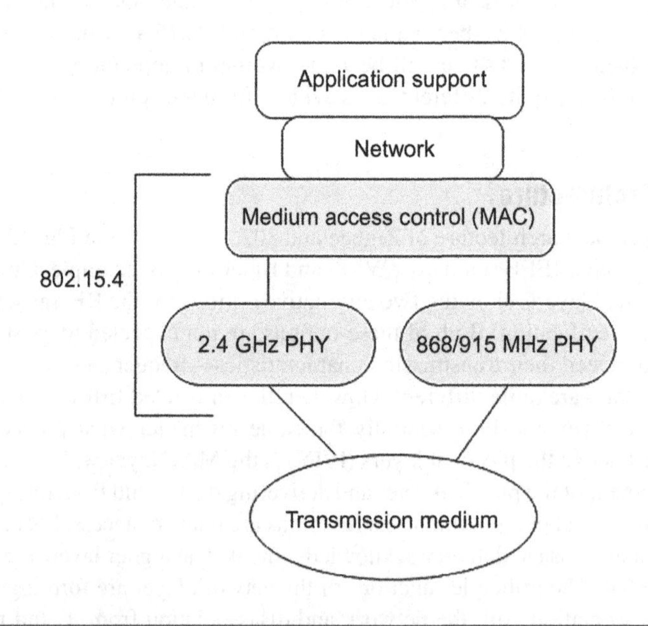


**FIG. 12.15**

802.15.4/Zigbee architecture.

Application software is not a part of the IEEE 802.15.4 specification. The Zigbee Alliance prepares profiles, or programming guidelines and requirements for various functional classes in order to assure product interoperability and vendor independence. These profiles define network formation, security, and application requirements while keeping in mind the basic Zigbee features of low power and high reliability.

### 12.3.2 Communication characteristics

In order to achieve high flexibility of adaptation to the range of applications envisioned for Zigbee, and other application classes, operation is specified primarily for three unlicensed bands—2.4 GHz, 915 MHz, and 868 MHz; the latter two being included in the same physical layer. Those two bands are generally mutually exclusive, their use being determined by geographic location and regional regulations. The following 28 transmitting channels are defined for those bands:

| Channel number | Center frequency range | Channel width |
|---|---|---|
| 0 | 868.3 MHz | 600 kHz |
| 1 to 10 | 906 to 924 MHz | 2 MHz |
| 11 to 27 | 2405 to 2480 MHz | 5 MHz |

In both physical layers, the modulation is DSSS (direct sequence spread spectrum) using either BPSK or O-QPSK (offset quadrature phase shift keying). The spreading parameters are defined to meet communication authority regulations in the various regions as well as desired data rates. For example, the chip rate of 600 kbps on the 902-928 band with BPSK modulation allows the transmission to meet the FCC paragraph 15.247 requirement of minimum 500 kHz bandwidth at 6 dB down for digital modulation. However the chip rate, and with it the data rate, has to be reduced on Channel 0 in order to meet the confines of the 868 to 868.6 MHz channel allowed under ERC recommendation 70-03 and ETSI specification EN 300-220. On the 2400 to 2483.5 MHz band, the bit rate of 250 kbps with O-QPSK allows a throughput, after considering the overheads involved in packet transmissions, to attain 115.2 kbps, a rate used for some PC peripherals for example.

O-QPSK is shown in Fig. 12.16. On the 2.4 GHz band sixteen different, almost orthogonal 32 chip long spreading sequences are available for transmission at 2 Mchips/s. In Fig. 12.16A, each consecutive sequence of four data bits, constituting a symbol, determines which of the sixteen spreading sequences is sent. Each chip is shaped as a half sign wave, as shown in Fig. 12.16B. On reception, the receiver can identify the spreading sequence, for example using a bank of matched filters, and thus decode the four data bits. O-QPSK with half-sine wave pulse shaping is similar to a frequency shift keying variant, MSK (minimum shift keying). It is fairly easy to generate and has relatively low spurious components, although a somewhat larger main lobe bandwidth compared to regular QPSK. Its main advantage over QPSK

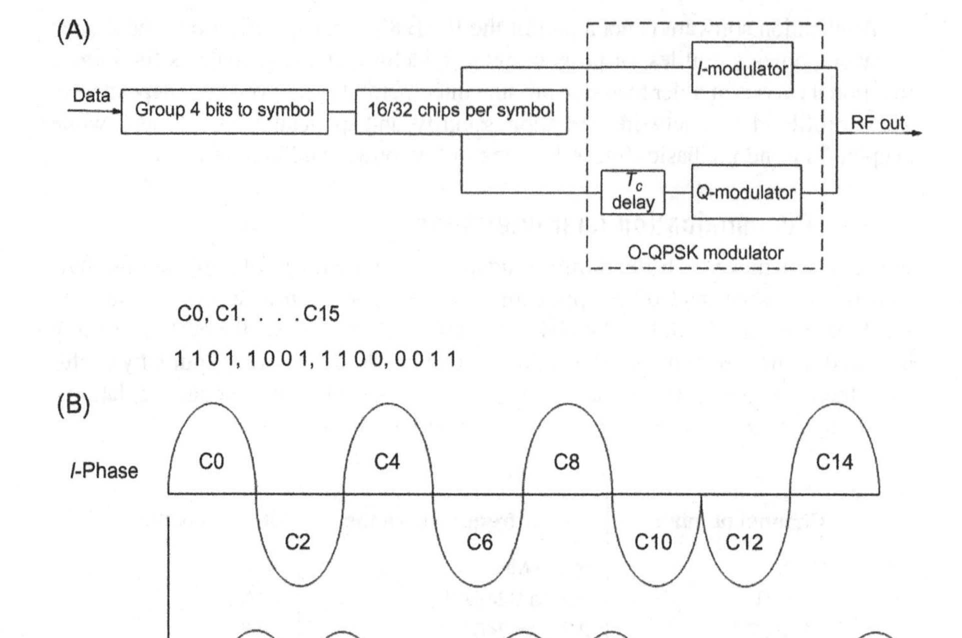


**FIG. 12.16**

O-QPSK modulation. (A) Spreading and modulation. (B) First 16 chips of spreading sequence with half sine wave shaping and quadrature offset.

is a constant time-domain signal envelope due to smooth phase transitions which never go through zero. Constant envelope simplifies transmitter power amplifier requirements as high linearity is not needed and higher efficiency can be achieved. O-QPSK modulation is defined as an option on the 915 and 868 MHz bands where 16 chip spreading sequences are used.

A multitude of other bands and a variety of modulation schemes are described in Ref. [5] from 160 MHz through 10 GHz. Specifications are classified according to frequency bands, user categories and modulation. Table 12.4 gives examples of RF characteristics of some of the 19 physical layers defined in the specification which demonstrate the range of possibilities of basing short-range low power wireless communication systems on IEEE 802.15.4.

The devices must be capable of radiating at least –3 dBm although output may be reduced to the minimum necessary in order to limit interference to other users. Maximum power is determined by the regulatory authorities. While much higher powers

**Table 12.4** Physical layers (PHY) and parameters

| PHY | Frequency bands (MHz) | Chip rate (if DSSS) (kcps) | Data rate (kbps) | [a]Minimum receiver sensitivity (dBm) |
|---|---|---|---|---|
| O-QPSK (orthogonal phase shift keying) | 2450, 2380 | 2000 | 250 | −85 |
| | 915, 780 | 1000 | 250 | −85 |
| | 868 | 400 | 100 | −85 |
| BPSK (binary phase shift keying) | 915 | 600 | 40 | −92 |
| | 868 | 300 | 20 | -92 |
| CSS (chirp spread spectrum) | 2450 | | 1000, 250 (optional) | -85, -91 |
| HRP (high rate pulse repetition frequency) UWB | 249.6 to 749.6 and 3.1 GHz to 4.8 GHz and 6 GHz to 10.6 GHZ | Pulse modulation | 0.11 Mbps to 27.24 Mbps | |
| LRP (low rate pulse repetition frequency) UWB | 3 bands between 6.2896 and 9.1856 | Pulse modulation | 31.25, 250, 1000 | |
| MSK (minimum shift keying) | 433 | | 31.25, 100, 250 | |

[a] *These limits are for a packet error rate of 1%.*

are allowed, it may not be practical to transmit over, say 10 dBm, because of absolute limits on spurious radiation and the general objective of low-cost and low-power consumption.

## 12.3.3 Device types and topologies

Two device types, of different complexities, are defined. A full function device (FFD) implements the full protocol set and acts as a network coordinator. Devices capable of minimal protocol implementation are reduced function devices (RFD). Due to the distinction between device types, networks in which most members require only minimum functionality, such as switches and sensors, can be made significantly cheaper and have lower power consumption than if all devices were constrained to have maximum capability.

Flexibility in network configuration is achieved through two topologies—star and peer-to-peer that are depicted in Fig. 12.17. A network may have up to 64k members, one of which is a PAN (personal area network) coordinator. The function of the PAN coordinator, in addition to any specific application it may have, is to initiate, terminate, or route communication around the network. It also provides synchronization

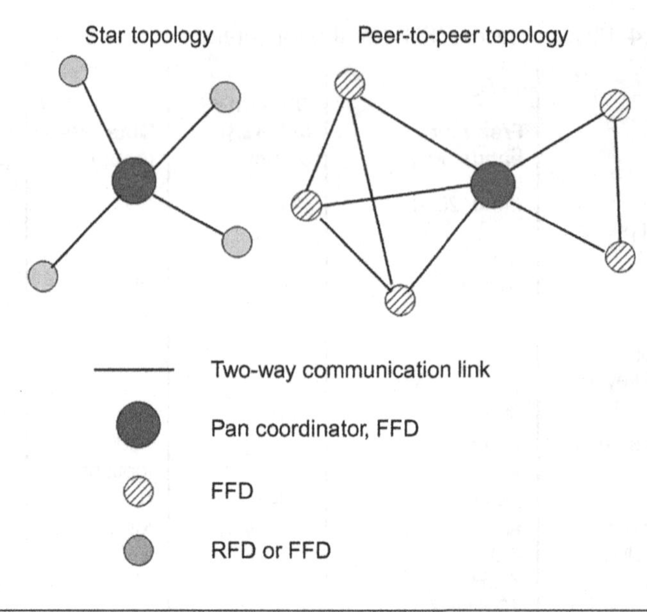


**FIG. 12.17**

Network topologies.

services. In a star network, each device communicates directly with the coordinator. The coordinator must be a FFD, and the others can be FFDs or RFDs. Relatively simple applications, like PC peripherals and toys, would typically use the star topology.

In the peer-to-peer topology, any device can communicate with any other device as long as it is in range. RFDs cannot participate, since an RFD can only communicate with a FFD. More complicated structures can be set up as a combination of peer-to-peer groups and star configurations. There is still just one PAN coordinator in the whole network. One example of such a structure is a cluster-tree network shown in Fig. 12.18. In this arrangement devices on the network extremities may well be out of radio range of each other, but they can still communicate by relaying messages through the individual clusters.

## 12.3.4 Frame structure, collision avoidance, and reliability

802.15.4 frame construction and channel access are similar to those of WLAN 802.11 (Wi-Fi) but are less complex. The transmitted packet is shown in Fig. 12.19. The purpose of the preamble is to permit acquisition of chip and symbol timing. The PHY header, PHR, which is signaled by a delimiter byte SFD (Start-of-frame deliminator), notifies the baseband software in the receiver of the length of the subsequent data. The PSDU (PHY service data unit) is the message that has been passed down through the higher protocol layers. As shown, it can have a maximum of 127 bytes although monitoring and control applications will typically be much shorter. Included in the PSDU are information on the format of the message frame, a sequence number, address information, the data payload itself, and at the end, two bytes that serve as a frame check

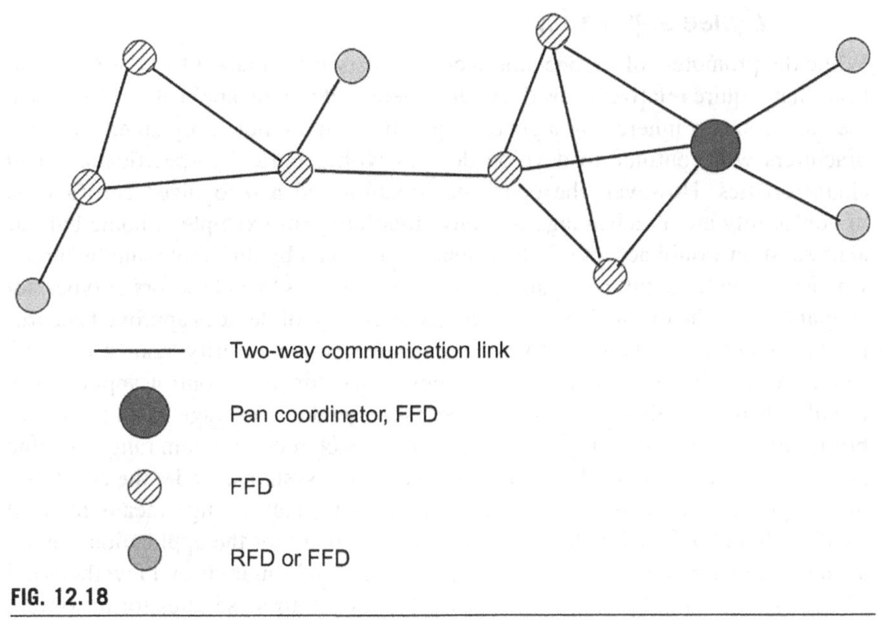


**FIG. 12.18**

Cluster-tree topology.

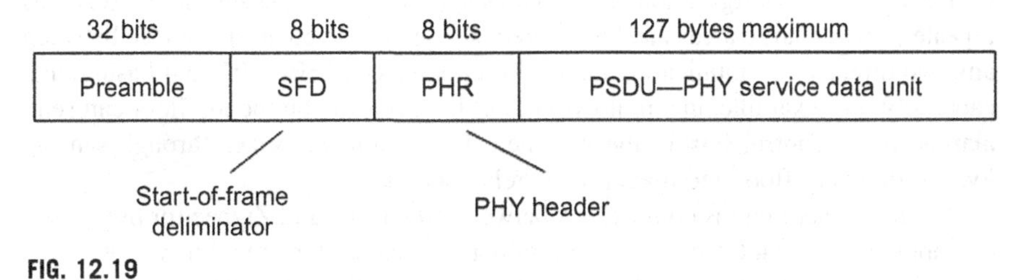


**FIG. 12.19**

Transmission packet.

sequence. Reliability is assured since the receiver performs an independent calculation of this frame check sequence and compares it with the value received. If any bits have been changed by interference or noise, the numbers will not match. Only when a match occurs, the receiving side returns an acknowledgement to the originator of the message. Lacking an acknowledgement, the transmission will be repeated until it is successfully received.

In order to avoid two or more stations trying to transmit at the same time, a carrier sense multiple access with collision avoidance (CSMA-CA) routine is employed, similar to that used in Wi-Fi, IEEE 802.11. The 802.15.4 receiver monitors the channel and if it is idle it may initiate a transmission. If the channel is occupied, the terminal must wait a random back off period before it can again attempt access. Acknowledgement messages are sent without using the collision avoidance mechanism.

### 12.3.5 Zigbee applications

While the promoters of Zigbee aim to cover a very large market for those applications that require relatively low data rates, there will remain applications for which the compromises inherent in a general specification are not acceptable, and manufacturers will continue to develop devices with proprietary specifications and characteristics. However, the open specification and a recognized certification of conformity are an advantage in many situations. For example, a home burglar alarm system would accept wireless sensors produced by different manufacturers, which will facilitate future expansion or allow installers to add sensors of types not available from the original system manufacturer. Use of devices approved according to a recognized standard gives the consumer some security against obsolescence. Although Zigbee claims to be appropriate for most control applications, it will not fit all of them, and will not necessarily take advantage of all the possibilities of the unlicensed device regulations. Its declared maximum range of some 50–75 m will fall short of the requirements of many systems. Under the constraint of low power output for low power consumption, greater range means reduced bandwidth and reduced data rate. In fact, a great many of the applications envisaged by Zigbee can get by very well with data rates of hundreds or a few thousand bits per second, and by matching receiver bandwidth to these rates for maximum sensitivity, ranges of hundreds of meters can be achieved.

One partial answer to the range question is the deployment of the Zigbee network in a cluster-tree configuration, as previously described. Adjacent nodes serve as repeaters so that large areas can be covered, as long as the greatest distance between any two directly communicating nodes does not exceed IEEE 802.15.4 basic range capability. For example, in a multi-floor building, sensors on the top floor can send alarms to the control box in the basement by passing messages through sensors located on every floor and operating as relay stations.

No doubt that there is competition between Bluetooth and Zigbee for use in certain applications, but the overall extent of deployment and the reliability of wireless control systems are higher because of it. The proportion of wireless security and automation systems has increased because adherence to the 802.15.4 standard provides high reliability, security, and reduced development costs compared to many proprietary solutions.

## 12.4 Other wireless personal area networks

### 12.4.1 6LoWPAN

IPv6 over low-power personal area networks (6LoWPAN) is an adaptation protocol layer for efficiently transporting internet protocol version 6 (IPv6) packets end-to-end over a WPAN physical layer such as IEEE 802.15.4. It takes advantage of over 30 years of IP technology development with open standards and interoperability proven through very widespread use of the internet by billions of users. 6LoWPAN

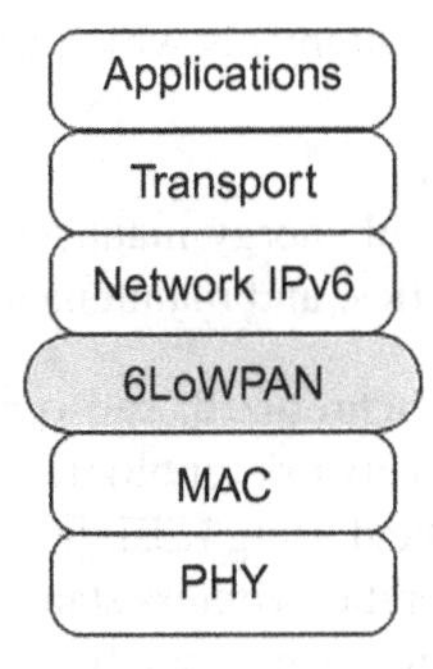


**FIG. 12.20**

6LoWPAN stack.

is an open standard defined in document RFC6282 by the Internet Engineering Task Force (IETF), the standards body that defines many of the open standards used on the internet. While originally conceived for 802.15.4 on 2.4 GHz, it is being adapted to Bluetooth LE and sub-1GHz low-power formats [6].

The level of 6LoWPAN in a 802.15.4 or other WPAN device protocol stack is just above MAC in the data link layer, as shown in Fig. 12.20. It serves as an adaptation layer to enable transmission of IPv6 datagrams over IEEE 802.15.4 radio links. Its principal function is compression of the headers from higher layers so that they can fit into 802.15.4 packets. Complementary to that function is fragmentation and reassembly of IPv6 data frames in order to transport them over the relatively short frames of up to 127 bytes that are specified for 802.15.4.

### 12.4.2 Thread network protocol

Thread is a mesh network protocol built according to IEEE 802.15.4 on 2.4 GHz and 6LoWPAN for routing IPv6 datagrams. It encompasses the IP stack transport layer using UDP (User Datagram Protocol), the network layer and the 6LoWPAN adaption layer. It was created by the Thread Group, an industry alliance whose mission is to focus on education, marketing and promotion of the Thread protocol as well as certification to insure interoperability among products of different manufacturers. A Thread network can securely connect more than 250 devices in a low-power wireless mesh network that gives direct internet and cloud access for every device. Thread is designed to support a wide variety of products for the home, including appliances, access control, climate control, energy management, lighting, safety and security [7].

### 12.4.3 WirelessHART

WirelessHART is a global standard that specifies a self-organizing mesh technology used for industrial control and monitoring. Some of its uses are [8]

- Process monitoring and measurements which are remote and uneconomical for wired monitoring
- Equipment health monitoring
- Environmental monitoring and energy management
- Asset management, diagnostics, and maintenance

The protocol utilizes a time synchronized, self-organizing, and self-healing mesh architecture, forming full mesh network topologies. The protocol currently supports operation in the 2.4 GHz ISM Band using IEEE 802.15.4 standard radios [9]. Developed as a multi-vendor, interoperable wireless standard, WirelessHART was defined specifically for the requirements of process field device networks. The standard began in 2004, and made its first appearance on the market in late 2007.

WirelessHART is based on the PHY layer specified in IEEE 802.15.4 for 2.4 GHz and specifies new data-link (MAC), Network, Transport and Application layers. All devices are time synchronized and use pre-scheduled fixed length time-slots. All WirelessHART devices must have routing capability—there are no reduced function devices like in Zigbee. Reliability is an utmost consideration and the protocol includes several mechanisms to insure successful coexistence in the 2.4 GHz band, including frequency hopping, clear channel assessment and so-called blacklisting of certain interfering channels, as well as path diversity to provide redundancy and self-healing properties [10].

### 12.4.4 Z-Wave

The Z-Wave protocol is an interoperable, wireless, RF-based communications technology designed specifically for control, monitoring and status reading applications in residential and light commercial environments. Like Thread, Z-Wave is a wireless network protocol aimed at home automation. Unlike Thread and Zigbee, it operates on sub-1 GHz bands, notably at 908.42 MHz in North America and 868.42 MHz in Europe. Different frequencies are specified for other countries. Physical and MAC sublayers are defined by the ITU (International Telecommunications Union) G.9959 standard [11]. The industry Z-Wave Alliance promotes the development and extension of Z-Wave for smart home and business applications and administers a produce testing program for certification in order to achieve interoperability regardless of manufacturer [12].

### 12.4.5 DASH7

DASH7 is designed for extremely low power applications including active RFID and WSN on sub-GHz ISM bands. The DASH7 Alliance was "formed to foster the existence and the further development of the DASH7 protocol specification (based on ISO 18000-7). It is the intent of the Alliance to enhance the technology beyond its current capabilities and physical boundaries to enable security, automation and control systems for a multitude of environments" [13].

DASH7 operates in the 433, 868, and 915 MHz bands. Data rates are 9.6, 55.55, and 166.667 kbps, modulation is GFSK and channel spacing is 25 and 200 kbps. Following are DASH7 Alliance protocol characteristics [14]:

- Bursty: data transfer is in the form of bursts, and does not include streaming content such as video, audio, or other isochronous (constant time periods) forms of data.
- Packet sizes are limited to 256 bytes.
- Asynchronous: Communication is by request-response. There is no periodic network handshaking or synchronization between devices.
- Stealth: Discovery beacons are not used and end nodes can choose to respond only to pre-approved devices.
- Mobility: Devices are inherently mobile and operation is upload based, so devices do not need to be managed extensively by a fixed infrastructure of base stations.

The protocol specification encompasses application, transport, network, data link and physical layers. It has well established security algorithms and protocols for authentication and encryption.

### 12.4.6 ANT/ANT+

ANT/ANT+ is a priority WSN technology managed by ANT Wireless, a division of Dynastream Innovations Inc. and promoted by the ANT+ Alliance in order to ensure interoperability between products of different companies. It is especially used in products in the sport, fitness and health sectors [15]. Designed for ultra-low power, ease of use, efficiency and scalability, ANT easily handles peer-to-peer, star, connected star, tree and fixed mesh topologies. It operates on the 2.4 GHz band, and uses time division multiple access (TDMA).

## 12.5 Ultra-wideband

A different approach to short-range communications with unique technical characteristics is UWB signal generation and detection. UWB adds applications and users to short-range communication without impinging on present spectrum use. Additionally, it has other attributes including range finding and high power efficiency that are derived from its basic principles of operation. We present the main features of UWB communication and an introduction to how it works.

### 12.5.1 Ultra-wideband technology

UWB technology creates a radio spectrum that is spread over a very wide bandwidth—much wider than the bandwidth used in the spread-spectrum systems previously discussed. UWB transmissions are virtually undetectable by ordinary radio receivers and therefore can exist concurrently with existing wireless communications without demanding additional spectrum or exclusive frequency bands.

These are some of the advantages cited for UWB technology:

- Low spectral density—very low probability of interference with other radio signals over its wide bandwidth
- High immunity to interference from other radio systems
- Low probability of interception/detection by other than the desired communication link terminals
- High multipath immunity
- Many high data rate UWB channels can operate concurrently
- Fine range-resolution capability
- Relatively simple, low-cost construction, based on nearly all-digital architectures.

Transmission and reception methods are unique, and are described briefly below.

There are two basic signal generating technologies for UWB devices: impulse radio (IR) and multiband OFDM (MB-OFDM).

*Impulse radio* is the legacy method of achieving wide bandwidth, dating back to the origins of radio. It is based on creating a sequence of short pulses, modulated by pulse amplitude modulation (PAM) or pulse position modulation (PPM). An example of a IR UWB "carrier" is a Gaussian monopulse, shown in Fig. 12.21 [16]. Its power spectrum is shown in Fig. 12.22. If the time scale in Fig. 12.21 is in nanoseconds, then the width of the pulse is 0.5 ns and the 3 dB bandwidth of the power spectrum is approximately 2.4 GHz with maximum power density at 2 GHz.

*MB-OFDM* uses an inverse FFT to place data in optimally spaced discrete subchannels to create a flat-topped wideband spectrum. Further widening is achieved by frequency hopping over a number of consecutive OFDM bands.

In order to pass information over a UWB impulse radio communication link, trains of pulses must be transmitted with some characteristic of a pulse or group

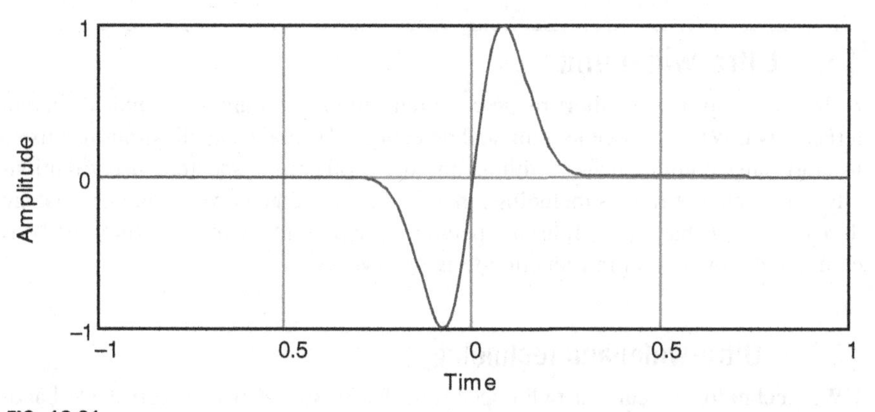


**FIG. 12.21**

UWB monopulse.

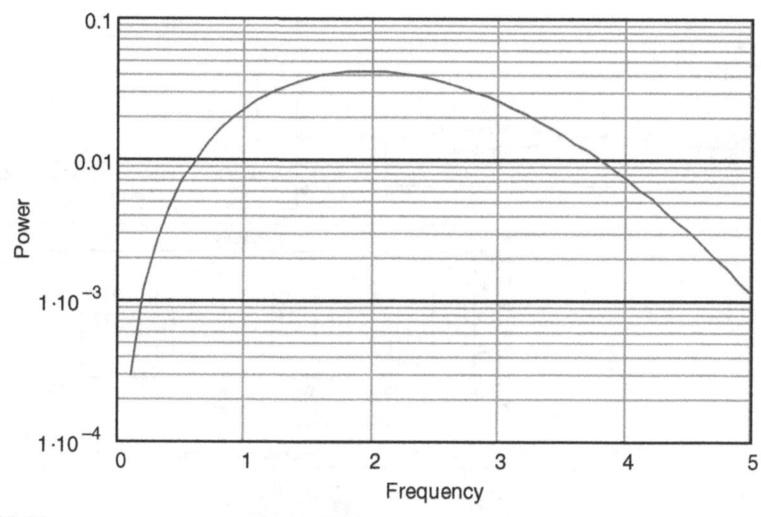


**FIG. 12.22**

Spectrum of monopulse.

of pulses varied in order to distinguish between "0" and "1." The time between consecutive pulses should be determined in a pseudo-random manner in order to smooth the energy spikes in the frequency spectrum. Reception of the transmitted pulse train is done by correlating the received signal with a similar timed sequence of pulses generated in the receiver. A large number of communication links can be maintained simultaneously and independently by using different pseudo-random sequences for each link.

A pulse similar to that of Fig. 12.21 can be generated by applying an impulse, or perhaps more conveniently a step-voltage or current, to a linear band limited network. Fig. 12.23 is a simulation of a sequence of UWB pulses created by stimulating a bandpass filter with a pseudo-randomly spaced sequence of impulses. The figure also shows the power spectrum of that sequence. The network that creates the individual UWB pulses includes the transmitter antenna, the propagation channel, and the receiving antenna, whose characteristics, in terms of impulse response or amplitude and phase vs. frequency must be known and accounted for in designing the system.

There are several ways of representing a UWB pulse as "1" or "0." One method is to advance or retard the transmitted pulse with respect to the expected time of arrival of the pulse in the receiver according to the agreed pseudo-random time sequence. This is PPM. Another method is to send the pulse with or without inversion, which is binary phase shift keying (BPSK). In both cases the correlation of the received pulse with a "template" pulse generated in the receiver will result in a different polarity, which depends on whether a "1" or a "0" was transmitted. A third way is on-off keying (OOK), where a pulse is transmitted for a "1" level and no pulse for a "0" level. In

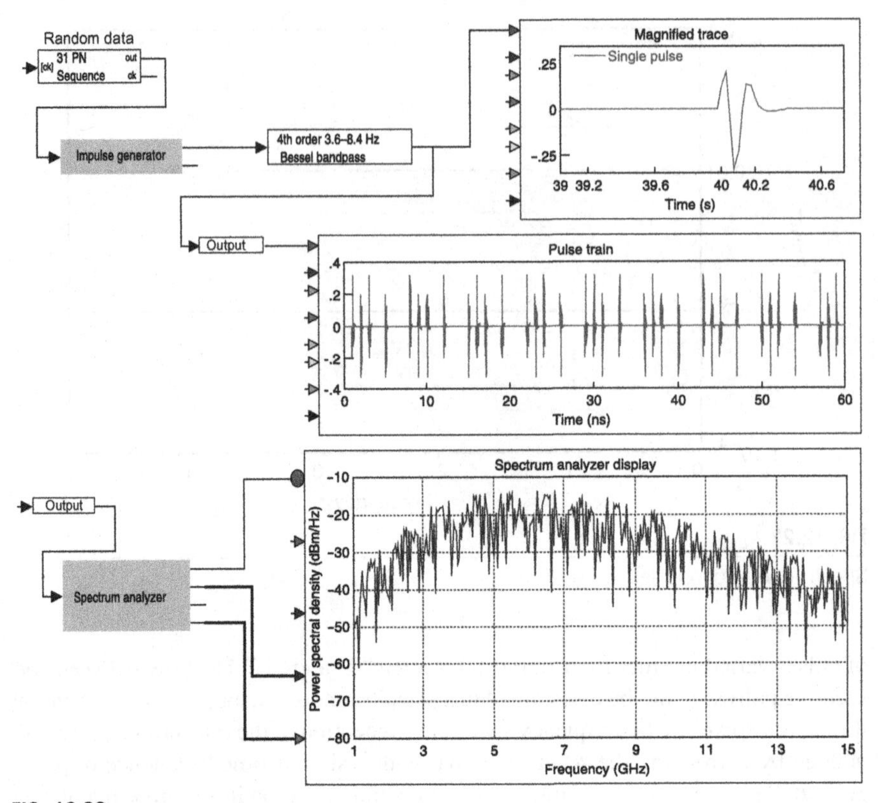


FIG. 12.23

Simulated sequence of UWB Pulses.

this case, a data bit level should be represented by a sequence of two bits of opposing levels, in order to preserve a constant average zero pulse stream level.

Detection of UWB bits is illustrated in Fig. 12.24. A "1" monopulse is represented by a negative line followed by a positive line, and a "0" monopulse by the inverse—a positive line and a negative line. The synchronized sequence generated in the receiver is drawn on the second line and below it the result of the correlation operation $\int f(t)g(t)dt$ where $f(t)$ is the received signal and $g(t)$ is the locally generated sequence. By sampling this output at the end of each bit period and then resetting the correlator, the transmitted sequence is reconstructed in the receiver. This is shown as the "Logic Output" of Fig. 12.24 which is delayed by one symbol period.

As mentioned above, an individual bit can be represented by more than one sequential monopulse. Doing so increases the processing gain by the number of monopulses per bit. Processing gain is also an inverse function of the pulse duty cycle. This is because, for constant average power, the power in the pulses contributing to each bit must be raised by (1/*duty-cycle*). By gating out the noise except during the interval of the expected incoming pulse, the signal-to-noise ratio will only be a function of the power in the pulse, regardless of the duty cycle. An example may make the explanation

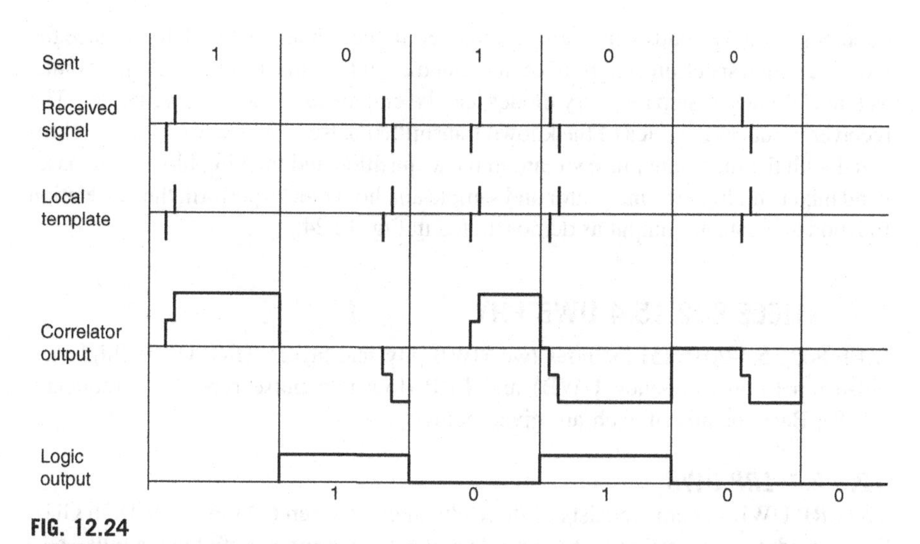


FIG. 12.24

Detection of UWB bit sequence.

clearer. Let's say we are sending data a rate of 10 Mbps. A UWB pulse in the transmitter is 200 picoseconds wide; 20 pulses represent one bit. The time between bits is $1/(10 \times 10^6) = 100$ ns. So the time between pulses is (100 ns)/20 pulses = 5 ns. The duty cycle is (200 ps)/(5 ns) = 0.04. Now the processing gain attributed to the number of UWB pulses per bit is 10 log(20) = 13 dB. That due to the duty cycle is 10 log(25) = 14 dB. Total processing gain is 13 + 14 = 27 dB.

A block diagram of a UWB system is shown in Fig. 12.25. The pulse generator block creates short impulse or step functions with rise times on the order of tens or at the most

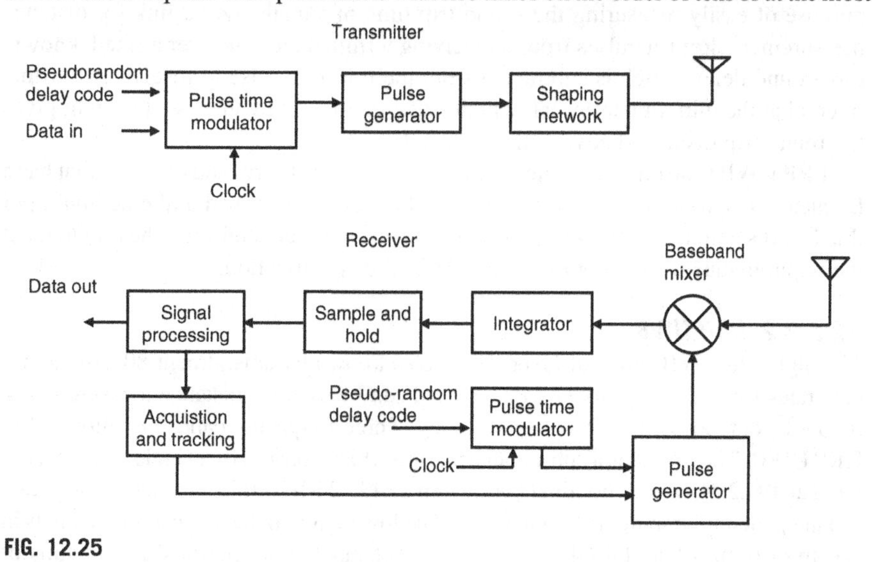


FIG. 12.25

UWB simplified block diagram.

hundreds of picoseconds which are conditioned in the shaping network to produce the required output spectrum. High speed integrated circuits or special circuit elements such as tunnel diodes or step recovery diodes, can be employed in the pulse generator. The receiver produces a replica of the known transmitted pulse sequence which is synchronized with the incoming pulse stream in the acquisition and tracking block. The baseband mixer (multiplier), integrator and sample and hold blocks perform the correlation function and data are output as demonstrated in Fig. 12.24.

### 12.5.2 IEEE 802.15.4 UWB PHY

IEEE 802.15.4-2016 [5] includes two UWB physical layers: HRP UWB (high rate pulse repetition frequency UWB) and LRP (low rate pulse repetition frequency UWB). Basic details of each are given below.

#### 12.5.2.1 LRP UWB

The LRP UWB version consists of three channels between 6.2896 and 9.1856 GHz. Three modes are specified. A base mode and extended mode both have a pulse repetition frequency (prf) of 1 MHz and use OOK modulation. The base mode data rate is 1 Mbps with one pulse per symbol. The data rate of the extended mode is 250 kbps. The lower data rate is achieved by convolution encoding with three code bits for each data bit. Thus, it has four pulses per symbol, and since the prf is constant the symbol period is four times the length in the base mode. A third long-range mode has a prf of 2 Mbps, Manchester encoding and a data rate of 31.25 kbps using 64 pulses per symbol.

All modes can optionally append a UWB location enhancing information postamble (LEIP) which is a sequence of pulses at the defined prf for each mode for the purpose of easily measuring the round trip time of signals over a link for distance measurement. Return pulses from a receiving terminal are sent after a fixed, known, turnaround delay which is subtracted from the transmit pulse to receive pulse time interval at the initiating terminal, which when multiplied by the speed of light gives the round trip distance between the terminals.

LRP UWB transmitters create a base band impulse of form and duration that has a frequency response fitting the appropriate channel, as described at the beginning of this UWB section. An RFID tag, for example, can be produced very cheaply to meet the requirements of the low rate UWB 802.15.4 specification.

#### 12.5.2.2 HRP UWB

The high rate UWB physical layer, HRP, also known by amendment 802.15.4a, has data rates from 120 kbps to 27.24 Mbps and can be used for distance measurements, indoor or outdoor, with sub-meter accuracy. Three frequency bands are allocated to HRP UWB. The lowest is a sub-gigahertz band which consists of a single channel centered at 499.2 MHz with nominal bandwidth of 499.2 MHz. It is intended for wall and ground penetration imaging applications. The low gigahertz band has four channels in spectrum from 3.1 to 4.8 GHz. Three of the channels have nominal bandwidths of

499.2 MHz and one channel, centered on 3993.6 MHz, has a nominal bandwidth of 1331.2 MHz. Eleven channels occupy frequencies from 6 to 10.6 GHz in the high GHz band. Eight of them have 499.2 MHz bandwidths. Three, with center frequencies at 6489.6, 7987.2, and 9484.8 MHz, have nominal bandwidths of 1081.6, 1331.2, and 1354.97 MHz.

The transmitted signal spectrum is created by multiplying a baseband pulse with a carrier at the channel center frequency. HRP UWB doesn't specify the exact pulse shape to use, but it does define a reference pulse for each bandwidth, shown in Fig. 12.26 for a pulse width of 2 ns (bandwidth 499.2 MHz). The cross correlation of the transmitted pulse has to be at least 0.8 for a period of one fourth of the pulse width.

Pulses like the one of Fig. 12.26 are transmitted in bursts. Modulation is a combination of the polarity of the series of pulses in a burst (BPSK) and the position of the burst within a symbol period (burst position modulation, BPM), which conveys 2 bits per symbol. The data rate is set by the number of pulses in a burst. User data rates are selected from 0.11, 0.85, 6.81, and 27.24 Mbps. User data rates are the raw data rates times the FEC coding rate.

Fig. 12.27 shows the structure of a HRP UWB PHY symbol. BPM is done by transmitting a burst within the first quarter of the symbol period (logic 0) or the third quarter (logic 1). The number of pulses (chips) per burst is $N_{cpb}$. The polarity of a

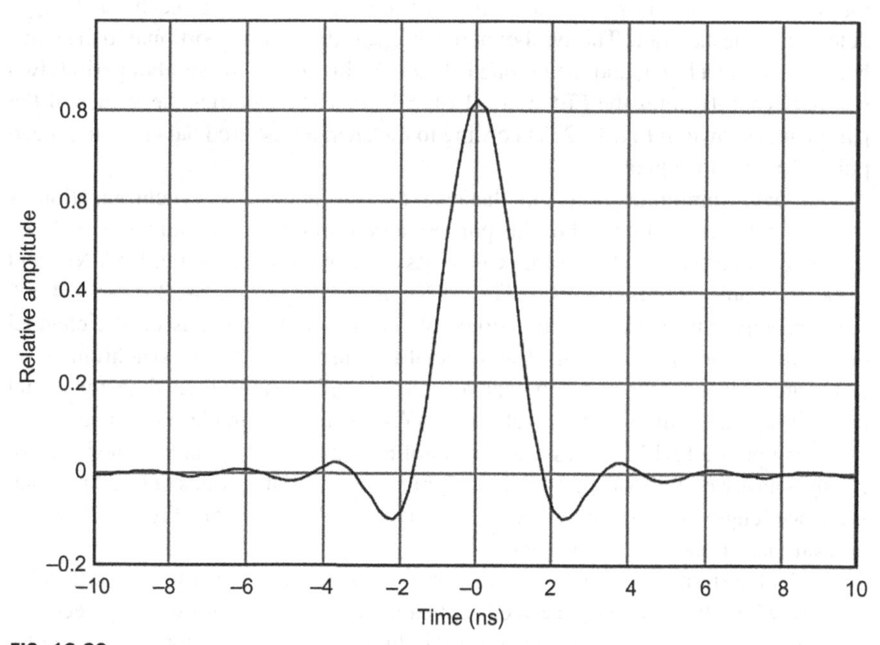


**FIG. 12.26**

Reference UWB pulse in 802.15.4 for 499.2 MHz bandwidth.

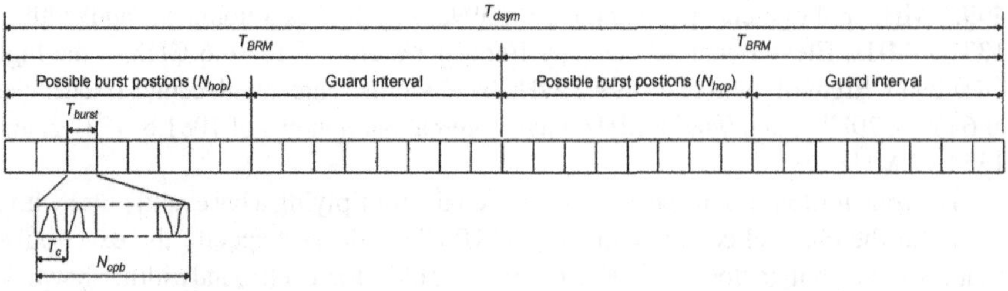

**FIG. 12.27**

802.15.4 UWB symbol.

pulse is determined for each symbol by a pseudo random scrambling function which provides interference suppression and spectral smoothing. $N_{\text{hop}}$ is the number of burst positions each of duration $T_{\text{burst}}$ in each half $T_{\text{BPM}}$. Eight are shown in the figure. The position occupied by the burst, the third is shown in the figure, changes pseudo-randomly from symbol to symbol. The pseudo-random code is known at the receiver and creates a form of time hopping spread spectrum. The use of time hopping pulses reduces the probability of collisions between multiple users on the channel and also smoothes out the signal spectrum. The second and fourth quarters of the symbol are "quiet" guard intervals with no bursts, They prevent intersymbol interference from multipath reflections. The number of pulses in each burst, $N_{\text{cpb}}$, determines the data rate. The symbol period $T_{\text{dsym}}$ is directly proportional to the number of pulses in a burst, and the bit rate is twice the inverse of the symbol period (two bits per symbol) times the FEC rate. IEEE 802.15.4 [5] specifies the values of the parameters shown in Fig. 12.27 according to the channel, desired data rate and mean pulse repetition frequency.

For data communication and distance measuring precise synchronization is required between terminals. For this purpose a synchronization preamble is prefixed to a HRB UWB frame. The preamble consists of a synchronization field (SYNC) and a start-of-frame delimiter (SFD). The SYNC may contain 16, 64, 1,024, or 4,096 identical repeating symbols. The choice of SYNC length depends on the channel delay profile (impulse response) and signal-to-noise ratio. Large synchronization fields are preferred for non-coherent receivers to allow additional time for signal acquisition and frame synchronization. A SYNC symbol is made up of a sequence of 31 or optionally 127 pulses. Each symbol may have one of three states: plus or minus polarity or zero value. There are eight ternary (3 state) codes defined for each sequence length and they are allocated among the UWB channels. Fig. 12.28 shows an example of one of these sequences.

Each of the ternary sequences has perfect autocorrelation, that is, it has a peak value of unity when the receiver generated replica is perfectly lined up with the received sequence and no sidelobes when the replica is displaced in time by more than one pulse width. Perfect correlation is retained when the input sequence is squared, as it is in a non-coherent receiver. The receiver synchronizes its local reference sequence to the

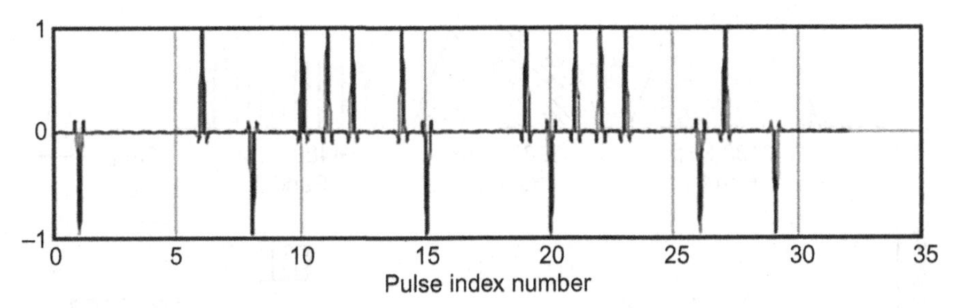


**FIG. 12.28**

Synchronizing sequence for HRB UWB in IEEE 802.15.4.

SYNC symbols of the received signal. It must then identify the epoch in the packet where the time of arrival measurement is made—at the end of the synchronizing preamble and the start of the PHY (physical layer) header. It does this by searching for and identifying the SFD, which directly follows the SYNC symbols. The SFD may contain 8 or 64 symbols. The default short sequence has 8 symbols and the long, optional 64-symbol SFD is used with the low data rate of 110 kbps. Each symbol is the same as the basis symbol of the SYNC, one of which is depicted in Fig. 12.28. After the receiver demodulates, or despreads, the SFD by correlating with the known sequence signal, it can recognize precisely the end of the last symbol in the SYNC preamble and can then interpret the remaining fields in the packet.

The FCC regulations pertaining to UWB transmissions are described in Chapter 10. Conditions for using UWB in Europe are specified by the European Union. In order to prevent interference with vital wireless services in the wide bandwidths covered by UWB radiation, spectral density limits allowed by the FCC and permitted in the European Union are relatively low, on the order of spurious radiation limits for conventional unlicensed transmissions. However, as we have seen, UWB has high processing gains and communication ranges on the order of tens or hundreds of meters can be attained.

### 12.5.3 OFDM UWB

Another method for creating an UWB signal that meets FCC requirements and has features attractive to short-range distance measurement and location is orthogonal frequency division multiplex (OFDM). OFDM is particularly efficient for capturing the total energy in multipath channels. It has high spectral efficiency due to its spectrum shape, which is almost completely flat over its bandwidth between extremities of 10 dB down. The properties of OFDM make it possible to exclude from the spectrum explicit frequency ranges—an advantage since UWB frequency coverage is not identical around the world. Frequency hopping gives additional spreading of the basic OFDM signal for UWB. Multiband OFDM (MB-OFDM) is the system in which consecutive OFDM UWB packets are displaced in frequency over adjacent

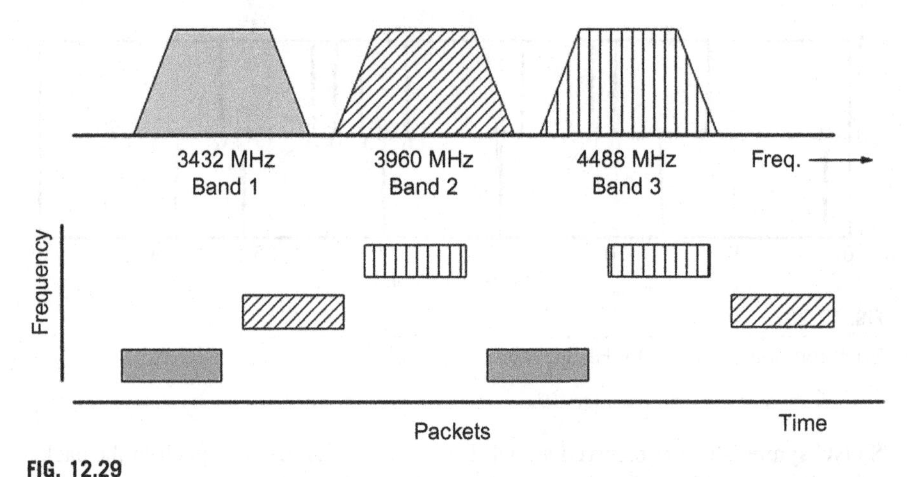


**FIG. 12.29**

Multiband OFDM UWB.

bands. The process is illustrated in Fig. 12.29. The upper part of the diagram shows the frequency spectrum spread over three OFDM bands. On the bottom we see how consecutive packets hop with time over the three frequency ranges.

Standard ECMA-368, issued by Ecma International, a European association for standardizing information and communication systems, specifies a MAC sublayer and a physical layer (PHY) for a high rate personnel area network [17]. It is based on a multiband OFDM technique that was proposed for IEEE 802.15.3a, an alternate PHY for the high rate WPAN standard IEEE 802.15.3. IEEE 802.15.3a was not completed due to lack of agreement on the UWB physical layer technology. Included in ECMA-368 are provisions to support ranging measurements between devices using two-way time transfer techniques.

ECMA-368 supports specific data rates from 53.3 through 480 Mbps. Its UWB spectrum of 3.1 to 10.6 GHz is divided into 14 bands, each with a bandwidth of 528 MHz. Twelve of these bands are grouped into units of three over which frequency hopping is conducted, as in the example of Fig. 12.29. A total of 110 subcarriers (100 data carriers and 10 guard carriers) are used per band to transmit the information. In addition, 12 pilot subcarriers allow for coherent detection. Frequency-domain spreading, time-domain spreading, and FEC coding are used to vary the data rates. MB-OFDM chips have been built into cameras, video monitors, TV sets with HDMI, docking stations, and military communications [18].

## 12.6 Summary

Several WPAN are described in this chapter. Bluetooth consists of two versions—BR/EDR, which we refer to as legacy Bluetooth, and BLE. BR/EDR is used mostly for streaming data applications like wireless headphones and loudspeakers. BLE, with its four defined roles and device capabilities, serves control and monitoring functions as well as one-way beacons, where very low energy consumption for long battery life is a necessity. Both versions operate on world-wide 2.4 GHz unlicensed frequency channels using frequency hopping medium access. Problems and solutions for coexistence between Bluetooth and IEEE 802.11 devices are discussed.

The IEEE 802.15.4 standard and Zigbee implementations for personal area networks are designed for short-range network topologies and use in low cost, low power devices. Several frequency ranges and modulation techniques are defined to suit sub-1 GHz regulations in different countries and regions, in addition to operation on the 2.4 GHz band.

Other WPAN standards for specific application requirements are discussed. They include 6LoWPAN, Thread, WirelessHART, Z-Wave, DASH7, and ANT.

UWB is a WPAN technology that stands apart from the others from the point of view of operating frequencies, bandwidth, data rates and consequently its applications. Two signal generation techniques are explained—impulse radio and multiband OFDM. Due to spreading signal energy over a wide bandwidth, UWB does not interfere with narrow band communications over the same frequency range. Also, its wide bandwidth makes it particularly suitable for short range high precision wireless distance measurement.

## References


[1] Specification of the Bluetooth System, Core Package Version 5.0, Bluetooth Special Interest Group, 2016.

[2] M. Woolley, Bluetooth 5 Go Faster. Go Further, White Paper. Bluetooth Special Interest Group, 2018.

[3] N. Hunn, Essentials of Short-Range Wireless, Cambridge University Press, 2010.

[4] A. Bensky, Bluetooth and 802.11 WLAN Coexistence, https:/www.researchgate.net/publication/268404004_Bluetooth_and_80211_WLAN_Coexistence, 2003. Accessed 24 October 2018.

[5] IEEE Standard for Low-Rate Wireless Networks, IEEE Std 802.15.4™, 2015.

[6] J. Olsson, 6LoWPAN Demystified, Texas Instruments, 2014.

[7] The Thread Group, https:/www.threadgroup.org/. Retrieved 24 October 2018, 2018.

[8] System Engineering Guidelines IEC 62591 WirelessHART, Emerson Process Management, 2016.

[9] Basics of wirelessHART Network Protocol, https:/automationforum.co/basics-wirelesshart-network/, Retrieved 26 October 2018, 2018.

[10] T. Lennvall, S. Svensson, A comparison of WirelessHART and ZigBee for industrial applications, in: IEEE International Workshop on Factory Communication Systems, Dresden, Germany, 21–23 May, 2008.

[11] ITU-T Recommendation G.9959, Short Range Narrow-Band Digital Radiocommunication Transceivers—PHY, MAC, SAR and LLC Layer Specifications (01/2015), 2015.

[12] Z-Wave Alliance, https:/z-wavealliance.org. Retrieved 24 October 2018, 2018.

[13] DASH7 ALLIANCE, www.dash7-alliance.org. Retrieved 24 October 2018, 2018.

[14] Wireless Sensor and Actuator Network Protocol VERSION 1.1, DASH7 Alliance, 2017.
[15] ANT/ANT+, https:/www.thisisant.com/company/. Retrieved 24 October 2018, 2018.
[16] A. Petroff, P. Withington, Time modulated ultra-wideband (TM-UWB) overview, in: Presented at Wireless Symposium/Portable by Design, Feb 25, 2000, San Jose, CA, 2000. http:/www.time-domain.com. Accessed 24 October 2018.
[17] ECMA Standard ECMA-368, High Rate Ultra Wideband PHY and MAC Standard, third ed., 2008.
[18] L. Frenzel, UWB for IoT? Microwaves & RF, 2017.

CHAPTER

# Radio frequency identification (RFID)

13

## 13.1 Introduction

Radio frequency identification (RFID) deserves its separate classification among short-range wireless technologies. A basic difference between RFID and the applications discussed in previous chapters is that RFID devices are not two-way communication devices per se but involve interrogated transponders. Instead of having two separate transmitter and receiver terminals, an RFID system consists of a reader that sends a signal to a passive or active tag and then receives and interprets a modified signal reflected or retransmitted back to it. Fig. 13.1 shows an RFID system. One or more tags or smart cards are interrogated by the reader to obtain the ID of each and the data are sent to a host application. Some types of tags/cards may be written to as well. Other tag devices include sensors whose outputs are read to the host.

The range of wireless links that are considered to be RFID is demonstrated by these typical applications:

- Aircraft identification. A terrestrial terminal, radar for example, transmits a beam in the direction of an aircraft. A transponder in the airborne vehicle responds by transmitting an identification code to the initiating terminal, which can be used to mark a radar screen blip, and, in wartime, signify friend or foe.
- Roadway toll collection. A signal from an overhead transmitter on a toll road activates a transponder mounted on a vehicle windshield which transmits the car ID to a toll station receiver for billing purposes.
- Inventory control. Tags on product cartons in a warehouse each reflect a signal to a handheld reader identifying the contents of the box.
- Animal husbandry. Embedded tags in cows are read by RFID readers when the animals come for milking for the purpose of acquiring statistics on the milk output of each individual cow.

The reader has a conventional transmitter and receiver having similar characteristics to the devices already discussed. The tag itself is special for this class of applications. It may be passive, receiving its power for retransmission from the signal sent from the reader, or it may have an active receiver and transmitter and embedded long-life battery. In contrast to other wireless communication terminals, an RFID tag has no signal inputs or outputs, but just transmits a response of internal information upon receipt of a trigger transmission. The tag may incorporate an external sensor element



Short-range Wireless Communication. https://doi.org/10.1016/B978-0-12-815405-2.00013-0
© 2019 Elsevier Inc. All rights reserved.

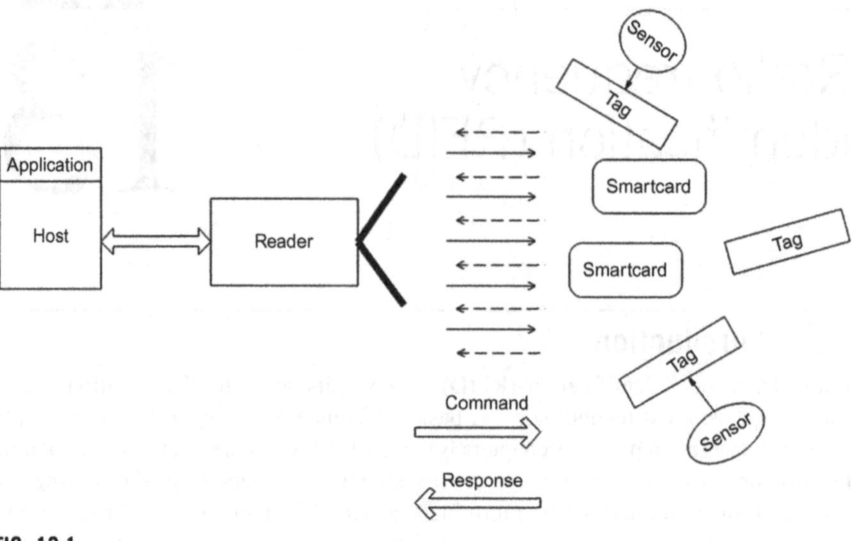


**FIG. 13.1**

Basic RFID system.

whose measurement values can be read in a similar manner to the tag's internal memory. A read-write tag has memory that can be modified by the reader.

In its most common form of operation, an RFID system works as follows. The reader transmits an interrogation signal, which is received by the tag. The tag may alter the incoming carrier signal in some unique manner and reflect it back, or its transmitting circuit may be triggered to read its ID code residing in memory and to transmit this code back to the receiver. Active tags have greater range than passive tags. If the tag is moving in relation to the receiver, such as in the case of toll collection on a highway or parts moving on a conveyer belt, the data transfer must take place fast enough to be complete before the tag is out of range.

These are some design issues for RFID:

- Tag orientation is likely to be random, so tag and reader antennas must be designed to give the required range for any orientation.
- Multiple tags within the transmission range can cause return message collisions. One way to avoid this is by giving the tags random delay times for response, which may allow several tags to be interrogated at once.
- Tag data must be coded to prevent misreading.

## 13.2 Frequencies

Ranges of RFID may be several centimeters up to tens of meter. The most common operating frequencies range from 125 kHz up to 2.45 GHz. Frequencies are usually those specified for unlicensed applications. Higher data rates demand

**Table 13.1** RFID frequencies

| Frequency | Typical range | Near field/far field |
|---|---|---|
| 125-134 kHz | Under 1 m | Near |
| 13.56 MHz | Up to 1.5 m | Near |
| 433 MHz | 4-9 m | Far |
| 868-928 MHz | 2-5 m | Far |
| 2.45 GHz | 1-2 m | Far |

higher frequencies. Table 13.1 shows spot frequencies and frequency bands typically occupied by RFID devices and whether their range is in the near field or far field [1].

Inductive coupling is used in the near field, while far field propagation is used in the far field.

## 13.3 RFID operation and classifications

Three classifications of RFID devices are passive, semi-passive, and active. These are further divided into five tag classes by EPCglobal [2]:

- Class 0—Read only tags. Passive, no batteries
- Class 1—Identity tags. Passive, read only
- Class 2—Higher functionality, read and write memory
- Class 3—Semi-passive, read and write memory, built in battery for increase read range
- Class 4—Active tags, active communication, allows tag networking
- Class 5—Active tags, allows communication with class 4 and other class 5 tags as well as other devices.

Passive devices have no local source of power. The electronic chip in which is stored object identification and modulation circuitry gets its power from the signal it receives from the reader. The antenna voltage resulting from the received wave is rectified and may be stepped up by a charge pump so that it can power the tag circuitry. Considering that the power transmitted by the reader is restricted by the regulating authority, the energy received is very low and the tag circuits must have low power consumption to present a small load to the equivalent source. Since the reader transmitter powers the tag, the reader must contend with the low reflected signal from the tag at the same time that it is transmitting on the same frequency. Because of their simplicity, passive devices can be made very cheap and compact.

A semi-passive RFID tag similarly modulates the reflected wave from the reader but in contrast to a fully passive device, power for the tag circuit is provided by a self contained battery. When there is no reader signal, power consumption is very small.

Operation is triggered by the incoming reader signal but the modulating circuits do not take power from that signal. Advantages of the semi-passive mode compared to fully passive are:

- longer range, since received signal threshold is lower than that needed by a passive device,
- reader transmitter power may be less,
- consequently, interference in the reader from its own transmission is reduced.

Disadvantages of semi-passive tags compared to passive are higher cost and size due to the battery, and the fact that the battery must be replaced periodically.

Active tags contain a complete radio transceiver and operate as a transponder. Range is considerably greater than that of passive tags, since the return signal to the reader is generated in the tag and not reflected.

Another categorization of RFID systems is near field and far field. Table 13.1 shows the frequency ranges of each. Transmission in the near field typically uses the mode of inductive coupling over short sub-meter ranges. Far field transmissions propagate under conditions described in Chapter 2.

RFID systems can also be distinguished as being read only or read-write. A read only tag has its identification number permanently written in the electronic chip during manufacture and cannot be changed. A read-write tag can have its ID determined at the time of deployment or changed by the reader transmissions according to necessity during operation. A tag whose readable contents can be changed must have a security protocol to insure that unauthorized persons cannot alter the device's memory.

### 13.3.1 Near field RFID

The near electromagnetic field, through which communication takes place, is conveniently taken to be up to a distance of $\lambda/2\pi$ when antenna dimensions are considerately smaller then wavelength $\lambda$, although this boundary is not abrupt. As indicated in Table 13.1, maximum range is around 1 m, whereas the near field boundary at 125 kHz is 382 m and at 13.56 MHz it is 3.5 m. Energy in the near field is transferred through induction or capacitance, and not radiation as it is in the far field. Inductive communication between reader and tag is conducted through opposing wire coils as in a transformer. The coils are loosely coupled over the communication range, and signals at both terminals may be relatively weak.

Fig. 13.2 shows a reader and passive tag coupled by induction. Both terminals have coils which are equivalent to primary (reader) and secondary (tag) windings of a transformer. The nature of the coils depend on the carrier frequency. Low frequency windings, at 125 kHz for example, have a large number of turns on a ferrite core. High frequency devices have circular or rectangular loops with one or a few turns, often printed on a circuit board. The tag is passive, with a diode rectifier supplying DC to the chip from the RF that appears across the tag resonant tank of $L_T$ and $C_T$. When the voltage powering the chip crosses a threshold, typically 4 V peak-to-peak, the chip comes to life and sends its identification [3]. It does this by activating a transistor switch across the tag tuned circuit in accordance with the data, which is

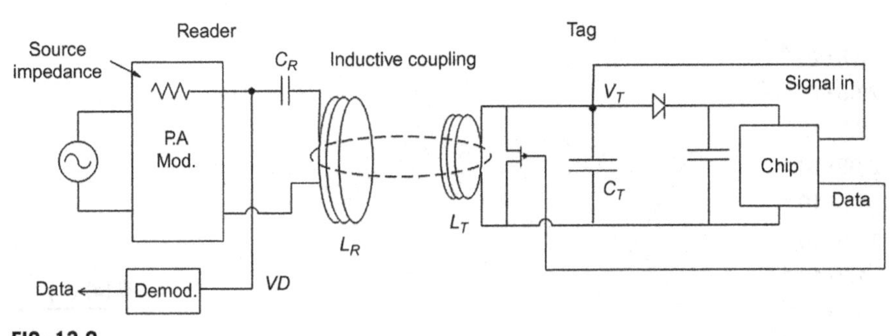


**FIG. 13.2**

Inductive coupling.

seen by the reader as a change in impedance at its coil. The resulting combined amplitude and phase shift keying is demodulated resulting in a data output. The "Signal in" from the reader to the tag chip is a polling signal to identify the tag from which a response is expected.

There are two considerations in designing the transmission parameters of a near field RFID system. On the one hand, the carrier signal at the tag has to be strong enough to power the chip at the maximum required range. Signal strength depends critically on tag orientation towards the reader antenna so a worst case situation has to be defined. On the other hand, the degree of mutual coupling between the tag and reader coils affects the depth of modulation of the tag signal at the reader. It is possible that both conditions, sufficient signal to power the chip and detection ability of the tag message at the reader, are not satisfied at the maximum specified range.

The signal strength from the reader is expressed as the magnetic field density $B$ in units of Tesla. In Fig. 13.3,for a circular loop antenna of radius $r$ meters, magnetic field density $B$ at point $A$, a distance $x$ in meters on a line going through its center and perpendicular to its plane, is

$$B = \frac{\mu_0 N I r^2}{2(x^2 + r^2)^{3/2}} \tag{13.1}$$

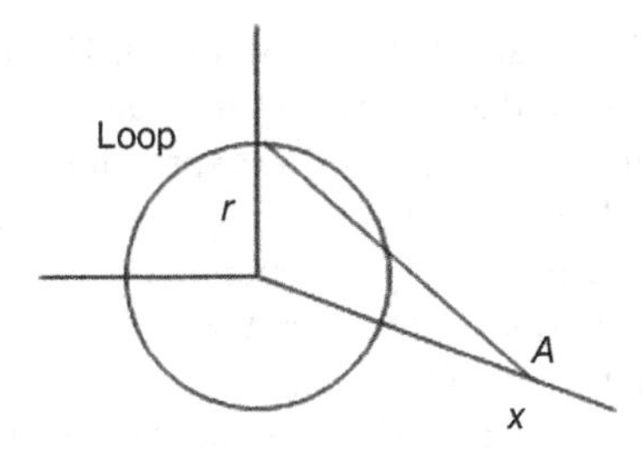


**FIG. 13.3**

Near field transmitter loop antenna.

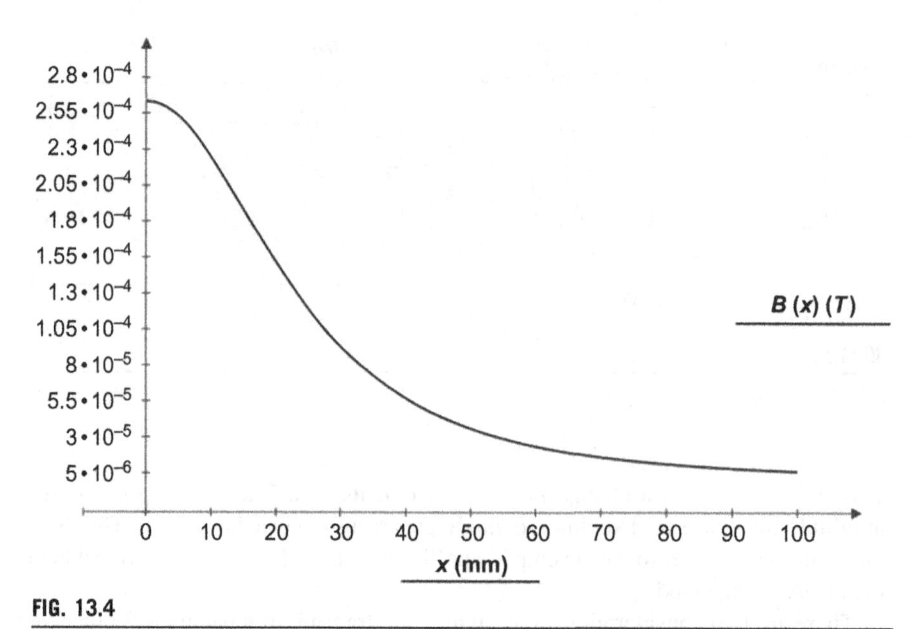


**FIG. 13.4**

Field intensity in Teslas from a 2 turn circular loop as a function of distance. Loop radius is 30 mm and loop current is 1 A.

where $\mu_0 = 4\pi \times 10^{-7}$ Webers/Amp-meter is the permeability of free space, $N$ is the number of turns in the loop, and $I$ Amperes is the current in the loop [4]. The formulation of $B$ takes into account the distance, $\sqrt{x^2+r^2}$, from a point on the loop to point $A$. Fig. 13.4 shows $B$ as a function of $x$, attaining an attenuation proportional to $x^3$ as $x$ increases beyond $r$. For a given distance of $x$, there is a radius $r$ that produces maximum intensity $B$, as shown in Fig. 13.5. This radius is $\sqrt{2}\cdot x$.

The voltage $v$ in the tag coil is found from Faraday's Law:

$$v = -\frac{d}{dt}\Phi = -\frac{d}{dt}\int \mathbf{B}\cdot\mathbf{dA} \tag{13.2}$$

That is, the voltage equals the rate of change of flux $\Phi$. The minus sign means that the polarity of the voltage is such that the current it creates in the coil results in a magnetic field that opposes the change in flux. The expression on the right indicates, from the dot product of vectors **B** and **dA**, that $v$ is maximum when the reader and tag loops are facing each other and in general is proportional to the cosine of the angle between the direction of **B** and the normal of the area **A** which is in the plane of the tag loop. When $C_T$ resonates with the tag coil inductance $L_T$ (Fig. 13.2) the voltage $V_T$ across $C_T$ at the input to the rectifier is $Q\cdot v$ where the Q of the circuit is the coil reactance divided by the series equivalent of the coil losses and the loading of the chip: $Q = X_L/RL$. $V_T$ is also proportional to the number of turns in the tag loop.

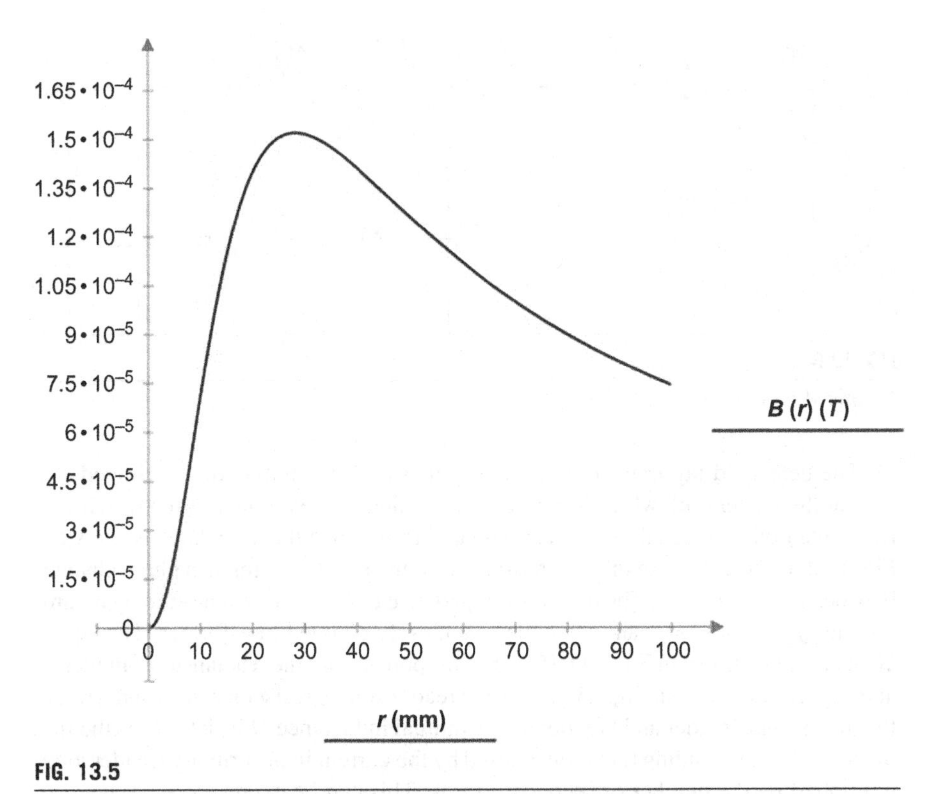


**FIG. 13.5**

Field intensity in Teslas from a 2 turn circular loop as a function of loop radius. Distance is 20 mm and loop current is 1 A.

We make a simplifying approximation that the flux density $B$ is uniform over the tag loop area $A_T$. Then maximum $V_T$ when the coils face each other is

$$V_T = Q_T B A_T N_T 2\pi f \tag{13.3}$$

where the subscript $T$ indicates referral to the tag and $f$ is the carrier frequency, which enters the expression due to the derivative in Eq. (13.2).

## Example 13.1

A near field RFID system operates on 13.56 MHz. The round reader loop has a radius of 3 cm and 6 turns of 1 mm diameter wire. The current in the reader coil is 10 mA. The 10 turn tag loop radius is 2 cm and the coil $Q$ is 20. Find the voltage at the input to the tag rectifier when the tag is 4.5 cm from the reader and facing it.

*Solution*

**(a)** Estimate the inductance of the reader coil. You can use Worksheet **Inductance.mcd** for a round loop. The result is $L_R = 5.7$ μH. Reactance is $X_R = 2\pi f L_R = 483\ \Omega$.
**(b)** From Eq. (13.1), $B = 2.1 \times 10^{-7}$ T (Tesla).
**(c)** From Eq. (13.3), $V_T = 4.6$ V.

This and similar problems can be solved with Worksheet **RFID_NF.mcdx**.

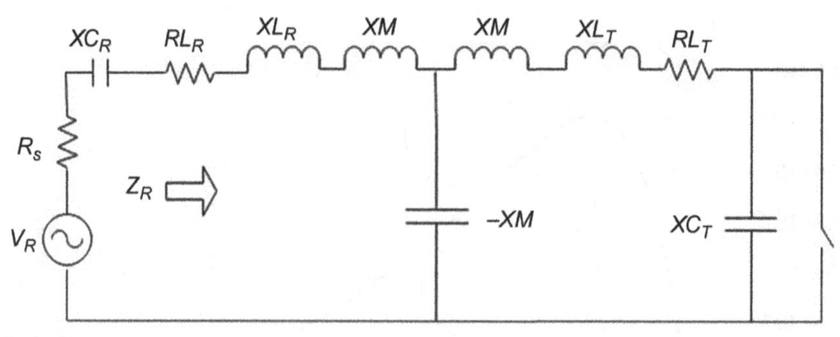


**FIG. 13.6**

Near field equivalent circuit.

The near field tag transmits its identity message by modulating the impedance seen at the reader coil, which in turn amplitude and phase modulates the reader carrier wave giving a signal that is demodulated to produce the chip data, as shown in Fig. 13.2. Coherent demodulation must be used in order to distinguish phase changes between the logic states. The degree of impedance change can be analyzed by regarding the system of reader and tag coils as a loosely coupled transformer. The equivalent circuit is shown in Fig. 13.6 [5]. The components are the reactances of inductors and capacitors shown in Fig. 13.2. $R_S$ is the reader source resistance. $RL_R$ and $RL_T$ are the losses in the reader and tag loops. The mutual inductance $M$ is defined as the flux in the secondary winding (tag loop) caused by the current in the primary (reader loop) multiplied by the number of secondary turns. This can be expressed as

$$M = \frac{BA_T N_T}{I} \tag{13.4}$$

$XM$ in Fig. 13.6 is the reactance of $M$.

An example shows how the modulation works.

## Example 13.2

The system is the same one defined in Example 13.1. Find the voltage into the demodulator (Fig. 13.2) when the transistor switch in the tag is open, VD1, and when it is closed, VD2. What is the peak modulation voltage?

*Solution*

**(a)** Estimate the inductance of the reader tag loop and reactances of both loops. The inductance can be estimated using worksheet **Inductance.mcdx** or worksheet **RFID_NF.mcdx**.

$$L_T = 9.5\,\mu H, XL_T = 807\,\Omega, XL_R = 483\,\Omega.$$

**(b)** Find the loss resistances of the reader and tag coils, $RL_R$ and $RL_T$. $Q$ of both loops is 20.

$$RL_R = 483/20 = 24, RL_T = 807/20 = 40$$

**(c)** Find the mutual inductance and reactance

From Eq. (13.4), $M = 270$ nH. $XM = 23\ \Omega$.

**(d)** Find resonating reader and loop capacitive reactances, $XC_R$, $XC_T$. Exact resonance varies with tag distance and consequently $M$. Assume the capacitive reactances for resonance are the negative of the corresponding reactances of the reader and tag loops. Reader source voltage is 10 V.

$$XC_R = -XL_R = -483\ \Omega, XC_T = -807\ \Omega$$

**(e)** Using Fig. 13.6 solve for the impedance seen at the source resistance for the cases where the switch is open, $Z1_R$ and is closed, $Z_{2R}$.

$$Z1_R = 24.3 - j1.6, Z2_R = 24.2 - j0.65$$

**(f)** Solve for the open switch and closed switch voltages, $VD1$ and $VD2$, respectively, at the junction of $C_R$ and $R_S$.

$$VD1 = (4.937 - j0.16)\ \text{V}$$

$$VD2 = (4.916 - j0.067)\ \text{V}$$

**(g)** The peak modulation voltage is $|VD1 - VD2| = 0.1$ V.

Near field modulation calculations are shown on Worksheet **RFID_NF**.

Fig. 13.7 shows a near field RFID reader for selective access through a turnstile, and a corresponding tag on a key ring. A short range is desirable for this application and the system pictured detects tags at a distance not exceeding 5 cm.

### 13.3.2 Far field RFID

The far field of a radio wave is considered to exist at distances greater than $2D^2/\lambda$, where $D$ is the largest antenna dimension. In the far field, the ratio of the electric field to the magnetic field is constant—$120\pi \approx 377\ \Omega$. Typical RFID far field frequencies are in the UHF bands, and operating range is usually greater than that in the near field. For half-wave dipole antennas, the far field at 433 MHz starts at around 35 cm from the antenna and 6 cm at 2.5 GHz.

Passive and semi-passive far field RFID devices are similar to radar in that the tags reflect back to the interrogator (reader) a portion of the energy that was radiated in their direction. The difference is that the tags responding to an interrogating signal modulate the antenna cross section, resulting in an amplitude and/or phase modulated reflection that is received at the reader receiver.

A UHF passive RFID system is shown in Fig. 13.8. The reader transmits a CW signal to the tag, periodically modulated with an interrogation message. The resulting voltage at the tag antenna is used to power up the tag, which has no other source of power. The tag demodulates the received signal and on command modulates the impedance seen by the tag antenna with its ID or other data, causing a modulated backscattered signal that is received at the reader. Since the reader is transmitting at the same time it is receiving, a circulator passes the weak received signal directly

FIG. 13.7

Near field RFID reader and tag on key ring.

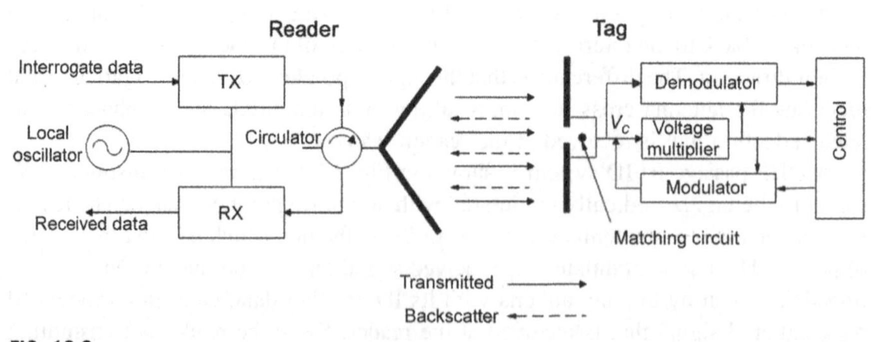


FIG. 13.8

UHF RFID reader and tag.

to the receiver input and significantly attenuates the feedback of the transmitted signal to reduce blocking out or desensitizing reception. The local oscillator is common to the TX and RX sections of the reader, facilitating coherent demodulation, because transmitted and received signals are inherently frequency locked since they are both from the same source.

While low power UHF communication is conducted over ranges of tens and hundreds of meters, the distance between readers and tags is limited because of two considerations: the signal strength at the passive tag needed to provide DC supply power, and the backscattered signal strength and modulation efficiency at the reader. The signal at a passive tag must be considerably stronger than that required for demodulation since it must have enough power to supply the tag's active circuits. The power at the tag antenna, $P_T$, assuming free space, is given by the Friis equation:

$$P_T = \frac{P_R G_R G_T \lambda^2}{(4\pi)^2 d^2} \tag{13.5}$$

where $P_R$ and $G_R$ are reader transmitter power and antenna gain, $G_T$ is the tag antenna gain, $\lambda$ is the wavelength and $d$ is the distance. Under the condition of perfect antenna matching, where the tag input impedance is the conjugate of the antenna impedance, which may be referred to the output of a matching circuit (Fig. 13.8), the peak chip input voltage is

$$V_C = \sqrt{2P_T R_C} \tag{13.6}$$

where $R_C$ is the input resistance of the chip. $V_C$ is stepped up by a voltage multiplier to the voltage required by the tag circuits. The chip power, of course, depends on the current supplied to those circuits and is what determines the range limit as far as concerns the reader to tag link. In addition to low circuit current, the efficiency of the voltage multiplier is also a factor in the power requirement. Minimum chip power is typically in the range of 50 to 500 μW for conventional silicon RF integrated circuits [6]. An exception was achieving a chip power input of 30 μW which allowed gaining an RFID distance of 4.5 m from reader power of 500 mW at 869 MHz [7]. The improved performance was obtained due to relatively high rectifier efficiency using Shottky diodes and optimizing backscatter modulation.

The power at the reader, $Pbs_R$, resulting from modulated reflected, or backscattered, radiation from the tag is expressed similarly to the radar equation, where the power is proportional to the fourth power of the distance:

$$Pbs_R = \frac{P_R {G_R}^2 {G_T}^2 \lambda^4}{(4\pi d)^4} M \tag{13.7}$$

where $M$ is the modulation factor:

$$M = \frac{1}{4} |\rho_1 - \rho_2|^2 \tag{13.8}$$

and

$$\rho_1 = \frac{Z_C{}^1 - \overline{Z}_T}{Z_C{}^1 + Z_T}$$
$$\rho_2 = \frac{Z_C{}^2 - \overline{Z}_T}{Z_C{}^2 + Z_T} \tag{13.9}$$

$Z_T$ is the source impedance from the tag antenna at the output of the matching circuit (Fig. 13.8). $Z_C$ is the tag chip input impedance and $\rho$ is the reflection coefficient. The subscripts on $\rho$ and superscripts on $Z_C$ designate the two logic levels of the modulating data from the chip.

The backscatter signal is modulated by switching the value of the chip impedance $Z_C$ that is seen from the tag antenna. This causes a change in the amplitude, phase, or both, of the reflected signal which is demodulated in the reader. The match or mismatch of the chip impedance to the antenna is indicated by the reflection coefficient $\rho$. For example, if $Z_C$ is a short circuit, $\rho = -1$, if it is an open circuit, $\rho = 1$, and if $Z_C$ matches $Z_T$ making $\rho = 0$, there is maximum power transfer to the chip from the tag antenna and minimum backscatter. Antenna mismatch creates a larger backscatter signal. However, it also reduces the power into the chip by reducing the voltage at the input to the voltage multiplier. The average of this power, a function of the reflection coefficient for each logic state 1 and 2 and the fraction of time in each state, $p1$ and $p2$, is the power available at the chip antenna $P_T$ times a power efficiency factor $PE$ [7]:

$$PE = p1\left(1 - |\rho_1|^2\right) + p2\left(1 - |\rho_2|^2\right) \tag{13.10}$$

For equal time in each logic state $p1 = p2 = 0.5$:

$$PE = 1 - 0.5|\rho_1|^2 - 0.5|\rho_2|^2$$

The choice of loads switched for each data logic state is a trade-off between the modulation factor and the power efficiency factor. Table 13.2 shows reflection coefficient pairs and the resulting $M$ and $PE$. Ref. [7] gives a detailed discussion on load switching considerations. Worksheet **RFID_FF.mcdx** does the calculations in Eqs. (13.5)–(13.10).

**Table 13.2** Far field RFID load trade-offs

| Rho 1 | Rho 2 | *M* | *PE* | Description of states |
|---|---|---|---|---|
| 0 | 1 | 0.25 | 0.5 | Perfect match and perfect mismatch (load shorted) |
| 1 | -1 | 1 | 0 | Load shorted and load open |
| $0.2 - j0.4$ | 0.8-j0.2 | 0.133 | 0.564 | Resistance switching with constant reactance example |
| $0.002 - j0.05$ | 0.9-j0.2 | 0.235 | 0.52 | Reactance switching with constant resistance example |
| $-(0.8 + j0.16)$ | $0.8 - j0.016$ | 0.66 | 0.33 | Ideal case. No power wasted. $M + PE \approx 1$ |

## 13.4 Communication protocols

The RFID communication protocol defines [8]

- modulation details, data rate, the way information is formatted
- medium access, dealing with collisions
- data definitions and organization

Only basic modulation types—amplitude or phase modulation—are used with passive tags, not spectrum efficient methods like QAM (quadrature amplitude modulation).

RFID communications are specified in ISO/IEC 18000, an international standard developed, published and sold by the International Organization for Standardization. It is organized in parts as follows:

- Part 1: Reference architecture and definition of parameters to be standardized
- Part 2: Parameters for air interface communications below 135 kHz
- Part 3: Parameters for air interface communications at 13.56 MHz
- Part 4: Parameters for air interface communications at 2.45 GHz
- Part 6: Parameters for air interface communications at 860 to 960 MHz general and four subparts according to types
- Part 7: Parameters for active air interface communications at 433 MHz

Part 5 relating to 5.8 GHz was withdrawn.

There exist a considerable number of standards for RFID, which apply to specific applications and protocol layers. This chapter deals with standards relating to the physical layer or air interface of RFID systems. For more comprehensive coverage of RFID standards in general, see Ref. [9].

### 13.4.1 UHF protocols

EPCglobal promotes a UHF protocol for passive RFID tags used in supply-chain tracking, which was approved as ISO 18000-6C [10]. It includes a unique electronic product code (EPC) identifier which is 64 or 96 bits long. Similar in purpose to a barcode, EPC additionally contains a serial number specific to each individual tag.

The EPC is stored on the transponder as a string of bits. It consists of a header with variable length and a number of data fields whose length, structure and function are determined by the header value. Currently, EPCs with a total length of 64 bits and 96 bits are specified [9].

The EPCglobal lGen2 Specification includes an operating procedure for an interrogator-talks-first RFID system operating in the 860-960 frequency range [11].

#### *13.4.1.1 Class 1 generation 2 (ISO 18000-6C) interrogator to tag communications*

Binary reader symbols that are transmitted to tags are shown in Fig. 13.9. The encoding method is PIE—pulse interval encoding. It provides a high average RF power to keep passive tags energized while the reader sends data. The tag distinguishes between "0" and "1" by the symbol length whose value is conveyed to the tag by

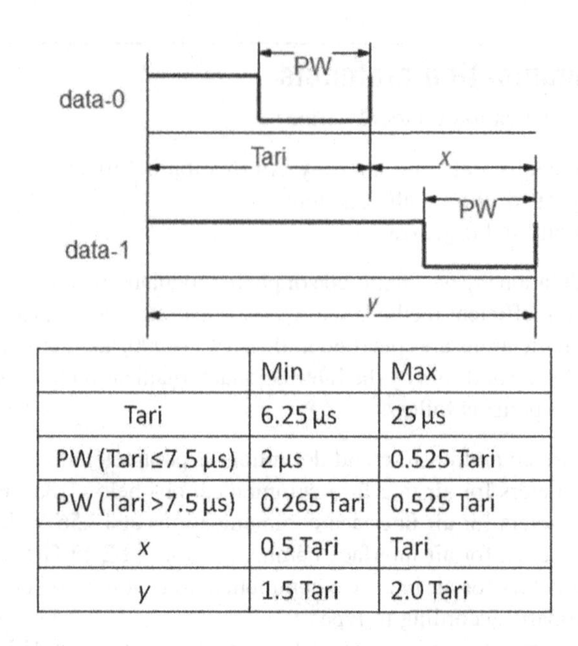


| | Min | Max |
|---|---|---|
| Tari | 6.25 μs | 25 μs |
| PW (Tari ≤7.5 μs) | 2 μs | 0.525 Tari |
| PW (Tari >7.5 μs) | 0.265 Tari | 0.525 Tari |
| $x$ | 0.5 Tari | Tari |
| $y$ | 1.5 Tari | 2.0 Tari |

**FIG. 13.9**

Pulse interval encoding (PIE) symbols from reader.

*After EPC™ Radio-Frequency Identity Protocols Generation-2 UHF RFID Specification for RFID Air Interface Protocol for Communications at 860–960 MHz Version 2.0.1 Ratified, EPCglobal Gen2 Specification, April 2015.*

a preamble in terms of "Tari" which is an abbreviation for Type A Reference Interval, defined in ISO/IEC 18000-6 part A. The value of Tari can be between 6.25 μs and 25 us. Modulation is ASK (amplitude shift keying) or PR-ASK (phase reversal-amplitude shift keying). Average data rate is between 27 and 107 kbps, depending on the value of Tari.

#### *13.4.1.2 Class 1 generation 2 (ISO 18000-6C) tag to interrogator communications*

Two methods are specified to encode the data that is backscattered from the passive tag. The encoding type is specified by the interrogator.

The *FM0 baseband method* (bi-phase space) is shown in Fig. 13.10 along with example sequences. For a logic "0" there is an inversion in the middle of a symbol period and for a logic "1" there is no inversion in the middle of a symbol period. There is always an inversion on the boundary between periods. FM0 signaling always ends with a dummy data −1 bit at the end of a transmission.

The second encoding method for a tag is *Miller-modulated subcarrier*. In this method basis functions for levels "0" and "1," shown in Fig. 13.11, are multiplied by a subcarrier containing $M = 2$, 4, or 8 cycles per data symbol, as specified in a command from the interrogator. The subcarrier bit rate remains constant over different $M$'s, so the symbol rate is divided by M. Fig. 13.11 shows Miller sequences

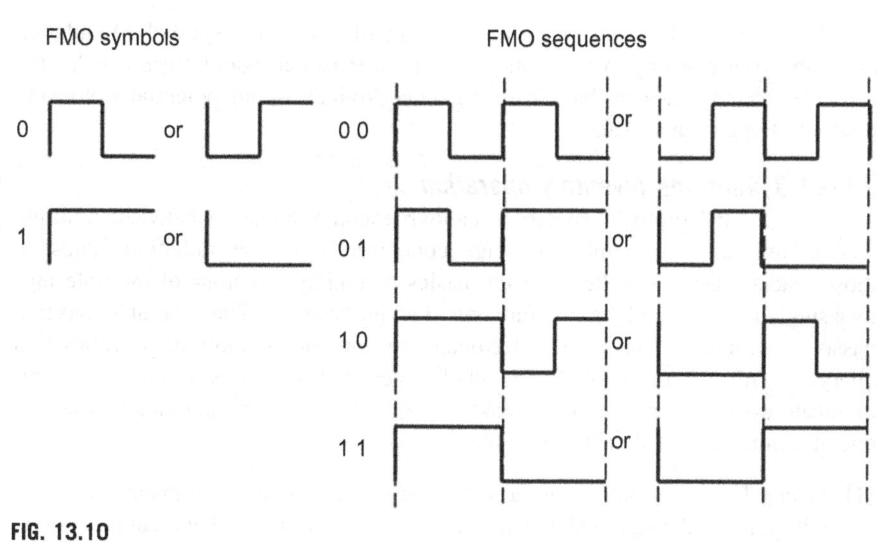


**FIG. 13.10**

Tag FM0 symbols.

*After EPC™ Radio-Frequency Identity Protocols Generation-2 UHF RFID Specification for RFID Air Interface Protocol for Communications at 860–960 MHz Version 2.0.1 Ratified, EPCglobal Gen2 Specification, April 2015, Fig. 6.9.*

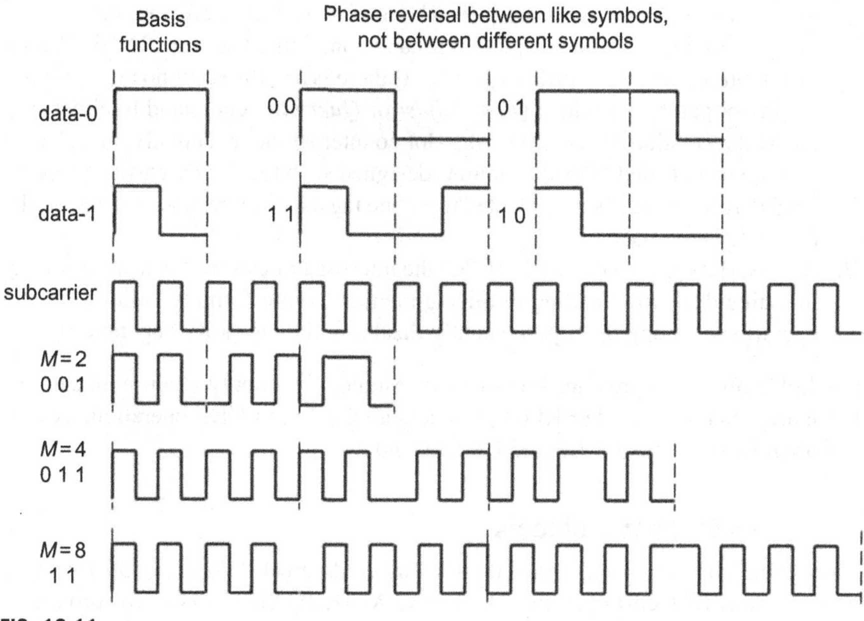


**FIG. 13.11**

Miller-modulated subcarrier method.

with $M = 2, 4$, and 8. A sequence is terminated with a dummy 1 symbol. Modulation of a subcarrier can improve reception quality at the interrogator since it helps the reader to separate the weak backscattered signal from the strong potentially blocking reader-to-tag carrier leakage.

#### 13.4.1.3 Multi-tag inventory operation

The Gen 2 EPC protocol specifies read/write communication between multiple readers and a large number of tags concurrently in the radiation fields of those readers. Described here are the basics of taking inventory of multiple tags by a single reader, which has the function of an interrogator. The general idea is that passive tags powered up by an interrogator respond at pseudo-random times to a *query* to identify themselves, after which the interrogator chooses which tags to communicate with. The problem is to avoid backscatter collisions and to read/write tags one at a time. This is done through the following steps [11].

**(1)** *Select*. The interrogator sends a command or a sequence of commands to all tags within range through which it sets flags in a subset of tags that meet the criteria for an inventory session.

**(2)** *Inventory*. Interrogator issues a *Query* command to initiate an inventory round and subsequent commands maintain it. The *Query* command sent by the interrogator contains a slot-count parameter Q. Participating tags pick a random value in the range 0 to ($2^Q - 1$) and load this value into a slot counter. A tag or tags with zero in the counter respond immediately by backscattering a random (or pseudo random) 16 bit number, RN16. If there is no collision, the interrogator replies with an ACK command containing the same RN16. The tag then replies with its identification EPC. If there is a collision or no tag response, the interrogator transmits a *QueryAdjust* or *QueryRep* command to either adjust the $Q$ value or decrement the slot counter so that eventually, all selected tags are inventoried. The algorithm is designed such that if $Q$ is chosen properly considering the number of selected tags, one tag only will respond to most of the *QueryRep* commands.

**(3)** *Access*. Having now the tag ID (EPC) the interrogator can choose to access a tag, initiating data flow, writing to the tag memory, or performing various other operations, among them permanently disabling the tag (a Kill operation).

The EPCglobal standards are specifically designed for supply chain applications. There are a number of other RFID protocols applicable to UHF operation, as well as numerous standards for LF and HF tags and readers.

### 13.4.2 13.56 MHz protocols

Three standards describing contactless smart cards are ISO/IEC 10536 for close coupled cards (0-1 cm) operating on 4.9152 MHz, ISO/IEC 14443 for proximity cards (up to 10 or 15 cm) and ISO/IEC 15693 for vicinity cards (up to 1 m). We discuss here the latter two, which operate on 13.56 MHz.

#### 13.4.2.1 Proximity-coupling smart cards

Standard ISO/IEC 14443 describes two different data transfer modulation methods, type A and type B. [12]. Readers must support both types, while a ISO/IEC conforming smart card needs support only one.

##### ISO/IEC 14443 type A

For type A cards 100% ASK modulation with modified Miller coding is used for data transmissions from the reader to the card at a rate of 106 kbps. A blanking interval in the middle of a bit period indicates level "1" and sequences of "0" are interrupted by a similar interval at the beginning of a bit period. A short interval of 2 to 3 us is used to guarantee maintenance of power to the passive tag from the reader transmission.

Card data are transferred to the reader through load modulation by a subcarrier of 13.56 MHz/16 = 847 kHz. Manchester coding modulates the subcarrier by the card data using on-off keying at a bit rate of 13.56 MHz/128 = 106 kbps. For a logic bit "1" the carrier is modulated with the subcarrier for the first half of the bit duration. For a logic bit "0" the carrier is modulated with the subcarrier for the second half of the bit duration. Reader and card transmissions are shown in Fig. 13.12.

The type A protocol uses a dynamic binary search tree algorithm to select individual cards for data transfer when multiple cards are within reader range and collisions between card responses can occur. The principle behind this algorithm is that collisions between Manchester encoded data from cards whose frames are synchronized reveal to the reader the positions of bits that differ. The algorithm works as follows.

**(1)** The reader sends a *request* command containing the highest serial number of the card group. Cards that are powered up by the reader transmission that meet the command criterion respond with their ID number.

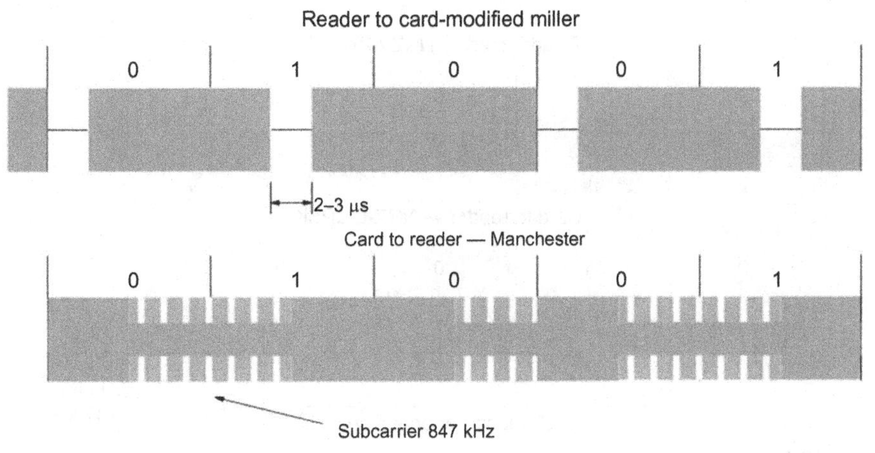


**FIG. 13.12**

ISO/IEC 14443 type A modulation.

**(2)** If only one card responds, the reader sends a *select* command with the ID of the lone responder and then can proceed with data exchange. If two or more cards respond, their transmissions collide. Bits of the ID's that are common will be received correctly. Bits that differ will be recognized by having a constant level, which contradicts the Manchester code where the signal level inverts in the middle of each bit.

**(3)** The reader sends a new *request* command with an upper limit serial number chosen from the knowledge of common bits learned from the previous iteration. As in step 2, one card can be selected, or if there are collisions, more common bits will be learned.

**(4)** Steps 2 and 3 are continued until all cards have been selected.

Because of the nature of the algorithm, it's not necessary for all serial number bits be communicated for all iterations. In an optimizing *dynamic* algorithm, the reader sends a parameter NVB (number of valid bits) in the *request* command, thereby reducing the amount of data transferred and the total time needed for the process.

## ISO/IEC 14443 type B

Transmission from reader to cards uses 10% ASK with NRZ coding in type B. Data are transferred from card to reader by load modulation with a 847 kHz subcarrier, which is modulated by BPSK by a NRZ-L (nonreturn to zero-level) coded data stream. Data rate is 106 kbps in both directions. Reader and card transmissions are shown in Fig. 13.13.

The anti-collision algorithm used in type B is a *dynamic slotted ALOHA* procedure. As in type A, reader responses have to be synchronized. After being powered up by the reader transmission, the cards in range wait for a reader *request-B* command. This command contains a parameter that gives the number of subsequent time slots that are available for a card response, as well as an application family identifier

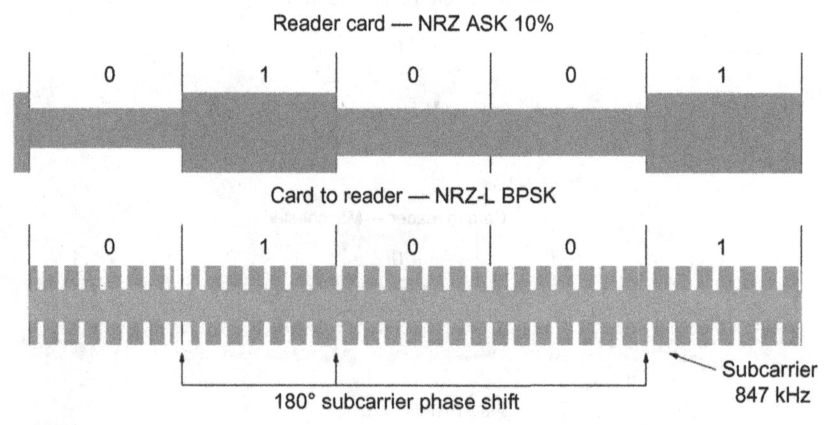


**FIG. 13.13**

ISO/IEC 14443 type B modulation.

for the search group. Each card that is in the indicated group randomly chooses a slot within those available for responding with its identification number. If the reader detects a single card response in one of the allocated slots, it initiates data transfer with that card. If there are no responses, or all slots have collisions, the reader issues a new *request-B* command, possibly with a new number of available time slots, and the card response procedure repeats. This is a stochastic process and since allocation of the number of time slots is dynamic, the time for scanning all cards can be optimized.

#### *13.4.2.2 Vicinity-coupling smart cards [9]*

Smart cards with detection range of up to around 1 m are described in ISO/IEC standard 15693 which covers functioning and operating parameters of contactless vicinity-coupling smart cards. These are passive near field cards operating on 13.56 MHz.

The vicinity-coupling card readers can use both 10% and 100% modulation index ASK modulation for transmissions from a reader to a card. Data are encoded by a pulse position method. There are two data coding modes: "1 of 256" and "1 of 4."

In the "1 of 256" procedure a time span of 4.833 ms of the carrier wave is divided into 256 slots of 18.88 μs. The value of a byte is indicated by an amplitude modulating pulse of 9.44 μs positioned in the second half of one of the slots, numbered from 0 to 255. "1 of 256" coding is shown in Fig. 13.14A. There are 8 bits in a span

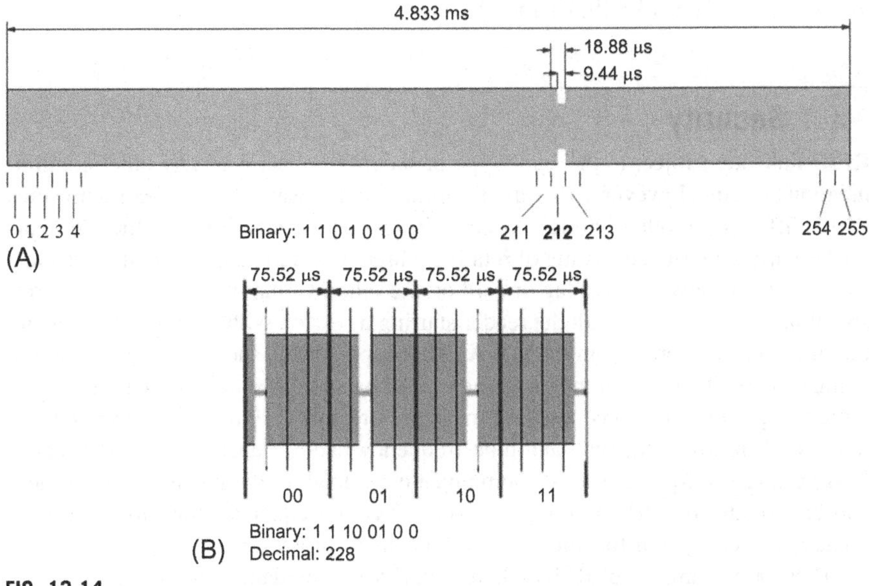


**FIG. 13.14**

ISO/IEC 15693 vicinity cards. (A) 1 of 256 coding mode and (B) 1 of 4 coding mode.

**Table 13.3** Smart card modulation and data rates

| | ASK | FSK |
|---|---|---|
| Subcarrier frequency | 423.75 kHz | 432.75/484.28 kHz |
| Divider ratio from $f_c = 13.56$ MHz | $f_c/32$ | $f_c/32$; $f_c/28$ |
| Long-distance data rate | 6.62 kbps | 6.62/6.68 kbps |
| Fast data rate | 26.48 kbps | 26.48/26.72 kbps |

of 4.833 ms, so the bit rate is 8/4.833 ms = 1655 bps. Data frames are bounded by start-of-frame (SOF) and end-of-frame (EOF) signals.

The "1 of 4" procedure is similar. A span of four pairs of 9.44 us slots is used to indicate a symbol value of 0, 1, 2, or 3 depending on whether the second, fourth, sixth or eighth slot is modulated. This is shown in Fig. 13.14B. There are two bits per symbol. Bit rate is $2/(8 \times 9.44\ \mu s) = 26483$ bps. SOF and EOF symbols delineate data frames.

The combination of modulation depth (10% or 100%) and coding method affects the maximum distance between reader and cards. The long distance mode favors 10% and "1 of 256" at the expense of data rate whereas 100% and "1 of 4" is best for high data rates but at shorter distances.

Data transfer from card to reader uses load modulation by a subcarrier that is ASK or FSK modulated with a Manchester coded data stream. The subcarrier modulation type is determined by a flag bit in the header of the reader transmission. Two data rates are also supported. The variations of subcarrier modulation type and data rate are shown in Table 13.3 [9, Chapter 9].

## 13.5 Security

RFID links are subject to the same type of security concerns as any other communication system. However, there are differences in the way they are dealt with since RFID differs from other wireless systems in several aspects. For one, the terminals are very asymmetric, consisting of relatively high powered readers on one end and a passive or low power active tag or card on the other. Communication is also correspondingly asymmetric, with the reader starting a session with an inquiry command that triggers a response from the tag. Another aspect is the short range of operation, being no more than several meters in the case of passive or semi-passive tags. Also, a large sub-group of RFID is based on magnetic or capacitive near field communication (NFC) in low frequency and high frequency bands, generally below 20 MHz. Passive and semi-passive tags communicate by load switching in the near field and backscatter on UHF frequencies which has consequences both for purposely breaching security and for implementing means of protection.

These are examples of RFID where security is a fundamental concern.

- Smartcards and tags used for banking and ticketing
- Entrance systems

- Automobile immobilizers and passive entry
- Supply chain product tracking
- Container identity and tracking
- Passport electronic ID

Security threats to RFID systems have been categorized as follows [13]:

- Spoofing identity: gaining unauthorized access by masquerading as someone else
- Tampering with data: modifying, adding or deleting data by an unauthorized person
- Repudiation: denying the truth or validity of an operation on an item
- Information disclosure: releasing information from an item to someone who cannot rightly gain access
- Denial of service: interruption of an authorized person's access to a system
- Elevation of privilege: gaining access to more privileges by a system user than he is authorized to receive

Threats to an RFID system may be carried out by unauthorized access to tags, use of rogue and cloned tags, and by infiltrating the communication link.

Three aspects of communication security in general are *confidentiality, integrity* and *authenticity*. They are described in Chapter 11, Section 11.6, along with other security concepts. RFID systems, in common with other communication links, must deal with service reliability, especially by protecting against denial of service attacks which thwart the functioning of the system.

While security threat countermeasures have been proven effective for wireless systems of the types discussed in Chapters 11 and 12, they cannot economically and functionally be used in all RFID applications. The scope of RFID is very wide, and while state-of-the-art security techniques can be incorporated in chips implanted in passports or in smartcards used for banking, they are not at all practical for inexpensive tags affixed to inventory items or pass tags for entrance to a gym, for example, which do not have the computing capability for sophisticated encryption and integrity algorithms. Much of the work on RFID security focuses on authentication, which can prevent tag duplication and use of rogue readers to gain access to stored data in tags. Authentication schemes for RFID aim to remove complexity from the tag to the reader, host, or other data base system to limit security related computation on the tag chip to simple operations [14].

## 13.6 Near field communication (NFC)

We have discussed ways to carry out communication for RFID in the near field, where the distance between reader and tag is significantly less than a wavelength. NFC also has another connotation as signifying NFC capabilities incorporated in smartphones and tablets [15]. Some uses of NFC in this context, using a smartphone, are

- Checking out at a grocery store
- Getting on a subway

- Secure payments
- Getting local exhibit information while visiting a museum
- Initiating file transfer between smartphones

The basic standard for NFC is ISO/IEC 14443 described above, in common with readers and tags in general that operate on 13.56 MHz. The NFC Forum develops protocol specifications and test mechanisms, and maintains a certification program to insure interoperability between certified devices. The program encompasses handsets, tags and readers.

NFC devices operate in three modes:

(1) Reader/writer mode: enables devices to read information stored on inexpensive NFC tags embedded in smart posters and displays. They can also write demanded data. Examples are reading timetables and getting special offers.
(2) Peer-to-peer mode: enables two devices to communicate for the purpose of exchanging information and sharing files. It works by a simple initiation act, like tapping two devices together, which triggers information transfer over a Bluetooth or Wi-Fi link.
(3) Card Emulation Mode: enables NFC devices to act like smart cards so that users can very simply make purchases, get tickets and access to mass transportation facilities. In this mode, the NFC device communicates with an external reader much like a traditional contactless smart card.

NFC devices have the same security vulnerabilities as described in the previous section for RFID equipment in general. Among the most serious problems, particularly connected with mobile payment arrangements or an access application in card emulation mode, is relay attacks. While NFC phones can be victims of a relay attack, they can also be used to carry one out against smart cards. The principle of a relay attack is shown in Fig. 13.15 [16, 17]. It is based on using a fake smartcard device, which can be a NFC phone, in proximity to the payment system reader. In normal operation, the reader triggers the client smartcard to transmit its ID and begin an authenticating procedure which, when successful, results in payment. The fake (emulated) smartcard, of course, does not have a recognized ID, but when it is in the reader's magnetic field, it will take the reader's interrogation message and relay it, through a Wi-Fi or other connection to a proxy reader that is in range of the victim NFC device. The proxy reader gets the response from the victim device, which may be encrypted, and transmits it back to the fake smartcard emulator which in turn backscatters it to the legal reader. Authentication and decryption processes can be carried out over the illegal relay link which will allow a fraudulent transaction or physical access to take place.

Two methods of counteracting the relay attack are based on detecting the excess delay required for using the relay link as compared to the response delay of a true proximity card. One method uses the frame waiting time (FWT) that ISO/IEC 14443 defines as the maximum time for the card to respond after a reader interrogation. Since the fraudulent relay path introduces additional delay, the fake smartcard response may not be valid. However, it has been pointed out [17] that when two NFC

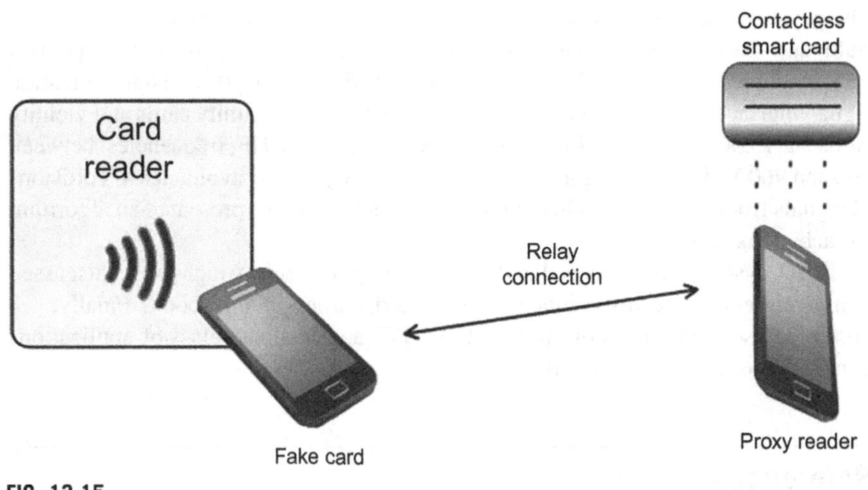


**FIG. 13.15**

Relay attack using NFC phones.

mobile phones are used for the attack there is no timing limitation because the devices are active, not passive as is the case for ISO/IEC 14443. The other method is called distance-bounding where round trip interrogation and response times are measured to find time-of-flight (TOF) and thereby distance between card and reader. One system using this method uses a UWB link which has the bandwidth for the necessary TOF resolution [18]. This method does not appear to be practical for card emulation mode NFC devices. A practical, cheap and effective way to prevent a relay attack is to keep the smart card in a radio frequency shielded covering, for example a wallet made of radiation impervious material, when it is not being used.

While relay attacks have the potential for compromising contactless NFC transactions and controlled access, the actual risk may not be high due to the short near field distances involved and the fact that the effort needed to carry out an attack would not in most cases justify the benefit that would accrue to the attacker. Also, the convenience of use and the fact that the smart card/NFC phone doesn't ever leave the hands of its owner may outweigh any concerns about ultimate security [19].

## 13.7 Summary

The roles of RFID systems over the frequency range of 125 through 2.45 GHz were discussed in this chapter. All of them have a configuration of a reader terminal that forms a communications link with a tag or (smart) card that serves to convey the identification of an article or human user. Two classifications were described: according to energy source of the tag/card—"passive" or "active," and type of propagation link—near field or far field with consequential frequency band divisions.

The physical principles of communication by way of induction coupling in the near field and backscattering over short distances in the far field were explained with examples and Mathcad worksheet backup. We explained the communication signal interfaces specified in international standards for proximity cards and vicinity cards at 13.56 MHz, and for backscattering tags on UHF frequencies between 860 and 960 MHz. Multi-tag inventory operations require the avoidance of collisions of signals from multiple tags in the range of a reader, and we presented an algorithm for achieving this.

RFID systems have specialized security requirements which were discussed while referencing security concepts presented earlier in this book. Finally, we presented uses and modes of operation of NFC, a special sub-class of applications generally associated with smart phones.

## References

[1] F. Hussien, D. Turker, R. Srinivasn, et al., Design considerations and trade-offs for passive RFID tags, in: Analog & Mixed-Signal Center, Texas A&M University, SPIE Europe International Symposium Microtechnologies for the New Millennium, Sevilla, Spain, May 2005, 2005.

[2] S. Smiley, UHF RFID Tag Communications: Protocols and Standards, RFID Insider, 2016. https://blog.atlasrfidstore.com/uhf-rfid-tag-communications-protocols-standards.

[3] Y. Lee, Antenna Circuit Design for RFID Applications, AN710, Microchip Technology Inc, 2003.

[4] W. Saslow, Electricity, Magnetism, and Light, Academic Press, 2002.

[5] W.W. Lewis, C.F. Goodheart, Basic Electric Circuit Theory, The Ronald Press Company, 1958.

[6] J.D. Griffin, G.D. Durgin, Complete link budgets for backscatter-radio and RFID systems, IEEE Antennas Propag. Mag. 51 (2) (2009).

[7] U. Karthaus, M. Fischer, Fully integrated passive UHF RFID transponder IC with 16.7-W minimum RF input power, IEEE J. Solid State Circ. 38 (10) (2003).

[8] D.M. Dobkin, Enigmatics, "The RF in RFID: Physical Layer Operation of Passive UHF Tags and Readers", http://www.enigmatic-consulting.com/Communications_articles/RFID/RFID_protocols.html, 2012. Accessed 24 October 2018.

[9] K. Finkenzeller, RFID Handbook, third ed., Wiley, 2010.

[10] Gen 2 EPC Protocol Approved as ISO 18000-6C, RFID Journal, 2006.

[11] EPC™ Radio-Frequency Identity Protocols Generation-2 UHF RFID Specification for RFID Air Interface Protocol for Communications at 860–960 MHz Version 2.0.1 Ratified, EPCglobal Gen2 Specification, April 2015.

[12] International Standard ISO/IEC 14443-2, Identification Cards—Contactless Integrated Circuit(S) Cards—Proximity Cards—Part 2: Radio Frequency Power and Signal Interface, first ed., 2001.

[13] S. Vattem, T. Anjali, Complete RFID security solution for inventory management systems, in: International Conference on Advances in Computing, Communications and Informatics (ICACCI), Manipal, Mangalore, India, 13–16 September 2017, 2017.

[14] A. Maarof, Z. Labbi, M. Senhadji, M. Belkasmi, Security analysis of low cost RFID systems, in: 5th Workshop on Codes, Cryptography and Communication Systems (WCCCS), El Jadida, Morocco, 27–28 November 2014, 2014.

[15] NFC Forum, https://nfc-forum.org/. Accessed 24 October 2018.

[16] T. Kasper, D. Carluccio, C. Paar, An embedded system for practical security analysis of contactless smartcards, in: Workshop in Information Security Theory and Practices, Heraklion, Crete, Greece, 2007.

[17] D. Cavdar, E. Tomur, A practical NFC relay attack on mobile devices using card emulation mode, in: MIPRO 2015, 25–29 May, Opatija, Croatia, 2015.

[18] G.P. Hancke, M.G. Kuhn, An RFID distance bounding protocol, in: Proceedings of IEEE/Create-Net Secure Comm 2005, Athens, Greece, September 5–9, 2005.

[19] J. Jumic, M. Vukovic, in:Analysis of Credit Card Attacks Using the NFC Technology, MIPRO 2017, May 22–26, Opatija, Croatia, 2017.

CHAPTER

# Technologies and applications

# 14

This concluding chapter covers networks and technologies that are not defined by a specific device standard yet encompass the principles, standards and regulations that were dealt with in previous chapters. In addition, we include a briefing of short-range wireless technologies that are outside the realm of the VHF-UHF devices and systems commonly considered as short-range communication.

First we describe ad hoc networks and wireless mesh networks (WMN) and then proceed to wireless sensor networks (WSN) and the Internet of things (IoT), all of which are inherently connected. While local and personal area networks (PAN) were discussed in previous chapters, here we come down in range to body area networks and look at its implementation. Location awareness is an important adjunct of wireless communication in general and mobile terminals in particular, so we will explain aspects of its integration in short-range systems. Finally, we take an overview of millimeter wave and optical short-range communication.

## 14.1 Wireless ad hoc networks

A *wireless ad hoc network* is a decentralized type of wireless network. The network is ad hoc because it does not rely on a preexisting infrastructure, such as fixed routers in wired networks or access points in centrally managed (infrastructure) wireless networks. Instead, in an ad hoc network each node participates in routing by forwarding data through other nodes as required to reach a destination, and so the determination of which nodes forward data is made dynamically based on the network connectivity.

In a "pure" ad hoc network all devices have equal status and each can associate with any other network device either directly when range permits, or over multiple hops. This is shown in Fig. 14.1.

Wireless ad hoc networks can be further classified by their application: mobile ad hoc networks (MANET), WMN, and WSN.

In order to support communication between any two devices in the network, intermediate nodes serve as relay points and forward messages from one adjacent node to another until the destination is reached. To do this, the nodes must maintain routing tables, whose entries vary dynamically as nodes move and turn on and off. If a path is interrupted, an alternate path must be found to deliver packets to the end point.



Short-range Wireless Communication. https://doi.org/10.1016/B978-0-12-815405-2.00014-2
© 2019 Elsevier Inc. All rights reserved.

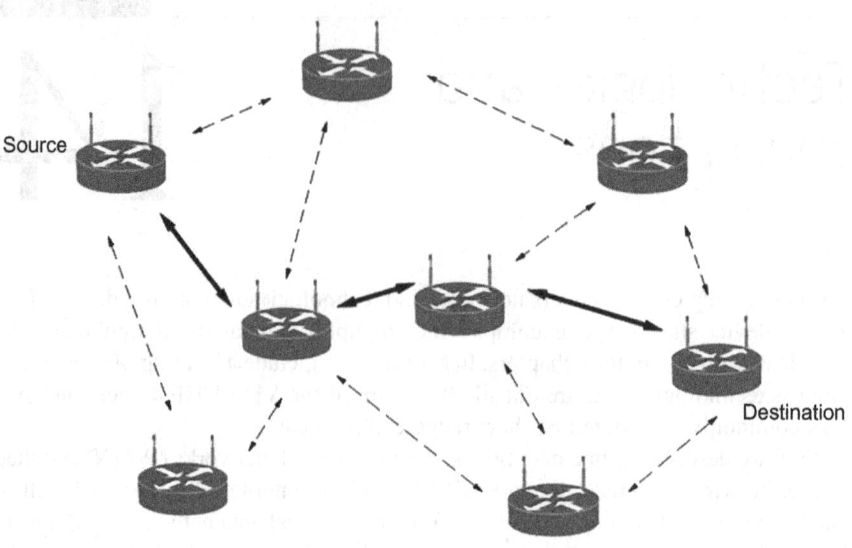


**FIG. 14.1**

Wireless ad hoc network. The heavy solid lines indicate the dynamic route between source and destination.

### 14.1.1 Routing protocols

An *ad hoc routing protocol* is a convention, or standard, that controls how nodes decide which way to route packets between computing devices in a mobile ad hoc network.

In ad hoc networks, nodes cannot know in advance the network topology. Instead, they have to discover it. The basic idea is that a new node announces its presence and listens for announcements broadcast by its neighbors. Each node learns about nodes nearby and constructs a map with the paths to reach them.

A routing protocol must discover new routes while keeping track of the validity of old ones. Various protocols are available and chosen depending on circumstances. Even a static ad hoc network must have dynamic routing tables, since nodes can fail and alternative routes must be discovered.

Routing protocols are classified in three categories:

- Proactive: current routing tables for the network are maintained continuously.
- Reactive: routes are established on demand, when there are data to transfer.
- Hybrid: combine features of proactive and reactive protocols.

Proactive and reactive protocols are described below.

#### *14.1.1.1 Proactive routing (table driven)*

In table driven routing protocols each node maintains one or more tables containing routing information to every other node in the network. All nodes keep on updating these tables to maintain the latest view of the network. Routes are available quickly (low latency) but table maintenance is slow and involves a lot of data and communication overhead. Optimized Link State Routing (OLSR) is an example of a proactive routing protocol where the routes are always immediately available when needed [1].

#### 14.1.1.2 On demand or reactive protocols

In these protocols, routes are created as and when required. When a transmission is needed from source to destination, it invokes the route discovery procedure. The route remains valid until destination is achieved or until the route changes or is no longer needed. This class has high latency and can cause network clogging.

Examples are dynamic source routing (DSR) and ad hoc on-demand distance vector (AODV) routing. They work by broadcasting route request messages from node to node (flooding), then waiting for the destination mode to return a reply message. In DSR node addresses are stored in the transmitted packets whereas AODV creates temporary tables in the route nodes. When the source finally gets a reply that a path to the destination is available, it can start to transmit its data [1].

Protocols differ by the message overhead used to support the routing, congestion from frequent flooding, and measures taken to maintain the routes, which change due to node mobility and failure of radio communication.

A widely used protocol for MANET is AODV. A simplified account of route discovery and repair is as follows, referring to Fig. 14.2 [2].

AODV defines three message types: route request (RREQ), route reply (RREP), and route error (RERR). Nodes are identified by their addresses, which may be IP v4 or IP v6 as assumed in the reference specification [2], or a different address format according the standards of the devices forming the network, for example 802.11 (Wi-Fi) or 802.15.4 (Zigbee). These are the stages of setting up a route on demand:

**(1)** In Fig. 14.2A, the Source, 2, initiates network flooding by broadcasting a RREQ which is received by the nodes in range, 1 and 4. RREQ contains source and destination addresses and a hop count field that starts at zero and is updated on each hop.

**(2)** Nodes 1 and 4 make note in a table of the addresses in the RREQ and the node from which it was received, then rebroadcast the RREQ.

**(3)** Similarly, nodes in range of these broadcasts act as in stage (2), until the destination node or an intermediate note which already has a route to the destination, receives the RREQ. A node receiving the RREQ a second time, node 4 for example which hears the message from node 3, will ignore it because it has a record of receiving it earlier.

**(4)** The destination node, 5 in this case (or an intermediate node with a route to the destination), recognizes its address and responds to the RREQ with a RREP to a node from which it was received. In Fig. 14.3A destination node 5 receives RREQ from nodes 3, 4, and 6. It will reply to the node from which the RREQ was received first. In our case this will most likely be node 4, since the number of hops from the source is the least (2 hops as compared to 3 hops from 3 and 4 hops from 6).

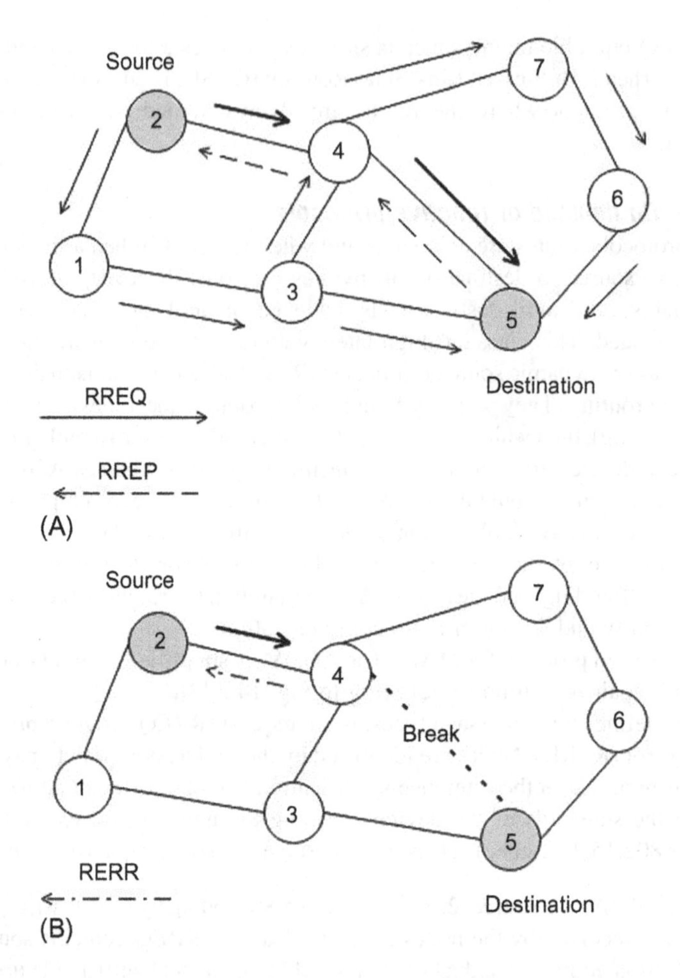


**FIG. 14.2**

Ad hoc network using AODV routing. (A) Setting up a route. (B) Error reporting.

**(5)** The RREP will continue to progress back to the source through the previous hop nodes over which the RREQ passed and which are written in the temporary route tables. In Fig. 14.2A RREP passes through node 4 to the source node 2.

**(6)** When the source receives the RREP, routing tables have been established, hop count (indication of distance to destination) is recorded and messages can be sent back and forth between the source and the destination.

Fig. 14.2B is used to show what happens when a break occurs on the route between communicating nodes 2 and 5.

**(1)** Node 4 realizes that its link to node 5 is broken, possibly due to movement of the terminals putting them outside radio range. It then initiates a RERR and sends it backwards on the route to the source, node 2.

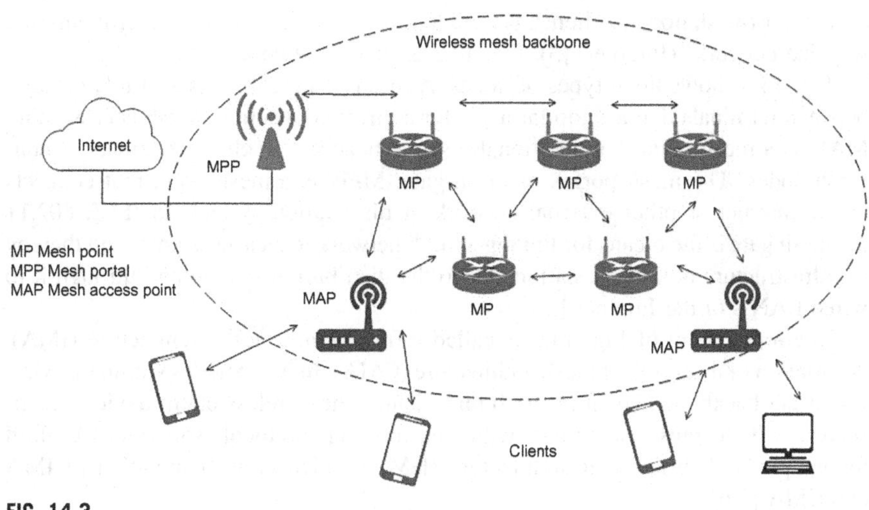


**FIG. 14.3**

Two tier wireless mesh network.

**(2)** On receiving the RERR, node 2 initiates a new route discovery. The new route will likely be along nodes 2, 4, 3, and 5 or 2, 1, 3, and 5.

AODV has other features not mentioned above. For one, it associates a sequence number with the source and destination addresses. Sequence numbers are incremented when new routes are formed and when faults are discovered. They prevent the possibility of a message going into an infinite loop and indicate when a route is stale and needs to be updated. Routes that are not used over a period of time are erased. Their validity can be extended by sending "Hello" messages to keep them open and reduce the overhead of renewing them. Also, the extent of the initial flooding of the network for route discovery can be limited and increased gradually as needed in order to reduce power and interference when source and destination are not too far apart in the network.

## 14.2 Wireless mesh networks

WMN based on ad hoc connections between wireless terminals provide an alternative to a fixed infrastructure for communication between the client devices themselves and between the clients and an external network or networks such as the Internet. Such a network provides flexibility of deployment in the case of mobile terminals. Moreover, multiple redundant wireless routes are able to bypass network faults to self-heal. The basis of the WMN is a wireless ad hoc network or mobile ad hoc network (MANET) shown in principle in Fig. 14.1. A wireless connection between a node and an adjacent neighbor is a mesh link. A two tier mesh network, consisting of a backhaul tier (ad hoc network such as shown in Fig. 14.1) and an access tier (mesh node to client), is used to move data between client terminals and wireline networks (Internet) [3]. Fig. 14.3 is such a network.

Fig. 14.3 shows three types of nodes. A mesh point, MP, acts as a relay station between terminals that are too far apart for a direct connection. Mesh access point, MAP, is a mesh point that additionally supports access to client terminals or non-mesh nodes. The mesh portal, or mesh gate, MPP, is a mesh point that connects to the Internet or other external network or distribution system. In IEEE 802.11 the mesh gate is the means for linking a mesh network to an access point and thereby to infrastructure non-mesh stations, or to the distribution system and through it to wired LAN's or the Internet [4].

The topography of Fig. 14.3 is called infrastructure mesh architecture (IMA). A variation of this is client mesh architecture (CMA). In it, the nodes within the wireless mesh backbone shown in the figure include the wireless client devices themselves, which must be fitted with the routing protocol software. A third topography is hybrid mesh architecture (HMA) which is a combination of IMA and CMA [5,6].

Three challenges that mesh networks must meet to provide optimum bandwidth over the coverage area are [3]:

- The efficient use of limited resources of capacity and time.
- Protection of resources, notably securing data and conserving power for long-term operation of mobile devices.
- Providing fairness by eliminating spatial bias—assurance that throughput of clients connected to mesh nodes closer to gateway nodes is not greater than those connected to mesh nodes further (more hops) from the gateways.

WMN have a wide range of applications and scale, among them gated communities for sharing Internet connections and other services among residences and community buildings, health systems for transferring medical data, and as part of large scale Wi-Fi networks where a wired (ethernet) infrastructure is not practical or reliable [5]. Another adoption of WMN is in the home environment for Internet access, where wired LAN is not available and coverage of all use areas, including the back yard, for example, is not achieved with a single Wi-Fi router. A system called HomeMesh has been described, which has some distinguishing features [7].

The architecture of HomeMesh is similar to that shown in Fig. 14.3. The mesh points (MP) have the roles of providing security and performing load balance, in addition to their forming the backbone for connectivity between the Internet and the client devices. In this network, the mesh routers, MP's and MAP's, may be normal desktop computers or laptops, and the MPP is a standard access point. Each mesh router has an additional off-the-shelf wireless interface card, allowing it to operate on two non-interfering Wi-Fi channels. One channel is used in what is called a managed mode for association with the access point or other routers for Internet access, while the second channel serves in a master mode for association with clients directly or through other routers. He et al. [7] claims that the use of two channels may improve throughput by a factor of 6 or 7 compared to a single channel.

In HomeMesh, mesh routing protocol software is installed in the routers while the access point and clients are non-modified devices. The wireless mesh backbone is transparent to them. The routers continuously monitor network traffic and divert data flow from a congested path to a lighter loaded one, achieving load balancing and optimal throughput. The routers periodically broadcast messages to maintain connectivity and update routing tables. Two metrics are used for selecting routes—hop count and what is called ETX. ETX is a function of the number of transmissions, including retransmissions, for sending a packet between nodes. Use of the two metrics for routing results in higher throughput than using hop count alone [7].

## 14.3 Wireless sensor networks

A WSN is a network of nodes, each of which is provided with a sensing device, that send information about their physical surroundings to an external client for a unified purpose. The basic layout of a WSN is shown in Fig. 14.4. Sensor nodes are spread over a geographical area called a *sensor field* and communicate between each other by point-to-point wireless links using ad hoc association, forming a network through which sensor data can flow from any node to an outlet point. Three types of nodes are defined [8]. *Common nodes* collect physical or environmental data from their associated sensors. *Sink nodes* receive and process the data from multiple common nodes and forward the information over a *gateway node* to the Internet or other external distribution network where it is accessed by an *observer*, the destination of the information. WSN nodes may also include actuators that perform control functions in the monitored area.

Examples of physical parameters or conditions that sensors detect or measure are light, sound, humidity, pressure, temperature, soil composition, air or water quality, and attributes such as size, weight, position, speed and direction [9].

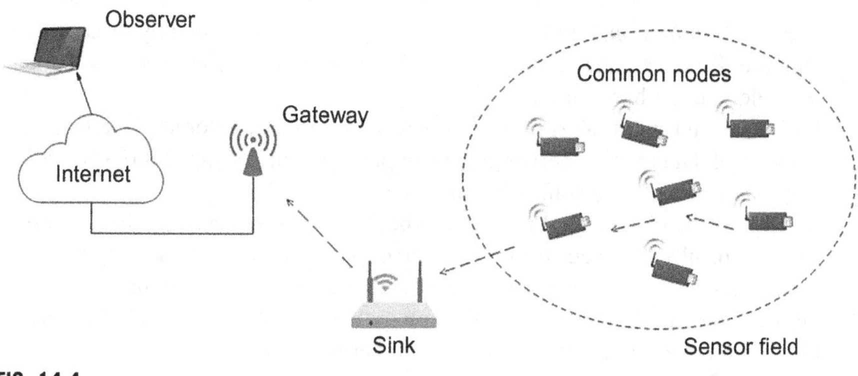


**FIG. 14.4**

Wireless sensor network.

A number of terms are commonly used to describe features of an WSN [8]. When all modes have the same hardware and capabilities the network is *homogeneous*; otherwise it is *heterogeneous*. Grouping of nodes within the network to facilitate or improve communication makes it *hierarchical*. In a *flat* network, all of its nodes communicate on the same level with the sink. A *static* WSN has stationary nodes whereas in a *dynamic* network the nodes are mobile. Topology may be dynamic in a static WSN since nodes may fail or be removed and new ones may be added. In a *symmetric* WSN each transceiver has the same communication range, otherwise it is *asymmetric*. A sensor network is called *continuous* when its sensors autonomously communicate their data and it is an *on demand* network when nodes respond to queries from the observer. A *reactive* sensor network reports data on the basis of events occurring in the environment whereas it is *programmed* when data is collected according to application defined conditions. A hybrid network combines two or more of the data transmission characteristics.

WSN have distinguishing characteristics [9]:

- High node density. Sensor nodes are deployed more densely than nodes in other ad hoc networks. Consequently, the data sensed by multiple nodes typically have a relatively high level of correlation and redundancy.
- Battery power. Sensor nodes are usually power by a battery. In many environments, it is difficult are impossible to replace or recharge batteries.
- Sensor nodes are limited in computation and storage capabilities
- Self organizing. The nodes are often randomly deployed without control of their exact location. They must autonomously configure themselves into a communication network.
- Self healing. Nodes may be deployed in hostile environments and be prone to damage and operational impairment. The network must reorganize itself as required to maintain communication in case of node failure. As a result, the network topology may change frequently.

These are some metrics that are considered for evaluating WSN [8]:

*Energy consumption*. Sensors may not be accessible conveniently or at all after they are dispersed, and so operation lifetime depends on low power and when possible, energy harvesting.

*Latency*. The time it takes from sensor response to a monitored phenomenon to delivery of data to the observer may be important in many applications. Late data may be of no value in some circumstances.

*Accuracy*. The exactness of a result can be defined as the fraction of valid results from all results obtained. It may be a function of environmental conditions.

*Fault tolerance*. There are several ways nodes may fail, including power loss, physical destruction, and communication problems. The network should continue to function even when some nodes no longer operate.

*Goodput*. This term is defined as the ratio of the total number of packets received by the observer to the total number of packets sent by all the sensors over a given period of time.

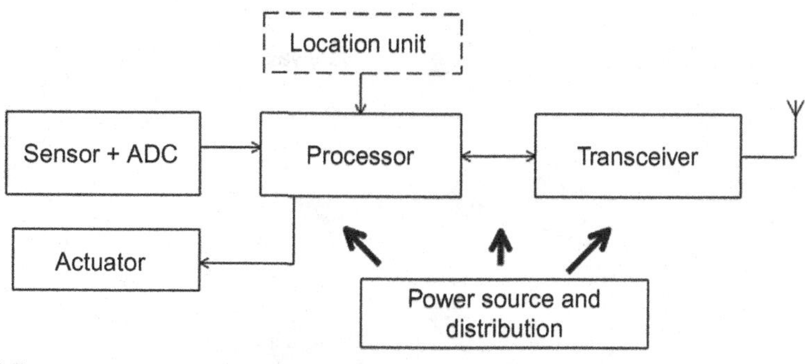


FIG. 14.5

WSN node hardware.

WSN have many varied applications [9]. Some examples are:

- Environmental monitoring: air and water quality, earthquake detection, forest fires, and floods.
- Military: track movement of forces, intrusion protection, and guidance to targets.
- Health care: disabled patient tracking, behavior monitoring, and monitoring of body functions.
- Industrial process monitoring and control.
- Security and surveillance.
- Home intelligence: smart home and remote metering.

Sensor hardware may have various forms, including general purpose computers, dedicated sensor nodes and system-on-chip nodes. Hardware organization of sensor nodes is shown in Fig. 14.5. Localization facilities are optional but are often needed in conjunction with the sensor measurement reporting and also for assisting in routing. Radio communication is typically conducted over UHF ISM bands, but some networks use optical or infrared communication which is virtually interference free but has the limitation of requiring line-of-sight.

A WSN may have various topologies but a typical example is the hybrid star-mesh network shown in Fig. 14.6 [10]. In this heterogeneous configuration, the sensor nodes do not have message forwarding capability and can be made for low power and cost, whereas the mesh nodes each support multiple sensors for multi-hop routing. These nodes may have a mains power supply. In another example, tree topology which is shown in Fig. 14.7, sensor nodes are configured in a hierarchy of groups. Data from each sensor node progresses through routers on the tree, eventually arriving at the PAN coordinator, or sink node.

### 14.3.1 Protocols and algorithms

Of prime importance in WSNs is the routing of data from the sensor nodes to a sink. Routing protocols must consider the special characteristics and needs of this type of network. There are power and resource limitations, latency considerations and the

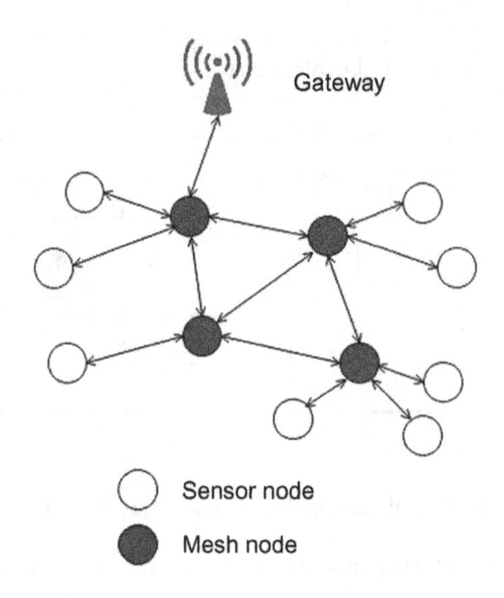


**FIG. 14.6**

Hybrid star-mesh topology.

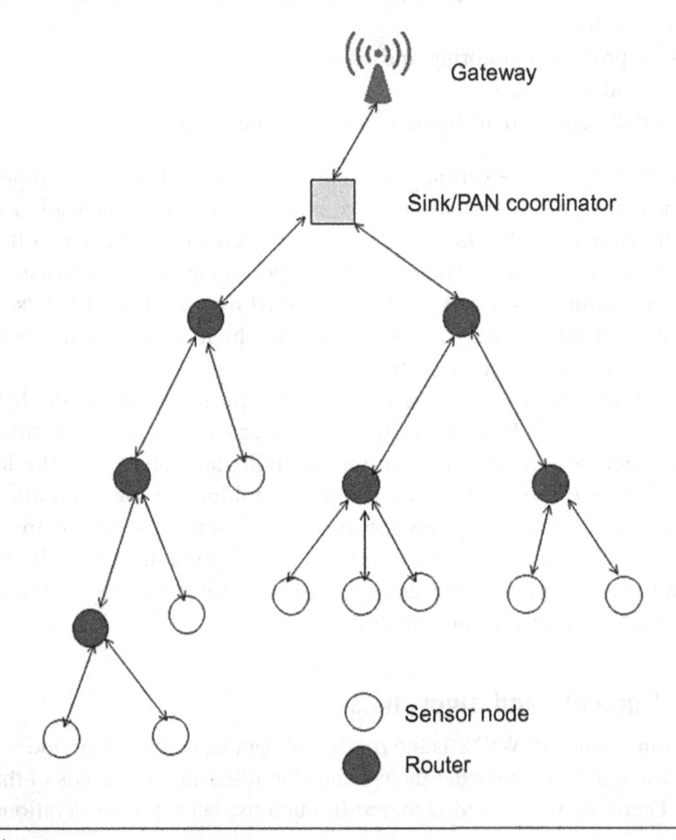


**FIG. 14.7**

Tree topology.

time-varying quality of the wireless channel which can cause packet loss and delay. Routing protocols may be divided into several classes [10]. One class is designed for a flat topology in which all nodes are considered peers. Protocols for a flat network architecture have the advantage of minimum overhead to maintain the infrastructure and to discover alternate routes between node endpoints in the case of faulty intermediate nodes. Another class of routing protocols aims to achieve energy efficiency, stability and scalability. To accomplish this nodes are organized in clusters in which a device with higher energy capability serves as a cluster head. The cluster head coordinates the activity of its component nodes and forwards data between clusters. Clustering can reduce energy consumption and extend the network lifetime. Still another class of protocols uses location based routing, which is useful where a source node query relates to sensor node position within the sensor field. The query may specify an area of interest or a specific point in the network environment.

An example of a WSN routing protocol based on clusters is LEACH (low energy adaptive clustering hierarchy) [11]. It is designed to collect data from sensor nodes that are organized in groups and deliver it to a sink, typically a base station. Its main objectives are extension of network lifetime, reduced energy consumption at each sensor node, and use of data aggregation to reduce the number of communication messages. LEACH has a hierarchical approach to organize the network into a set of clusters, each managed by a cluster head. Such a network is shown in Fig. 14.8. The nodes are

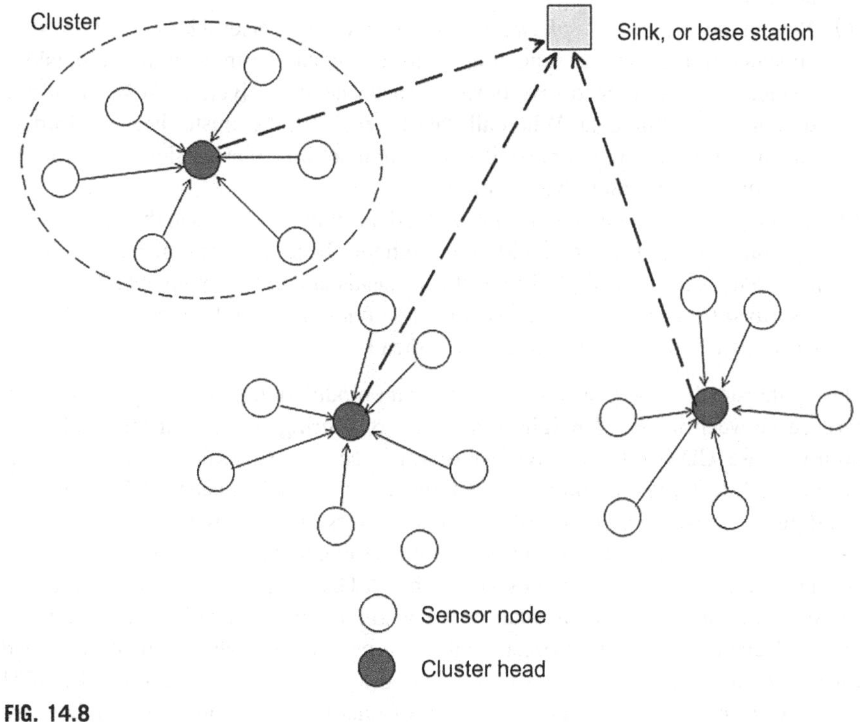


**FIG. 14.8**

WSN clusters.

homogenous, clusters are formed dynamically, and the role of cluster head is passed on in a random manner to all members of the network in order to average out the energy needed to communicate with the remote base station among all nodes.

The LEACH operation is carried on in *rounds*, where each round begins with a set-up phase when the cluster is organized, followed by a steady-state phase where sensor nodes send data to the cluster heads which aggregates it and forwards the information to a base station. The operation is as follows in each round.

**(a)** Initially, in the *set-up phase*, each node decides whether or not to become a cluster-head for the current round. Its decision is based on a randomized algorithm that is a function of a predetermined percentage of cluster heads among all nodes in the network.

**(b)** The self-determined cluster heads broadcast an advertisement message to all of the other nodes. These nodes decide on which cluster to join based on the received signal strength of the broadcast. In order to prevent interference among the multiple broadcasts, the cluster heads transmit using a carrier sense multiple access (CSMA) protocol, based on listening for a quiet channel before transmitting.

**(c)** Each node informs its chosen cluster head that it will be a member. Again, CSMA is used to prevent interference.

**(d)** Based on the number of nodes that joined its cluster, the cluster head creates a TDMA (time division multiple access) schedule which is broadcast back to the nodes in the cluster.

**(e)** Now in the *steady-state phase*, nodes transmit their sensor data in the time slots allocated to them in the schedule. Transceivers can be in sleep mode outside of their slots in order to save power. Cluster head receivers must be on all the time to collect the data. When all data is received, the cluster head performs signal processing to compress it into a compact information field which it transmits to the remote base station.

**(f)** After a predetermined time during which data from all sensors in the network has been preprocessed and transmitted through the cluster heads to the base station, a new round begins, with different cluster heads and cluster composition as described from (a) above. The steady-state phase is made long compared to the set-up phase in order to minimize overhead.

During the steady-state phase, spread spectrum modulation is used to prevent interference between nodes of neighboring clusters. During set-up, cluster heads randomly chose CDMA (code division multiple access) spreading codes for their clusters. While TDMA eliminates interference within each cluster, CDMA restricts interference between the nodes of different clusters in the network.

Other protocols have been devised to improve existing protocols or to better suite special sensor network objectives [12]. The PEGASIS protocol aims to improve LEACH. Instead of cluster formation it constructs communication link chains of nodes. Each node selects its closest neighbor as next hop in a chain. Lower overhead for the chain communication set-up compared to the cluster formation in LEACH gives performance enhancement in terms of energy consumption, but delays are greater due to sequential communication over the chain. The TEEN protocol is more appropriate than LEACH for event based applications. In it, cluster heads occupy different levels in a hierarchy, each passing data to a higher level cluster head until the base station is reached. Data is collected from a sensor only if an event of interest occurs, based on a threshold of measured data. An improvement on TEEN is APTEEN, which provides periodic information retrieval. Another category of WSN protocols is geographical routing protocols which exploit location knowledge at each node to provide efficient and scalable routing.

## 14.4 The Internet of things

The ITU (International Telecommunications Union) defines IoT as "A global infrastructure for the information society, enabling advanced services by interconnecting (physical and virtual) things based on existing and evolving interoperable information and communication technologies." It also defines the word "thing": "With regard to the Internet of things, this is an object of the physical world (physical things) or the information world (virtual things), which is capable of being identified and integrated into communication networks" [13]. These all embracing definitions make it possible to include virtually all telecommunications networks, wireless and not, under the IoT roof. However, for practical purposes, specifically regarding standards and security considerations, a more restrictive and specific understanding of the term is in order: "IoT connects heterogeneous devices that provide sensing, control, actuation, and monitoring activities for smarter environments" [14]. Among entities covered in this book that relate to or are part of IoT are WLAN (Chapter 11), WPAN (Chapter 12), RFID (Chapter 13), and WMN and WSN as described above in this chapter and WBAN below. Fig. 14.9 is an example of IoT. Household appliances, in this case an air conditioner and a microwave oven,

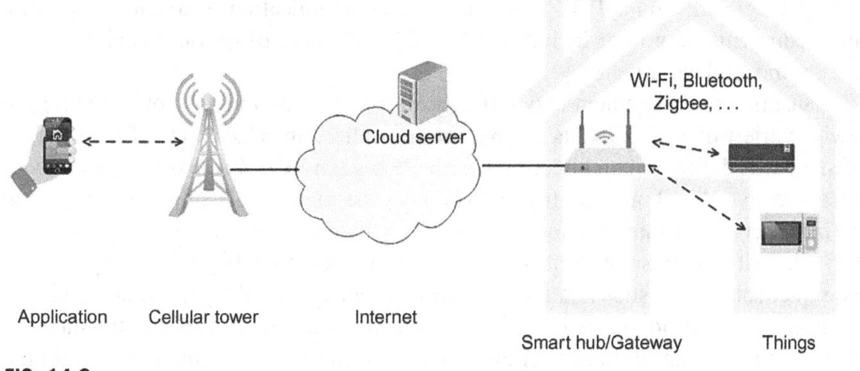


**FIG. 14.9**

Internet of things.

are fitted with sensors and actuators which are part of wireless nodes. These nodes are connected, typically through a wireless local network, to a smart hub acting as a sink, which in turn communicates through a router to the Internet, which is accessed through a cellular network by a remote smartphone. An app on the smartphone lets the user monitor and control devices and systems in his home while he is away. An external server connected to the Internet (cloud server) can store and process data from the things so that local nodes, which in many cases are battery powered, can be as simple as possible with minimum energy consumption. Among features which must be integral to such an arrangement are identification and addressing of the terminal things, and end-to-end security to prevent access to the things by unauthorized persons.

The concept of IoT has expanded the environments and applications where connectivity of devices and ubiquitous virtual access to information about the characteristics of all sorts of objects improves the quality of people's lives. The domains of IoT have been grouped as follows, with some examples [15]:

- *Transportation and logistics*. Traffic control, routing, depot management, status of transported goods, assisted driving, supply chain monitoring and management.
- *Healthcare*. Patient flow monitoring, objects and people tracking, patient and drug identification and authentication, access control, theft prevention, data collection, medical inventory management, remote patient health monitoring.
- *Smart environment*. Home and office comfort, heating, air conditioning, lighting, energy conservation, industrial plant part monitoring, machine vibration, machine-to-machine communication (M2M).
- *Personal and social*. Automatic social network event updates, person and thing location updates, theft detection.

### 14.4.1 Standardization

We discussed in previous chapters and sections standardization in respect to short-range wireless communication in general which applies as well to IoT. Standards originate in organizations such as IEEE, IETF (Internet Engineering Task Force), ITU, EPCglobal, and ETSI (European Telecommunication Standards Institute), and manufacturers groups including the Wi-Fi Alliance, Bluetooth Special Interest Group and Zigbee Alliance.

Sensors are an ingredient in practically all IoT systems and most of them work in a wide variety of environments that require specialized interfaces. The IEEE 1451 set of standards defines an architecture which allows sensor and actuator nodes to connect into live distributed control networks in the manner of "plug and play" and defines a common interface to give a network independent view of devices. IEEE 1451 describes how smart transducers can embed local intelligence to support, for example, self-diagnostics and local control and analytical algorithms, and can perform self-declaration to the network based on an electronic data sheet. Sub-standard IEEE 1451.2 defines a transducer electronic data sheet (TEDS) and standard digital interface and communication protocols to be used between transducers and a microprocessor [16]. As an example, an intelligent vehicle monitoring system for installation in buses was designed based on IEEE 1451.2 [17]. A single field programmable gate array (FPGA) chip was fabricated as an interface between various types of sensors and a microcontroller. The system collects video information, environmental information including temperature and humidity, and position information in real time through a GPS module. Processed data are transmitted to a server through 4G cellular where it can be accessed by client terminals. By taking advantage of the smart interface architecture, the system could be adapted to other types of sensors and IoT applications.

### 14.4.2 Security

IoT is particularly vulnerable to infractions for several reasons. Its components are often unattended and therefore open to physical attacks. Also, while there are multiple modes of protection for wireless transmissions, IoT devices have low computing resources and little energy to spare, so they cannot in many cases employ directly sophisticated authentication, data integrity and encryption facilities. Some security aspects can be handled by relegating protection methods to proxy devices in the communication link.

Research has been conducted to develop security algorithms which better fit the requirements of IoT. We give a brief description of a system that was designed for smart home automation through an Internet connected WSN, used, for example, for healthcare and environmental monitoring. Called triangle based security algorithm (TBSA), it uses an efficient key generation mechanism for data encryption. TBSA is claimed to provide the security requirements of authentication, confidentiality, integrity, replay protection, non-repudiation, and secure localization, among others [18]. The key used for authentication and encryption is generated for each sensor node from the sensor's identification number (ID) and time of collection of data. The algorithm for transforming the ID and time to an encryption key is similar to the calculations for finding a side of a non-right angle triangle from two known sides and their included angle. Security over a Wi-Fi communications channel is achieved from knowledge of the network SSID (service set identifier) and password. While TBSA is shown to demand reduced energy consumption for data encryption as compared to more conventional security mechanisms, its relative effectiveness in terms of security is not proved in Ref. [18]. However, the compromises between reduced energy consumption and level of security may be worthwhile in some applications, since lack of protective measures leave IoT systems open to access of unauthorized information and denial of service [19].

## 14.5 Optical wireless communication

Short-range communication using an optical medium is an alternative to radio waves for appropriate applications. While the radio spectrum from around 30 kHz to 300 GHz is progressively being used more efficiently and being able to handle much of the ever increasing data that are communicated, the optical spectrum that includes infrared, visible and ultraviolet wavelengths is able to absorb some of the communication load over short distances for network access and backhaul links. Special features of optical wireless communication (OWC) include ultra-high bandwidth, the absence of electromagnetic interference (EMI), spatial confinement (narrow beam width) for virtually unlimited reuse, and high physical security. Also, the optical spectrum is unregulated so there are no spectrum licensing fees [20].

OWC is carried out on three sub-bands: infrared (IR), visible (VL), and ultraviolet (UV). The radiating elements of visual light communication (VLC—390 to 750 nm wavelengths) are light emitting diodes, (LED), which can be pulsed at very high speeds. VLC is considered for WLAN, WPAN and vehicular networks, among others. Infrared communication has been in the mass consumer market for many years in the form of remote controls for home entertainment systems and wireless USB devices using IrDA (infrared data association) protocols. Terrestrial point-to-point systems known as free space optical systems (FSO) operate at near IR wavelengths of 750 to 1600 nm. They use laser transmitters and provide high data rates of 10 Gbps with potential for alleviating the backhaul bottleneck. Ultraviolet communication (UVC) on wavelengths of 200 to 280 nm has the advantage that it is immune to solar noise, since sunlight has negligible radiation on this sub-band at ground level. UVC may be employed outdoors to support low power short-range uses as in WSN.

OWC applications can be categorized according to range.

- Ultra short-range OWC is proposed for interconnections in integrated circuits, where high bandwidth and low latency, as well as convenience for multidimensional arrays and freedom from cross-talk give it advantages over wired connections.
- Short-range applications for use over tens of centimeters are viable for wireless body area networks (WBAN) with wearable sensors providing physical and bio-chemical information about a person. While RF communicating sensors are most often used for these purposes, optical communication from sensors to a sink node could be employed in hospitals and other medical facilities where RF radiation is restricted or prohibited because of EMI. Organic LED (OLED) technology makes it possible to integrate VLC transceivers into wearable devices and clothing as part of a WBAN.
- Medium range OWC, on the order of meters, has been investigated for use in WLAN's. Li-Fi is a term for a technology whose end use is similar to Wi-Fi, however it can be deployed in places where EMI cannot be tolerated. Actually, infrared was one of the media specified in the original IEEE 802.11 specification in addition to direct sequence spread spectrum (DSSS) and frequency hopping spread spectrum. Optical WLAN communication, through Li-Fi, may be making a comeback. A feature of OWC in this range, not previously available, is the use of general purpose lighting fixtures for transmission. LED lighting is replacing, worldwide, incandescent bulbs and fluorescent lamps, and we may expect to see dual purpose solid state devices used for illumination and for

communication, both on the scale of indoor WLAN as well as in city streets with high powered LED street lamps, traffic signs and advertising displays. Car headlights and taillights could also be used for vehicle to vehicle and vehicle to road side infrastructure communication.

An area where OWC is being developed is underwater communication. Acoustic communication in water can cover hundreds of meters, but at very small bandwidths and relatively high latency due to the speed of sound in water. While OWC cannot cover the same range, it is far superior to acoustic communication in both of these qualities. At visible light wavelengths of 450 to 550 nm, VLC may communicate underwater at data rates of hundreds of Mbps for ranges under one hundred meters [20].

## 14.6 Millimeter wavelength communication

The previous descriptions of RF communication in this book were based on the most used spectrum which is between around 300 MHz to 6 GHz, a span of more than 5 GHz which is not wholly used for mass communication. Of course, we deal here with short-range communication, which doesn't include cellular and other non-short-range networks that occupy the same frequency range. Considering the ever expanding volume of information that needs to be communicated wirelessly, much more spectrum has been put into use. We showed, particularly in Chapter 11, that significant throughput gains are achieved in the physical layer by higher modulation modes, increased channel bandwidth and MIMO. However, a much greater capacity increase to meet future needs is possible only through spectrum expansion. The answer provided by 5th generation cellular communication and several other services is the millimeter wavelength bands covering 30 to 300 GHz. Even "small" allocated portions of this range are roughly the same bandwidth as the whole spectrum generally used for mass communication through early 21st century. An example is the 59 to 64 GHz band that is allocated by the FCC for unlicensed operations.

Amendment IEEE 802.11ad became the directional multi-gigabit (DMG) PHY specification for millimeter wave communication in IEEE 802.11-2016. To make up for the high pathloss in the 60 GHz mm band, directional channel access was adopted. As a result, there are significant differences compared to physical and MAC layer specifications at sub-6 GHz frequencies, particularly as regards beam forming. The 60 GHz unlicensed band offers multi-GHz throughput with low modulation modes (BPSK for example) and reduced interference, but this is at the expense of adverse signal propagation characteristics, basically the high pathloss and reduced performance of NLOS operation with mm wavelengths. Consequently the communication range of mm wave signals is significantly reduced compared to low microwave for mobile ad hoc terminals at non-licensed band powers. The low penetration characteristic of the very short wavelengths is mitigated by high reflection from smooth surfaces, particularly metal, so mm wave communication is most suitable in indoor environments where there are suitable reflectors and blocking can be minimized. The features of 802.11 DMG (802.11ad) make it ideal for applications that need instant wireless synchronization, and high speed media file exchange between mobile devices without fixed network infrastructure, as well as wireless cable replacement, for example to connect wirelessly to high definition displays [21]. IEEE 802.11-2016 specifies a maximum data rate of 8085 Mbps with pi/2 64-QAM modulation (optional) and a maximum mandatory data rate of 1155 Mbps using pi/2 BPSK modulation.

Above 10 GHz, the path attenuation includes peaks around discrete frequencies due to radiation absorption by gas components in the atmosphere, notably oxygen and water vapor. This is apparent in Fig. 14.10. Note the sharp peak at 60 GHz. This attenuation is small at ranges of tens of meters and is actually advantageous in reducing in-channel and co-channel interference from larger distances, thereby improving performance.

In order to make up for high propagation losses, mm wave devices use directional antennas. Dynamic directional characteristics can be created by switching multiple directional antennas, or by creating directional antenna patterns through phase and amplitude shift networks connected to multiple radiating elements. The antenna array will be significantly smaller than that used with RF terminals that operate on the usual sub-6 GHz bands. Transmitter and receiver antenna patterns are created and adjusted to obtain maximum receiver input power, and with it signal-to-noise ratio. In DMG 802.11, the process of optimizing antenna pointing is called beamforming training.

The beamforming process is carried out using a robust modulation and coding scheme in what is called the DMG control mode, necessary since signal to noise ratio will be low before the terminal antenna patterns are adjusted. In the control mode, the data rate is 27.5 Mbps using differential BPSK and a code rate of ½. We describe the course beam pattern adjustment principle in four steps, demonstrated by an example in Fig. 14.11. Fig. 14.11A shows initiator and responder timing. Antenna scans and beam patterns are displayed in Fig. 14.11B.

Step 1: A terminal, the *initiator*, sends a sequence of frames, called a sector level sweep, each one using a different antenna sector pattern, with an identifying number in a field within the frame (Pattern A).

Step 2: The *responder* antenna pattern is quasi-omnidirectional, shown in Pattern B as a circle. The responder notes the strongest received sector, which is "3" in the example.

Step 3: The responder replies with feedback frames in a sequence of antenna sector beams (Pattern C). With each beam, it transmits the number of the strongest received initiator sector in Step 1, shown in the numerator of the displayed ratio, and the number of its transmitted sector, shown in the denominator.

Step 4: The initiator, not yet knowing its strongest sector, uses a quasi-omnidirectional pattern to listen to the responder transmissions. When the

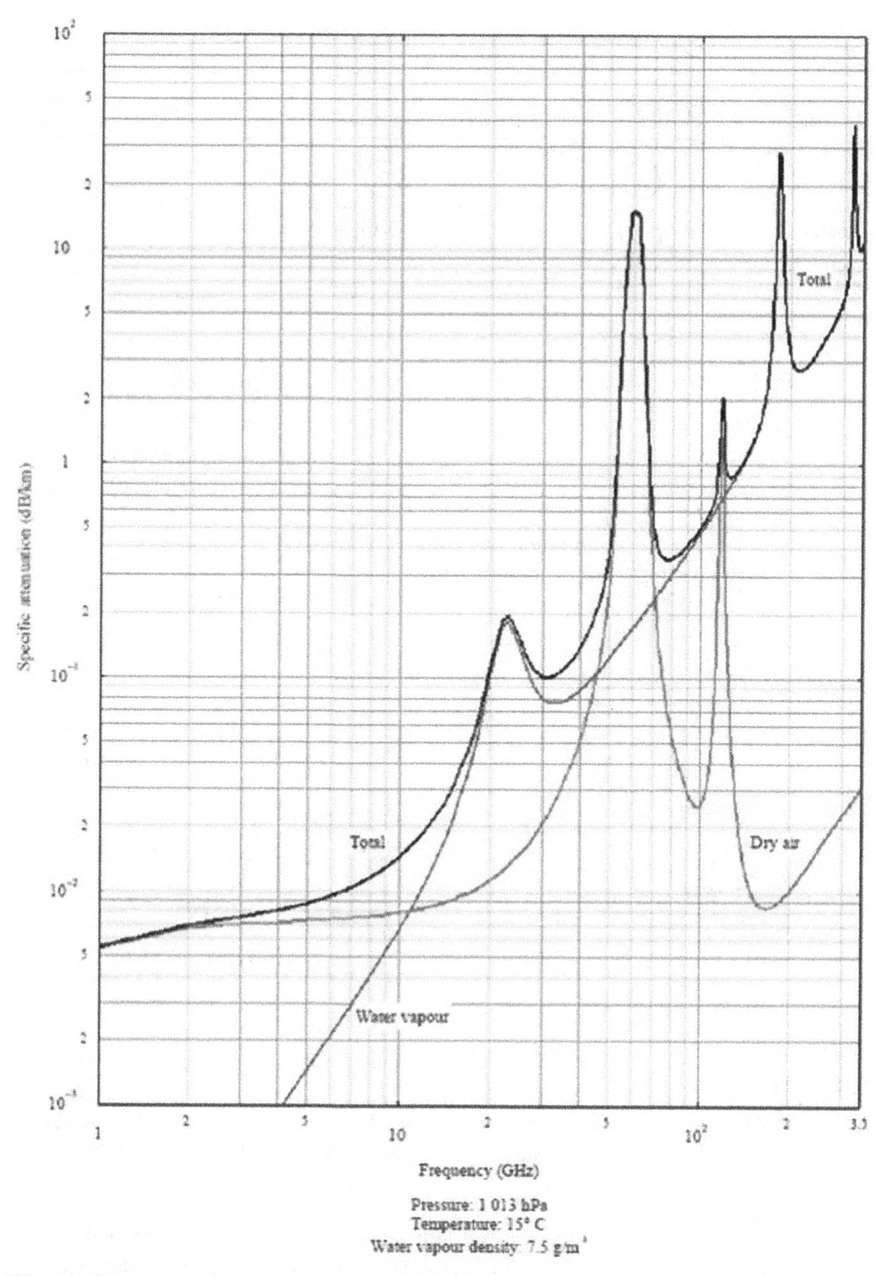


**FIG. 14.10**

Specific attenuation due to atmospheric gases (Ref. ITU-R P.676-9, with permission).

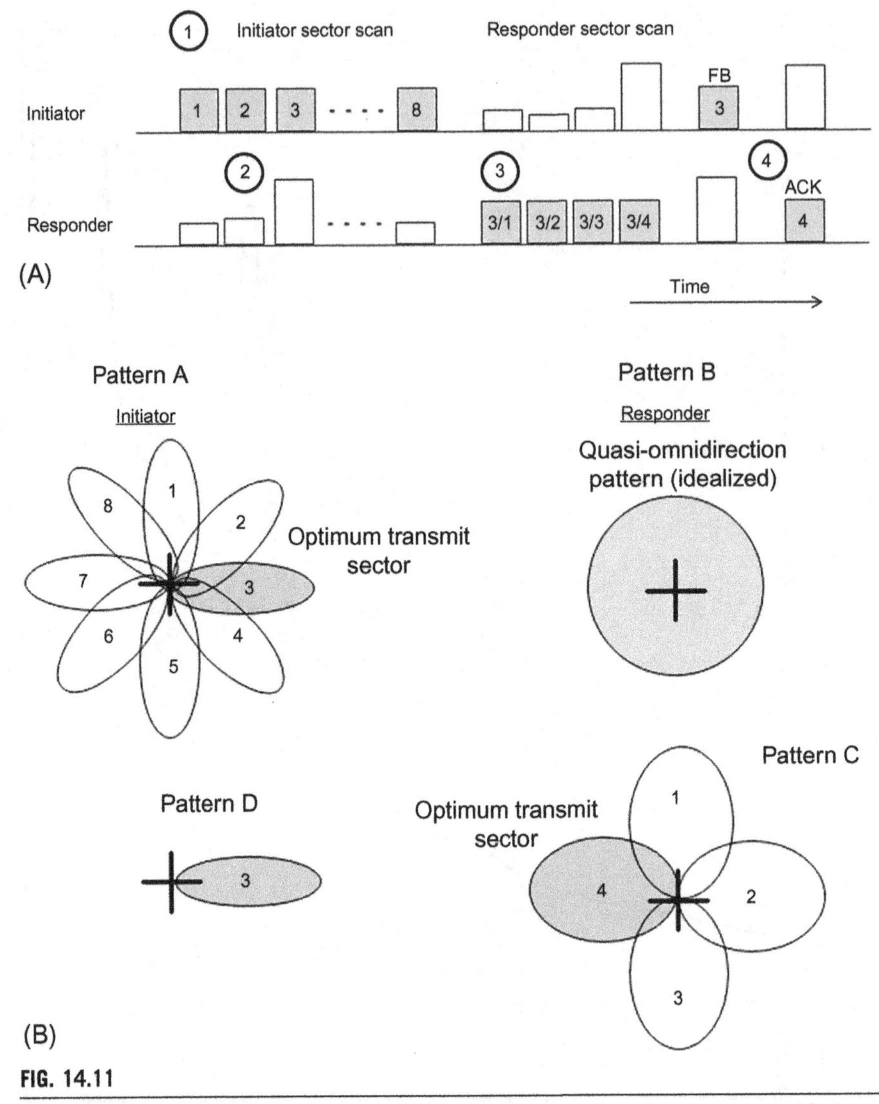


**FIG. 14.11**

DMG mm Wave beamforming concept. (A) Timing diagram. Gray boxes are transmissions, white boxes are received signals. Numbers in circles correspond to steps in the text explanation. (B) Antenna patterns.

responder scan is complete, the initiator sends a feedback message (Pattern D) specifying the strongest received responder sector, which is "4" in the example. The process is completed with an acknowledgement "ACK" from the responder.

Now the signal to noise ratio at both ends is significantly improved, allowing the use of modulation and coding at high data rates, which depend on the range, typically up to 10 m. Antenna pattern optimization can be further carried out in a subsequent optional beam refinement phase to fine tune the selected sectors [21].

Significant increases in data rate as well as MIMO and multi-user MIMO (MU-MIMO) capabilities is specified in an amendment to the DMG specification, IEEE 802.11ay, known as EDMG (enhanced DMG) [22]. It achieves improved performance through bonding or aggregating the basic 60 GHz channels of 2.16 GHz bandwidth, MIMO and MU-MIMO with up to eight data streams, and dual polarized antennas. The increased data rate, up to 100 Gbps, is considered for use in wireless backhauling and new applications such as augmented reality and virtual reality.

## 14.7 Location awareness in short-range networks

It seems that people like to know where they are, and where their family and friends happen to be at any particular time. That is location awareness. Location awareness is the essence of navigation. For hundreds of years, sailors found their way by measuring the position of stars. Aircraft navigators used celestial navigation up to the middle of the 20th century. Today, navigation is also carried out by observing objects in the sky, but now those objects are artificial satellites. In the past, observations were by sight—today they are through radio, and the location accuracy obtained couldn't be imagined only a few decades ago.

While GPS (global positioning system), or to use the more inclusive term GNSS (global navigation satellite system) is the best known and probably the most accurate location technology for its class of applications, it does have limitations and it alone does not satisfy location requirements in all situations. Navigation satellites are on the order of 20,000 km above the earth's surface, hardly a short-range as considered in this book. But location awareness is an essential facility in virtually all of the short-range networks we have been discussing—WLAN, WPAN, WMN, WSN, and IoT. While the location information often is obtained through associated independent satellite navigation receivers where the environment is suitable (outdoors, with a clear view of much of the sky), our interest here is location and positioning methods that are a facility of the short-range communication equipment itself.

First, a note on nomenclature. The term "location" answers "Where is it." "Positioning is a more inclusive term, which may refer to distance or angle measurement, or the attitude of an object. However, the two words are often used synonymously.

### 14.7.1 Proximity location

The simplest location method is proximity. The fact of reception of a transmission from an access point or beacon, where the coordinates of that terminal are known, indicates that the receiving device is nearby. The accuracy of its location is within the communication range of the wireless link. This depends on the radiated power—the lower the transmission power, the smaller the reception spatial volume around the transmitter. An improvement on the estimation of the device's location, or distance from the beacon, can be derived from the received signal strength indicator (RSSI). In Fig. 14.12 a client device that receives a signal from the beacon approximates his position as being within a circle around the beacon, whose location

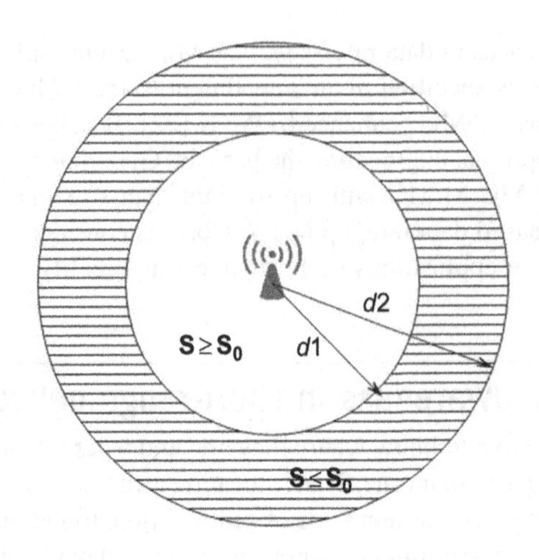


**FIG. 14.12**

Proximity location estimate using an RSS threshold. *s* is measured signal strength and $s_0$ is a threshold.

is known. If the signal strength *s* is greater than a threshold $s_0$, the client is within the white region. If *s* is weaker than $s_0$, it is in the hatched region. As mentioned, the size of these regions depend on the transmitter power. Systems may use for proximity only indications within the white circle, that is, signals that are above the threshold. The circles, of course, are just an approximation. Devices lying on a circle surrounding a beacon will not have constant signal strength because of propagation irregularities, notably reflection, refraction, scattering and blocking. Although the circle is presented as illustrative, the reception region is actually three dimensional. In free space it would be a sphere but on earth its shape depends on proximity to the ground and other features of the environment.

Reception by a client, or target device, of signals from more than one beacon can better approximate their position. This is shown in Fig. 14.13. In the figure, the client at position 1 has established connections with devices A and B whose coordinates are obtainable. Device C is out of range, or its signal is below a threshold. The client is located in the shaded region. Estimated position coordinates can be found by calculating the centroid of the region, or an approximation from the intersections of the constant radius circles that bound the region. When the client moves to position 2, it is in range only of devices B and C, and is located in the corresponding shaded region.

Fig. 14.14 illustrates the concept of locating and tracking personnel in an office space. Bluetooth low energy (BLE) devices are particularly suited for this purpose. Each beacon transmits message frames that include an identification number. The receiving device carried by a person finds the locations of the beacons within range from a lookup table and can estimate its location. The dashed circles in the figure indicate roughly the coverage area of each beacon, the size of which depends on

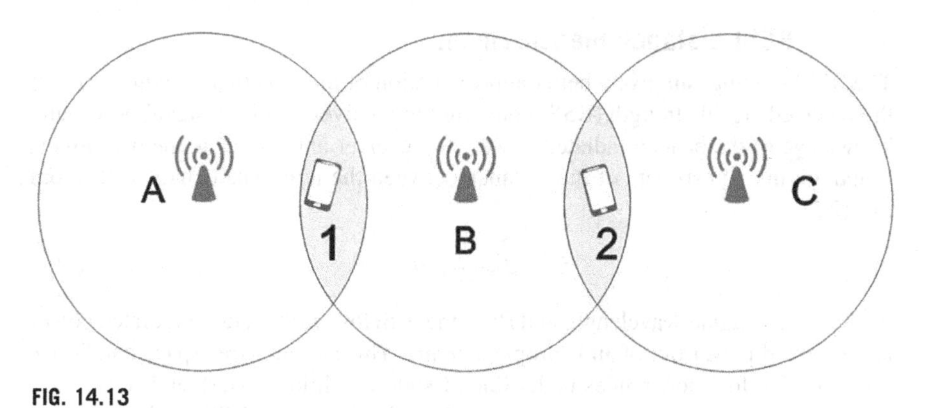


**FIG. 14.13**

Location estimation from reception from multiple beacons.

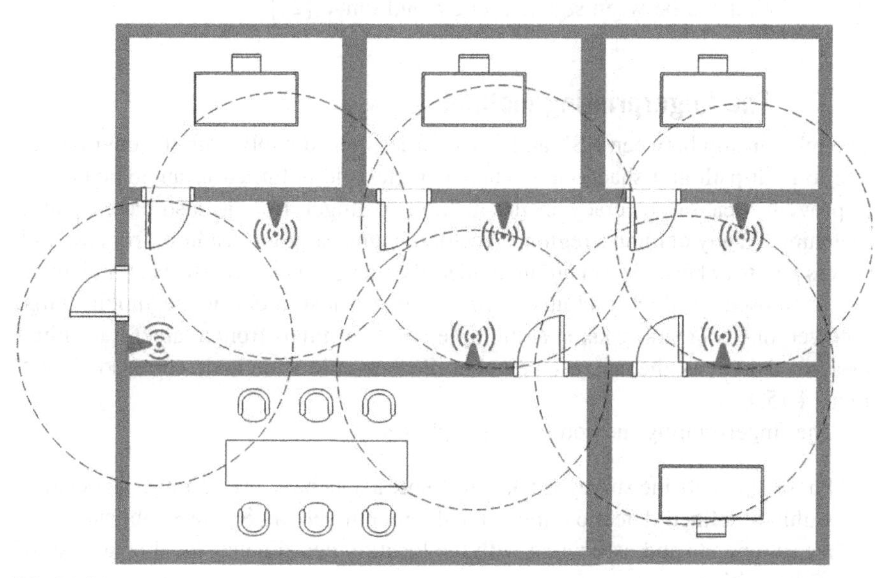

**FIG. 14.14**

Multiple location beacons in an indoor environment.

the beacon radiated power. The beacons are battery powered, and their energy consumption is reduced by sending short, low duty cycle signals. The low duty cycle insures that there will be a very low probability of collision between the emissions of the beacons. The target receiving device includes a transceiver, Wi-Fi for example, and can pass its identification, as well as the ID's of the beacons to a central processor that can locate and track people in the office. The device itself may be a smartphone, which contains both Bluetooth and Wi-Fi radios.

### 14.7.2 RSSI distance measurement

The RSSI reading can give a better approximation of location than proximity. From the received signal strength (RSS) converted to receiver input RF signal power and knowledge of the beacon radiated power and receiver antenna gain, the path loss is found, giving an estimate of the distance between the terminals (Chapter 2). From Eq. (2.5)

$$d = \frac{\lambda}{4\pi}\sqrt{PL} \tag{14.1}$$

where $\lambda$ is the signal wavelength and $PL$ is the path loss as the ratio of radiated power and received power out of an isotropic antenna. This assumes free space conditions and is usable for short ranges under line of sight conditions. Mesh and sensor networks with low cost battery powered nodes will likely use RSSI and expression (14.1) over links between adjacent devices as a step in estimating node location. A better estimation is possible using a more sophisticated algorithm than the free space relationship between signal strength and range [23].

### 14.7.3 The fingerprinting method

The relationship between RSS and range varies considerably with the environment due to multipath and shadowing. This characteristic is turned to an advantage for improving location accuracy in the method of fingerprinting, also called pattern matching. At any point in a region, a vector of signal strength readings from multiple access points or beacons can uniquely identify a target location. The vector elements are composed of RSS readings from multiple transmitters at the mobile target receiver, or RSS reading taken at multiple fixed terminals from a target transmitter. Described here is the nearest neighbor RSS location method, corresponding to Fig. 14.15.

The fingerprinting method has three phases:

Phase 1: An offline survey is taken to create a data base. At a number of positions within a defined detection area, signal strength vectors $\mathbf{S_n}$ are established by measurements and associated with the location coordinates, local or global, of each position.

Phase 2: To locate a target's position, real-time online signal strengths are recorded between it and all access points in range.

Phase 3: The real-time vector is compared with the vectors in the data base. The data base coordinates at the point whose vector is closest to the online measurement vector is taken to be the location of the target. "Closest" is considered to be the minimum Euclidean distance, expressed in vector notations as

$$D_n = |\mathbf{S_T} - \mathbf{S_n}| \tag{14.2}$$

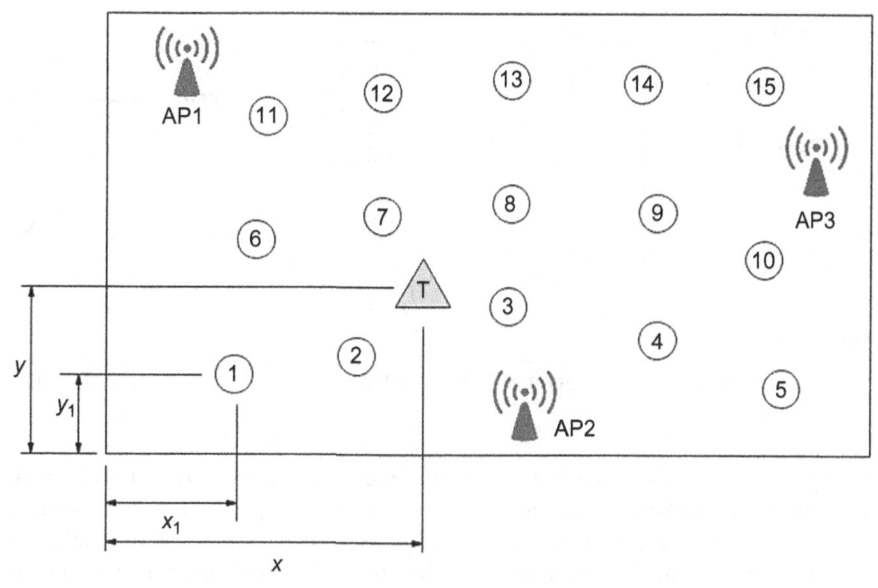


**FIG. 14.15**

Nearest neighbor fingerprint method. The circles are data base positions with known coordinates. $(x,y)$ are the coordinates of the target $T$, whose position is estimated.

where $D_n$ is the Euclidean distance at point "$n$" in the data base between the target signal strength vector $\mathbf{S_T}$ and the vector $\mathbf{S_n}$. Finer accuracy is achieved by taking the target location as the average of the coordinates of several points with minimum distances. In Fig. 14.15, points 2, 3, and 7 are the nearest neighbors and the location of target "T" is estimated as having the average of their coordinates [24].

## 14.7.4 Time-of-flight methods

While RSS provides best accuracy using fingerprinting, suitable only for a specific region and requiring advance preparation and updating of a data base, time-of-flight (TOF) methods used to measure distance and indirectly location can provide higher accuracy without such limitations. TOF uses wave propagation time to determine distance, or distance difference, through time or phase measurements.

### *14.7.4.1 Time of arrival (TOA)*

Distance between two terminals, $d$, is found by subtracting the emission time, $t_1$, at one terminal of a recognizable instance on a signal, called an epoch, from the time of arrival (TOA) of the epoch, $t_2$, at the second terminal, and multiplying the result by the speed of light:

$$d = (t_2 - t_1)c \tag{14.3}$$

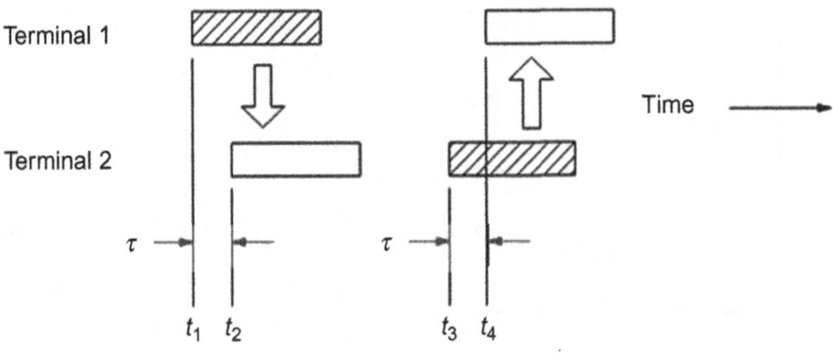


**FIG. 14.16**

Two-way time of arrival. Hatch patterned frames are transmission; white frames are reception.

Using this expression directly requires that the clocks at both terminals be synchronized, an unlikely situation particularly when low-power low cost devices are involved. Using two-way transmissions, as shown in Fig. 14.16, each side measures arrival and departure times of the epoch. The distance between the terminals is

$$d = \tau \cdot c = \frac{(t_4 - t_1) - (t_3 - t_2)}{2} \tag{14.4}$$

where $\tau$ is TOF of the frame. $t_1$ and $t_4$ are measured at terminal 1 and $t_2$ and $t_3$ are measured at terminal 2. Terminal 2 sends its time measurements to terminal 1 if terminal 1 performs the distance calculation.

The resolution of the clocks in the devices must be appropriate for the required accuracy, and the receiver bandwidth must be high compared to the inverse of the time resolution. Bandwidth and clock resolution can be traded for measurement time by making multiple time measurements and averaging the results.

Two way TOA with IEEE 802.15.4 UWB PHY is described in Ref. [25]. UWB high bandwidth and correspondingly short pulses can be used to measure distances to accuracies of well under 1 m in high multipath indoor environments. Because of the high bandwidth, individual reflected signals in the multipath response can be identified and the TOA of the earliest arriving symbol is then used to estimate the true line of sight distance.

Wi-Fi Certified Location by the Wi-Fi Alliance brings sub-meter accuracy to indoor location estimation. It is based on the fine timing measurement (FTM) protocol from IEEE 802.11-2016 which is applicable to very high throughput, wide bandwidth devices. Location is based on distance measurements obtained through two way TOA shown in Fig. 14.16 and Eq. (14.4).

Location coordinates of a target are determined from distance measurements to fixed reference stations whose locations are known. The setup is shown in Fig. 14.17. A least three reference, or base stations, are needed for estimating unambiguously the target location. Target coordinates are found from the intersection of circles around

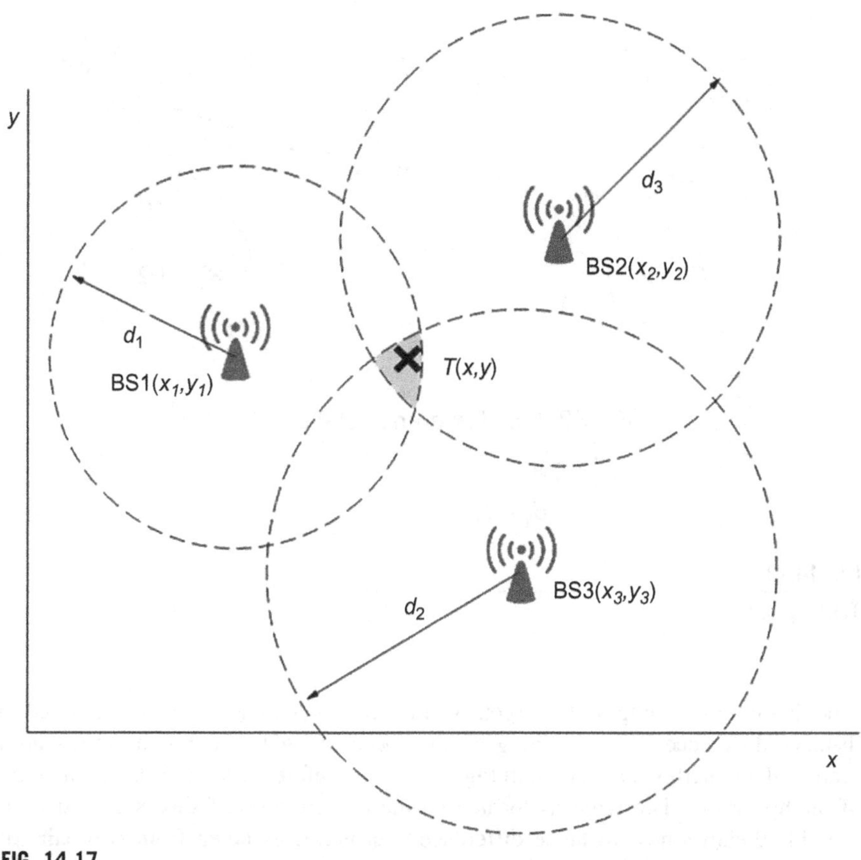


**FIG. 14.17**

Target location based on TOA distance measurements.

each base station, whose radii are the estimated distances to the target. They are obtained by solving these simultaneous equations for $(x, y)$:

$$\begin{aligned}(x_1 - x)^2 + (y_1 - y)^2 = d_1{}^2 \\ (x_2 - x)^2 + (y_2 - y)^2 = d_2{}^2 \\ (x_3 - x)^2 + (y_3 - y)^2 = d_3{}^2\end{aligned} \tag{14.5}$$

Because the distances $d_1$, $d_2$, and $d_3$ are subject to errors, a least squares solution can be applied which estimates the target coordinates as those that give the minimum of the sum of the errors of the three equations.

#### *14.7.4.2 Time difference of arrival (TDOA)*

TDOA is a TOF method that doesn't require a transceiver in the target device. It is based on TOA difference measurements taken between the target and a pair of fixed stations. Transmissions from the fixed reference stations must be synchronized in

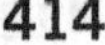



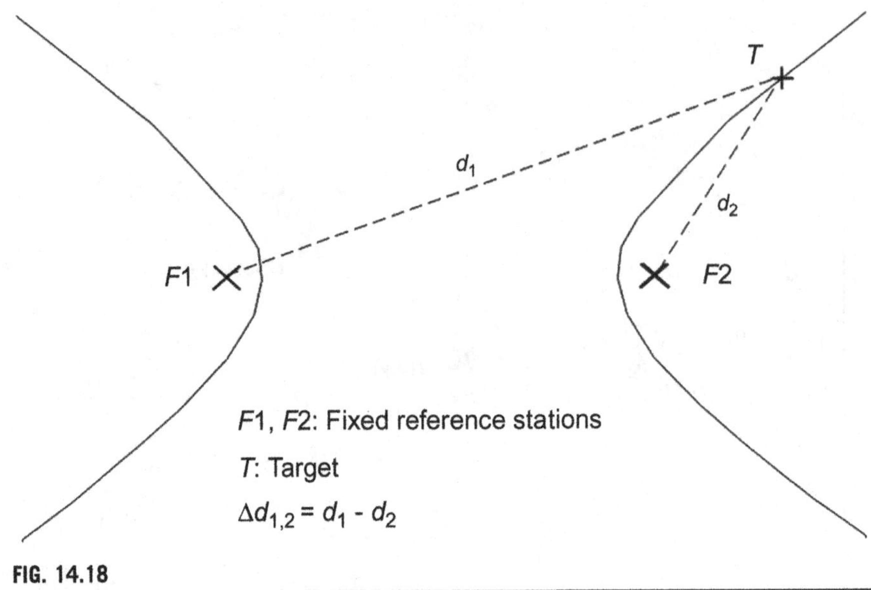


**FIG. 14.18**

TDOA geometry.

time, but not the timing in the target. A time difference measurement, converted to distance difference by multiplying by the speed of light, situates the target on a branch of a hyperbola, as shown in Fig. 14.18. The reference stations are at the focii of the hyperbola. The target is located on the branch whose focus is closest to it. Fig. 14.19 shows how distance difference measurements taken from two pairs of fixed stations among three stations result in two hyperbolas which intersect at the location of the target. Target coordinates $(x, y)$ are found by solving the expressions for the distance differences

$$\begin{aligned} \Delta d_{1,2} &= d_1 - d_2 = \sqrt{(y_1 - y)^2 + (x_1 - x)^2} - \sqrt{(y_2 - y)^2 + (x_2 - x)^2} \\ \Delta d_{3,2} &= d_3 - d_2 = \sqrt{(y_2 - y)^2 + (x_2 - x)^2} - \sqrt{(y_3 - y)^2 + (x_3 - x)^2} \end{aligned} \tag{14.6}$$

where $(x_1,y_1)$, $(x_2,y_2)$, $(x_3,y_3)$ are the coordinates of the three reference stations. A fourth reference station and another independent pair of distance differences are required to be sure of an unambiguous solution.

### *14.7.4.3 Phase of arrival (POA)*

A third TOF positioning method is phase of arrival (POA). In common with TOA and TDOA the estimated distance with POA is a function of speed of propagation, but the measured metric is relative phase, not time. This is shown in Fig. 14.20 in which the transmitting waveform and the receiving waveform are compared [24] The distance

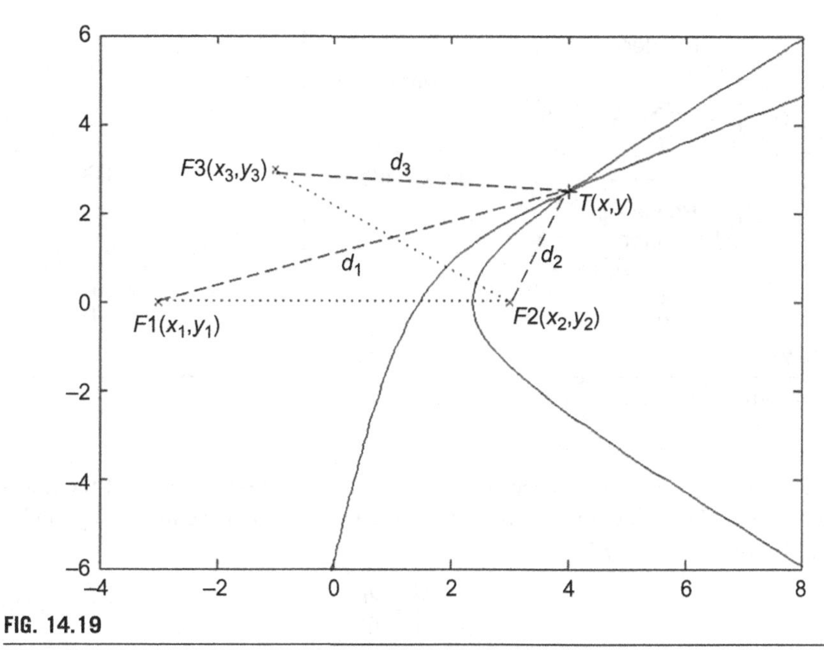


**FIG. 14.19**

TDOA location. Dotted lines connect the focii of each hyperbola.

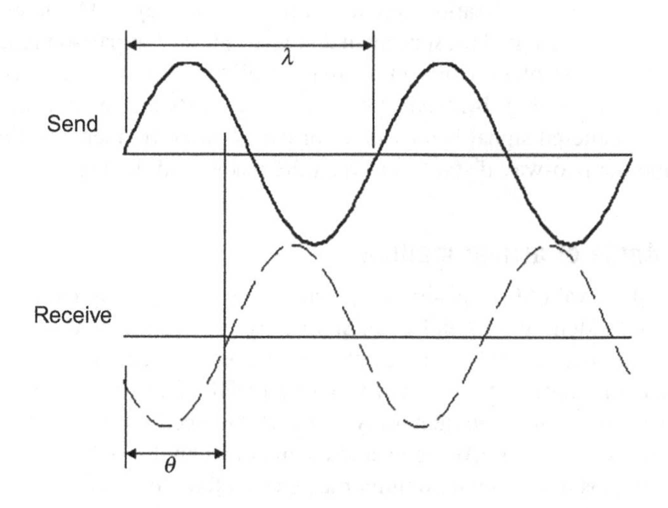


**FIG. 14.20**

Phase of arrival distance measurement.

$d$ over which the wave propagates is proportional to the phase $\theta$, $0 \leq \theta \leq 2\pi$, but it is ambiguous by the whole number of wavelengths between sender and receiver, $n$:

$$d = \lambda\left(\frac{\theta}{2\pi} + n\right) = \frac{c}{f}\left(\frac{\theta}{2\pi} + n\right) \tag{14.7}$$

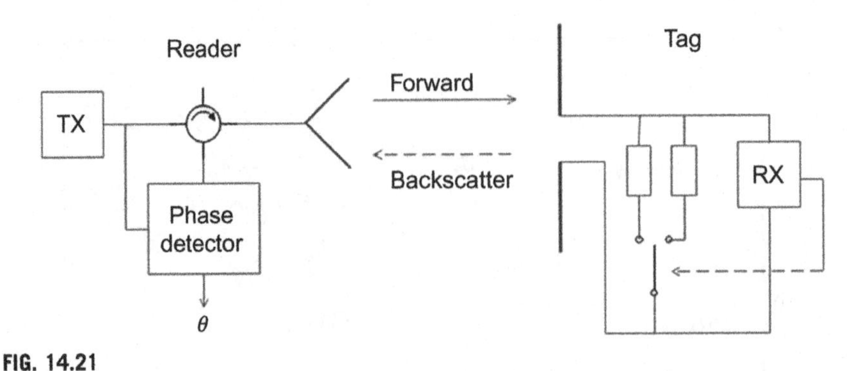


**FIG. 14.21**

Phase of arrival method applied to RFID distance measuring.

where $\lambda$ is the wavelength, $c$ is the speed of light and $f$ is the frequency of the transmitted signal. The ambiguity is removed by measuring $\theta$ at two frequencies, $f_1$ and $f_2$ that differ by less than $c/d$:

$$d = \frac{c}{2\pi} \cdot \frac{\theta_2 - \theta_1}{f_2 - f_1} \tag{14.8}$$

Phase errors can be averaged out by measuring on more than two frequencies.

Just as clock synchronization is required for the one-way TOA method, a phase reference of the send signal must be available at the phase comparison terminal. This requirement is no problem with back scattered RFID for two-way distance measurement at the reader. In Fig. 14.21, the phase difference between the transmitted signal and the backscattered signal is measured for two or more frequencies. Eq. (14.8) is used to find the two-way distance between the reader and the tag.

## 14.7.5 Angle of arrival method

The angle of arrival (AOA) positioning method has not been applied to short-range location to the extent of RSS and TOF. Short-range device antennas are for the most part omnidirectional, and also their electrical size, on the order of a half wavelength, is not suitable for narrow beam patterns in small UHF band devices. However, with the proliferation of multi-element arrays for MIMO in cellular networks and Wi-Fi, and the introduction of mmWave as a mainstream technology that uses directional antennas, AOA is becoming more important as a method for adding location awareness to communications systems.

Two ways of using AOA to estimate location are shown in Fig. 14.22. Fig. 14.22A combines AOA with a distance measuring method—RSS or TOA—to estimate location using only one reference terminal. In Fig. 14.22B the direction lines of the directional antennas of two terminals cross at the target location. Target coordinates are found from the known fixed terminal coordinates and the antenna beam angles in relation to a common reference direction.

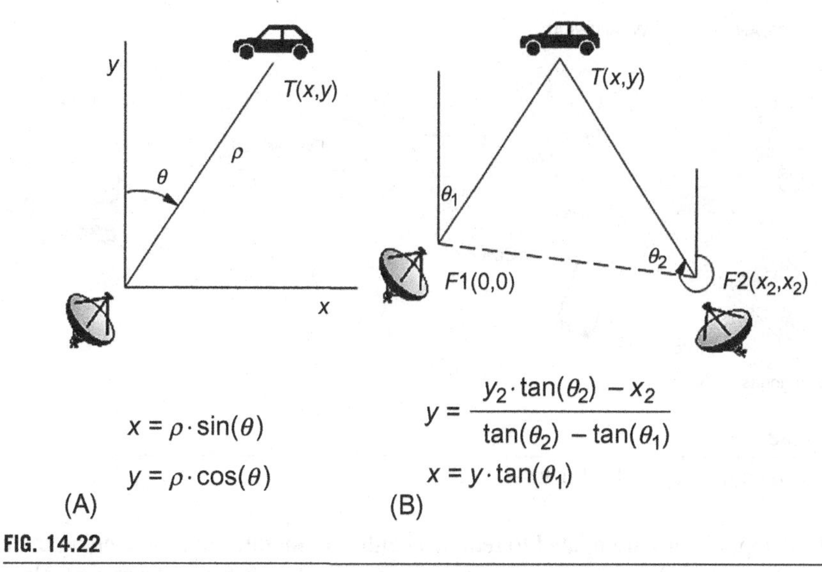


**FIG. 14.22**

Using AOA to find location. (A) Rho-theta, (B) theta-theta, or triangulation.

## 14.8 Wireless body area networks

Wireless Body Area Networks (WBAN) look like a subset of both WSN and IoT. They are composed of wireless sensor nodes that communicate with aggregating sink nodes, and overall provide autonomous connections between things—sensors—and storage devices or human analyzers and responders. However, in addition, WBANs have some very distinctive characteristics. WBAN radio signals propagate in close vicinity of a living body and as a result path loss follows very different rules as compared to propagation in a WSN. Antenna design must take into account the very close proximity of the body which can significantly affect antenna efficiency and size. Whereas WSNs may have a very large number of sensors distributed over a wide area, and provision for redundancy and node failure, WBANs have a maximum dimension of around 3 m and no more than tens of nodes. Its nodes must be highly reliable since failure can be consequential to the users health or even life. Node radio transmission power must be very low to minimize heating of and otherwise affecting adjacent body tissues. Also, wireless communication links change dynamically with the user's movements, and the body mounted nodes must not encumber his activity.

Body area network nodes are situated in, on or very near a living body, usually a human but not necessarily so. Implantable biosensors, such as a pacemaker or glucose sensor, are surgically installed inside the body or under the skin. Wearable sensors can be contained in a wrist watch, in a hearing aid, or generally in or on clothing. Actuators that interact with the user may also be part of nodes, similarly to WSNs. WBANs are used for medical and non-medical applications Examples of

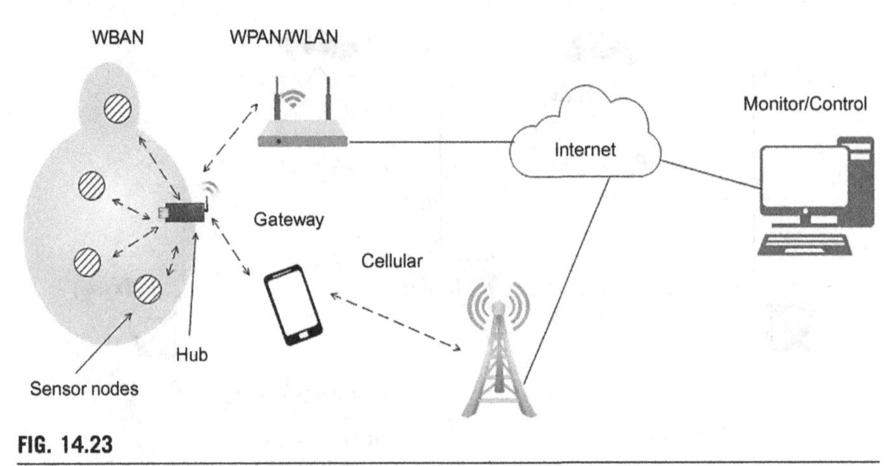


**FIG. 14.23**

WBAN communication infrastructure.

medical applications are related to remote health care monitoring, for keeping track of heart beats, blood pressure, temperature and physical movement of a patient. Non-medical applications are used for sports and pedestrian navigation. There are also numerous military uses for WBANs for remote monitoring of soldiers' physical conditions on the battlefield and could include cameras and navigation sensors.

The basic architecture of a WBAN and its communication links is shown in Fig. 14.23. Sensor nodes communicate directly with a hub, which is a control center worn on the body that relays data from the WBAN to a gateway. The gateway may be a Wi-Fi access point or the master node in a Bluetooth or Bluetooth LE network which moves the WBAN data over the Internet. A smartphone can act as a gateway for connection to the Internet through the cellular network. The destination connection for the WBAN is a monitor/control point which could be for example in a doctor's office or the smartphone of a relative of a monitored person, where a decision on the data and a response if necessary will be made [26–28].

### 14.8.1 IEEE 802.15.6

IEEE 802.15.6 is an international standard for short-range wireless communication in the vicinity of, or inside, a human body (but not limited to humans). It uses existing industrial scientific medical (ISM) bands as well as frequency bands approved by national medical and/or regulatory authorities.[29]. It supports data rates up to 12.6 Mbps. The standard specifies three types of physical layers: narrow band, ultra wideband, and human body communication (HBC).

#### *14.8.1.1 Network topology*

Nodes and hubs are organized into logical sets referred to as body area networks (BAN). A BAN may have only one hub and up to 64 nodes. The topology is star, and nodes may connect to their hub by one hop, or by two hops through a relay node (one hop only in MICS band—see Table 14.1) Relay nodes can also exchange their

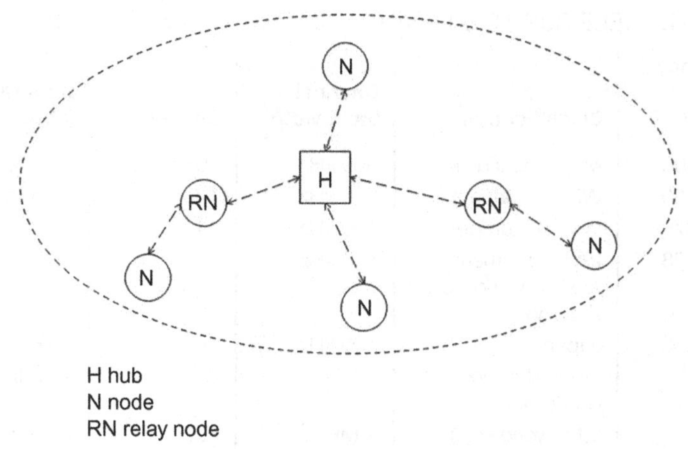


FIG. 14.24

BAN network topology.

own frames with the hub. This is shown in Fig. 14.24. The standard details several access mechanisms based on polling and random access which take into account priorities to meet quality of service (QoS) requirements. Provisions are made for preventing collisions between neighboring BANs by offsetting access time allocations within regularly repeated beacon periods or superframes, and by frequency hopping.

#### *14.8.1.2 Narrow band PHY*

The standard supports a number of narrow band UHF frequency bands as shown in Table 14.1 [29].

Modulation type for six of the bands in Table 14.1 is differential phase shift keying (DPSK), with options for the number of symbol states: $M = 2$ ($\pi/2$-DBPSK), 4 ($\pi/4$-DQPSK), and 8 ($\pi/8$-D8PSK). GMSK (Gaussian minimum shift keying) modulation is used in the 420 to 450 MHz band. Multiple data rates, and robustness are achieved through the different number of states per symbol, code rate (ratio of data bits per total number of bits in a code block), and spreading factor—the repetition of each bit two or four times. The lowest data rate for each frequency band is used for the PLCP (physical layer convergence protocol) header in the transmitted frame while higher rates are used for the payload data. In the 402 to 405 MHz MICS band, specified transmitter EIRP in the low power low duty cycle mode (LP/LDC) as defined in applicable regulations is maximum −40 dBm. In the non-LP/LDC mode, a transmitter shall be capable of transmitting at most −16 dBm EIRP. On the other narrow band PHY frequency bands a transmitter shall be capable of transmitting at least −10 dBm EIRP. The standard states that "Devices should transmit lower power when possible in order to reduce interference to other devices and systems and to protect the safety of the human body. The maximum transmit power is limited by local regulatory bodies" [29].

**Table 14.1** IEEE 802.15.6 frequency bands for narrow band PHY

| Frequency band (MHz) | Classification | Channel bandwidth | Channels | Data rate (kbps) |
|---|---|---|---|---|
| 402 to 405 | MICS[a] (worldwide) | 300 kHz | 10 | 57.5 to 455.4 |
| 420 to 450 | WMTS[b] (Japan) | 320 kHz | 12 | 57.5 to 187.5 |
| 863 to 870 | WMTS (Europe) | 400 kHz | 14 | 76.6 to 607.1 |
| 902 to 928 | ISM[c] (N. America, Australia, and New Zealand) | 400 kHz | 60 | 76.6 to 607.1 |
| 950 to 958 | Japan | 400 kHz | 16 | 76.6 to 607.1 |
| 2360 to 2400 | Medical device (worldwide) | 1 MHz | 39 | 91.9 to 971.4 |
| 2400 to 2483.5 | ISM (worldwide) | 1 MHz | 79 | 91.9 to 971.4 |

[a] *MICS: Medical Implant Communications Service.*
[b] *WMTS: Wireless Medical Telemetry Service.*
[c] *ISM: Industrial Scientific Medical.*

#### *14.8.1.3 Ultra wideband PHY*

UWB has several advantages for WBAN:

- optional high data rates
- robust performance
- low complexity
- ultra low power, on the order of that used in the MICS band
- low interference to other devices

The UWB physical layer is used for communication between on-body devices, and between a hub and off-body devices. IEEE 802.15.6 includes two different UWB technologies—impulse radio (IR-UWB) and wideband frequency modulation (FM-UWB). There are two modes of operation: default mode, used for medical and non-medical operations, and high QoS mode which is destined for high-priority medical applications. The two technologies share the same operating frequency bands and bandwidth. There are three channels in the low band, with center frequencies of 3494.4, 3993.6, and 4492.8 MHz. Support for the 3993.6 MHz channel is mandatory for devices operating on this band. Eight channels are specified for the high band, with center frequencies from 6489.6 to 9984.0 MHz. Devices operating on the high band are required to support at least the channel with center frequency 7987.2 MHz. Bandwidth for all channels is 499.2 MHz.

The default mode PHY technology is IR-UWB whereas FM-UWB is optional. Mandatory modulation is on-off signaling (see Chapter 12, Section 12.5). An advantage of this type of modulation is that a relatively simple non-coherent receiver architecture can be used. Data has error correction coding, and six coded bit rates are specified ranging from 394.8 kbps to 12.636 Mbps with corresponding minimum receiver sensitivity of −91 to −76 dBm for a packet error rate (PER) less than 1% in added white Gaussian noise (AWGN). Mandatory data rate is 487.5 kbps. The same duty cycle (pulse burst duration over symbol duration) is maintained for all data rates.

A second modulation type for IR-UWB is DPSK, where phase changes between consecutive phase shift keying symbols convey the information. The high QoS mode must support DPSK and a data rate of 487.5 kbps. The error correction code has a higher redundancy, that is a higher ratio of code bits to data bits, for this modulation type as compared to on-off keying, which makes it more reliable (at the expense of data rate). Differential binary phase shift keying and differential quadrature phase shift keying are specified. Included is an option for spreading the DBPSK/DQPSK symbols with a seven bit sequence in order to enhance interference refection. Eight coded bit rates between 243 kbps and 7.8 Mbps are specified, differentiated by DBPSK, DQPSK and spreading [29].

The wideband FM-UWB technology is an optional PHY in the default mode targeting low data rate medical body area networks. Information data phase modulates a 1.5 MHz subcarrier which in turn frequency modulates a carrier frequency using a high modulation index to obtain the ultra-wideband signal. Features of the technology are

- constant envelope—peak power equals average power
- flat spectrum with steep spectral roll-off
- after wideband FM demodulation (despreading) without down conversion, the signal is narrow band continuous phase binary FSK, for relatively simple synchronization and detection
- high processing gain—low complexity with robustness against interference and multipath.

The FM-UWB data rate is 202.5 kbps. Mandatory low band frequency is 3993.6 MHz and mandatory high band frequency is 7087.2 MHz. Bandwidth is approximately 500 MHz at 10 dB down. Minimum receiver sensitivity is −85 dBm for a bit error rate (BER) equal to or below $10^{-6}$.

#### *14.8.1.4 Human body communications PHY*

The third physical layer alternative for WBAN is human body communications (HBC). Actually, it doesn't involve a wireless link but rather a data connection through the body. Some advantages of HBC are [30]:

- Little signal leakage outside the body—high security. The signal is confined to the surface of the body with little energy radiated to the surroundings. This avoids the possibility of eavesdropping and also of interference from external sources.
- Low signal attenuation—low transmission power. Low path loss through body conduction means reduced transmitter power and lower power consumption of the system as a whole.

- Low carrier frequency. The relatively low frequency—tens of MHz—reduces transmitter and receiver complexity, size and power.
- No antenna required. Coupling to the signal path (the body) is through electrodes.

The HBC PHY uses electric field communication (EFC) technology. The frequency band of operation is centered at 21 MHz and the bandwidth at 3 dB down is 5.25 MHz. Information data rates are 164 kbps (mandatory), 328 kbps, 656 kbps, and 1.3125 Mbps. Modulation is accomplished by spreading the data to 42 Mcps (Megachips per second). The spreading operation is similar to DSSS, described in Chapter 4. A difference is that the spreading sequence in DSSS is a pseudo random code, whereas in HBC the spreading code is an alternating sequence of logical zero and one, clocked at 42 Mcps. This gives a fundamental frequency of 21 MHz, which is filtered to fit the transmit spectral mask determining the bandwidth.

#### 14.8.1.5 WBAN security

Secure communication is an important issue in all wireless communications, but it has special significance for body area networks, primarily in medical applications, because of its potential contribution to a user's physical well-being. Security mechanisms in WBANs relate to these matters [26]:

(1) Security key distribution among nodes.
(2) Insuring availability of patient information to a physician at all times.
(3) Authentication of data to prevent falsification by an adversary.
(4) Data integrity to prevent modification, by physical aberrations such as noise, or on purpose.
(5) Confidentiality—prevention of data disclosure.
(6) Data freshness—prevention of replay and reuse.

Secure communication requires a node and a hub to follow the simplified state diagram shown in Fig. 14.25. Note that the master key (MK), the shared secret between node and hub, is not used directly for encrypting frames, but is used to create a pairwise temporal key (PTK). The PTK encrypts frames for a period of time constituting a session, after which a new PTK is created from the MK [29].

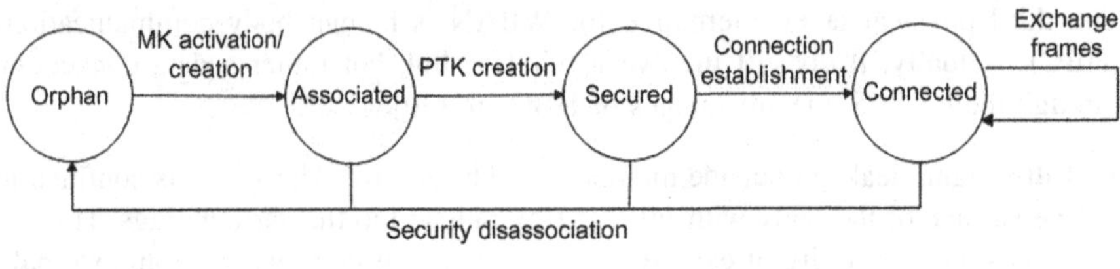


**FIG. 14.25**

Simplified security state diagram.

- Orphan state—initial state to activate or create a shared MK, and for authentication.
- Associated state—creation of a PTK which is used for secure frame transmissions in the connected state.
- Secured state—with the PTK confirmed, this state serves for requesting and establishing a connection before exchanging information data frames.
- Connected state—secure frames are exchanged between hub and node in accordance with the channel access arrangement in the BAN.

In the case of unsecured communication (level 0, defined below), the nodes go directly from orphan state to connected state.

All nodes and hubs chose among three security levels, selected in the associated state:

- Level 0—unsecured communication. All messages are transmitted in unsecured frames.
- Level 1—authentication but not encryption. Message authenticity and integrity are provided, but not confidentiality and privacy.
- Level 3—authentication and encryption. Authenticity, integrity, confidentiality and privacy, and replay defense are provided.

A technique for managing security keys that is specific to WBANs utilizes biometrics. It uses information from the body sensors to create and securely distribute cryptographic keys. Examples are generating and distributing session keys from the ECG (electrocardiograph) signal, and using the timing information of heartbeat for authentication or secure cipher key distribution [26].

## 14.9 Energy harvesting

Devices in networks discussed in this chapter, notably WSN, IoT, and WBAN, are generally battery powered and their unattended lifetime depends on battery life. In many applications battery replacement is particularly inconvenient or even impossible. For very low power devices, energy harvesting, or energy scavenging, is an answer. Natural energy sources exist in different environments, both outdoors and indoors, that may be converted to the form needed to power electronic circuits. Examples are light, wind, heat differential, vibration and ambient radio signals.

### 14.9.1 Solar and artificial lighting

Solar panels are the most familiar transducer for energy harvesting. Composed of photovoltaic cell arrays, they are almost always used together with storage batteries to regulate the non-constant energy provided by a light source, be it the sun or indoor lighting. Solar cell power depends on the surface area of the photovoltaic cell array, the orientation towards the source of light and the spectral composition of the light, as well as time of day and latitude in the case of sunlight. Outdoor deployed WSN would be the most likely to benefit from solar cells. In one study, a solar panel of area $18 \times 8.1\ cm^2$ provided on average approximately 600 mW to a IEEE 801.15.4 wireless sensor node, which included transceiver, sensor, and microcontroller power. This result was obtained from tests in sunlight during the morning [31]. Another report was based on a study of the use of solar energy to extend smartphone battery life. It found that a solar panel with an area of $103\ cm^2$, the approximate size of a smartphone, provided around 12 mW when illuminated by simulated indoor incandescent lighting of 500 lux. LED and fluorescent sources of the same energy density provided less power. The report concluded that reasonable indoor lighting coupled to a solar energy harvester the size of a smartphone cannot provide enough power (stated as 30 mW) to charge the device during the idle state, although it can increase the time between mains driven charges [32].

### 14.9.2 Thermoelectric energy

Thermoelectric energy generator (TEG) harvesters exploit the Seebeck effect by which a voltage is created between two sides of a pair of junctions of two dissimilar metals, forming a thermocouple, that are exposed to a temperature differential. An array of series connected thermocouples, called thermopile, provides the voltage required for the electronic device. The power output is

$$P = \frac{(S \cdot \Delta T)^2}{4 \cdot R_{load}} \tag{14.9}$$

where $S$ is the coefficient of voltage vs. temperature (a function of the metals making up the thermopile), $\Delta T$ is the temperature differential, and $R_{load}$ is the load resistance for maximum power. The thermal efficiency, which is reflected in $S$ in Eq. (14.8) is

$$\eta \leq \frac{T_h - T_c}{T_h} \tag{14.10}$$

where $T_h$ is the hot side temperature and $T_c$ is the cold side temperature, both in degrees Kelvin.

TEGs have been considered to support the use of vacuum insulation panels in order to increase the thermal efficiency of homes and lower heating and cooling energy costs. These panels have high thermal resistance in comparison with foam insulation, but their efficiency depends on maintaining a vacuum between the two sides of the panels. The vacuum can leak, and maintenance can be performed by monitoring absolute pressure in the panels which gives an indication of the leakage. The output of pressure sensors is transmitted by a wireless sensor node, powered by TEGs that use the temperature difference across the panels. In this application, power output varies considerably with season and time of day, in accordance with the inside/outside temperature differences. As reported, an average output of 28.7 μW was measured over the course of a summer day. In a simulation experiment, a sensor node consisting of a temperature sensor, microcontroller and wireless transmitter

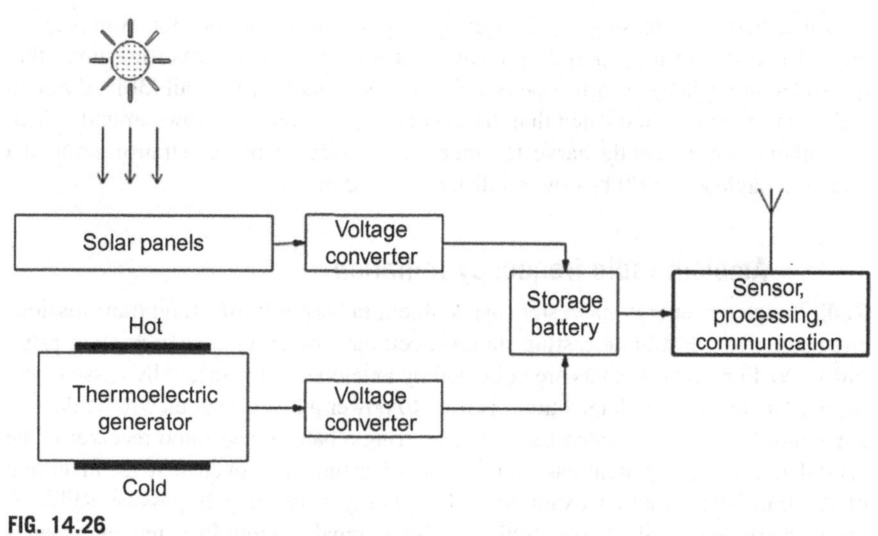


**FIG. 14.26**

Multiple source energy harvesting.

was powered intermittently by a TEG module. With a $\Delta T$ of 15°C, data transmission occurred every 2 s, representing delivered power of 109 μW [33].

Another energy harvesting application for thermoelectric generators together with solar panels is in a WSN that is designed to be self-sustaining, without support by replaceable batteries, and floats on water—lakes, rivers, ponds or pools. Its sensors collect environmental data including temperature, humidity, and water quality in the form of pH and transmits it to a gateway on land up to 500 m away. Solar panels provide most of the operating power but their output is dependent on time of day and weather. A TEG device gets its temperature differential from contact through heat sinks with the surrounding air and warming of sunlight on one side and water flow on the other. In the reported application, the maximum temperature difference was 6°C and an array of seven TEG units provided maximum power of 0.92 W. Output voltage varied between 0.1 and 1.2 V, and was stepped up by a boost converter circuit to 5 V to charge a battery pack. The multiple source energy harvesting concept is shown in Fig. 14.26 [34].

### 14.9.3 Kinetic energy

Piezoelectric transducers generate electricity from stress or vibration, which is called kinetic energy harvesting when used to power electronic circuits. There are many applications for piezoelectric transducers in sensor networks. Some examples are transducers located on bridges subject to vibrations from moving traffic, transducers in floors which respond to human traffic on them, and transducers mounted in the soles of shoes which power body area network nodes [35]. Harvesting energy from a vibrating cantilever beam coated with piezoelectric paint that was made to vibrate by a user activity, pressing a button or opening or closing a door for example, has been described. In a proposed application, a kinetic energy harvesting node that transmits only when someone opens a closet was explained in detail [36]. Although the simulated results indicated that the created energy was quite low, around 4.4 μJ, the authors showed that the harvested energy could directly power a transmission of a one byte payload at 100 bps over a distance of 10 m.

### 14.9.4 Ambient radio frequency radiation

Radio frequency energy harvesters use ambient radiation from strong transmissions emanating from FM broadcasting stations, cellular towers and, when in close proximity, Wi-Fi routers. Signals are collected by antennas and detected by passive rectifiers. The resulting voltage output is used to power a low power electronic device. An example of radio broadcast signals powering a battery less radio receiver is the crystal detector radios of the early days of broadcasting, used even up to the invention of the transistor. Another example of RF energy harvesting is passive RFID. A research article described how ambient cellular signals on multiple frequency bands were harvested to power a wireless sensor node. It claims that 40 μW DC power could be obtained from RF input power between −10 and −5 dBm in LTE, GSM, and ISM bands. A low node power duty cycle is required while a storage capacitor is charged during standby intervals [37].

### 14.9.5 Energy harvesting management circuit

The voltage and current output over time of energy harvesting devices rarely fits the requirements of the systems they have to supply, so an energy management circuit is necessary for efficient utilization. Its functions include voltage conversion and energy storage to match system supply power needs with the vagaries of the energy source. The energy available from harvesting, taking into account the conversion efficiency, should at least equal the required energy of the user device, taken over a period of time that includes power during full operation and the time the device is in idle or sleep mode. In order to obtain optimum performance, the current supplied by the harvester to the storage element has to have a value that depends on the type of harvester and its power input. This is called maximum power point tracking (MPPT). For example, solar cells provide maximum power when their output current is obtained at a voltage of around 80% of the open circuit value. A thermoelectric generator may need an output current at 50% of its open circuit voltage for highest efficiency. Other roles of the management circuit are prevention of over and under voltage conditions to the destination device, and protection of the storage element from overcharging and from excessive discharge.

A diagram of an energy harvester management system is shown in Fig. 14.27. The output of a harvester transducer, which may be a fraction of a volt in the case of individual cells or small arrays, or more than ten volts for large solar panels for example, is transformed in the voltage converter block to the voltage required by the

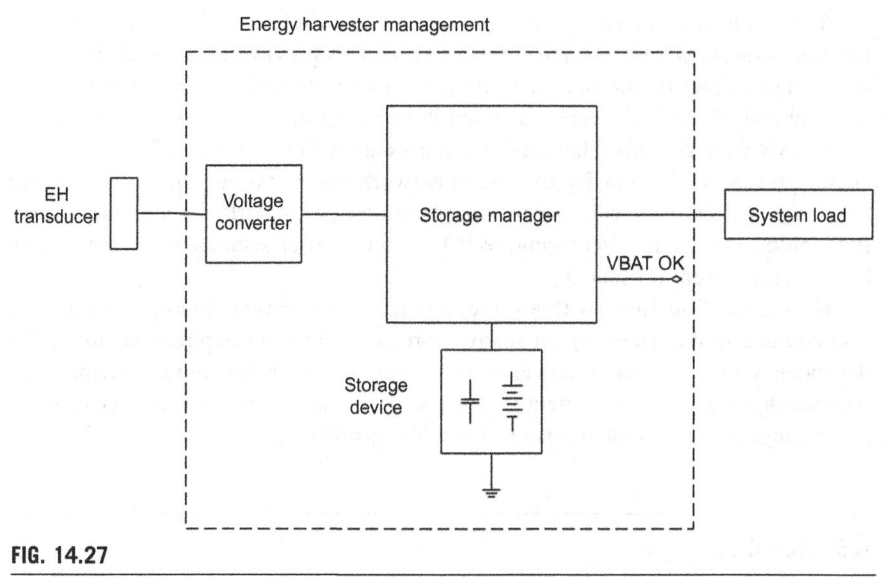


**FIG. 14.27**

Energy harvesting management.

user system. The storage manager switches this voltage to a storage device—rechargeable battery or a capacitor—when the storage device's voltage is low or when the user device is in a low power idle or sleep mode. The charging current is determined by the MPPT circuit for maximum efficiency. An indicator output, labeled VBAT OK in the figure, notifies the user that operating voltage is available and facilitates orderly shutdown when the storage voltage is below the operating threshold. The energy management system itself needs power to operate, which it must also receive from the harvester transducer. For example, the Texas Instruments bq25504 battery manager integrated circuit turns on when the transducer voltage $V_{IN}$ is at least 330 mV and can continue to harvest energy down to $V_{IN} = 80$ mV [38].

## 14.10 Summary

A discussion of wireless ad hoc networks in general was followed by an explanation of the features of WMN and WSN. Routing protocols of proactive and reactive variants were presented and functional details of their operation were described with specific examples. Then the IoT was discussed. IoT is composed of systems and networks that are primary subjects of this book, including WLAN and WPAN, as well as the infrastructure-less networks covered earlier in the chapter. Other short-range wireless systems exist which do not use the VHF-UHF frequencies on which mass communication wireless devices are based. We showed how the special characteristics of optical communication and millimeter wavelength networks expand the scope of short-range wireless communication resources.

As an adjunct to the communication function of wireless devices, we introduced location awareness based on the short-range devices previously covered. The three physical principles of distance measurement—signal strength, propagation TOF and angle-of-arrival methods were explained in some detail.

WBAN were described has having features apart from the type of networks discussed previously, both in deployment of network nodes, extreme limitations on use of energy, and on the special types of applications, particularly for measuring body functioning for health monitoring. WBAN entails strict security measures whose basic features were presented.

The chapter concludes with a survey of energy harvesting, an important area of research due to the necessity for many short-range wireless applications to reduce dependency on replaceable power sources. Several examples were described and then we showed how requirements for an energy management regime that can apply to the range of natural energy sources are being fulfilled.

## References

[1] S.A. Ade, P.A. Tijare, Performance comparison of AODV, DSDV, OLSR and DSR routing protocols in mobile ad hoc networks, Int. J. Inf. Technol. Knowl. Manag. 2 (2) (2010) 545–548.

[2] C. Perkins, Ad Hoc On-Demand Distance Vector (AODV) Routing, RFC 3561, The Internet Society, 2003.

[3] J.D. Camp, E.W. Knightly, The IEEE 802.11s extended service set mesh networking standard, IEEE Commun. Mag. 46 (8) (2008) 120–126.

[4] IEEE Standard 802.11-2016, IEEE Computer Society, 2016.

[5] K. Rathan, S.E. Roslin, A survey on routing protocols and load balancing techniques in wireless mesh networks, in: 2017 International Conference on Intelligent Computing and Control (I2C2), Coimbatore, India, 23–24 June, 2017.

[6] Y. Liu, K.F. Tong, et al., Wireless mesh networks in IoT networks, in: Y. Liu, K.-F. Tong, X. Qiu, Y. Liu, X. Ding (Eds.), Wireless Mesh Networks in IoT Networks, 2017, 2017 International Workshop on Electromagnetics: Applications and Student Innovation Competition, London, 2017, pp. 183–185.

[7] T. He, S.H.G. Chan, C.F. Wong, HomeMesh: a low-cost indoor wireless mesh for home networking, IEEE Commun. Mag. 46 (12) (2008) 79–85.

[8] L.B. Ruiz, J.M. Nogueira, A.A.F. Loureiro, MANA: a management architecture for wireless sensor networks, IEEE Commun. Mag. 41 (2) (2003) 116–125.

[9] J. Zheng, A. Jamalpour, Introduction to wireless sensor networks, in: Wireless Sensor Networks: A Networking Perspective, Wiley-IEEE Press, 2009 (Chapter 1).

[10] M. Matin, M. Islam, Overview of wireless sensor network, in: Wireless Sensor Networks—Technology and Protocols, IntechOpen, 2012 (Chapter 1).

[11] W.R. Heinzelman, A. Chandrakasan, H. Balakrishnan, Energy-efficient communication protocol for wireless microsensor networks, in: Proceedings of the 33rd Hawaii International Conference on System Sciences, 2000.

[12] I. Akyildiz, M. Vuran, Wireless Sensor Networks, Wiley, 2010.

[13] Overview of the Internet of things, in: Recommendation ITU-T Y. 2060, International Telecommunication Union, 2012.

[14] L. Farhan, S.T. Shukur, A.E. Alissa, et al., A survey on the challenges and opportunities of the Internet of things (IoT), in: The 11th International Conference on Sensing Technology, 2017.
[15] L. Atzori, A. Iera, G. Morabilo, The Internet of Things: a survey, in: Computer Networks 54, Elsevier, 2010.
[16] P. Conway, D. Heffernan, B. O'Mara, P. Burton, T. Miao, IEEE 1451.2: an interpretation and example implementation, in: Proceedings of the 17th IEEE Instrumentation and Measurement Technology Conference, vol. 2, 2000, pp. 535–540.
[17] S. Wang, Y. Hou, F. Gao, X. Ji, A novel IoT access architecture for vehicle monitoring system, in: 2016 IEEE 3rd World Forum Internet Things, WF-IoT 2016, 2017, pp. 639–642.
[18] S. Pirbhulal, H. Zhang, M.E. Alahi, H. Ghayvat, S. Mukhopadhyay, Y.-T. Zhang, W. Wu, A novel secure IoT-based smart home automation system using a wireless sensor network, Sensors 17 (1) (2016) 69.
[19] D. Geneiatakis, I. Kounelis, R. Neisse, I. Nai-Fovino, G. Steri, G. Baldini, Security and privacy issues for an IoT based smart home, in: 2017 40th International Convention on Information Communication Technology, Electronics and Microelectronics, 2017, pp. 1292–1297.
[20] M. Uysal, M.H. Nouri, Optical wireless communications—an emerging technology, in: 2014 16th International Conference on Transparent Optical Networks (ICTON), 2014.
[21] T. Nitsche, C. Cordeiro, A. Flores, et al., IEEE 802.11ad: directional 60 GHz communication for multi-Gbps Wi-Fi, IEEE Commun. Mag. 52 (12) (2014).
[22] Y. Ghasempour, C.C. da Silva, E.W. Knightly, IEEE 802.11ay: next-generation 60 GHz communication for 100 Gbps Wi-Fi, IEEE Commun. Mag. 55 (12) (2017).
[23] Y. Zhuang, J. Yang, Y. Li, et al., Smartphone-based indoor localization with Bluetooth low energy beacons, Sensors Mag. 15, 596 (2016) 20.
[24] A. Bensky, Wireless Location Technologies and Applications, second ed., Artech House, 2016.
[25] Applications of IEEE Std 802.15.4, IEEE 802.15 Document 15-14-0226-00-0000, 2014.
[26] S. Movassaghi, M. Abolhasan, J. Lipman, et al., Wireless body area networks: a survey, 1658 IEEE Commun. Surv. Tutor. 16 (3) (Third Quarter) (2014).
[27] R. Negra, I. Jemili, A. Belghith, Wireless body area networks: applications and technologies, Procedia Comput. Sci. 83 (2016) 1274–1281.
[28] M.T. Arefin, M.H. Ali, A.K.M.F. Haque, Wireless body area network: an overview and various applications, J. Comput. Commun. 5 (2017) 53–64.
[29] IEEE Std 802.15.6™-2012, IEEE Computer Society, 2012.
[30] J.F. Zhao, X.M. Chen, B.D. Liang, Q.X. Chen, A review on human body communication: signal propagation model, communication performance, and experimental issues, Hindawi Wirel. Commun. Mob. Comput. 2017 Article ID: 5842310. (2017) 15.
[31] O.N. Samijayani, H. Firdaus, A. Mujadin, E. Al, Solar energy harvesting for wireless sensor networks node, in: 2017 International Symposium on Electronics and Smart Devices, 2017, pp. 30–33.
[32] N. Jain, X. Fan, W.D. Leon-salas, A.M. Lucietto, Extending battery life of smartphones by overcoming idle power consumption using ambient light energy harvesting, in: 2018 IEEE International Conference on Industrial Technology, 2018, pp. 978–983.

[33] M. Yun, E. Ustun, P. Nadeau, A. Chandrakasan, Thermal energy harvesting for self-powered smart home sensors, in: 2016 IEEE MIT Undergraduate Research Technology Conference, 2016, pp. 1–4.
[34] W.K. Lee, M.J.W. Schubert, B.Y. Ooi, S.J.Q. Ho, Multi-source energy harvesting and storage for floating wireless sensor network nodes with long range communication capability, IEEE Trans. Ind. Appl. 54 (3) (2018) 2606–2615.
[35] S.K. Dewangan, A. Dubey, Design & implementation of energy harvesting system using piezoelectric sensors, in: 2017 International Conference on Intelligent Computing and Control Systems (ICICCS), 2017, pp. 598–601.
[36] P. Lynggaard, A self-supporting wireless IoT node that uses kinetic energy harvesting, in: Internet Things Business Models, Users, and Networks, IEE 2017, 2017, pp. 1–6.
[37] U. Muncuk, K. Alemdar, J.D. Sarode, K.R. Chowdhury, Multi-band ambient RF energy harvesting circuit design for enabling battery-less sensors and IoTs, IEEE Internet Things J. 5 (4) (2018) 2700–2714.
[38] bq25504 Ultra Low-Power Boost Converter With Battery Management for Energy Harvester Applications, Texas Instruments, 2015.

# Index

Note: Page numbers followed by *f* indicate figures, *t* indicate tables and *b* indicate boxes.

## C

## D

## F



## G

## N

## O



## P

## Q



## R

## S

## Z

Printed in the United States
By Bookmasters